3D Printing of Concrete

3D printing (or additive manufacturing) of concrete has opened new possibilities for automated, efficient, and design-flexible construction. This book brings together contributions from leading experts worldwide to present the scientific foundations and emerging applications of 3D-printed concrete, covering its properties, materials, modeling, and selected case studies from construction practice.

This book reviews the fresh and hardened properties of 3D-printed concrete, as well as existing reinforcement strategies; discusses novel printable materials and mixture design approaches for improved sustainability and performance; presents advances in computational modeling and artificial intelligence technologies; and examines recent construction applications and future opportunities.

This book is primarily intended for researchers and professionals working in concrete materials, additive manufacturing, and digital construction.

Modern Concrete Technology Series

A series of books presenting the state-of-the-art in concrete technology.

Series Editor
Geert De Schutter
Department of Structural Engineering and Building Materials
Faculty of Engineering and Architecture
Ghent University
Ghent
Belgium

Sustainability of Concrete
P. C. Aïtcin and Sidney Mindess

Concrete Surface Engineering
B Bissonnette, L Courard and A Garbacz

Textile Reinforced Concrete
Ava Peled, Arnon Bentur and Barzin Mobasher

Durability of Concrete: Design and Construction
Mark Alexander, Arnon Bentur and Sidney Mindess

Concrete Permeability and Durability Performance: From Theory to Field Applications
Roberto J. Torrent, Rui D. Neves and Kei-ichi Imamoto

Shotcrete: Materials, Performance and Use
M. Jolin and D.R. Morgan

Active Rheology Control of Cementitious Materials
Geert De Schutter and Karel Lesage

Fundamentals of Durable Reinforced Concrete – 2nd edition
Mark G. Richardson

Temperature Development in Early Age Concrete Elements: Fundamentals, Prediction and Control
Yunus Ballim

Aggregates in Concrete – 2nd edition
Mark Alexander, Sidney Mindess and Mike Otieno

3D Printing of Concrete: Properties, Materials, and Modeling
Zhanzhao Li and Aleksandra Radlińska

For more information about this series, please visit: https://www.routledge.com/Modern-Concrete-Technology/book-series/MCT

3D Printing of Concrete
Properties, Materials, and Modeling

Edited by
Zhanzhao Li
Aleksandra Radlińska

CRC Press
Taylor & Francis Group
Boca Raton London New York

CRC Press is an imprint of the
Taylor & Francis Group, an **informa** business

Designed cover image: Shutterstock

First edition published 2027
by CRC Press
4 Park Square, Milton Park, Abingdon, Oxon, OX14 4RN

and by CRC Press
2385 NW Executive Center Drive, Suite 320, Boca Raton FL 33431

CRC Press is an imprint of Taylor & Francis Group, LLC

British Library Cataloguing-in-Publication Data
A catalogue record for this book is available from the British Library

ISBN: 978-1-032-54898-2 (hbk)
ISBN: 978-1-032-56301-5 (pbk)
ISBN: 978-1-003-43486-3 (ebk)

DOI: 10.1201/9781003434863

Typeset in Times
by Apex CoVantage, LLC

Contents

Preface

The idea of 3D printing of concrete – layer by layer, with precision and efficiency – did not originate on Earth. It was born from a bold question posed by space agencies and engineers alike: how can we build habitats on the Moon or Mars using only the materials we find there? In the harsh environments of space, where transporting traditional construction materials is costly and impractical, additive manufacturing offered an elegant solution: automated, resource-efficient construction using local regolith and novel binders.

What began as a visionary concept for extraterrestrial living has rapidly found footing on Earth. As global demand for housing and infrastructure grows – alongside the urgent need to reduce carbon emissions, construction waste, and labor-intensive practices – 3D printing offers a transformative path forward.

Yet, among all major industries, construction has been the slowest to adopt automation. While manufacturing lines have long embraced robotics and digital workflows, buildings are still too often assembled brick by brick, by hand, on-site. Now, robotic construction and digital fabrication promise a paradigm shift: faster builds, greater geometric freedom, reduced material usage, and enhanced design integration.

Such a promise is not without its challenges. Realizing this vision requires a deep understanding of materials that behave differently when printed than when cast. It demands new models for rheology, setting kinetics, interface bonding, and long-term durability. The technology must also grapple with regulatory, safety, and scalability questions before it can reshape our built environment.

This book was written to illuminate the rapidly evolving field of concrete 3D printing – from the science of materials and mix designs to the mechanics of deposition and the digital models that drive them. By bringing together insights from engineering, materials science, and automation, we hope to contribute to building not only smarter structures but also a smarter way to build – whether here on Earth or far beyond it.

Author and Contributor Biographies

Mohammed Alnaggar is a senior research scientist at Oak Ridge National Laboratory in the United States. He obtained his PhD in structural engineering from Northwestern University in 2014, where his research focused on physics-based models of aging and deterioration of concrete materials, particularly alkali–silica reaction modeling coupled with creep and shrinkage. He received his MSc and BSc degrees in civil engineering from Zagazig University, Egypt, in 2009 and 2002, respectively. Prior to joining Oak Ridge National Laboratory in 2021, he served as an assistant professor in the Department of Civil and Environmental Engineering at Rensselaer Polytechnic Institute from 2014 to 2021. His research interests include multi-scale multi-physics computational modeling of cementitious materials, load transfer mechanisms in reinforced concrete, and 3D printing of concrete structures. He is a member of ACI committees 209 (Creep and Shrinkage in Concrete), 209–0D (Numerical Methods and 3D Analyses), 446 (Fracture Mechanics of Concrete – Joint ACI-ASCE), 239 (UHPC), and 239-0C (Structural Design on UHPC).

Adewumi John Babafemi is a professor in the Department of Civil Engineering at Stellenbosch University, South Africa. He earned his undergraduate and master's degrees from Obafemi Awolowo University, where he also lectured between 2005 and 2019. He obtained his PhD from the Department of Civil Engineering at Stellenbosch University, South Africa, in 2015. He was also a postdoctoral research fellow at the university between 2016 and 2017. He joined the department as a senior lecturer in 2019. He is a member of the Unit for Construction Materials (UCM) and was the division head of structural engineering and civil engineering informatics from 2020 to 2025. His research focuses on sustainable concrete and 3DCP.

Kasra Banijamali is a PhD candidate in the Bert S. Turner Department of Construction Management at Louisiana State University (LSU). His research focuses on automated quality control and real-time monitoring in construction 3D printing, with a particular emphasis on machine vision, sensor integration, and machine-learning-driven process optimization. He holds a BSc in Civil Engineering from the University of Tehran and an MSc in Computer Science from LSU, where his work included developing embedded sensing systems, automated inspection methods, and process automation for 3D-printed concrete. Meanwhile, he has published several articles in top-tier journals addressing challenges in real-time geometrical inspection, automation, and process optimization for large-scale construction 3D printing. His research interests include robotics, programming, and applied science, where he approaches models and algorithms as simplified abstractions of complex systems.

Tathagata Bhaduri is a lecturer in the Department of Civil and Environmental Engineering at Rensselaer Polytechnic Institute. He obtained his PhD in civil engineering from Rensselaer Polytechnic Institute in 2021, focusing on computational modeling of ultra-high-performance concrete (UHPC), multi-scale numerical simulation, and flow-induced fiber anisotropy. He received his Master of Technology from the Indian Institute of Technology, Kanpur, in 2016, and his Bachelor of Engineering from the Indian Institute of Engineering Science and Technology, Shibpur, in 2014. His research interests include computational modeling of advanced concrete materials, smart vibration control systems, and structural bracing. He is an active member of the ACI committees 239 and 564.

Sven G. Bilén is a professor of engineering design, electrical engineering, and aerospace engineering at Penn State, with over 30 years of experience developing systems for extreme environments – from space to the Arctic. He applies a systems design approach to translate early-stage needs into validated, deployed technologies. His work spans satellites, wireless sensor networks, software-defined radios, space propulsion systems, and robotic concrete printing. Sven co-founded the AddConLab following the success of PennStateDen@Mars in NASA's 3D-Printed Habitat Challenge and later helped launch X-Hab 3D to bring research into real-world applications. He is a registered professional engineer in Pennsylvania, a senior member of IEEE, an associate fellow of American Institute of Aeronautics and Astronautics (AIAA), and a member of American Geophysical Union (AGU), American Society for Engineering Education (ASEE), International Council on Systems Engineering (INCOSE), and Sigma Xi.

Hanbin Cheng is a PhD candidate in the Department of Civil and Environmental Engineering at the Pennsylvania State University. He received his BEng in Civil Engineering from Huaqiao University and MEng in Civil Engineering from Harbin Institute of Technology, China. His research interests lie in cement and concrete materials, specifically in the areas of 3D concrete printing (3DCP), concrete durability, recycled aggregates, seawater–sea sand concrete, data-driven concrete science, and structural health monitoring. He has served as a voting member of American Concrete Institute (ACI) Committees 565–00 "Lunar Concrete" and an associate member of ACI Committees 135–00 "Machine Learning-Informed Construction and Design" and ACI Committees 564–00 "3-D Printing with Cementitious Materials."

Dhrutiman Dey is the director and chief operating officer of Stratify 3D Private Limited, a pioneering company in scalable 3DCP technologies and services. He earned his bachelor's degree in production engineering from Veer Surendra Sai University of Technology, Burla, Odisha, followed by a master's degree in materials science and engineering from IIT Gandhinagar. He subsequently worked at IIT Kharagpur as a project staff, contributing to the development of high-performance materials for the Defence Research and Development Organisation (DRDO). Soon after this, he moved to IIT Guwahati for his doctoral research and graduated in 2025. His PhD focused on the development of 3D-printable concrete materials utilizing industrial waste, along with their mechanical and thermal performance characterization. Over the course of his research, he authored 17 papers in reputed journals and conferences and published three patents. Beyond academic contributions, Dr. Dey played a key role in the conceptualization and execution of large-scale 3D printing projects in collaboration with the Indian Army and the Institute of Planning and Maintenance (IPM) team at IIT Guwahati. His work bridged laboratory research with field applications, earning recognition through the Best Exhibitor Award and Best Paper Presentation Award at various national and international platforms.

Weijian Ding is currently a lecturer in the School of Construction Engineering at Shenzhen Polytechnic University. His research interests include concrete health monitoring, mechanisms and applications of piezoelectric/magnetoelectric multifunctional cementitious composites, intelligent sensing, and self-healing concrete. He obtained his PhD degree in the Department of Mechanics and Aerospace Engineering at Southern University of Science and Technology, where he proposed an innovative strategy for monitoring the in-service condition of concrete based on an integrated sensing element (ISE) concept.

José P. Duarte is the Stuckeman Chair in Design Innovation and professor of architecture and landscape architecture at Penn State, where he directs the Stuckeman Center for Design Computing. With over 30 years of research experience in architecture, computation, and digital fabrication, his work focuses on mass customization, housing design, and additive construction. He holds a professional degree in architecture from Lisbon Technical University and master's and PhD degrees from the Massachusetts Institute of Technology. Duarte co-founded the Additive Construction Laboratory (AddConLab) and the start-up X-Hab 3D, which develop large-scale 3D printing systems for construction. He led the team that was awarded the second place in NASA's 3D-Printed Habitat Challenge and co-edited *Mass Customization and Design Democratization.*

Michael R. Fiske is a technical fellow of advanced materials and manufacturing at the Amentum Space Exploration Division, located at NASA's Marshall Space Flight Center in Huntsville, Alabama. He obtained bachelor's and master's degrees in metallurgical engineering from the University of Alabama in 1984 and 1986, respectively. He has 39 years of experience in design, development, fabrication, and test of unique space-based materials processing hardware. This experience encompasses ground and spaceflight science hardware for the Shuttle *Middeck*, Spacelab, and ISS, J-2X rocket engine assembly and test, AM, and, most recently, additive construction. Mr. Fiske has led Marshall Space Flight Center's (MSFC) additive construction technology development since 2004. He is responsible for some of the world's earliest 3D-printed concrete structures and has focused on the development of flight systems for planetary surface infrastructure and habitat architecture. In addition, he has worked extensively on the development of in situ resource utilization (ISRU)-based cementitious materials with a focus on lunar launch/landing pads. He is active in ACI, AIAA, Minerals, Metals and Materials Society (TMS), and the Project Management

Institute (PMI). He was also the chair of the AIAA Space Processing Technical Committee in 1999–2000. He is a member of the Industrial Advisory Board for the Department of Metallurgical Engineering at the University of Alabama and has been an adjunct faculty member for Roger Williams University in Bristol, Rhode Island, where he taught classes in Introduction to Manufacturing Methods and Introduction to Engineering Materials.

Shashank Gupta is a PhD candidate in civil and environmental engineering at Princeton University and a member of the Architected Materials and Additive Manufacturing (AM^2) Lab. He earned his MSc in civil engineering (structures) at Politecnico di Milano in 2021 and BE (Hons.) in civil engineering from BITS Pilani in 2018. His research focuses on bio-inspired architected cementitious materials with improved toughness and ductility, integrating experimental, theoretical, and computational fracture mechanics with digital fabrication. He is a recipient of the Charlotte Elizabeth Procter Fellowship, a campus-wide honorific fellowship, at Princeton University. He has also been awarded the merit-based Silver scholarship at Politecnico di Milano. He is an associate member of ACI 564 (3D Printing with Cementitious Materials) and ACI 236 (Material Science of Concrete), and a young member of RILEM.

Maryam Hojati is an assistant professor in the Department of Civil, Construction, and Environmental Engineering at the University of New Mexico, where she established the Advanced Sustainable Construction (ASC) Lab in 2019. She earned her PhD in civil engineering with a focus on construction materials and additive manufacturing. Her research advances sustainable infrastructure technologies, including 3D printing of cementitious and earthen materials, and the integration of autonomous systems into construction workflows. Dr. Hojati's expertise includes the rheology of fresh and printable concretes, advanced material characterization, microstructural analysis of novel binders, and circular economy applications in the built environment. Her work has deepened the understanding of material–process–performance relationships in extrusion-based additive manufacturing, with publications on geopolymer systems, engineered cementitious composites (ECCs), earth and material durability under harsh environmental conditions. Through the ASC Lab, she leads interdisciplinary projects combining experimental methods with automation and advanced manufacturing to address challenges of resilience, speed, and sustainability in construction. Actively collaborating across engineering disciplines, from materials science to robotics, Dr. Hojati is committed to training the next generation of engineers in Construction 4.0 and advancing innovation in infrastructure development through research-informed education and inclusive, technology-driven solutions.

Yohan Jacquet, with a background in civil and mechanical engineering, earned a PhD, in France, in underwater 3DCP, mainly focusing on rheology, process design, and the durability of materials printed and cured in water. His research explores the fluid–solid transition in printable materials, work recognized with several distinctions, including the ACI's Wason Medal in 2022. After completing his PhD, he taught as a building information modeling (BIM) specialist and expanded his investigations to clay-based and alternative binder systems for 3D printing. In 2024, at the Technical University (TU) of Berlin, he advanced a new technique called injection 3DCP. Granted the Marie Skłodowska-Curie postdoctoral fellowship, in June 2025 he began an independent project on the numerical simulation and life-cycle assessment of novel 3DCP processes, conducted between the Technical University of Denmark and ETH Zurich. His current work combines advanced numerical and experimental modeling of additive construction and pioneers new concepts, including powder rheology, to propose innovative approaches for digital construction innovation.

Ali Kazemian is an assistant professor in the Bert S. Turner Department of Construction Management at Louisiana State University (LSU). Since 2014, his research has focused on robotic construction-scale 3D printing, and the design and investigation of novel automated quality control and inspection methodologies to enable higher levels of construction automation and robustness. His contributions have been recognized by the community on multiple occasions, including being named a Kavli fellow by the National Academy of Sciences, winning of the prestigious "ASTM Young Professional in Additive Manufacturing" national award, and having his research featured in *The New York Times* and *MIT Technology Review* and on the Fox44 TV channel. He earned his PhD in civil engineering and a master's degree in computer science, both from the University of Southern California. Prior to joining LSU in 2020, he worked for 3 years as a senior R&D engineer at a robotic construction company in California. He currently also serves as an associate editor for *Progress in Additive Manufacturing* (Springer) and the *Journal of Architectural Engineering* (ASCE).

Zhanzhao Li is a Rice Academy Postdoctoral Fellow in the Department of Civil and Environmental Engineering at Rice University, USA. He received his PhD (2024) and BEng (2019) in civil engineering from the Pennsylvania State University, USA, and Harbin Institute of Technology, China, respectively. His research interests include hydration, durability, materials informatics, and digital fabrication of cementitious materials. He served as the co-chair of ACI Committee S805 "Student Leadership Council" (2023–2024) and is a voting member of ACI Committees 135 "Machine Learning-Informed Construction and Design," 201–0H "Aggregate Reactions," and 565 "Lunar Concrete." He is also an active member of RILEM Technical Committees 301-ASR "Risk Assessment of Concrete Mixture Designs with Alkali-Silica Reactive (ASR) Aggregates" and 315-DCS "Data-Driven Concrete Science." He received the ACerS Stephen Brunauer Award (2026) and was recognized as an ACerS Rising Star (2024) and an ACI Emerging Leaders Alliance Member (2023).

Xiqiang Lin holds a PhD in the Department of Civil and Hydraulic Engineering, College of Civil Engineering and Architecture, Zhejiang University. He is currently chief engineer at Hangzhou Guanli Intelligent Technology Co., Ltd., and a core research member at the Institute of Intelligent Construction, Zhejiang University. His research focuses on construction-scale additive manufacturing, including the rheology and mix design of printable concrete, intelligent construction robotics and equipment, and end-to-end printing workflows. From 2012 to 2021, he served as a senior R&D engineer at the Technical Center of China State Construction Engineering Corporation (CSCEC), where he led or contributed to major 3D-printing initiatives and helped deliver China's first on-site 3D-printed two-story office building (Guangzhou, 2019). Since 2021 at Guanli, he has led the R&D and field deployment of materials, equipment, and processes, advancing the intelligentization, industrialization, and internationalization of 3DCP; his team's technologies have been implemented on jobsites in multiple countries. Representative projects include the world's first 3D-printed mosque (Middle East, 2024), a reinforced-concrete pedestrian bridge in Zhejiang (2024), a scenic-area rest pavilion in Anhui (2024), a traditional *siheyuan* residence in Hebei (2023), and the "Mars Habitat" hotel in Gansu (2021). He has authored more than 20 papers, served as an editor or contributor to Chinese industry standards, co-authored the monographs *Concrete 3D Printing Technology and Applications* and *Intelligent Construction of Concrete Structures*, and holds ten granted invention patents.

Inka Mai studied industrial and civil engineering at Technical University in Braunschweig, Germany; Chalmers University of Technology, Gothenburg, Sweden; and the University of Rhode Island, USA. After obtaining an MSc, she became a PhD candidate at the Institute of Building Materials, Concrete Construction and Fire Safety in Braunschweig. After her PhD, she continued as a postdoctoral researcher and was in charge of the working group on rheology and digital fabrication. A research stay at the University of South Brittany, Lorient, France, enabled her to deepen her knowledge in analytical modeling and 3DCP. In 2023, she became a junior professor at TU Berlin, Institute of Civil Engineering, Chair of Robot-Assisted Manufacturing of the Built Environment. She has received several research awards, such as Klaus-Dyckerhoff Prize for Young Scientists and the research prize of the governing mayor of Berlin for young scientists.

Nicholas A. Meisel is an associate professor of engineering design in the School of Engineering Design and Innovation at Penn State and an affiliate faculty in the Department of Mechanical Engineering. He graduated from Virginia Commonwealth University in 2010 with his BS in mechanical engineering and received his PhD in mechanical engineering from Virginia Tech in 2015. Nick has the privilege of directing the Made By Design Lab, which conducts research in design for additive manufacturing (AM), including investigation into design concepts that drive innovation, limits on the manufacturability of designs, and the impact of AM on student design and manufacturing learning. Nick has received several awards for his work in the realm of design for AM, including the National Science Foundation (NSF) CAREER Award in 2021 and the International Outstanding Young Researcher in Freeform and Additive Manufacturing Award in 2023.

Ali M. Memari is a professor and Bernard and Henrietta Hankin Chair in Residential Building Construction in the Departments of Architectural Engineering and Civil and Environmental Engineering at Penn State, and director of the Pennsylvania

Housing Research Center. During his 40 years of teaching and research experience, he has taught various courses related to structural and architectural engineering and has focused his research on full-scale laboratory testing, characterization, and evaluation of residential and commercial buildings with respect to various structural and environmental loading systems. He has over 350 publications, including papers in journals and conference proceedings, book chapters, edited books, and research reports.

Jyotirmoy Mishra is a postdoctoral researcher in the Department of Civil Engineering at Stellenbosch University, South Africa. He obtained his undergraduate and master's degree (dual degree) from KIIT University, India, specializing in structural engineering (2016). He continued his passion for research in geopolymer materials by pursuing a PhD, which he completed in 2023 at Veer Surendra Sai University of Technology, India. During his PhD journey, he explored the feasibility of ferrochrome ash, a waste from ferroalloy industries, as a precursor material for geopolymer synthesis. Soon after completing his PhD, he worked as a research associate at the same institution on a project until 2024. In 2024, he was employed as a postdoctoral researcher at Stellenbosch University and is a member of the UCM.

Reza Moini is an assistant professor of the Civil and Environmental Engineering Department at Princeton University, NJ, USA. He earned his PhD from the Lyles School of Civil Engineering at Purdue University in 2020. He is an associated faculty member of the Princeton Materials Institute, Princeton Center for Computational Science and Engineering (PICSciE), and Princeton Robotics Initiative. He is the recipient of several awards, most recently, the Walter P. Moore Jr., Faculty Achievement Award from ACI, the Howard B. Wentz Jr. Junior Faculty Award from the School of Engineering and Applied Science at Princeton University, and the National Science Foundation early CAREER award. His research interest lies at the interface of materials mechanics and advanced design enabled by advanced manufacturing and robotics.

Dr. Moini's overarching research is in the areas of engineering mechanics and autonomous manufacturing and robotics of construction materials. He has pioneered the field of architected cement-based materials and structures, engineered with damage-resilient and functional characteristics unattainable with conventional compositions and construction methods. His research is motivated by the intellectual challenges in understanding the fracture and statistical mechanics of architected and disordered materials. His group uses experiments, simulation, and theory to engineer the fracture mechanics of intrinsically brittle and quasi-brittle cement-based materials through a materials-by-design approach, and to engineer toughening mechanisms in cement-based materials, often inspired by biological systems. His group has developed numerical frameworks such as the coupled phase-field (FP) cohesive zone model (CZM) to understand the interfacial energy dissipation in additively manufactured concrete and hard–soft layered cementitious composites. His lab has spearheaded several advanced manufacturing methods that enable several new mechanics and functions in construction materials, including multi-material additive manufacturing with concrete and polymers (soft hyperelastic or hard elastic), laser processing for the tabulation of brittle materials, pressure-based feedback control in extrusion-based robotic additive manufacturing, in situ carbonation during additive manufacturing for carbonate cements, and autonomous robotic additive manufacturing with concrete. His group has also developed open-source path generation and control algorithms and workflows that enable a more accessible design of advanced cement-based materials.

Aimane Najmeddine is an associate research scholar in the Department of Civil and Environmental Engineering at Princeton University. He earned his PhD in civil engineering from Virginia Tech in 2022, where his research focused on the long-term thermos-chemo-mechanical behavior of polymeric and elastomeric materials – research that has become increasingly vital, given the growing concerns about microplastic proliferation in our environment. He also holds both MS and BS degrees in civil engineering. Dr. Najmeddine has authored several publications, including peer-reviewed journal articles and conference papers in leading journals such as *Journal of the Mechanics and Physics of Solids*, *International Journal of Mechanical Sciences*, *International Journal of Solids and Structures*, and *Engineering Fracture Mechanics*.

At Princeton, Dr. Najmeddine's research has evolved to focus on understanding the mechanical responses and fracture behavior of architected hard–soft multi-material assemblies, leveraging recent advances in additive manufacturing technologies. His work integrates theoretical analysis, computational modeling, and experimental validation to investigate fracture mechanics and explore multi-functionality in multi-material architected composites. This work aims to design materials that combine exceptional strength with toughness, essential properties for developing more robust, sustainable, and resilient built infrastructure.

Shadi Nazarian is the H. Ralph Hawkins, FAIA Endowed Chair and a professor of architecture at the University of Texas at Arlington, where she founded and directs the Architectural Robotic Construction Laboratory. Previously, she taught at Penn State, where she co-founded the AddConLab, directed the Transitional Material Interface Lab, co-led her award-winning team in the NASA Mars 3D-Printed Habitat Challenge competition, and co-founded the start-up X-Hab 3D. She holds a post-professional degree in architecture from Cornell University, a professional degree in architecture, and a Bachelor of Environmental Design from the University of Minnesota. Her scholarship is focused on design, fabrication, and post-processing of functionally graded materials and structures with varying structural and insulative performance, and optical transparency. Her work on seamless and graded transition between glass and geopolymer concrete reveals new architectural and spatial possibilities, advancing the role of structural ceramics in contemporary design.

Jinlong Pan is a full professor at School of Civil Engineering, Southeast University, China. He obtained his bachelor's (1999) and master's (2002) degrees in civil engineering from Zhejiang University, China, and received his PhD (2005) in civil engineering from the Hong Kong University of Science and Technology. His research interests include high-performance construction materials, prefabricated and composite structures, and 3DCP technology. Prof. Pan has published more than 180 refereed papers in top-ranked journals (h-index = 43) and holds 35 authorized invention patents. He has contributed to seven national and provincial standards, authored one monograph, and co-authored six textbooks. Prof. Pan is the recipient of numerous awards, including the Elsevier Highly Cited Paper Award (2009), the First Prize of the Ministry of Education Science and Technology Progress Award (2011), the Second Prize of the National Science and Technology Progress Award (2012), and the Second Prize of the Ministry of Education Science and Technology Progress Award (2019). In recognition of his research impact, he was listed among Elsevier's "Highly Cited Chinese Researchers" in 2024.

Biranchi Panda is an assistant professor in the Department of Mechanical Engineering at the Indian Institute of Technology (IIT) Guwahati. He brings over 9 years of pioneering research in concrete 3D printing, with more than 60 peer-reviewed publications in high-impact journals, an h-index of 46, and over 11,500 citations – establishing his global recognition in the field. Dr. Panda's core expertise lies in the formulation of printable cementitious mixtures using industrial by-products, and its performance characterization for several applications. At IIT Guwahati, Dr. Panda heads a fully equipped 3DCP laboratory, from which five doctoral candidates have already graduated and eight more are currently pursuing their PhDs under his supervision. He has successfully translated laboratory innovations into real-world applications, including the construction of defence infrastructure in Arunachal Pradesh for the Indian Army and a sustainable 3D-printed security post on the IIT Guwahati campus. As principal investigator, he has completed two funded projects under Science Engineering Research Board (SERB) and Department of Science and Technology (DST) and is currently leading three ongoing projects with a combined worth of approximately ₹130 crores. In addition to his academic contributions, he actively provides technical advisory to start-ups and major industry leaders such as Larsen & Toubro (L&T) on challenges related to 3DCP.

Aleksandra Radlińska is a professor in the Department of Civil and Environmental Engineering at Penn State and a member of AddConLab. Her research focuses on developing innovative materials and construction methods to enhance infrastructure durability and reduce CO_2 impact for terrestrial and extraterrestrial applications, including the analysis of geopolymer concrete mixed and hardened on the International Space Station. Dr. Radlińska is the chair of the newly established ACI 565 Lunar Concrete Committee, past chair of the Concrete Research Council that helps transform academic research into industrial applications, and past chair of ACI Committee 123 Research and Current Developments. She served as a chair-elect of the American Ceramic Society (ACerS) Cements Division in 2017 and as a meeting chair in 2018, which was held on Penn State's University Park campus. Dr. Radlińska is a fellow of the ACI.

Xiaoyan Sun, PhD, is an associate professor at Zhejiang University. She received her PhD from Dalian University of Technology in 2004, completed postdoctoral research at Tsinghua University (2004–2006), and joined the College of Civil Engineering and Architecture at Zhejiang University in 2006. From 2009 to 2010, she conducted collaborative research at Stanford University and was selected for Zhejiang University's "Rising Star Program" and Zhejiang Province's "151 Talent Program." As principal investigator, she has led and completed 30+ projects, including the National High-Tech R&D Program (863), the NSFC program, the Zhejiang Natural Science Foundation project, China–US collaborative projects, the Ministry of Education Doctoral Fund, and multiple provincial programs. She has received a Provincial Science and Technology Progress Second Prize, holds 10+ authorized patents and one software copyright, and has published 100+ papers, with 60+ indexed by SCI/EI. Her research focuses on 3D-printed building structures and processes and off-world (extraterrestrial) 3D-printing intelligent construction, covering printable materials and process coupling, structural performance, and studies specific to extraterrestrial 3D printing.

Fei Teng is a PhD candidate at the Department of Building and Real Estate, the Hong Kong Polytechnic University. His current research mainly focuses on multi-material 3DCP, sustainable strain-hardening cementitious composites (SHCC), and automation in construction. He received his bachelor's and master's degrees in civil engineering in 2017 and 2020, respectively, from Southeast University, China.

Guanfei Tian, PhD, research fellow, is the chairman of Hangzhou Guanli Intelligent Technology Co., Ltd. He graduated from the School of Civil and Hydraulic Engineering at Tsinghua University in 2006 and worked at the China Academy of Building Research from 2006 to 2016, successively serving as the director of the Center for Concrete Testing Technology & Instruments; deputy general manager of Beijing Jianyan Kunlun Technology Co., Ltd.; deputy chief engineer of the Institute of Building Materials; and director of the Waterproofing Technology Center. He has received three provincial/ministerial second prizes and five third prizes for scientific and technological progress. In 2017, he founded Jianyan Huace (Hangzhou) Technology Co., Ltd., and Hangzhou Guanli Intelligent Technology Co., Ltd., focusing on high-end concrete testing instruments and construction 3D-printing materials/equipment. He has led or participated in one Zhejiang key project and two Zhejiang "Pioneer/Leading Goose" projects and holds ten invention patents, over 30 utility models, and more than 20 design patents. His research focus is on intelligent construction equipment for 3DCP, such as gantry systems and large-reach robotic arms; materials and process optimization; and the Moli intelligent control software. He has advanced the engineering deployment of multiple large-scale 3D-printed architecture and landscape projects, improving print accuracy and on-site automation.

Christopher Vreeland completed his bachelor's degree in civil engineering from the University of New Mexico in 2024 and continued to receive an MS in structural mechanics by 2025. His interest was in elevating the performance of materials, building methods, and structures. His research centered on composites for construction applications using natural fibers in a geopolymer matrix. Through this project he developed low-carbon, bio-inspired, functionally graded materials that offer both structural and insulating abilities. Following his time in materials science, Chris shifted focus to the field of structural design, where he hopes to incorporate advanced construction products and techniques in future work.

Hailong Wang, PhD, is a professor and PhD supervisor at Zhejiang University and director of the Institute of Intelligent Construction and Management. He received his PhD from Tsinghua University in 2006 and carried out visiting research at Queen's University Belfast and the University of California, Berkeley. He was selected for Zhejiang University's "Rising Star Program" and Zhejiang Province's "151 Talent Program," and holds multiple academic positions such as board member of the Building Materials Division of the Architectural Society of China. He has led more than 40 national and provincial projects, including the National High-Tech R&D Program (863), the NSFC, and Zhejiang's Key R&D Program; has received the Second Prize of the National Science and Technology Progress Award and five provincial or ministerial first/second prizes; holds over 20 authorized patents; and has published more than 100 papers. His research focuses on 3D-printed structures: low-carbon printable cementitious materials and reinforcement–process co-design, interlayer interface mechanics and structural performance, fiber/steel reinforcement, durability and service-life assessment, early-age rheology, buildability and geometric control, as well as in situ/online monitoring with digital twins and related standardization. In addition, he has led and participated in multiple engineering projects on 3D-printed structures, covering structural design, process–material integration, and performance verification, thereby promoting the large-scale application of intelligent construction research outcomes in building and transportation infrastructure.

Li Wang is a professor at Hebei University of Technology (China) and director of the Hebei Provincial Engineering Research Center for 3D Printing in Construction. He is a recipient of the Excellent Young Scientists Fund of the National Natural Science Foundation of China (NSFC) and has established himself as a leading researcher in 3D-printed cementitious materials. Prof. Wang has authored over 80 peer-reviewed papers in high-impact journals, including *Composite Part B: Engineering*, *Cement and Concrete Research*, and *Additive Manufacturing*. His work has garnered more than 6,400 citations and an h-index of 40. His research focuses on mix design optimization, interfacial bonding, structural performance, and sustainability assessment of 3D-printed concrete. He has served as the principal investigator for multiple projects, including a subproject of the Ministry of Science and Technology's 14th Five-Year National Key R&D Program, four NSFC projects, and several provincial-level grants. Beyond academia, Prof. Wang plays an active role in professional societies, serving as deputy secretary-general of the 3D Printing Branch of the China Concrete and Cement-based Products Association and as a member of various national expert committees on construction robotics and infrastructure 3D printing, with 43 invention patents (including American and Japanese patents), contributions to Chinese national standards, and recognition through prestigious awards such as the First Prize of Provincial Science and Technology and a Gold Medal at the Geneva International Exhibition of Inventions.

Nathan D. Watson is a mechanical engineer specializing in AM and 3DCP. He holds both a BS in mechanical engineering and an MS in AM and design from Penn State. As VP of engineering at X-Hab 3D, he leads the development of mobile concrete printing systems and has served as principal investigator on Defense Advanced Research Projects Agency (DARPA)-funded research. Watson has presented at national conferences, contributed to peer-reviewed publications, and played a key role in NASA's 3D-Printed Habitat Challenge, earning top honors. His work bridges engineering, architecture, and civil infrastructure, advancing autonomous construction technologies.

Yiwei Weng is an assistant professor at the Department of Building and Real Estate, the Hong Kong Polytechnic University. Dr. Weng received his dual bachelor' degrees (2014) in material science engineering and financial engineering from the University of Science and Technology Beijing, and a master's degree (2015) in mechanical and aerospace engineering from Nanyang Technological University (NTU). He received his PhD in civil and environmental engineering from NTU in 2020. Before joining the Department of Building and Real Estate as an assistant professor in 2021, he was a research fellow at the NTU (2020–2021). His research focuses on digital construction, specializing in additive manufacturing, construction materials, and robotics for building and construction.

Zhengyu Wu obtained his master's degree in civil engineering from the University of Florida in 2019 and his PhD in architectural engineering from the Pennsylvania State University in 2024. During his doctoral studies, he joined the Additive Construction Laboratory (AddConLab) at Penn State, where he worked with Prof. Ali M. Memari and Prof. José P. Duarte on reinforcement strategies for 3D-printed concrete structures. His research involved comprehensive experimental and analytical studies on the development length and bond behavior of steel rebar in printed concrete. He has authored and co-authored multiple peer-reviewed journal papers on reinforcement in 3D-printed concrete. He is currently a structural engineer at LERA Consulting Structural Engineers (New York, USA), where he contributes to the structural design and analysis of high-rise building projects while gaining practical experience with modern design practices and workflows in the structural engineering industry.

Muhammad Saeed Zafar is a civil engineering researcher with expertise in 3DCP, rheology optimization, and sustainable cementitious materials. He earned his BSc in civil engineering from Swedish College of Engineering and Technology and MSc in structural engineering from National University of Sciences and Technology, Pakistan. He further obtained his PhD in civil engineering from the University of New Mexico, USA, in 2024, where his work integrated machine learning and design of experiments (DoE) to optimize the fresh properties of 3D-printed cementitious composites, including biopolymer- and fiber-reinforced systems. His research contributions include the development of self-reinforced ultra-ductile printable materials, evaluation of printability and buildability, and rheological characterization using advanced analytical techniques. Dr. Zafar has published in high-impact journals such as *Additive Manufacturing*, *Construction and Building Materials*, *Case Studies in Construction Materials*, *Journal of Building Engineering*, and *Sustainability*, and is co-inventor of a patent on 3D-printed ECCs. He currently serves as research engineer-II and lab manager at the University of New Mexico, Gerald May Department of Civil Construction and Environmental Engineering, where he oversees materials and structural testing laboratories and collaborates on interdisciplinary research in infrastructure resilience and digital construction.

Mingzhong Zhang is an associate professor in sustainable infrastructure materials in Department of Civil, Environmental and Geomatic Engineering at University College London (UCL), UK. He received a PhD in civil engineering from Delft University of Technology, the Netherlands, in 2013. Following this, he worked as a postdoctoral research associate at the University of Manchester. He then joined UCL as a lecturer in 2015 and was promoted to associate professor in 2019. He specializes in the multiscale and multi-physics characterization and modeling of advanced civil engineering materials and structures, including nature-inspired materials, sustainable cement and concrete, 3D-printed/digital concrete, and fiber-reinforced composites. As of August 2025, he has published one academic book, six book chapters, and more than 150 peer-reviewed papers (h-index = 55) in top-ranked journals on construction materials, structural engineering, and computational mechanics. He is the receipt of the Royal Society Kan Tong Po International Fellowship and has been listed among the world's top 2% scientists by Stanford University since 2020. He serves as an associate editor/editorial board member of several ISI journals, including *Journal of Sustainable Cement-Based Materials*, *Construction and Building Materials*, and *Scientific Reports*. He is a member of EPSRC Peer Review College, Institute of Concrete Technology (ICT), Institution of Civil Engineers (ICE), and RILEM.

Binrong Zhu is an associate professor in civil engineering and an academic staff of the Jiangsu Provincial College Key Laboratory of Intelligent Bridge Construction and Safe Operation & Maintenance in School of Civil Engineering at Nanjing Forestry University, China. He obtained his PhD in civil engineering from Southeast University, China, in 2022. He has published more than 40 papers in high-impact journals, including 13 papers in Science Citation Index (SCI)-indexed journals as the first or corresponding author. He is a member of the 3D Printing Academic Committee and Solid Waste Subcommittee of the Chinese Ceramic Society. His research interest includes 3DCP, fiber-reinforced cementitious composites (e.g., ECCs and UHPC), and sustainable construction materials.

Part 1

Properties of 3D-Printed Concrete

1 Fresh Properties of 3D-Printed Concrete

Hanbin Cheng and Aleksandra Radlińska

1.1 INTRODUCTION: BACKGROUND AND DRIVING FORCES

3D concrete printing (3DCP) is an innovative construction technique that has gained significant recognition and interest from both the industry and the research community due to its unique characteristics. This technology eliminates the need for traditional formwork construction (Xiao, Ji et al. 2021), allowing architects and engineers to realize complex shapes and aesthetically appealing concrete elements. Additionally, it can reduce construction material waste on-site by 30%–60% (Zhang, Wang et al. 2019). The high productivity and economic construction time associated with 3DCP have the potential to significantly lower overall construction costs (Batikha et al. 2022). Numerous concrete structures have already been fabricated using 3DCP for both civil and military purposes (Khan and McNally 2023; Růžek et al. 2023). Furthermore, this technology is being explored as a promising method for extraterrestrial construction on the Moon and Mars (Kloft et al. 2020; Giwa et al. 2024). However, the widespread application of 3DCP continues to face several challenges. Unlike conventional concrete, 3D-printable concrete must possess fresh properties that are compatible with the specific requirements of the printing process at various stages. Therefore, it is crucial to consider these fresh properties in relation to the transportation and stiffening needs of the printing process.

The desired fresh properties of printable concrete involve several key considerations, including flowability (or pumpability), extrudability, buildability, and open time (Li et al. 2020). Flowability, or pumpability, refers to the consistency of the printable concrete and assesses how smoothly the material can be pumped from the mixing chamber to the printing nozzle (Paritala et al. 2023). Extrudability denotes the ability of the fresh concrete to be transported through a hopper and pumping system to the nozzle, where it must be extruded as a continuous filament with a stable shape (Andrew Ting et al. 2022). A mixture is considered to have "excellent" extrudability if it can be extruded successfully without any blockages in the delivery system or extruder. Buildability is the ability to 3D print by continuously adding concrete layers to the required height without significant deformation or collapse of the freshly printed component (Chang, Chen et al. 2023). Finally, open time is defined as the time period during which the fresh material maintains a level of workability that

DOI: 10.1201/9781003434863-2

allows it to be pumped and extruded through the nozzle without blockages, ensuring the production of a specimen with good quality (Li et al. 2020).

The distinct approach required for fabricating concrete structures through 3D printing presents unique challenges for the fresh properties of concrete, setting them apart from those of conventional concrete. During the printing process, the concrete must first be easily pumped to the printhead, requiring adequate flowability. Following this, the material needs to be smoothly and continuously extruded through the printer nozzle, demonstrating good extrudability. Once extruded, the concrete must maintain the intended shape of the deposited layers, exhibiting stability against gravitational and other forces without the use of formwork. Additionally, achieving sufficient open time is essential for successful printing and ensuring a strong bond between each layer. However, it is important to balance this, as an excessively long open time can compromise the shape stability and buildability of the printed concrete. Therefore, the fresh properties of 3D-printable concrete must be carefully tailored to meet the specific demands of pumpability, extrudability, and buildability – fundamental performance criteria that are critical to the success of 3DCP.

The rheology of a mixture, encompassing static yield stress, dynamic yield stress, plastic viscosity, and thixotropy, primarily governs the desired fresh properties (Souza et al. 2020). For 3D-printed concrete (3DPC) to meet the dual demands of pumpability and extrudability, while also ensuring sufficient buildability, it must demonstrate both appropriate thixotropy and satisfactory static yield stress. Thixotropy, in this context, refers to the ability of fresh concrete's yield stress to decrease from static to dynamic levels as the flocculation in the cementitious system is disrupted under shear force (Zhang, Zhang et al. 2019). However, once the external shear force is removed, the yield stress can recover over time, potentially exceeding its initial value. While the structural breakdown due to high shear forces facilitates pumping and extrusion, the shape stability of extruded concrete filaments and the structural stability of printed elements are predominantly determined by the static yield stress and its development (Vallurupalli et al. 2021).

These four properties are dependent on mixture design as well as mixing and printing systems. The influent factors regarding these properties and the development of the corresponding rheological properties are discussed in the following sections.

1.2 FLOWABILITY AND PUMPABILITY

Flowability and pumpability are critical parameters that define the flow behavior of fresh materials during their transportation. In 3DCP, the mixed concrete is initially pumped through a hose or pipe system to reach the designated printing endpoint (Xiao, Ji et al. 2021, Andrew Ting et al. 2022, Khan and McNally 2023, Paritala et al. 2023, Růžek et al. 2023, Giwa et al. 2024). Ensuring sufficient flowability and pumpability is vital, as it prevents blockages within the hose or pipe during the pumping process.

1.2.1 Experimental Techniques

The flowability of mixtures can be evaluated using several straightforward testing methods, such as the slump flow test (Tay, Qian et al. 2019), V-funnel test (Mu et al. 2022),

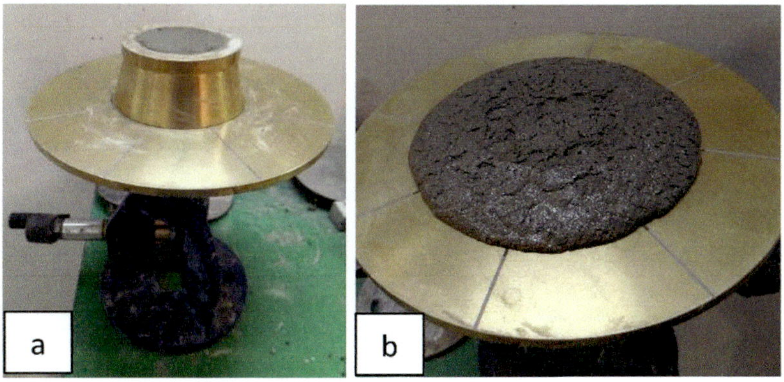

FIGURE 1.1 Slump-flow test experimental setup: (a) before the mold is removed; (b) the spread of the material after dropping the table for 25 times (Tay, Qian et al. 2019).

and squeeze flow test (Tao et al. 2022). Additionally, viscosity and buildability can be preliminarily assessed through the V-funnel and squeeze flow tests. These methods are simple yet effective and are routinely employed for the preliminary assessment of a mixture's flow behavior during the pumping stage. Among these methods, the slump flow test remains the most widely adopted, as evidenced by current research (Rahul et al. 2019; Tay, Qian et al. 2019; Vallurupalli et al. 2021; Tao, Ren et al. 2022; Chang, Chen et al. 2023). The slump test procedure, in accordance with ASTM C230, employs a conical mold. The procedure begins with the filling of the conical mold with mortar. Upon removal of the mold, the flow table is dropped 25 times, and the spreading diameter of the mortar is meticulously documented, as shown as in Figure 1.1.

Concrete mixtures demonstrating flowability within the range of 14–19 cm (Kim, Koh et al. 2016; Zhang, Hou et al. 2019; Bong et al. 2021, Muthukrishnan et al. 2021; Jiang, Liu et al. 2022; Zhang, Deng et al. 2022), contingent on the specific printing technology and conditions, generally satisfy printability standards. Consequently, the design of printable concrete mixtures must guarantee that the requisite flowability is achieved. A significant study on the influence of flowability and pumpability on printing quality was conducted by Tay et al. (Tay, Qian et al. 2019). They quantified the flowability and pumpability of various mixtures through the application of the flow table test and a pumping flow rate test. Their findings revealed that the pumpability index, defined as the ratio of a mixture's flow rate to that of water, exhibits a direct proportionality to the slump flow. They observed that a significant reduction in both slump value and pumpability index leads to increased roughness in the printed concrete, large cracks, and potential reductions in interlayer bond strength or structural integrity, despite enhanced shape retention capacity. Conversely, when the slump and pumpability index are excessively high, the printed specimen exhibits enhanced smoothness and reduced voids but experiences diminished buildability and potential instability due to lowered yield stress. Thus, concrete mixes with a high slump and pumpability index, when within the acceptable range, are ideal for surface quality and suitable for printing. Their findings can be summarized as a printable region using slump and slump-flow results (Figure 1.2).

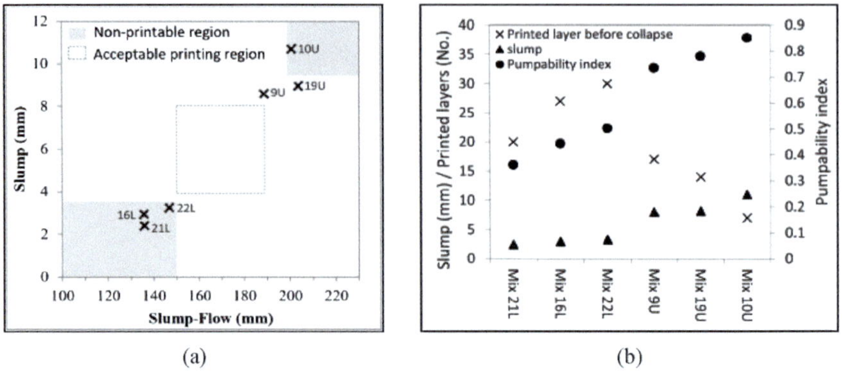

FIGURE 1.2 (a) Printable and non-printable regions; (b) correlation of slump, pumpability index, and maximum printed layer achieved (Tay, Qian et al. 2019).

The flowability and pumpability of mixture can be assessed using a rheometer test from the perspective of rheological behavior. This behavior of a fresh concrete closely follows the Bingham model, which relates shear stress (τ) and plastic viscosity (μ) using the equation

$$\tau = \tau_0 + \mu\dot{\gamma}. \tag{1.1}$$

For a Bingham fluid, the minimum stress required to maintain flow is known as yield stress (τ), while plastic viscosity (μ) represents the flow resistance. The Bingham model is typically fitted to the down curve to determine the dynamic yield stress, whereas the maximum point on the upcurve provides the static yield stress, as illustrated in Figure 1.3.

For concrete to be viable for 3D printing, it is essential that the dynamic yield stress and plastic viscosity remain low to facilitate optimal flowability and pumpability, while the static yield stress must be sufficiently high post-extrusion to enhance buildability (Chang, Chen et al. 2023). Adequate thixotropy within the fresh paste of the print filament is crucial for supporting buildability (Lee et al. 2022). Although thixotropy facilitates a reversible structural build-up, precise control over curing kinetics is vital to attain the requisite fresh strength, which is critical for buildability and irreversible. This can be assessed by analyzing the envelope area, as shown in Figure 1.4.

Gosselin et al. emphasized a pivotal requirement for printability: an extended dormant period to optimize pumpability, followed by rapid stiffening to guarantee buildability (Gosselin et al. 2016). Tran et al. conducted a study investigating the rheological properties of printable concrete with varying percentages of polypropylene fibers. Based on their findings, they introduced a "printable rheology box" for 3D-printable fiber-reinforced concrete mixtures, characterized by plastic viscosity ranging from 22 to 60 Pa•s and dynamic yield stress ranging from 250 to 500 Pa (Tay, Qian et al. 2019).

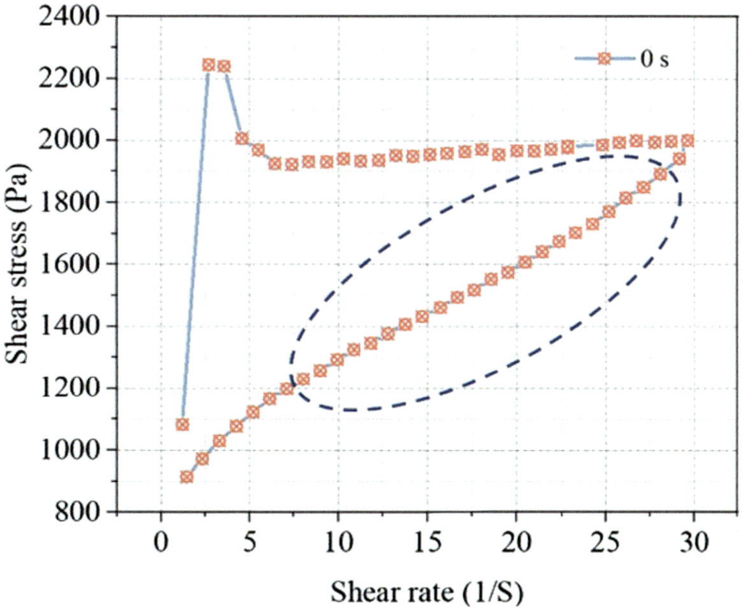

FIGURE 1.3 Bingham curve of cement mortar (Cheng, Radlińska et al. 2024).

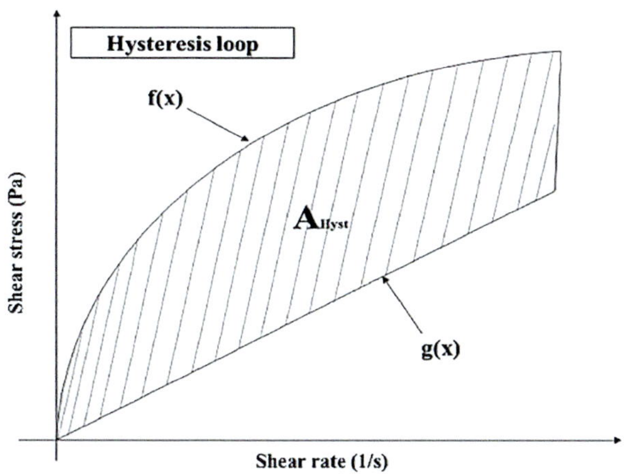

FIGURE 1.4 Hysteresis loop (Lee, Lee et al. 2022).

1.2.2 Influencing Factors

The flowability and pumpability of a printable mixture are primarily determined by its design, which encompasses factors such as cement content, binder compositions, fiber types, aggregate content, shape and gradation, chemical admixtures, and the type of hose employed for pumping.

1.2.2.1 Use of Supplementary Cementitious Materials and Fibers

Generally, various types of supplementary cementitious materials (SCMs), such as fly ash, silica fume, and slag, are used in concrete for better performances, including durability, mechanical properties, and workability (Cheng 2019; Zou, Cheng et al. 2019; Zou, Du et al. 2019; Cheng, Liu et al. 2021; Xiao, Cheng et al. 2024). Besides, the use of SCMs also modifies fresh properties of concrete. An adequate paste content is crucial for enhancing the flowability and pumpability of concrete during the pumping process. The formation of a sufficient slip layer around the hose facilitates pumping and prevents pressurized bleeding under high pumping pressure (Yuan et al. 2024). To achieve the desired flowability and pumpability, various SCMs are employed to replace a portion of the cement content. For example, fly ash, which possesses a lower specific gravity than cement, is typically used to replace cement by the same weight. This increases the volume of paste while proportionally reducing the cement concentration, which in turn reduces the connections among flocculated cement particles. The "balling effect," resulting from the spherical geometry and smooth surface of fly ash, facilitates particle sliding and reduces friction among particles (Bentz and Ferraris 2010). The incorporation of fly ash also exerts a filling effect on the mixture, enhancing flowability by increasing the packing density of the paste, thereby reducing water entrapment within the particle flocs (Ramamurthy and Harikrishnan 2006; Bentz and Ferraris 2010; Mehdipour et al. 2013). It has been reported that increasing the fly ash content leads to lower resistance to flow and a decrease in thixotropy (Jiang, Fall et al. 2020). However, it is important to note that different types of fly ash exhibit varying effects on flowability and pumpability due to the presence of unburnt carbon. Residual carbon typically absorbs superplasticizer and consumes water, resulting in a reduction in flowability (Chandara, Sakai et al. 2010). A similar phenomenon was observed in the research conducted by Choi et al. (Choi, Park et al. 2015). Their study showed that fly ash containing approximately 2.88% by weight of unburnt carbon actively adsorbs superplasticizer, leading to a reduction in flowability.

Silica fume is another commonly utilized SCM in concrete. Studies have shown that the incorporation of silica fume, along with additional water, increases the concentration of fines content in concrete, while simultaneously reducing the plastic viscosity of the mixture. This results in an improvement in both flowability and pumpability to a certain extent (Mechtcherine et al. 2014). The effects of silica fume on the pumping performance of concrete have been extensively investigated. The addition of silica fume has been found to significantly enhance the flow rate of pumped concrete. When zirconia silica fume was utilized, the flow rate was observed to double compared to that of concrete containing ordinary silica fume (Jae Hong Jeong et al. 2016). It was reported that the incorporation of silica fume resulted in a 5% reduction in plastic viscosity and yield stress. However, with further replacement, both yield stress and plastic viscosity began to increase. This increase is attributed to the absorption of superplasticizer by the silica fume, which enhances the reactive process due to its high specific surface area and fine particle size (Jiang, Fall et al. 2020). To further enhance the flowability of printable concrete, the incorporation of a mixed composition of nano clay and silica fume has been identified as a viable alternative. According to research (Zhang, Zhang et al. 2019), a mixture containing

2% nano clay and 2% silica fume enhances early flowability within minutes of mixing, thereby facilitating easier pumping and extrusion. However, reduced flowability was observed at later stages, leading to congestion and disruptions during printing. Furthermore, poor printability was associated with an increased number of voids within the concrete filaments. A favorable yield stress range for pumping a printable mix with a layer thickness of approximately 10 mm is 0.6–1 kPa, as measured by a flow table test (Panda and Tan 2018).

The incorporation of slag is another way to enhance flowability and pumpability of 3D-printable concrete because it has a smaller specific gravity than cement, yielding more paste volume for a given mixture design (Rashad 2018). According to research (Jiang, Fall et al. 2020), with the incorporation of slag, the plastic viscosity and yield stress of concrete have reduction. Thus, the flowability and pumpability were enhanced. The main reason is that slag can occupy the voids left by the coarser mortar and concrete particles, reducing matrix internal frictions. Based on other findings (Choi, Park et al. 2015), a replacement ratio of 60% was identified as the optimal one, which shows a noticeable improvement in pumpability and flowability. However, the reduction in viscosity and yield stress poses a threat to the buildability and shape rentability of 3D-printable concrete.

Fiber reinforcement is an effective strategy for enhancing the structural integrity of 3D-printable concrete; however, it negatively impacts flowability and pumpability. The incorporation of fibers generally reduces the flowability of 3DPC (Jiang, Liu et al. 2022). In contrast, Rushing et al. found that the addition of 0.46 vol.% short reinforcing fibers (0.5-inch-long steel or nylon fibers) to the printed concrete slightly increased flowability compared to the control mixture, with the steel fiber-reinforced mixture exhibiting better flowability than the nylon fiber-reinforced one (Rushing et al. 2017).

1.2.2.2 Physical Properties of Aggregates

Aggregate is the primary component of printable concrete, and its morphology significantly influences the flowability and pumpability of the material. For instance, concrete made with rounded aggregates typically requires less pumping pressure than that made with crushed aggregates. The primary reason is that crushed aggregates possess a larger specific surface area, necessitating more paste to coat the particles and form a slip layer. In the research (Choi, Kim et al. 2013), the impact of aggregate shape on the pumpability of concrete was analyzed using the multi-model suspension viscosity model (Farris 1968) to simulate slip layer properties. Additionally, the analysis of concrete flow through pipes revealed that the particle concentration profiles of sand and gravel are related to the maximum particle concentration (Choi, Kim et al. 2013). Intrinsic viscosity is typically used to assess the effect of aggregate shape on pumpability, with a value of 2.5 being characteristic of spherical particles (Liu, Guo et al. 2020). Compared to rounded aggregates, crushed aggregates exhibit higher intrinsic viscosity, leading to increased resistance to particle displacement, and thus higher overall viscosity. The influence of aggregate size on the pumpability of concrete was thoroughly investigated (Choi, Kim et al. 2013). Aggregate size significantly influences pumping conditions; larger aggregates lead to increased yield stress, viscosity, and pumping pressure. For instance, the pumping pressure required for concrete containing 25 mm aggregates was nearly double that needed for a mixture with 10 mm aggregates (Sung Choi 2014). Research indicates that the pumpability of a mixture is influenced by the gradation of aggregates in concrete, with enhanced pumpability observed as the proportion of fine aggregates and binders increases and the content of coarse aggregates decreases. A mix with up to 30% volume replacement of coarse aggregates was noted to be extrudable without segregation or blockages (Rahul and Santhanam 2020). Weng et al. advocate the use of continuous sand particle gradation to meet the rheological requirements of 3DCP, specifically to achieve low viscosity and high yield stress in the mix. Their experimental work demonstrated that this mix achieves higher buildability at a given pumping pressure than those containing gap-graded or uniformly graded sand particles (Weng et al. 2018). The water absorption behavior of aggregates is another influential factor affecting the rheological performance of concrete (Westerholm et al. 2008; Ahmad et al. 2024; Cheng 2024; Cheng, Ahmad et al. 2025; Deng et al. 2025). As demonstrated by Zhuang et al. (Zhuang et al. 2025), theoretical analyses employing the compressible packing model were conducted to elucidate the relationship between aggregate characteristics and rheological behavior from the perspective of particle packing.

1.2.2.3 Use of Chemical Admixtures

Superplasticizer is a crucial component of printable concrete, providing adequate flowability and pumpability, as well as high strength, even at a low water-to-binder ratio. Malaeb observed that the characteristics of fresh concrete change gradually as the superplasticizer content increases, until a certain threshold is reached; beyond this point, further increases in superplasticizer dosage lead to increased flowability but a loss of buildability. They reported a critical superplasticizer dosage of 0.8%, although this value may vary depending on the overall flowability dictated by the material mix proportions (Malaeb et al. 2019). Therefore, determining the optimal superplasticizer dosage for a given printable concrete mix is essential. In a study on polypropylene-reinforced concrete with a binder comprising 63% cement, 29% silica fume, and 8% fly ash, it was found that increasing the superplasticizer dosage from 0.75% to 1% reduced void content but also induced greater filament deformation and a consequent loss of buildability (Paritala et al. 2023). Various thickening agents have been utilized in printable mixture designs to achieve optimal cohesion. For instance, hydroxypropyl methylcellulose (HPMC) has demonstrated the ability to enhance the yield stress and flowability of printable mixtures (Chen, Chaves Figueiredo et al. 2019; Chen, Li et al. 2019). Research revealed that a printable limestone calcined clay was developed with an optimal HPMC concentration of 0.24%. The inclusion of HPMC resulted in reduced pump pressure requirements and improved buildability (Chen, Li et al. 2019). For successful printing, an adequately long dormant time to ensure pumpability is crucial. Retarders can be employed to achieve this. The dosage of the retarder depends on ambient temperature, printing parameters, dimensions of printed elements, and the targeted strength (Shanableh et al. 2017). Retarders are typically used for fast-setting printable concrete, such as sulfoaluminate cement-based composites or magnesium cement-based composites.

1.2.2.4 Materials and Size of Hose and Nozzle

The pumping behavior of flowable concrete through pipes composed of different materials was investigated. It was found that a higher silica fume dosage led to a sticky mix, affecting the flow rate by increasing it in rubber pipes while decreasing it in steel and acrylic pipes (Jacobsen, Haugan et al. 2009). They also observed that the influence of pipe material is more pronounced in mixes without admixtures and fillers. The pressure required to pump concrete through the pipe, the kinematics of the print head, the cross-sectional opening of the nozzle, buildability, and the curvature of the printed geometry are all directly or indirectly influenced by the size of the printed element and printing duration. Paul et al. examined the effect of nozzle size and shape on overall printability and reported that as the nozzle size increases, the extrusion pressure must be increased to synchronize with the speed of the print head (Paul et al. 2018).

1.3 EXTRUDABILITY

Extrudability refers to the ability of a mixture to be smoothly extruded as a continuous filament from the nozzle at the print head, maintaining a consistent material flow (Andrew Ting, Noel Quah et al. 2022). It is primarily influenced by the composition of the mixture and the printing system (Chen, Chaves Figueiredo et al. 2019; Chen, Li et al. 2019; Li, Hojati et al. 2020; Andrew Ting et al. 2022; Jiang, Liu et al. 2022; Chang, Chen et al. 2023; Paritala et al. 2023).

1.3.1 Experimental and Numerical Techniques

Ram extruders are widely employed to assess extrudability for printing applications, and similar manual extruders have also been used for this purpose (Nerella et al. 2019). The penetration resistance method and squeezing test are also commonly used for the initial assessment of mixture extrudability. However, these testing methods cannot fully replicate the actual extrusion conditions encountered in 3D printing (Toutou et al. 2005; Paul et al. 2018; Gu and Khayat 2024). The direct assessing method is carrying out a single-layer printing test, shown as in Figure 1.5. An extrudability coefficient was proposed to optimize printing quality (Ma et al. 2018). The relationship among extrudability, extrusion rate, and printing speed was investigated. Rushing et al. developed a laboratory evaluation method using a Bailey Ceramics Standard 9 clay extruder assembly (Rushing et al. 2017). To ensure that extrudability characterization is independent of geometry-induced artifacts, it is essential that the chosen testing device operates on a principle similar to that of the machine used in printing applications. Extrudability can be also characterized by measuring rheological properties (Barra Bizinotto et al. 2017; Cheng, Radlińska et al. 2024). Panda and Tan used a rheometer to study the yield stress of materials and evaluate their extrudability (Panda and Tan 2018). High static yield stress results in inadequate extrudability and blockage problems (Rahul and Santhanam 2020).

Numerical simulation is another way to investigate the extrudability of 3D-printable concrete. Several numerical models were proposed to analyze the fluidity mechanisms

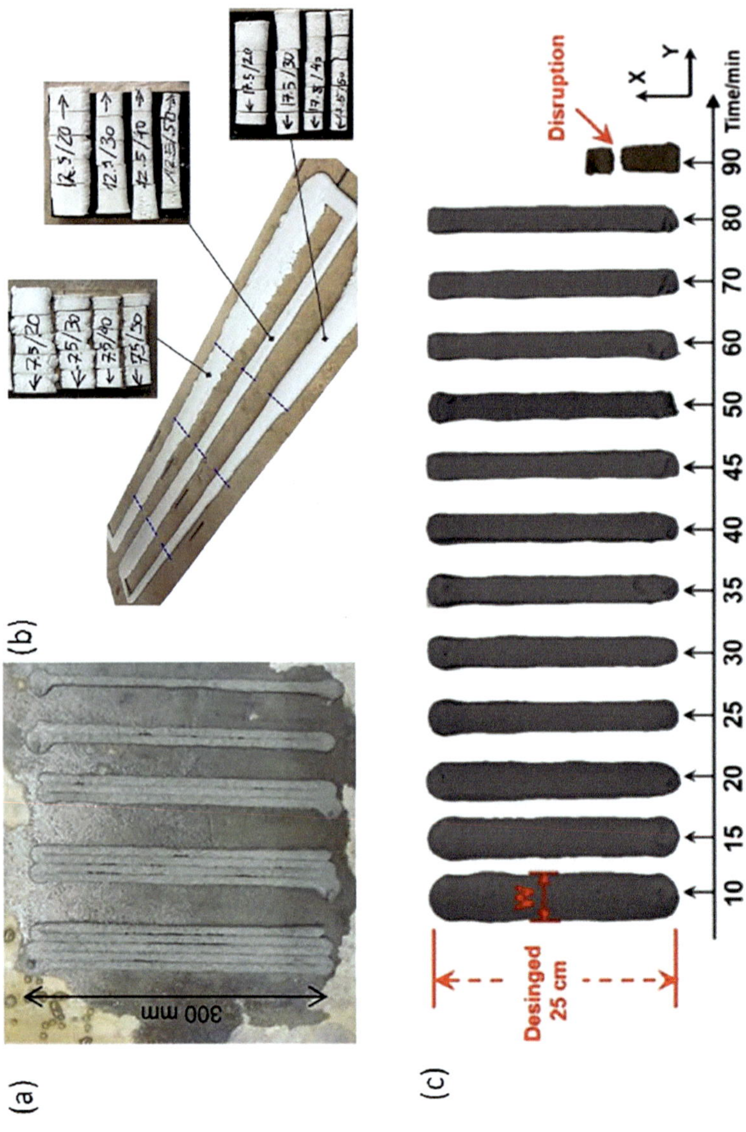

FIGURE 1.5 Test sample to evaluate extrudability: (a) continuous single-layer filament deposition over a designed printing length (Le et al., 2012); (b) dimensional measurement of printed filaments at selected locations to assess extrusion uniformity and shape stability (Comminal et al., 2020); and (c) time-dependent single-filament printing test showing continuous extrusion at different elapsed times and extrusion disruption at 90 min (Ma et al., 2018).

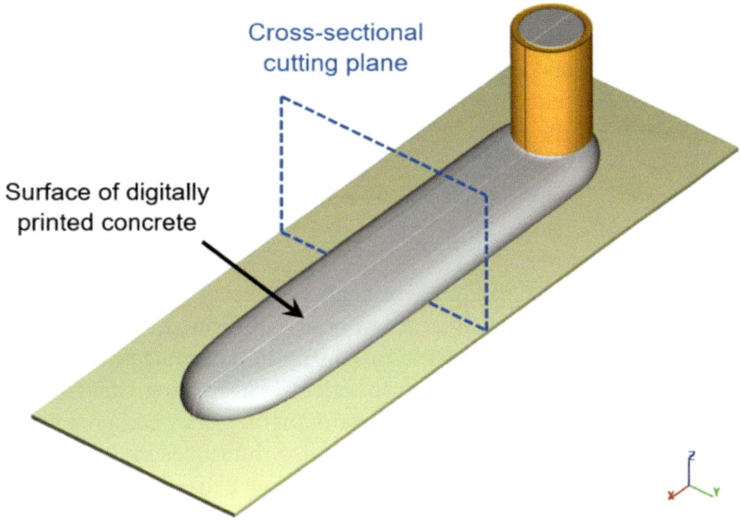

FIGURE 1.6 Example of simulated layer geometry based on CFD (Comminal et al. 2020).

of mortar. These models provide structural and microscopic insights into fluidity levels during the extrusion and deposition processes. For instance, Comminal et al. developed a model based on computational fluid dynamics (CFD) and demonstrated how the geometrical aspect ratio between nozzle size and filament height affects filament geometry (Comminal et al. 2020), as shown in Figure 1.6. Similarly, Liu et al. (Liu, Li, Weng et al. 2020) developed a support vector machine model to investigate the critical factors influencing the extrusion mechanism and deformation. The simulation results revealed that the deformation of extruded filaments depends on nozzle speed and yield stress, with higher nozzle travel speeds leading to significantly reduced filament deformation. Although the effectiveness and feasibility of numerical simulations have been confirmed by these studies, the accuracy of these simulations requires further improvement, particularly for printed specimens with longer printing processes (Iftekar et al. 2023). The primary reason for this limitation is that the dynamic effects of moisture and thermal influences were not incorporated into the numerical characterization. Accurately positioning the nozzle and correcting deviations in the printing plate, or more likely, the inhomogeneous rheological properties of the printed mortar and fluctuating extrusion frequency, may further complicate the simulation process (Roussel 2018).

1.3.2 Influent Factors

The binder content and aggregate proportions typically influence the extrudability of concrete. For example, research has shown that increasing the cement content while decreasing the sand content leads to improved extrudability (Ma, Li et al. 2018). This improvement is attributed to the increased paste volume, which provides better lubrication, similar to the effects observed on flowability and pumpability as previously discussed. Hambach and Volkmer (2017) found that the addition of superplasticizer ensures smooth and continuous extrusion, while the introduction of fibers in mixtures often causes frequent blockages during printing. To avoid extrusion issues, it is recommended that the fiber content not exceed 1% by volume (Panda, Chandra Paul et al. 2017).

Printing parameters significantly influence extrusion quality. As demonstrated by Wolfs et al. (Wolfs, Bos, van Strien et al. 2018), an increase in printing height often results in curling between two fresh layers. Conversely, reducing the height between the nozzle and the printed layer leads to squeezing, potentially causing an overload on the surface of the unhardened concrete. Consequently, an inappropriate nozzle-to-build distance can lead to instability in the printed filaments. A larger nozzle can cause collapse due to the insufficient load-bearing capacity of the filament, while a smaller nozzle can induce separation of mortar components (Joh et al. 2020). In addition to nozzle size, its geometry also influences the extrusion process. Roussel elaborated on the differences in material flow behavior between rectangular and conical nozzles: laminar flow occurs in rectangular nozzles, resulting in a stiff and non-deformed filament, while non-laminar flow occurs in conical nozzles, causing the material to expand upon deposition despite easier extrusion (Roussel 2018). Roussel et al. (Roussel, Spangenberg et al. 2020) concluded that a strong correlation exists between filament deformation and compressive stresses. The primary reason for this correlation is that when the nozzle travel speed is lower than the quotient of extrusion flow rate and cross-sectional area, compressive stresses are generated on the surface and inside the filament. Consequently, localized buckling observed during the printing process is attributed to these compressive stresses.

However, an increased nozzle travel speed can result in filament tearing, and in some cases, the filament may become discontinuous (Roussel, Spangenberg et al. 2020). Previous studies have demonstrated that filament swelling and corner rounding frequently occur at the corners of the printed path. It is commonly observed that more mortar is extruded to the outside of the nozzle during the corner deposition process, leading to this phenomenon (Liu, Li, Tay et al. 2020). Therefore, adjusting the nozzle's movement speed at corners and reducing the material delivery rate can mitigate this issue (Tay, Li et al. 2019).

1.4 BUILDABILITY

Buildability refers to the ability of printed concrete to resist deformation caused by the cumulative weight of successive layers as printing progresses. In the absence of a framework, it is crucial that the extruded filament maintains a stable shape, as this is essential for ensuring buildability. Sufficient early green strength is necessary for the printed layers to ensure that the first layer can support the weight of subsequent layers without distortion or failure. To achieve optimal mechanical performance, the fresh layers must exhibit strong bonding to prevent the formation of cold joints. A buildable mix should exhibit high plastic viscosity and yield stress, as discussed in Section 1.4.1.

FIGURE 1.7 Typical failure modes during the printing process: (a) plastic collapse dominant failure mode; (b) elastic buckling dominant failure mode; (c) failure mode between elastic buckling and plastic collapse (Chang, Chen et al. 2023).

A better understanding of buildability can help prevent printing failures. During the printing process, the printed structure may fail due to material yielding, structural instability, or a combination of both, as shown in Figure 1.7. Material yielding leads to the plastic collapse of the printed structure, while structural instability or imperfections can result in elastic buckling. Therefore, to ensure the buildability of 3DPC structures, both the material's strength and stiffness must develop adequately during the deposition process. Buildability is closely related to factors such as setting time, printing interval gaps, green strength, and the bonding capacity between layers.

1.4.1 Methods for Buildability Assessment

1.4.1.1 Experimental Methods

It is stated that printing trials are the most straightforward method for assessing buildability. These trials can consider various material properties and printing parameters. Hollow cylinders (Moini et al. 2022), square cylinders (Wolfs and Suiker 2019), and wall structures (Suiker 2018) are commonly printed for buildability quantification. Plastic failure, elastic buckling, and combination failures are often observed in these trials. Although the maximum height of printed concrete can be directly measured, these experimental trials are typically time- and labor-intensive. In this context, rheological assessments and compression tests have been adopted to evaluate buildability (Casagrande et al. 2020; Jo et al. 2020; Joh et al. 2020; De Vlieger et al. 2023). Static yield stress is a key parameter associated with the buildability of concrete. Through a series of tests, it is concluded that the suggested range for static yield stress is 0.3–0.9 kPa. If the static yield stress is below this range, i.e., 0.3 kPa, the fresh material is too soft, leading to segregation in the pipe–pump–nozzle system. The extruded material cannot maintain its shape due to excessive deformation. Conversely, if the static yield stress exceeds this range, i.e., 0.9 kPa, the cementitious material becomes difficult to extrude and print continuously (Le et al. 2012). The compression test is another straightforward method to quantify buildability. For instance, the buildability of cylindrical samples was evaluated (Ivanova et al. 2022).

The impact of mix design on build-up capacity was examined using a compression test (Kazemian et al. 2017). In these tests, the layers placed on top of one another are regarded as compressive loading applied to the specimen. A shape retention parameter has been proposed to assess the buildability of concrete. It has been reported that reducing the stand-off distance compromises the dimensional precision of the printed structure (Nerella, Krause et al. 2020).

1.4.1.2 Analytical Models

To achieve a more efficient and less labor- and time-intensive assessment of buildability, analytical models have been proposed, including rheological and solid mechanics models. Roussel (2018) laid the foundation for rheological models. In his research, a comprehensive study was presented on how material properties – such as yield stress, elastic modulus, viscosity, and structuration rate – affect buildability. The proposed criteria evaluate rheological requirements and assess the final geometrical dimensions of a single filament, taking into account factors such as surface cracking and structural instability. To measure the structural build-up rate, Roussel adopted the parameter G_0 (Roussel 2006) and static yield stress $(\sigma_{\circ})$ to predict material yielding using Eq. 1.2.

$$\tau_0(t_{rest}) = \sigma_{0,t=0} + A_{thix} t_{rest} \tag{1.2}$$

Here, the t_{rest} refers to the resting time after material deposition, and $\sigma_{0,t=0}$ refers to material initial yield stress. It should be pointed out that the sheared extruded material has a lower static yield stress than the unsheared one because of the breakdown of the connection among cement particles. A_{thix} is the yield stress development due to the structuration and flocculation. According to the time range considered, the material development can be described by a linear or nonlinear tendency (Perrot, Pierre et al. 2014).

As is stated in Kruger et al. (2019a, 2019b, 2021), it is necessary to consider re-flocculation of cement particle network when concrete is under high shear stress. He proposed a bi-linear model to describe the material yield stress. Re-flocculation (R_{thix}) and structuration (A_{thix}) are considered, shown as below:

$$\begin{aligned} \tau_s(t) &= \tau_{D,i} + R_{thix} t \quad for\, t \le t_{rf} \\ \tau_s(t) &= \tau_{s,i} + A_{thix}(t - t_{rf})\, for\, t \ge t_{rf} \\ t_{rf} &= \frac{\tau_{s,i} - \tau_{D,i}}{R_{thix}}, \end{aligned} \tag{1.3}$$

where $\tau_s(t)$ is the static or the apparent yield stress of the printed material after agitation. $\tau_{s,i}$ and $\tau_{D,i}$ stand for initial static and dynamic yield stress of printed material, respectively, which can be measured with rheological tests.

A more simper and direct model was proposed by Perrot (Perrot, Rangeard et al. 2015). He defined the optimal build-up rate for considering material yielding. The critical printing height can be estimated by using the strength of printed materials, explained as

$$H_{critical} = \frac{\sqrt{3}\tau_0(t_{pr})}{\rho g}. \tag{1.4}$$

If the strength of the bottom layer reaches the material strength, plastic collapse is assumed to occur.

Further development was made by Wangler (Wangler et al. 2019), who proposed the maximal horizontal printing velocity, as shown in

$$V_{r,\max} = \frac{\sqrt{3}LA_{thix}}{\rho g H_{layer}}, \tag{1.5}$$

where L refers to the contour length. Unlike other analytical models, this model examines the impact of structural geometry on predicting the dominant plastic collapse failure mode, as a geometric coefficient is introduced into the equation. However, this parameter is applicable only to typical wall geometries.

Another analytical model for the prediction of the critical printed height was proposed (Roussel 2018):

$$H_{c,buck} = (\frac{8EI}{\rho g A})^{1/3}, \tag{1.6}$$

where E and I refer to the material stiffness and second moment of inertia, respectively, and A is the horizontal rectangular cross-sectional area.

While the aforementioned analytical models are valuable for assessing buildability, they struggle to account for the influence of geometry and structural heterogeneity (Kruger et al. 2019a, 2019b; Zhang, Zhang et al. 2019). To address these limitations, Suiker introduced a mechanistic model that incorporates a range of critical parameters. These parameters include time-dependent material stiffness and strength, printing velocity, boundary conditions, structural imperfections, and non-uniform gravitational loading (Suiker 2018).

This model is particularly effective in analyzing the build-up performance of wall structures by considering both elastic buckling and plastic collapse as potential failure modes (Wolfs and Suiker 2019). As outlined in Suiker's model (Suiker 2018; Suiker et al. 2020), the elastic buckling or plastic collapse of a straight wall during the printing process can be systematically analyzed.

$$\frac{\bar{l}_{cr}}{\bar{l}_p} < \bar{\Lambda}: \text{elastic buckling}$$

$$\frac{\bar{l}_{cr}}{\bar{l}_p} > \bar{\Lambda}: \text{plastic collapse} \tag{1.7}$$

$$\text{with } \bar{\Lambda} = \left(\frac{h}{D_0}\right)^{1/3} \frac{\sigma_{p,0}}{(pg)^{2/3}}$$

In this model, the dimensionless parameter Λ serves as the "failure mechanism indicator." According to this model, the failure mechanism is determined by a combination of factors, including wall thickness (h), initial material yield strength ($\sigma_{p,0}$), initial bending stiffness (D_0) of the printed wall structure, material density (ρ), and gravitational acceleration.

1.4.1.3 Numerical Models

The aforementioned analytical models provide simple and time-efficient solutions for buildability assessment. However, their primary limitation lies in their applicability, which is restricted to specific printing geometries, with wall structures being the primary focus. To address this constraint, many researchers have turned to numerical simulations as a more versatile approach. In numerical models, two theoretical frameworks – fluid mechanics and solid mechanics – have been employed to simulate and analyze these behaviors.

CFD models, grounded in fluid mechanics, have been employed to explore the relationship between layer geometry and various printing parameters, such as printing velocity and nozzle height (Mollah et al. 2021). To date, most CFD models have concentrated on the extrusion process or the geometry prediction of a single printing layer. More recently, this method has also been used to simulate the stability and deformation of multiple layers. However, it is important to note that CFD models have not yet been utilized for buildability quantification.

A finite element method (FEM)-based model was proposed using the ABAQUS software package to investigate structural failure during the printing process (Wolfs, Bos, Salet et al. 2018; Wolfs and Suiker 2019), as shown in Figure 1.8. This method allows for the simulation of plastic collapse and elastic buckling. For plastic collapse simulation, time-dependent material properties and the Mohr–Coulomb failure criterion are employed. Uniaxial compression and shear tests are conducted to determine the required material properties at varying ages. To account for large deformations in structural analysis, a geometric nonlinear analysis is utilized. However, a significant quantitative discrepancy of nearly 60% was observed between the FEM simulations and the printing experiments. This disparity is primarily attributed to the omission of localized damage, stress redistribution, and nonuniform gravitational loading in the simulations. Regarding elastic buckling, a linear buckling analysis was proposed (Wolfs, Bos et al. 2019). An initial imperfection, based on the first-mode instability, was obtained and then incorporated into the numerical model for nonlinear buckling analysis. This approach yielded a good quantitative agreement with the printing tests. However, the buckling failure modes produced by the FEM-based method differed from those observed in the printing experiments. This difference is likely due to the exclusion of nonuniform gravitational loading in the 3D printing simulations. Building upon the FEM models described above, Ooms et al. proposed a parametric tool to create input files for 3D printing models without extensive manual modeling (Ooms et al. 2021; Vantyghem et al. 2021). The pre-processing step was automatically carried out for structural analysis. With this technique, the influence of nonuniform gravitation loading on failure height was explored. And how different interaction methods, tie constraints and contact-based interactions, affect structural deformation was explored. Their work shows that if the 3D printing model based

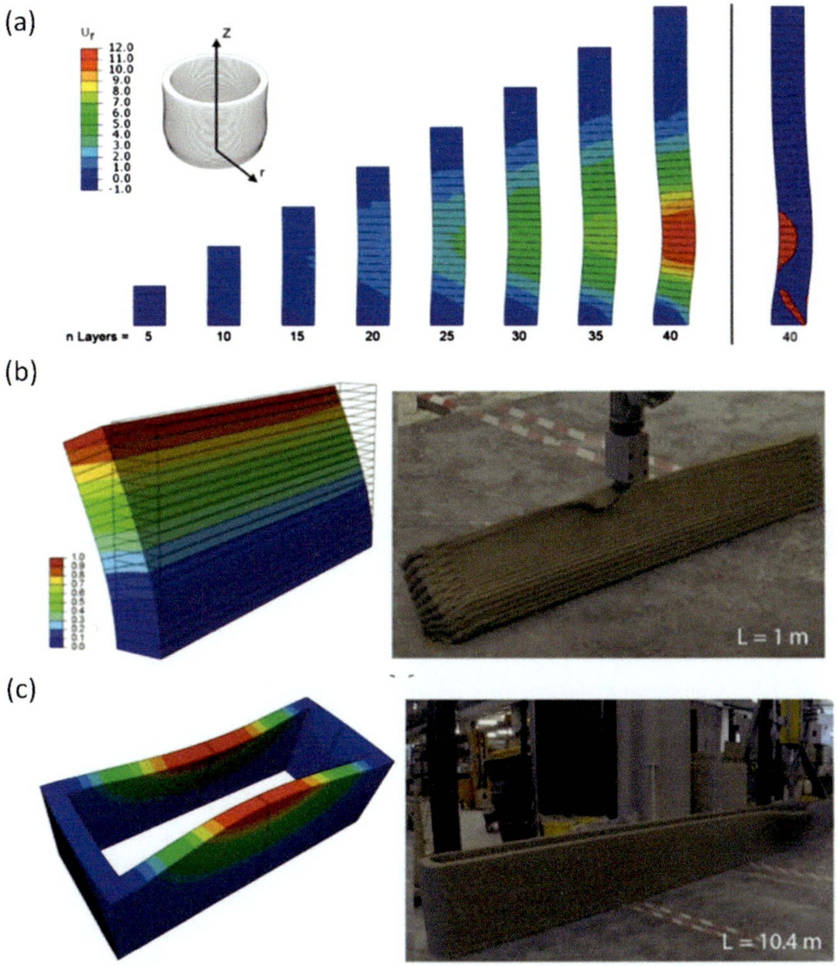

FIGURE 1.8 FEM-based failure simulations: (a) plastic collapse of a hollow cylinder structure (Wolfs and Suiker 2019); (b) elastic buckling of a wall; (c) elastic buckling of a rectangular structure (Chang, Chen et al. 2023).

on ABAQUS is used, the option (i.e., model change interaction with strain) is the only choice with regard to the printing process. A similar research was presented by Nguyen-Van (Nguyen-Van et al. 2021), wherein a novel computational framework to model the structural buildability of complex triply periodic surface via the given toolpath was proposed (Figure 1.9). This technique can present how printing speed influences the critical printed height and vertical deformation.

Numerical simulation models for 3DCP offer valuable insights for predicting deformation and assessing buildability.

However, current research predominantly focuses on the extrusion process. In terms of buildability assessment, these simulations often simplify the analysis by

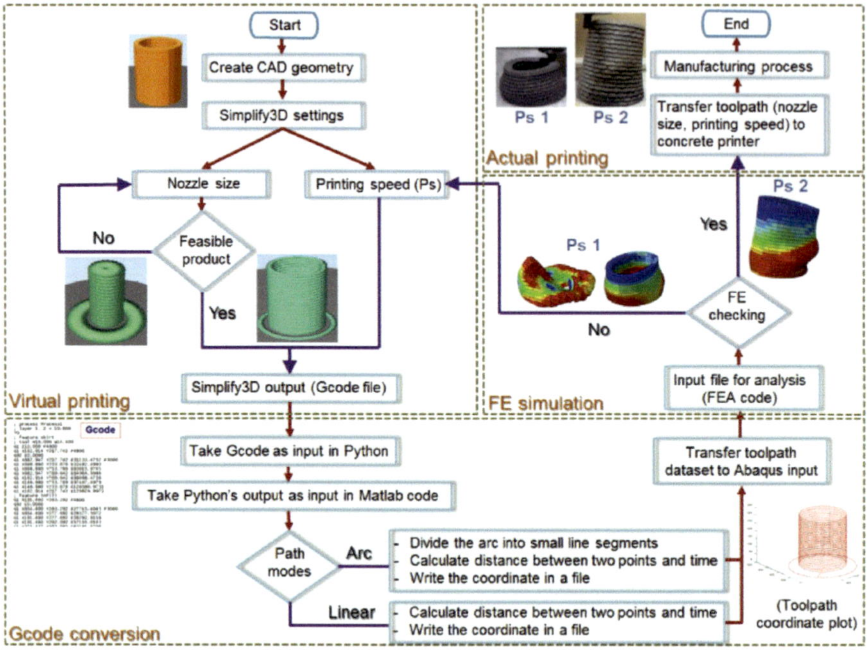

FIGURE 1.9 Computational framework for 3DCP (Nguyen-Van et al. 2021).

assuming that structural failure occurs when the printed structure reaches the material's strength. This assumption can lead to the underestimation of critical printing height, as it neglects stress redistribution and nonuniform stress distribution. Furthermore, 3D printing involves a layer-by-layer extrusion process, which can result in more variable mechanical properties – such as strength and stiffness – compared to traditional casting methods. Due to this material variation and nonuniform gravitational loading, localized damage may occur, potentially impacting structural buildability. Therefore, it is crucial to develop models that incorporate material heterogeneity, nonuniform gravitational loading, localized damage, and stress redistribution to more accurately quantify buildability in 3DCP.

Solid mechanics numerical models have been proposed in which the material is assumed to behave as an elasto-plastic substance. These models typically consider only instantaneous strain during the printing process, while time-dependent deformations are neglected. However, factors such as creep, plastic and autogenous shrinkage, and consolidation settlement under distributed compressive loading, which increase with printing time, are crucial considerations. According to published research, several early-age creep tests on 3D-printable mortar and paste have been conducted (Chen, Li et al. 2019; Esposito et al. 2021). These experiments indicate that early-age creep accounts for approximately 7% of the viscoelastic deformation in the tested samples. Given this significant contribution, it is essential to incorporate early-age creep into 3D printing models to better understand its impact on structural deformation prediction and buildability quantification.

The lattice simulation method has been proposed to investigate the impact of localized damage on buildability (Chang, Liang et al. 2022). This method also allows for the consideration of additional factors, such as printing velocity, time-dependent material stiffness and strength, and nonuniform gravitational loading. Recently, the early-age creep behavior of 3D-printable mortar, measured through cyclic compressive loading, has been incorporated into this model to study its impact on buildability quantification (Chang, Liang et al. 2023).

1.4.2 Influent Factors

Buildability is influenced by both mixture design and printing parameters. Zhang et al. studied printable mixtures incorporating nano clay and silica fume, finding that the addition of 2% nano clay and 2% silica fume improved buildability by 150% and 115%, respectively (Zhang, Zhang et al., 2018). Furthermore, the combined addition of nano clay and silica fume was shown to optimize buildability. Specifically, the inclusion of 0.5% nano clay increased the static yield stress by 300% (Dejaeghere et al. 2019). Other materials, such as rice husk (Muthukrishnan et al. 2021) and limestone (Jia et al. 2024), have also been demonstrated to enhance buildability.

For alkali-activated concrete, Panda et al. (2018) recommended using a reduced amount of slag, approximately 14% of the total binder content for 3DCP, as the setting time of fly ash-based geopolymer cement decreases with higher slag concentrations. Weng et al. (2018) investigated mixtures with different sand gradations. They reported that mixtures with continuous gradation could print a greater number of layers in the fresh state compared to gap-graded and uniformly graded mixtures.

Research has shown that nozzle size significantly influences the yield stress required to ensure buildability (Shakor et al. 2019). Larger nozzle sizes necessitate supporting a greater weight of subsequent layers. Various nozzle shapes, including square, rectangular, circular, and elliptical, are commonly used. While a circular nozzle offers ease in adjusting the extrusion angle when printing complex geometries, it can create voids and result in poor bonding between layers. In contrast, square or rectangular nozzle shapes provide a good surface finish and enhance buildability (Le et al. 2012; Shakor et al. 2019). Optimizing the concrete discharge rate, the print speed, and the time gap between subsequent layers is crucial for achieving optimal buildability. The time gap should be carefully balanced; it must not be so short that the structure collapses, nor so long that it compromises interfacial bonding (Liu, Li, Tay et al. 2020; Shakor et al. 2020). To optimize the quality of the printed structure, the fresh properties and print parameters must be tailored according to the final scale and geometry of the element. Tripathi et al. suggest that buildability in concrete printing can be enhanced through set-on-demand techniques (Tripathi et al. 2022), which can be achieved by employing methods such as accelerator intervention, magnetorheological control, ultrasonication at the print head, and microwave heating (Muthukrishnan et al. 2021).

1.5 OPEN TIME

Open time refers to the interval from the initial contact of the dry mix with water to the latest time when the mixture remains printable. During this period, the mixture should maintain good flowability, pumpability, and extrudability without causing any blockages during pumping. It is important to note that initial and final setting times are not reliable indicators for determining open time. Research has shown that discontinuous extrusion and nozzle blockages can occur well before the initial setting time, as reported by Kazemian et al. who strongly suggested that the setting time should not be used as a substitute for the open time (Kazemian et al. 2017).

An optimal open time is crucial for successful 3DCP. A short open time increases the risk of blockages and poor bonding between layers. Conversely, a long open time can negatively impact shape retainability and buildability. Therefore, balancing the open time is essential to achieve the desired quality and performance in 3DCP.

1.5.1 Experimental Techniques

The flow table test is a straightforward method for determining open time (Le et al. 2012). Rheology measurements are also widely utilized (Le et al. 2012; Panda, Paul, Hui et al. 2017; Panda et al. 2018, Panda and Tan 2018). Panda and colleagues evaluated the changes in intrinsic rheological properties, such as yield stress, viscosity, and thixotropy, over time and defined open time as the period when these rheological properties reach the critical values necessary for the smooth extrusion of fresh materials (Panda, Paul, Hui et al. 2017; Panda et al. 2018; Panda and Tan 2018). Additionally, a single filament printing test has been employed to determine open time (Ma et al. 2018). While this method is more direct, it is also time- and labor-intensive. In this approach, open time is determined by the moment when the first disruption in extrusion occurs.

1.5.2 Influent Factors

The primary factors influencing open time are closely related to mixture design, including cement content, binder composition, aggregate content, aggregate shape, size, and gradation, and the addition of superplasticizers, retarders, and accelerators. The analysis of these factors in relation to open time is similar to the discussion in Section 1.2.

In the context of 3D-printable geopolymer, the type and concentration of the activating agent and precursor significantly affect open time. For instance, Panda et al. designed a fly ash-based geopolymer activated with potassium silicate solution and observed that the open time of the geopolymer mortar was relatively short, around 20 minutes. They suggested that rapid-hardening materials should be mixed and printed simultaneously, rather than being mixed and stored in a container for an extended period before printing (Panda et al. 2018).

1.6 RHEOLOGY EVOLUTION

Based on the above discussions, it is evident that all fresh properties are closely related to the rheology of the mixture, with yield stress being of particular interest. During the pumping and extrusion process, a lower yield stress is desirable to ensure good flowability, pumpability, and extrudability. After deposition, yield stress and elastic modulus must increase sufficiently over time to maintain shape and ensure buildability. Figure 1.10 presents the evolution of the desired yield stress with respect to the 3D printing process (Marchon, Kawashima et al. 2018, Roussel 2018).

Stage 1: At the end of the mixing phase, cementitious materials exhibit an initial yield stress (τ_{c0}) and an initial elastic modulus (G_0). The evolution of these properties is influenced by the competition between flocculation and de-flocculation, caused by agitation during pumping and extrusion. In this stage, a low initial yield stress is favorable for better flowability and extrudability. The addition of superplasticizers and retarders, or the incorporation of silica fume, slag, and fly ash can help lower the yield stress.

Stage 2: Once extruded and deposited, cementitious materials begin to exhibit intrinsic structural build-up. At this stage, cement particles tend to flocculate and form a network of interacting particles as colloidal attractive forces dominate (Marchon et al. 2018). The evolution of yield stress and elasticity during this stage is minimal. The structural build-up rate after deposition can be modified by the use of flocculation or thixotropic agents, such as clays, which lower the initial yield stress but accelerate its evolution over time (Kim, Beacraft et al. 2010; Marchon et al. 2018; Roussel, Bessaies-Bey et al. 2019).

Stage 3: This stage represents a transition period with an almost constant yield stress, corresponding to the induction or dormant period. During this stage, hydration products nucleate at the pseudo-contact points between cement particles within the network. This nucleation process turns colloidal interactions into high-energy interactions, forming solid bridges that increase the elastic modulus on a macroscopic

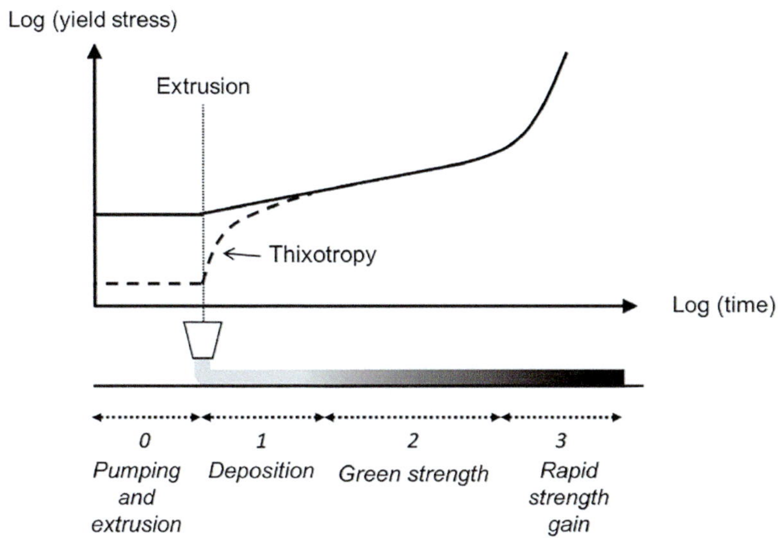

FIGURE 1.10 Development of static yield stress of 3D-printable concrete. The printing direction is from the right to left, with a gray degree indicating the age of the materials (Marchon et al. 2018).

scale. This process is reversible, as the interparticle connections can be broken by shear or remixing (referred to as reversible structural build-up). However, the material at this stage is not stiff enough to support subsequently deposited layers. At the end of this stage, the material is no longer flowable or extrudable, marking the end of the open time for printable materials. The duration of this stage (or open time) can be further tailored using water-reducing admixtures with retarding effects, retarders, or accelerators.

Stage 4: The sudden and irreversible increase in macroscopic yield stress and elastic modulus results from the growth in the size or number of solid bridges (irreversible structural build-up). The printed material, with increased yield stress, can support more weight and minimize deformation due to rapid strength gain. The buildability of printed materials is therefore determined by the evolution during this stage. As new layers are deposited, the stress in the underlying layers increases, necessitating that the evolution of yield stress outpaces gravity-induced stresses. It is important to note that even if gravity-induced stress remains below the yield stress and no spreading flow occurs, the additional strain from elastic behavior could threaten the geometric accuracy of the printed elements.

1.7 OUTLOOK

The future development of 3DCP will increasingly rely on the precise control of fresh-state properties and the establishment of robust multi-scale characterization methods. A key challenge remains achieving a dynamic balance among pumpability, extrudability, buildability, and open time, particularly as novel binders, nano-modifying agents, and functional admixtures are incorporated. Their effects on rheological parameters and early-age mechanical behavior require further systematic investigation. Moreover, the integration of experimental methods with numerical simulations will play a vital role in advancing the understanding of fresh property evolution. Multi-physics coupled models – combining fluid mechanics, solid mechanics, and chemical kinetics – are expected to improve the predictive accuracy of filament stability, interlayer bonding, buckling, and creep. In addition, artificial intelligence and data-driven mixture design approaches offer promising tools to optimize fresh-state performance across wide parameter spaces, enabling performance–process–structure integration. Looking ahead, as 3D printing expands into extreme environments such as cold, arid, or extraterrestrial conditions, future research should also emphasize material adaptability and sustainability. This includes promoting the use of low-carbon raw materials and developing life-cycle assessment frameworks to unify performance evaluation with environmental benefits. In sum, advancing the understanding of fresh-state behavior through interdisciplinary research will be essential for the large-scale and practical deployment of 3DCP.

REFERENCES

Ahmad, I., M. Shokouhian, H. Cheng and A. Radlińska (2024). "Enhancement of mechanical properties and freeze–thaw durability of recycled aggregate concrete using aggregate pretreatment." *Iranian Journal of Science and Technology, Transactions of Civil Engineering* **49**(4): 3605–3629.

Andrew Ting, G. H., T. K. Noel Quah, J. H. Lim, Y. W. Daniel Tay and M. J. Tan (2022). "Extrudable region parametrical study of 3D printable concrete using recycled glass concrete." *Journal of Building Engineering* 50.

Barra Bizinotto, M., F. Faleschini, C. G. Jiménez Fernández and D. F. Aponte Hernández (2017). "Effects of chemical admixtures on the rheology of fresh recycled aggregate concretes." *Construction and Building Materials* **151**: 353–362.

Batikha, M., R. Jotangia, M. Y. Baaj and I. Mousleh (2022). "3D concrete printing for sustainable and economical construction: A comparative study." *Automation in Construction* 134.

Bentz, D. P. and C. F. Ferraris (2010). "Rheology and setting of high volume fly ash mixtures." *Cement and Concrete Composites* **32**(4): 265–270.

Bong, S. H., M. Xia, B. Nematollahi and C. Shi (2021). "Ambient temperature cured 'just-add-water' geopolymer for 3D concrete printing applications." *Cement and Concrete Composites* 121.

Casagrande, L., L. Esposito, C. Menna, D. Asprone and F. Auricchio (2020). "Effect of testing procedures on buildability properties of 3D-printable concrete." *Construction and Building Materials* 245.

Chandara, C., E. Sakai, K. A. M. Azizli, Z. A. Ahmad and S. F. S. Hashim (2010). "The effect of unburned carbon in palm oil fuel ash on fluidity of cement pastes containing superplasticizer." *Construction and Building Materials* **24**(9): 1590–1593.

Chang, Z., Y. Chen, E. Schlangen and B. Šavija (2023). "A review of methods on buildability quantification of extrusion-based 3D concrete printing: From analytical modelling to numerical simulation." *Developments in the Built Environment* 16.

Chang, Z., M. Liang, Y. Xu, E. Schlangen and B. Šavija (2022). "3D concrete printing: Lattice modeling of structural failure considering damage and deformed geometry." *Cement and Concrete Composites* 133.

Chang, Z., M. Liang, Y. Xu, Z. Wan, E. Schlangen and B. Šavija (2023). "Early-age creep of 3D printable mortar: Experiments and analytical modelling." *Cement and Concrete Composites* 138.

Chen, Y., S. Chaves Figueiredo, C. Yalcinkaya, O. Copuroglu, F. Veer and E. Schlangen (2019). "The effect of viscosity-modifying admixture on the extrudability of limestone and calcined clay-based cementitious material for extrusion-based 3D concrete printing." *Materials (Basel)* **12**(9).

Chen, Y., Z. Li, S. Chaves Figueiredo, O. Çopuroğlu, F. Veer and E. Schlangen (2019). "Limestone and calcined clay-based sustainable cementitious materials for 3D concrete printing: A fundamental study of extrudability and early-age strength development." *Applied Sciences* **9**(9).

Cheng, H. (2019). *Study on the Deterioration Law and Uniaxial Compressive Constitutive Relationship of Concrete under Sulfate Attack*. Master of Engineering, Harbin Institute of Technology.

Cheng, H. (2024). *Improving Freeze–Thaw Resistance and Fatigue Resistance of Recycled Aggregate Concrete*. Center for Integrated Asset Management for Multimodal Transportation Infrastructure Systems (CIAMTIS) (UTC), U.S. DOT Region 3 University Transportation Center.

Cheng, H., I. Ahmad, M. Shokouhian and A. Radlińska (2025). "Enhancing flexural fatigue performance of recycled aggregate concrete under freeze–thaw cycles through aggregate and cement matrix modification: Effects, prediction models, and mechanisms." *Construction and Building Materials* 481.

Cheng, H., T. Liu, D. Zou and A. Zhou (2021). "Compressive strength assessment of sulfate-attacked concrete by using sulfate ions distributions." *Construction and Building Materials* 293.

Cheng, H., A. Radlińska, M. Hillman, F. Liu and J. Wang (2024). "Modeling concrete deposition via 3D printing using reproducing kernel particle method." *Cement and Concrete Research* 181.

Choi, M. S., Y. J. Kim and S. H. Kwon (2013). "Prediction on pipe flow of pumped concrete based on shear-induced particle migration." *Cement and Concrete Research* **52**: 216–224.

Choi, M. S., S. B. Park and S.-T. Kang (2015). "Effect of the mineral admixtures on pipe flow of pumped concrete." *Journal of Advanced Concrete Technology* **13**(11): 489–499.

Comminal, R., W. R. Leal da Silva, T. J. Andersen, H. Stang and J. Spangenberg (2020). "Modelling of 3D concrete printing based on computational fluid dynamics." *Cement and Concrete Research* 138. https://doi.org/10.1016/j.cemconres.2020.106256

De Vlieger, J., L. Boehme, J. Blaakmeer and J. Li (2023). "Buildability assessment of mortar with fine recycled aggregates for 3D printing." *Construction and Building Materials* 367.

Dejaeghere, I., M. Sonebi and G. De Schutter (2019). "Influence of nano-clay on rheology, fresh properties, heat of hydration and strength of cement-based mortars." *Construction and Building Materials* **222**: 73–85.

Deng, Q., Y. Zhuang, A. Nasr, H. Liu and Z. Duan (2025). "Assessing the effect of particle characteristics on the rheological properties of mortar with recycled fine aggregate." *Journal of Building Engineering* 107.

Esposito, L., L. Casagrande, C. Menna, D. Asprone and F. Auricchio (2021). "Early-age creep behaviour of 3D printable mortars: Experimental characterisation and analytical modelling." *Materials and Structures* **54**(6).

Farris, R. J. (1968). "Prediction of the viscosity of multimodal suspensions from unimodal viscosity data." *Transactions of the Society of Rheology* **12**(2): 281–301.

Giwa, I., M. Dempsey, M. Fiske and A. Kazemian (2024). "3D printed sulfur-regolith concrete performance evaluation for waterless extraterrestrial robotic construction." *Automation in Construction* 165.

Gosselin, C., R. Duballet, P. Roux, N. Gaudillière, J. Dirrenberger and P. Morel (2016). "Large-scale 3D printing of ultra-high performance concrete – A new processing route for architects and builders." *Materials & Design* **100**: 102–109.

Gu, Y.-c. and K. H. Khayat (2024). "Extrudability window and offline test methods to predict buildability of 3D printing concrete." *Cement and Concrete Research* 182.

Hambach, M. and D. Volkmer (2017). "Properties of 3D-printed fiber-reinforced Portland cement paste." *Cement and Concrete Composites* **79**: 62–70.

Iftekar, S. F., A. Aabid, A. Amir and M. Baig (2023). "Advancements and limitations in 3D printing materials and technologies: A critical review." *Polymers (Basel)* **15**(11).

Ivanova, I., E. Ivaniuk, S. Bisetti, V. N. Nerella and V. Mechtcherine (2022). "Comparison between methods for indirect assessment of buildability in fresh 3D printed mortar and concrete." *Cement and Concrete Research* 156.

Jacobsen, S., L. Haugan, T. A. Hammer and E. Kalogiannidis (2009). "Flow conditions of fresh mortar and concrete in different pipes." *Cement and Concrete Research* **39**(11): 997–1006.

Jae Hong Jeong, K. P. J., C. K. Park, S. H. Lee, and S. H. Kwon (2016). "Effect of admixtures on pumpability for high-strength concrete." *ACI Materials Journal* **113**(3): 323.

Jia, Z., L. Kong, L. Jia, L. Ma, Y. Chen and Y. Zhang (2024). "Printability and mechanical properties of 3D printing ultra-high performance concrete incorporating limestone powder." *Construction and Building Materials* 426.

Jiang, H., M. Fall, E. Yilmaz, Y. Li and L. Yang (2020). "Effect of mineral admixtures on flow properties of fresh cemented paste backfill: Assessment of time dependency and thixotropy." *Powder Technology* **372**: 258–266.

Jiang, Q., Q. Liu, S. Wu, H. Zheng and W. Sun (2022). "Modification effect of nanosilica and polypropylene fiber for extrusion-based 3D printing concrete: Printability and mechanical anisotropy." *Additive Manufacturing* 56.

Jo, J. H., B. W. Jo, W. Cho and J.-H. Kim (2020). "Development of a 3D printer for concrete structures: Laboratory testing of cementitious materials." *International Journal of Concrete Structures and Materials* **14**(1).

Joh, C., J. Lee, T. Q. Bui, J. Park and I. H. Yang (2020). "Buildability and mechanical properties of 3D printed concrete." *Materials (Basel)* **13**(21).

Kazemian, A., X. Yuan, E. Cochran and B. Khoshnevis (2017). "Cementitious materials for construction-scale 3D printing: Laboratory testing of fresh printing mixture." *Construction and Building Materials* **145**: 639–647.

Khan, M. and C. McNally (2023). "A holistic review on the contribution of civil engineers for driving sustainable concrete construction in the built environment." *Developments in the Built Environment* 16.

Kim, H., T. Koh and S. Pyo (2016). "Enhancing flowability and sustainability of ultra high performance concrete incorporating high replacement levels of industrial slags." *Construction and Building Materials* **123**: 153–160.

Kim, J. H., M. Beacraft and S. P. Shah (2010). "Effect of mineral admixtures on formwork pressure of self-consolidating concrete." *Cement and Concrete Composites* **32**(9): 665–671.

Kloft, H., M. Empelmann, N. Hack, E. Herrmann and D. Lowke (2020). "Reinforcement strategies for 3D-concrete-printing." *Civil Engineering Design* **2**(4): 131–139.

Kruger, J., A. du Plessis and G. van Zijl (2021). "An investigation into the porosity of extrusion-based 3D printed concrete." *Additive Manufacturing* 37.

Kruger, J., S. Zeranka and G. van Zijl (2019a). "3D concrete printing: A lower bound analytical model for buildability performance quantification." *Automation in Construction* 106.

Kruger, J., S. Zeranka and G. van Zijl (2019b). "An ab initio approach for thixotropy characterisation of (nanoparticle-infused) 3D printable concrete." *Construction and Building Materials* **224**: 372–386.

Le, T. T., S. A. Austin, S. Lim, R. A. Buswell, A. G. F. Gibb and T. Thorpe (2012). "Mix design and fresh properties for high-performance printing concrete." *Materials and Structures* **45**(8): 1221–1232.

Lee, K.-W., H.-J. Lee and M.-S. Choi (2022). "Correlation between thixotropic behavior and buildability for 3D concrete printing." *Construction and Building Materials* 347.

Li, Z., M. Hojati, Z. Wu, J. Piasente, N. Ashrafi, J. P. Duarte, S. Nazarian, S. G. Bilén, A. M. Memari and A. Radlińska (2020). "Fresh and hardened properties of extrusion-based 3D-printed cementitious materials: A review." *Sustainability* **12**(14).

Liu, G., X. Guo, W. Cheng, L. Chen and X. Cui (2020). "Investigating the migration law of aggregates during concrete flowing in pipe." *Construction and Building Materials* 251.

Liu, Z., M. Li, Y. W. D. Tay, Y. Weng, T. N. Wong and M. J. Tan (2020). "Rotation nozzle and numerical simulation of mass distribution at corners in 3D cementitious material printing." *Additive Manufacturing* 34.

Liu, Z., M. Li, Y. Weng, Y. Qian, T. N. Wong and M. J. Tan (2020). "Modelling and parameter optimization for filament deformation in 3D cementitious material printing using support vector machine." *Composites Part B: Engineering* 193.

Ma, G., Z. Li and L. Wang (2018). "Printable properties of cementitious material containing copper tailings for extrusion based 3D printing." *Construction and Building Materials* **162**: 613–627.

Malaeb, Z., F. AlSakka and F. Hamzeh (2019). 3D concrete printing. *3D Concrete Printing Technology*: 115–136.

Marchon, D., S. Kawashima, H. Bessaies-Bey, S. Mantellato and S. Ng (2018). "Hydration and rheology control of concrete for digital fabrication: Potential admixtures and cement chemistry." *Cement and Concrete Research* **112**: 96–110.

Mechtcherine, V., V. N. Nerella and K. Kasten (2014). "Testing pumpability of concrete using Sliding Pipe Rheometer." *Construction and Building Materials* **53**: 312–323.

Mehdipour, I., N. A. Libre, M. Shekarchi and M. Khanjani (2013). "Effect of workability characteristics on the hardened performance of FRSCCMs." *Construction and Building Materials* **40**: 611–621.

Moini, R., J. Olek, P. D. Zavattieri and J. P. Youngblood (2022). "Early-age buildability-rheological properties relationship in additively manufactured cement paste hollow cylinders." *Cement and Concrete Composites* 131.

Mollah, M. T., R. Comminal, M. P. Serdeczny, D. B. Pedersen and J. Spangenberg (2021). "Stability and deformations of deposited layers in material extrusion additive manufacturing." *Additive Manufacturing* 46.

Mu, J., Y. Li, C. Jin, Y. Liu, H. Li and J. Liu (2022). "Simulation of V-tunnel test for fresh concrete on the basis of lattice Boltzmann method." *Cement and Concrete Composites* 133.

Muthukrishnan, S., S. Ramakrishnan and J. Sanjayan (2021). "Technologies for improving buildability in 3D concrete printing." *Cement and Concrete Composites* 122.

Nerella, V. N., M. Krause and V. Mechtcherine (2020). "Direct printing test for buildability of 3D-printable concrete considering economic viability." *Automation in Construction* 109.

Nerella, V. N., M. Näther, A. Iqbal, M. Butler and V. Mechtcherine (2019). "Inline quantification of extrudability of cementitious materials for digital construction." *Cement and Concrete Composites* **95**: 260–270.

Nguyen-Van, V., B. Panda, G. Zhang, H. Nguyen-Xuan and P. Tran (2021). "Digital design computing and modelling for 3-D concrete printing." *Automation in Construction* 123.

Ooms, T., G. Vantyghem, R. Van Coile and W. De Corte (2021). "A parametric modelling strategy for the numerical simulation of 3D concrete printing with complex geometries." *Additive Manufacturing* 38.

Panda, B., S. Chandra Paul, L. J. Hui, Y. W. D. Tay and M. J. Tan (2017). "Additive manufacturing of geopolymer for sustainable built environment." *Journal of Cleaner Production* **167**: 281–288.

Panda, B., S. Chandra Paul and M. Jen Tan (2017). "Anisotropic mechanical performance of 3D printed fiber reinforced sustainable construction material." *Materials Letters* **209**: 146–149.

Panda, B., S. Chandra Paul, N. A. N. Mohamed, Y. W. D. Tay and M. J. Tan (2018). "Measurement of tensile bond strength of 3D printed geopolymer mortar." *Measurement* **113**: 108–116.

Panda, B. and M. J. Tan (2018). "Experimental study on mix proportion and fresh properties of fly ash based geopolymer for 3D concrete printing." *Ceramics International* **44**(9): 10258–10265.

Paritala, S., K. K. Singaram, I. Bathina, M. A. Khan and S. K. R. Jyosyula (2023). "Rheology and pumpability of mix suitable for extrusion-based concrete 3D printing – A review." *Construction and Building Materials* 402.

Paul, S. C., Y. W. D. Tay, B. Panda and M. J. Tan (2018). "Fresh and hardened properties of 3D printable cementitious materials for building and construction." *Archives of Civil and Mechanical Engineering* **18**(1): 311–319.

Perrot, A., A. Pierre, S. Vitaloni and V. Picandet (2014). "Prediction of lateral form pressure exerted by concrete at low casting rates." *Materials and Structures* **48**(7): 2315–2322.

Perrot, A., D. Rangeard and A. Pierre (2015). "Structural built-up of cement-based materials used for 3D-printing extrusion techniques." *Materials and Structures* **49**(4): 1213–1220.

Rahul, A. V. and M. Santhanam (2020). "Evaluating the printability of concretes containing lightweight coarse aggregates." *Cement and Concrete Composites* 109.

Rahul, A. V., M. Santhanam, H. Meena and Z. Ghani (2019). "3D printable concrete: Mixture design and test methods." *Cement and Concrete Composites* **97**: 13–23.

Ramamurthy, K. and K. I. Harikrishnan (2006). "Influence of binders on properties of sintered fly ash aggregate." *Cement and Concrete Composites* **28**(1): 33–38.

Rashad, A. M. (2018). "An overview on rheology, mechanical properties and durability of high-volume slag used as a cement replacement in paste, mortar and concrete." *Construction and Building Materials* **187**: 89–117.

Roussel, N. (2006). "A thixotropy model for fresh fluid concretes: Theory, validation and applications." *Cement and Concrete Research* **36**(10): 1797–1806.
Roussel, N. (2018). "Rheological requirements for printable concretes." *Cement and Concrete Research* **112**: 76–85.
Roussel, N., H. Bessaies-Bey, S. Kawashima, D. Marchon, K. Vasilic and R. Wolfs (2019). "Recent advances on yield stress and elasticity of fresh cement-based materials." *Cement and Concrete Research* 124.
Roussel, N., J. Spangenberg, J. Wallevik and R. Wolfs (2020). "Numerical simulations of concrete processing: From standard formative casting to additive manufacturing." *Cement and Concrete Research* 135.
Rushing, T. S., G. Al-Chaar, B. A. Eick, J. Burroughs, J. Shannon, L. Barna and M. Case (2017). "Investigation of concrete mixtures for additive construction." *Rapid Prototyping Journal* **23**(1): 74–80.
Růžek, V., P. Louda and V. Nosek (2023). "The military and civilian applications of geopolymers." *Advances in Military Technology* **18**(2): 259–273.
Shakor, P., S. Nejadi and G. Paul (2019). "a study into the effect of different nozzles shapes and fibre-reinforcement in 3D printed mortar." *Materials (Basel)* **12**(10).
Shakor, P., S. Nejadi, S. Sutjipto, G. Paul and N. Gowripalan (2020). "Effects of deposition velocity in the presence/absence of E6-glass fibre on extrusion-based 3D printed mortar." *Additive Manufacturing* 32.
Shanableh, A., Z. Jianchao, T. Zhang, M. Faried, C. Wengang, M. Maalej, S. Barakat, M. Omar, S. Al-Toubat, R. Al-Ruzouq and K. Hamad (2017). "3D printing cement based ink, and it's application within the construction industry." *MATEC Web of Conferences* 120.
Souza, M. T., I. M. Ferreira, E. Guzi de Moraes, L. Senff and A. P. Novaes de Oliveira (2020). "3D printed concrete for large-scale buildings: An overview of rheology, printing parameters, chemical admixtures, reinforcements, and economic and environmental prospects." *Journal of Building Engineering* 32.
Suiker, A. S. J. (2018). "Mechanical performance of wall structures in 3D printing processes: Theory, design tools and experiments." *International Journal of Mechanical Sciences* **137**: 145–170.
Suiker, A. S. J., R. J. M. Wolfs, S. M. Lucas and T. A. M. Salet (2020). "Elastic buckling and plastic collapse during 3D concrete printing." *Cement and Concrete Research* 135.
Sung Choi, M. (2014). "Numerical prediction on the effects of the coarse aggregate size to the pipe flow of pumped concrete." *Journal of Advanced Concrete Technology* **12**(8): 239–249.
Tao, Y., Q. Ren, K. Lesage, K. Van Tittelboom, Y. Yuan and G. De Schutter (2022). "Shape stability of 3D printable concrete with river and manufactured sand characterized by squeeze flow." *Cement and Concrete Composites* 133.
Tay, Y. W. D., M. Y. Li and M. J. Tan (2019). "Effect of printing parameters in 3D concrete printing: Printing region and support structures." *Journal of Materials Processing Technology* **271**: 261–270.
Tay, Y. W. D., Y. Qian and M. J. Tan (2019). "Printability region for 3D concrete printing using slump and slump flow test." *Composites Part B: Engineering* 174.
Toutou, Z., N. Roussel and C. Lanos (2005). "The squeezing test: A tool to identify firm cement-based material's rheological behaviour and evaluate their extrusion ability." *Cement and Concrete Research* **35**(10): 1891–1899.
Tripathi, A., S. A. O. Nair and N. Neithalath (2022). "A comprehensive analysis of buildability of 3D-printed concrete and the use of bi-linear stress-strain criterion-based failure curves towards their prediction." *Cement and Concrete Composites* 128.
Vallurupalli, K., N. Farzadnia and K. H. Khayat (2021). "Effect of flow behavior and process-induced variations on shape stability of 3D printed elements – A review." *Cement and Concrete Composites* 118.

Vantyghem, G., T. Ooms and W. De Corte (2021). "VoxelPrint: A Grasshopper plug-in for voxel-based numerical simulation of concrete printing." *Automation in Construction* 122.

Wangler, T., N. Roussel, F. P. Bos, T. A. M. Salet and R. J. Flatt (2019). "Digital concrete: A review." *Cement and Concrete Research* 123.

Weng, Y., M. Li, M. J. Tan and S. Qian (2018). "Design 3D printing cementitious materials via Fuller Thompson theory and Marson-Percy model." *Construction and Building Materials* **163**: 600–610.

Westerholm, M., B. Lagerblad, J. Silfwerbrand and E. Forssberg (2008). "Influence of fine aggregate characteristics on the rheological properties of mortars." *Cement and Concrete Composites* **30**(4): 274–282.

Wolfs, R. J. M., F. P. Bos and T. A. M. Salet (2018). "Early age mechanical behaviour of 3D printed concrete: Numerical modelling and experimental testing." *Cement and Concrete Research* **106**: 103–116.

Wolfs, R. J. M., F. P. Bos and T. A. M. Salet (2019). "Triaxial compression testing on early age concrete for numerical analysis of 3D concrete printing." *Cement and Concrete Composites* 104.

Wolfs, R. J. M., F. P. Bos, E. C. F. van Strien and T. A. M. Salet (2018). "A real-time height measurement and feedback system for 3D concrete printing." *High Tech Concrete: Where Technology and Engineering Meet*: 2474–2483.

Wolfs, R. J. M. and A. S. J. Suiker (2019). "Structural failure during extrusion-based 3D printing processes." *The International Journal of Advanced Manufacturing Technology* **104**(1–4): 565–584.

Xiao, J., G. Ji, Y. Zhang, G. Ma, V. Mechtcherine, J. Pan, L. Wang, T. Ding, Z. Duan and S. Du (2021). "Large-scale 3D printing concrete technology: Current status and future opportunities." *Cement and Concrete Composites* 122.

Xiao, S., H. Cheng, Z. Que, T. Liu and D. Zou (2024). "Enhancing marine anti-washout concrete: Optimal silica fume usage for improved compressive strength and abrasion resistance." *Construction and Building Materials* 428.

Yuan, S., Z. Xu and J. Liu (2024). "Insights and challenges of predicting concrete pumpability: A state-of-art review." *Journal of Building Engineering* 95.

Zhang, C., Z. Deng, C. Chen, Y. Zhang, V. Mechtcherine and Z. Sun (2022). "Predicting the static yield stress of 3D printable concrete based on flowability of paste and thickness of excess paste layer." *Cement and Concrete Composites* 129.

Zhang, C., Z. Hou, C. Chen, Y. Zhang, V. Mechtcherine and Z. Sun (2019). "Design of 3D printable concrete based on the relationship between flowability of cement paste and optimum aggregate content." *Cement and Concrete Composites* 104.

Zhang, J., J. Wang, S. Dong, X. Yu and B. Han (2019). "A review of the current progress and application of 3D printed concrete." *Composites Part A: Applied Science and Manufacturing* 125.

Zhang, Y., Y. Zhang, G. Liu, Y. Yang, M. Wu and B. Pang (2018). "Fresh properties of a novel 3D printing concrete ink." *Construction and Building Materials* **174**: 263–271.

Zhang, Y., Y. Zhang, W. She, L. Yang, G. Liu and Y. Yang (2019). "Rheological and harden properties of the high-thixotropy 3D printing concrete." *Construction and Building Materials* **201**: 278–285.

Zhuang, Y., Q. Deng, M. Hou, A. Nasr, Z. Duan and H. Zhang (2025). "Clarifying the role of aggregate characteristics in the rheological and mechanical properties of coal gangue aggregate concrete with a packing perspective." *Journal of Building Engineering* **111**: 113121.

Zou, D., H. Cheng, T. Liu, S. Qin and T.-H. Yi (2019). "Monitoring of concrete structure damage caused by sulfate attack with the use of embedded piezoelectric transducers." *Smart Materials and Structures* **28**(10).

Zou, D., C. Du, T. Liu, J. Teng and H. Cheng (2019). "Time-dependent deformations of concrete columns under different construction load histories." *Advances in Structural Engineering* **22**(8): 1845–1854.

2 Mechanical Properties of 3D-Printed Concrete

Fei Teng, Weijian Ding, and Yiwei Weng

2.1 INTRODUCTION

The construction industry has recently seen significant advancements with the emergence of additive manufacturing (AM) technologies, particularly extrusion-based 3D-printed concrete (3DPC) (Lu, Weng et al. 2019). In the conventional casting method, fresh concrete is poured into molds (formwork) and compacted through vibration to ensure uniformity. This process results in relatively homogeneous and isotropic concrete structures with well-understood mechanical properties. Quality control focuses on maintaining consistent concrete mixes, proper placement, and adequate curing conditions to achieve the desired strength and durability. In contrast, 3DPC constructs structures layer by layer using an extrusion process, eliminating the need for formwork and vibration (Ler et al. 2024). Figure 2.1 shows a typical 3D printing system and the relative components (Gomaa et al. 2024). This shift from casting to printing offers advantages such as enhanced design flexibility, reduced material waste, and accelerated construction times. However, new challenges are also introduced due to this innovative construction approach, which directly influence the mechanical performance of the final structure. Compared with cast concrete, the mechanical properties of hardened 3DPC are affected by more factors.

A primary concern with using 3DPC is the weak interlayer bond strength between sequentially deposited layers (Ding et al. 2023). Figure 2.2 shows the interlayer structure with the weak bonding during printing (Chen, Çopuroğlu et al. 2021), which exhibits more pores and voids at the interfacial regions. The increased porosity is caused by the absence of vibration and evaporation of surface moisture in 3DPC (Paul et al. 2018), and the pore shapes are also different compared to those of cast concrete. Figure 2.3 shows the typical comparison of pore shape in cast and printed concrete structures (Liu, Liu, Wu et al. 2022). Without vibration, entrapped air is inadequately removed, resulting in elongated voids. These interfaces become the weakest points in the structure, compromising the overall mechanical performance of the printed structure. Poorly bonded interfaces may also lead to delamination and cracking under load, further diminishing the load-bearing capacity of the structure and safety. The weak interfaces also lead to pronounced anisotropy, in which mechanical properties vary in different loading directions (Arunothayan et al. 2020). Existing research has highlighted the detrimental effects of anisotropy in 3DPC.

DOI: 10.1201/9781003434863-3

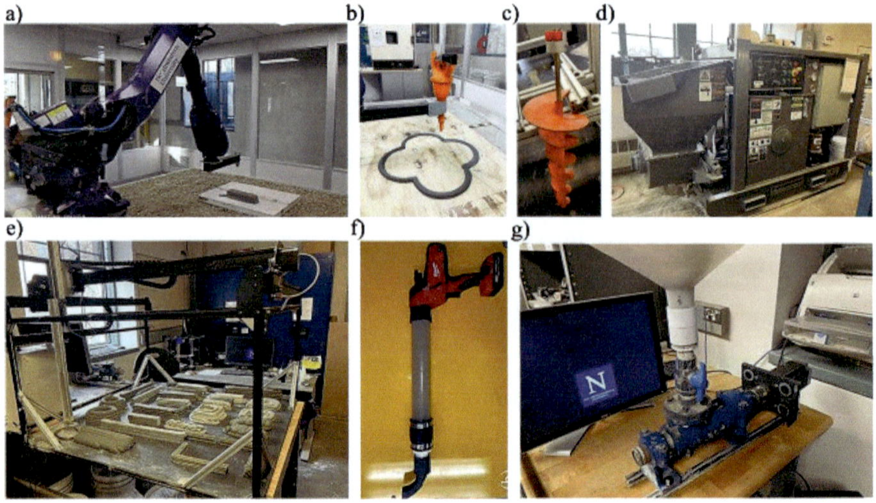

FIGURE 2.1 Different printing systems utilizing: (a) a six-axis robotic arm; (b) an auger system connected to the robotic arm; (c) the rotational auger; (d) a TK7 pump; (e) a three-axis gantry 3D printer; (f) a piston-type extruder; (g) a cavity pump (Gomaa et al. 2024).

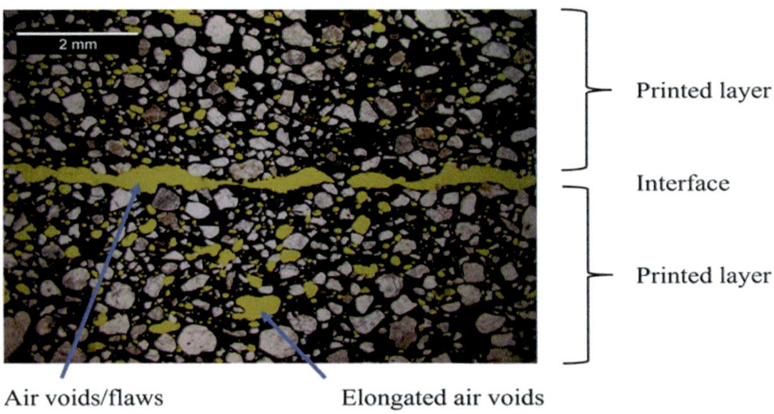

FIGURE 2.2 Interlayer structure of 3DPC (Chen, Çopuroğlu et al. 2021).

A typical experimental case is shown in Figure 2.4 (van den Heever et al. 2022), revealing the increased porosity, reduced compressive strength, and the anisotropic behavior of printed structures. Figure 2.4a illustrates the coring directions of the samples. The porosity of the two types of 3D-printed samples and cast samples is shown in Figure 2.4b, while the results of compressive strength comparison are shown in Figure 2.4c.

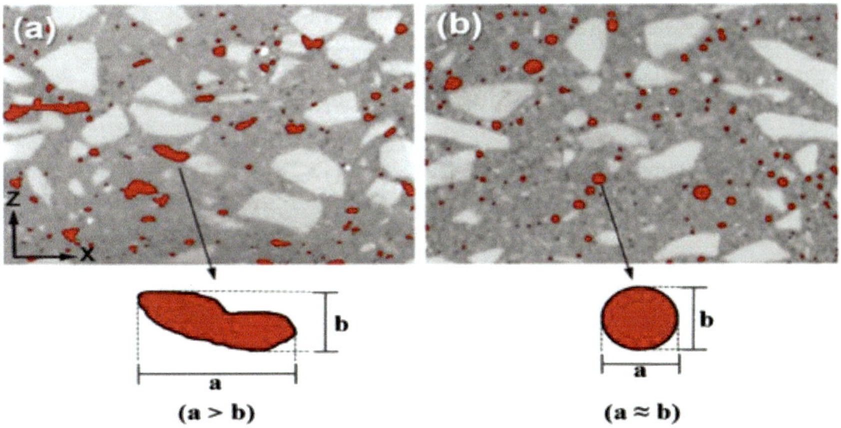

FIGURE 2.3 Pore shapes: (a) 3DCP; (b) casting concrete (Liu, Liu, Wu et al. 2022).

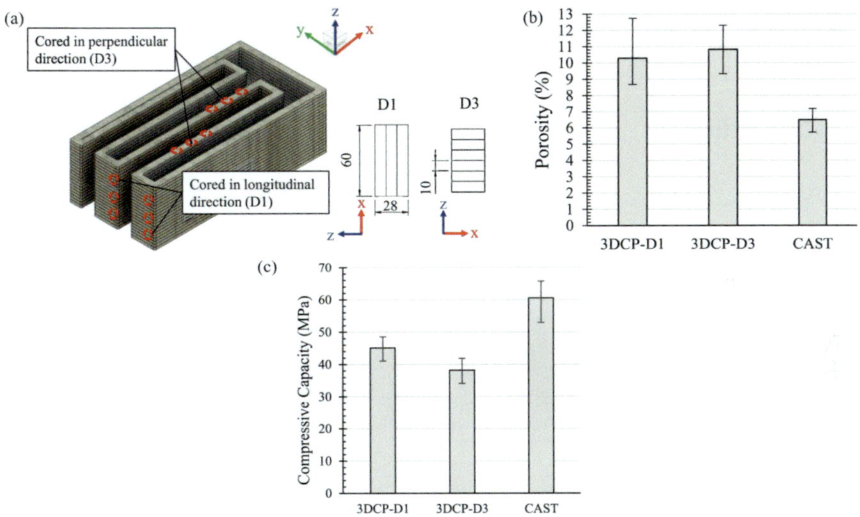

FIGURE 2.4 (a) Sample coring direction and location of sample extraction; (b) total porosity of evaluated samples; (c) compressive capacity of evaluated samples (van den Heever et al. 2022).

The extrusion process may also introduce variations in the volume flow rate, influenced by factors such as the rheological properties of material, the age of the mix, and the parameters of the printing equipment (Ji et al. 2022). These variations cause inconsistencies in layer dimensions, potential cracking, and bulging, leading to adverse impacts on the mechanical properties. Characterizing anisotropy is essential for assessing the mechanical properties of 3DPC. Unlike traditional cast concrete, which is typically considered isotropic and thus behaves consistently regardless of the loading direction, 3DPC exhibits unique mechanical performance that is heavily influenced by the direction of loading. As a consequence, conventional testing methods used for cast concrete cannot be directly applied to 3DPC (Nicolas et al. 2022). For example, the experimental results of Wolfs et al. (2019) clearly reveal the significant differences in compression, flexural, and tensile test results between cast and printed concrete, as shown in Figure 2.5. Generally, the mechanical properties of 3DPC are lower when measured in the direction perpendicular to the printed layers compared to that of the direction parallel to layers. Variations in the sampling height,

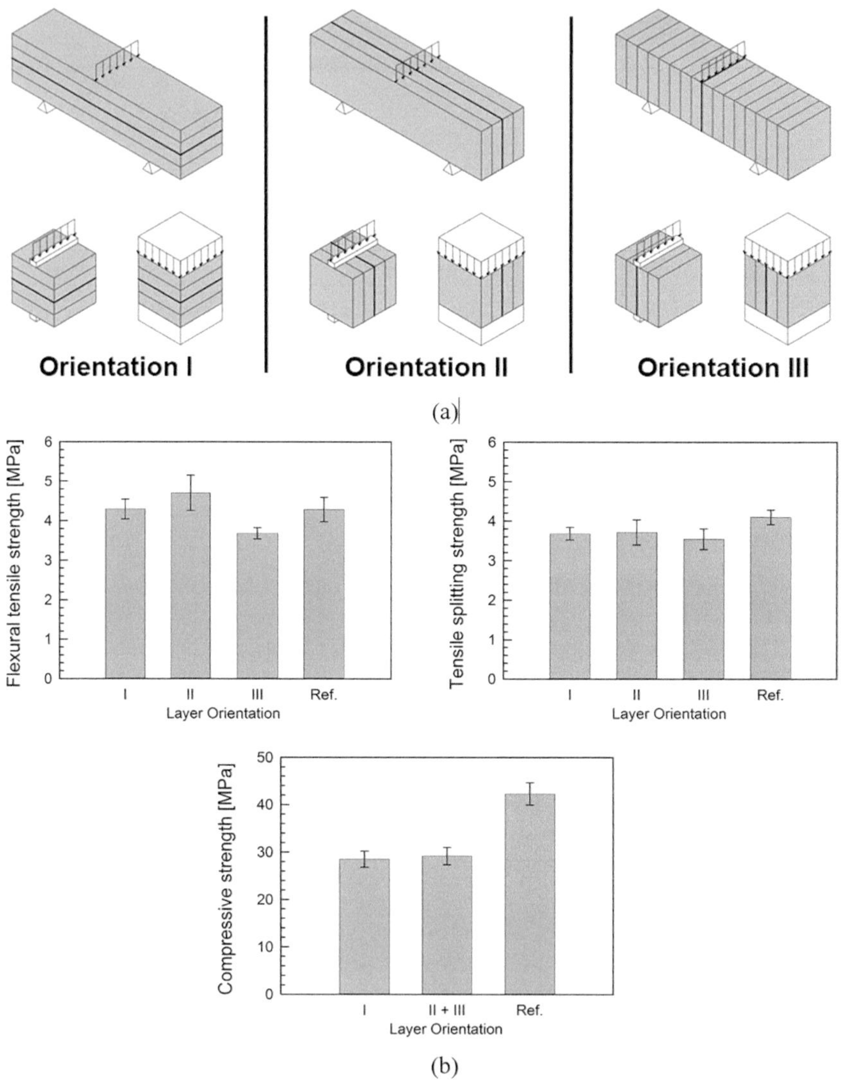

FIGURE 2.5 (a) Mechanical tests in different orientations; (b) results of flexural tension, tensile splitting, and compression tests for different layer orientations (Wolfs et al. 2019).

position, and shapes of specimens can also lead to differences in mechanical performance (Panda et al. 2018).

Currently, the anisotropy of the mechanical properties and weak interfaces of 3DPC pose challenges to the quality control of structures. This chapter focuses on the influence of mechanical testing methods, printing parameters, and material composition on the hardened mechanical properties of 3DPC. Specifically, it states the significance of reducing the adverse effects of anisotropy through strategies such as optimizing the printing process, optimizing material mixtures, controlling environmental conditions, and establishing standardized testing methods.

2.2 EVALUATION METHODS OF HARDENED MECHANICAL PROPERTIES

Hardened 3DPC exhibits unique mechanical properties characterized by anisotropy and weak interlayer bonding. This anisotropic behavior can be observed in compressive strength, flexural strength, and splitting tensile strength. In general, regardless of the ratio and type of printable concrete materials, compressive strength has a low level of mechanical anisotropy, followed by splitting tensile strength, while flexural strength shows the highest mechanical anisotropy (Wang et al. 2023). The directions used to characterize anisotropy should be defined based on the relationship between the loading direction and the printing direction (Zhang et al. 2024; Bos et al. 2016). For the convenience of description, this chapter uniformly uses the XYZ direction. The typical schematic of the directions for compressive strength and flexural strength is shown in Figure 2.6 (Arunothayan et al. 2022). Specifically, the X direction is parallel to the moving direction of the printing nozzle; the Y direction is perpendicular to the X direction and refers to the length direction of adjacent printed stripes; and the Z direction is oriented vertically to the XY plane, indicating the height of the printed layers. This spatial coordinate system facilitates a more comprehensive description and analysis of the mechanical properties of 3DPC.

2.2.1 Compressive Strength and Testing Methods

The uniaxial compressive strength testing samples are typically cubes or cylinders. When conducting compressive strength tests, it is often necessary to cut and/or polish the printed samples to meet the testing requirements due to the uneven surface of 3D-printed samples. The cutting and polishing processes can ensure that the geometric shape and surface quality meet the testing standards. However, this process may introduce additional stresses, potentially leading to errors in test results. During testing, compressive strength data in different directions should be collected and statistically analyzed. The compressive strength of 3DPC exhibits significant directional differences, with the strength in the X direction typically being higher than that in the Y and Z directions. Significant differences exist between the compressive strength of 3DPC and that of traditional cast concrete.

Figure 2.7 (Zhang et al. 2024) illustrates a comparison to evaluate the compressive strength of 3DPC in the X, Y, and Z directions versus that of cast concrete at 7 and 28 days. The results reveal that the ratios of compressive strength of 3DPC

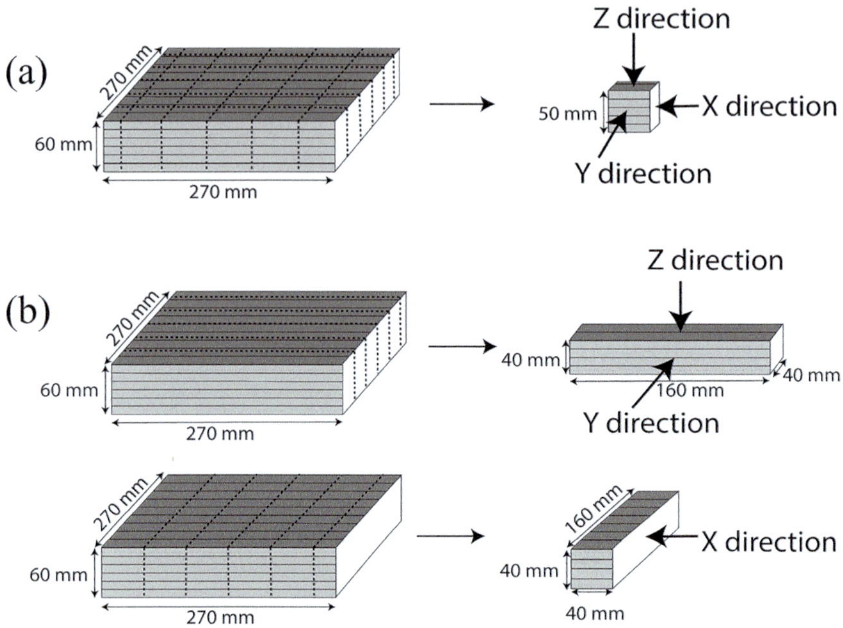

FIGURE 2.6 Test directions of (a) compressive strength; (b) flexural strength (Arunothayan et al. 2022).

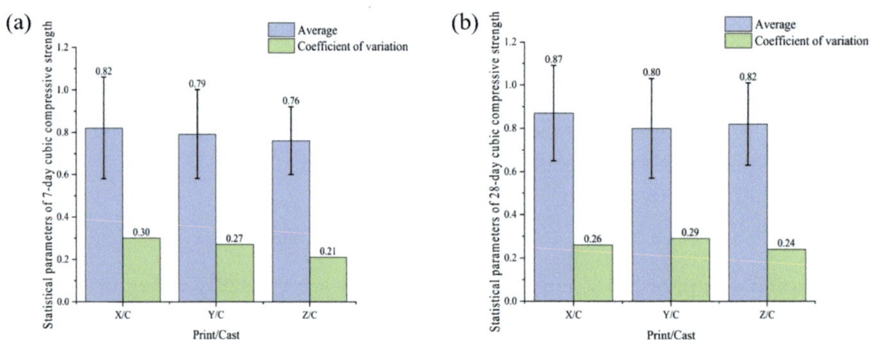

FIGURE 2.7 Anisotropy of compressive strength at (a) 7 days; (b) 28 days (Zhang et al. 2024).

compared to cast concrete are 0.82, 0.79, and 0.76, respectively, at 7 days, and 0.87, 0.80, and 0.82, respectively, at 28 days. All these values are below 1.0, indicating that the strength of 3DPC is lower than that of the cast concrete. The evaluation of the compressive performance of 3DPC can also be conducted by calculating the average compressive strength and the coefficient of variation (COV) in different directions. The study (Zhang et al. 2024) showed that the COVs for 3DPC at 7 and 28 days are in the ranges of 0.04–0.37 and 0.01–0.50, respectively, indicating a high level of variability. The COV for 3DPC is generally higher than that for cast concrete, with an increase ranging from 4.2% to 69%. This suggests that the compressive strength of 3DPC is significantly influenced by the printing process and material characteristics, leading to notable strength variability. The main factors affecting the variability of compressive strength in 3DPC include porosity, pore shape, and interlayer bond strength, which contribute to the microheterogeneity of the material during the printing process.

2.2.2 Flexural Strength and Testing Methods

Flexural strength is determined through three-point or four-point bending tests. The samples required for flexural testing are typically in the shape of rectangular prisms or beams. Similar to the compressive tests, the surface of the samples may need to be cut and polished to ensure that the geometric shape and surface quality meet testing requirements. A loading device applies force at a constant rate until the sample fails, while the load and mid-span deflection are recorded. Generally, the flexural strength of 3DPC is lower than that of traditional cast concrete. Research indicates that 3DPC exhibits pronounced anisotropic characteristics, with significant differences in flexural strength depending on the loading direction. The flexural strength in the X direction is typically higher than that in the Y and Z directions. For instance, Panda et al. (2017) analyzed the flexural strength of 3D-printed fiber-reinforced geopolymer mortar and found that the highest flexural strength occurred in the X direction, with strength variations of up to 50% between different loading directions, as shown in Figure 2.8. This anisotropy primarily arises from the interlayer interfaces and pore structures formed during the printing process.

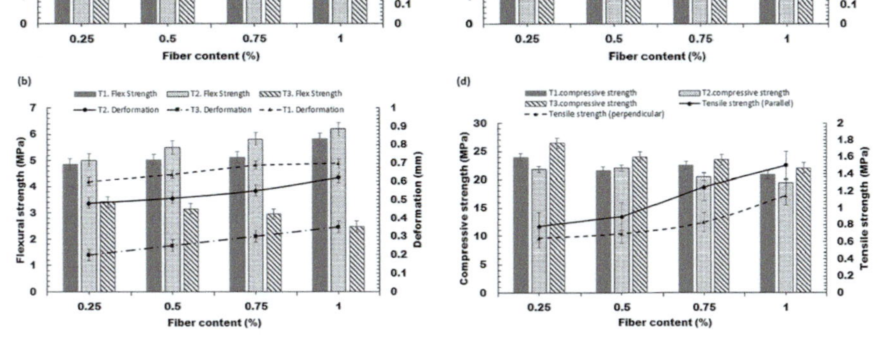

FIGURE 2.8 Flexural properties of short glass fiber/geopolymer composites with (a) 3 mm, (b) 6 mm, and (c) 8 mm fiber length; (d) compressive/tensile properties of 3 mm short glass fiber (Panda et al. 2017).

2.2.3 Interlayer Bond Strength and Testing Methods

Traditional casting produces a uniform material without interlayer interfaces, whereas 3D printing introduces significant anisotropy and weaknesses in bond strength between layers. Currently, there are no standardized methods specifically designed for evaluating interlayer bond strength in printed concrete. Existing testing approaches are largely adapted from conventional concrete testing methods. All interlayer interface testing techniques can generally be categorized into tensile strength tests (Figure 2.9) and shear strength tests (Figure 2.10) (Ding et al. 2023). Each testing method has its own advantages and limitations, with the test results influenced by numerous factors such as specimen geometry and loading conditions. Direct tensile testing is widely employed to measure interlayer bond strength. However, it has notable limitations, including the risk of adhesive failure and issues related to eccentric loading. Although modified direct tension tests have been introduced to evaluate interlayer bond strength, measurement accuracy is affected by factors such as stress concentrations at the crack tip and the presence of a multiaxial tensile stress state near the notch tip. Indirect tensile methods, such as splitting tensile tests, are also frequently utilized by researchers due to the standardized dimensions of the specimens and the simplicity of the required testing apparatus. To account for fracture energy, wedge-splitting tensile tests are sometimes employed. In the context of shear testing, researchers have recognized that achieving a state of pure shear without any

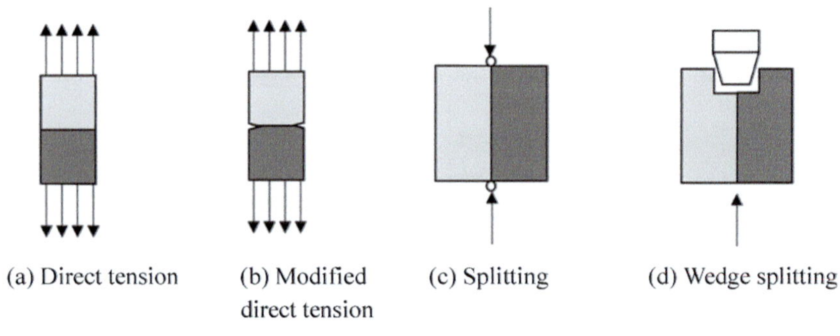

FIGURE 2.9 Tension test setups for concrete interfaces (Ding et al. 2023).

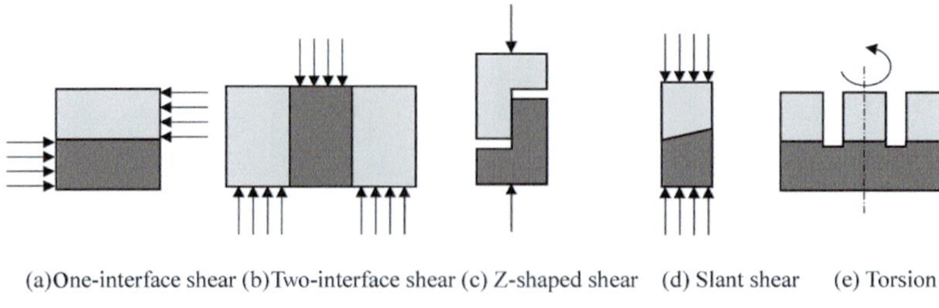

FIGURE 2.10 Shear bond test for concrete interfaces (Ding et al. 2023).

accompanying normal stress is practically impossible during experimental procedures. Nevertheless, direct shear tests remain the most commonly used method due to the simplicity of specimen preparation and testing procedures.

To address the challenges of load orientation and uneven stress distribution, some researchers (Silfwerbrand 2003) have developed the single-interface Z-shear test, which approximates a pure shear stress state within a localized region of the interface while mitigating the effects of stress concentration at the boundaries. However, preparation of specimens for the single-interface Z-shear test is relatively complex (Arunothayan et al. 2020). Additionally, oblique shear tests have been developed to assess interfacial bond strength under the combined influence of compression and shear stress.

The results of the various testing methods described above are often difficult to compare due to significant variations in specimen geometry and test setups, which profoundly influence the material properties under consideration. Consequently, there is an urgent need for collective efforts to establish universally accepted and standardized testing methods. The quality of interlayer bonding is critical to the overall mechanical performance of 3DPC and is influenced by factors such as moisture content, porosity, and printing parameters. During the 3D printing process, moisture migration plays a pivotal role in determining interlayer bonding quality. Specifically, the time interval between layer depositions enables moisture evaporation and the migration of free water, which impacts the humidity field at the interface and subsequently affects the hydration process of the cement. Moisture loss during this process can be divided into two stages: the physical stage, primarily driven by evaporation, and the chemical stage, which involves hydration reactions and the formation of early hydration products.

Another critical factor influencing interlayer bonding is the maximum operating time (MOT). If the printing interval exceeds the MOT, interlayer bond strength declines rapidly, adversely affecting both the material's printability and its bonding performance after curing (Pan et al. 2021). The layered structure inherent to the 3D printing process and the absence of vibration during layer deposition result in the formation of pores of varying sizes, primarily located in the interlayer regions. These interconnected voids can evolve into banded cracks, which significantly weaken interlayer strength.

2.3 EFFECT OF PRINTING PARAMETERS ON MECHANICAL PROPERTIES

Printing parameters are crucial factors that influence the mechanical properties of printed concrete. Researchers have focused on investigating parameters such as printing speed, layer height, interval time, printing path, nozzle dimension, lifting height, and printing environment. These factors directly or indirectly affect the printing quality of 3DPC, causing geometric defects and differences in physical and chemical properties, thereby impacting the final mechanical performance.

2.3.1 Influence of Printing Speed

Printing speed plays a critical role in determining the mechanical properties and print quality of 3DPC. When the printing speed is excessively high, insufficient concrete material is extruded, leading to narrower and thinner printed layers. This reduction in layer thickness decreases the contact area between adjacent layers, thereby weakening interlayer bonding and reducing bonding strength. Consequently, the overall structure exhibits diminished compressive and flexural resistance. Furthermore, rapid printing can introduce pores and defects between layers, reducing the density of the concrete and further compromising its mechanical performance.

Conversely, printing at an excessively slow speed can lead to over-extrusion and material accumulation, which results in uneven surfaces and shape distortions in the printed component. Prolonged printing times can exacerbate moisture loss in the previously deposited layers, thereby reducing their wettability and adversely affecting interlayer bond strength. Excessively slow speeds may also fail to provide sufficient structural support for upper layers, increasing the risk of deformation or even collapse.

Thus, selecting an appropriate printing speed is essential to improve the mechanical properties. Optimal speed ensures sufficient bond strength by avoiding poor adhesion and increased porosity associated with excessively fast speed, while also preventing material accumulation and structural instability caused by excessively slow speeds. By maintaining a balance, the integrity, density, and mechanical properties of 3DPC can be effectively optimized.

Many case studies have investigated the impact of printing speed. Singh et al. (2022) explored the effects of different printing speed on printing quality and mechanical properties. The study revealed that an excessively high printing speed (e.g., a horizontal speed of 25 mm/s) compromises the material's ability to maintain its shape at corners, leading to interlayer collapse and side material accumulation. These issues significantly reduce the overall printing quality, as illustrated in Figure 2.11a,b. Conversely, an excessively low printing speed results in over-extrusion and material accumulation, producing uneven surfaces and deformed shapes in the printed components, as depicted in Figure 2.11c.

An appropriate printing speed improves the overall structural performance of concrete. Pan et al. (2023) investigated the effect of printing speed on the compressive behavior of printed concrete and found that as printing speed increases, the number of initial defects within the specimens increases, compressive peak strain decreases, and compressive strength decreases significantly. Additionally, higher printing

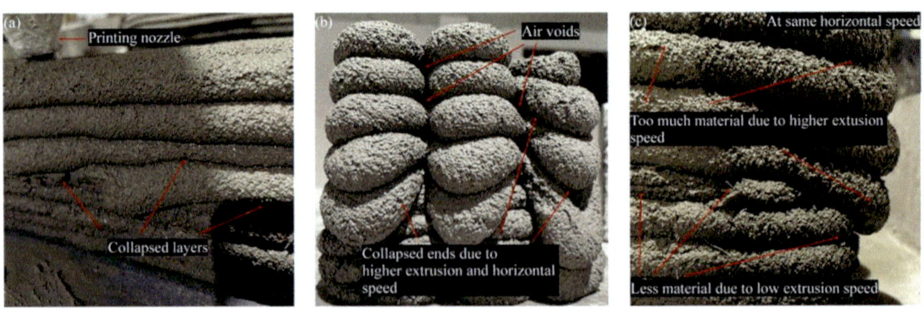

FIGURE 2.11 Failed specimens printed by varying the extrusion speed (Singh et al. 2022).

speeds may hinder adequate bonding between layers, thereby reducing interlayer bond strength and adversely affecting the mechanical performance of the overall structure. Al-Qutaifi et al. (2018) examined the influence of printing parameters on the mechanical properties of printed geopolymer concrete, suggesting that printing speed has a notable impact on flexural strength. Their findings demonstrated that an appropriate printing speed can improve flexural performance, whereas excessively high or low speeds can lead to a reduction in mechanical properties. Similarly, Chen et al. (2020) explored the effect of printing speed on the interlayer bond strength of printed limestone-calcined clay-based cement materials. They observed that excessively slow printing speeds increase the time interval between layer depositions, which results in water evaporation from previously deposited layers. This evaporation reduces the wettability between layers, preventing the newly deposited layer from fully bonding with the underlying layer, thereby forming a weak interface. This weak interface compromises the mechanical properties of the overall structure.

These findings further indicate that selecting an optimized printing speed not only improves the dimensional accuracy and surface quality of the printed structure but also indirectly enhances the mechanical properties of printed concrete by ensuring better interlayer bonding and structural integrity.

2.3.2 Influence of Layer Height

The influence of printing layer height on the mechanical properties of 3DPC is mainly reflected in the interlayer bond strength, structural anisotropy, and overall stability. Nair et al. (2021) investigated the effect of layer height on the bending and fracture response of ordinary and fiber-reinforced 3D-printed beams. Their study revealed that larger layer heights result in decreased mechanical properties due to poor interlayer bonding. Similarly, Geng et al. (2020) analyzed the interlayer interface characteristics of printed concrete and found that varying layer heights lead to the formation of a dual-layer structure, which directly impacts the micromechanical properties of the material. He et al. (2021) addressed this issue by designing a novel nozzle to improve interlayer strength, emphasizing the critical role of controlling layer height to enhance mechanical performance. Napolitano et al. (2021) further characterized the dynamic properties of the interlayer interface in printed concrete components and demonstrated that improper selection of layer height adversely affects the mechanical properties of the components. Ding, Xiao, Zou et al. (2020) studied the anisotropic behavior of 3DPC in bending and highlighted that variations in layer height contribute to anisotropy in mechanical properties. This study provides additional insights into the effects of layer height. For example, Figure 2.12 illustrates the relationship between layer height and porosity under specific mix ratios and printing parameters (Nair et al. 2021). When the layer height is small (5 mm, LH5), optical imaging shows fewer and smaller pores, which may be attributed to the higher pressure exerted by the print head during deposition. However, as the layer height increases to 10 mm (LH10) and 15 mm (LH15), larger and more uniformly distributed pores emerge, primarily located at the overlap regions of fibers in horizontal and vertical directions. This increase in porosity at higher layer heights can significantly weaken the mechanical properties of the material.

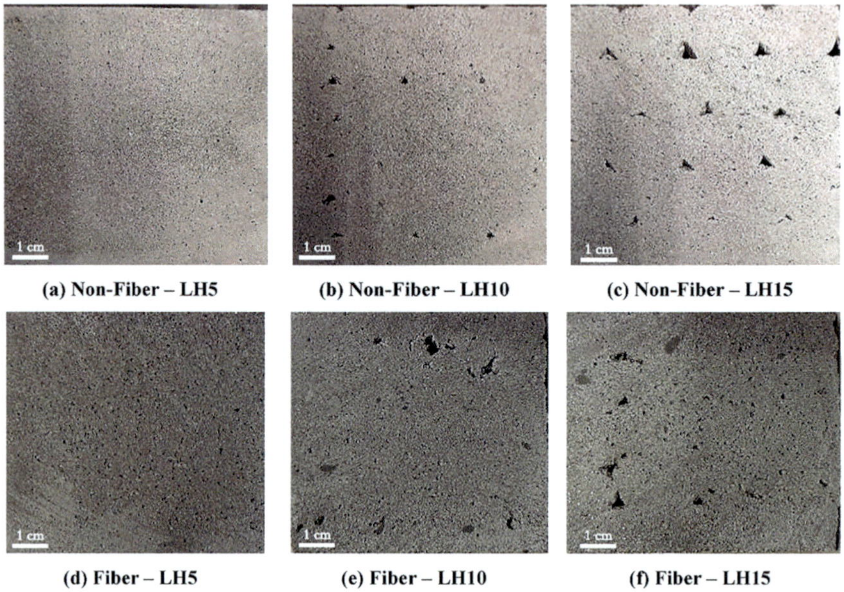

FIGURE 2.12 Optical images of beam cross sections showing typical void distribution at different layer heights in plain and steel fiber-reinforced mortars (Nair et al. 2021).

These findings emphasize the importance of optimizing layer height to minimize defects, reduce porosity, and enhance the mechanical performance of printed concrete components. Properly controlled layer height not only improves interlayer bonding but also mitigates the anisotropic effects observed in 3DPC structures.

2.3.3 Influence of Printing Path

The printing path can affect the mechanical properties of 3DPC by affecting the deposition mode of the material, the arrangement of the filling patterns, and the distribution of internal pores. Different printing paths will lead to different orientations of the interfaces. When the loading direction is parallel to the interfaces, the weak connection at the interface may be damaged first, reducing the mechanical properties of the material and causing the material to exhibit anisotropic characteristics. Complex printing paths may form discontinuous material deposition inside the material, resulting in an increase in pores and defects. Stress concentration occurs easily at these pores, weakening the overall strength and stiffness of the material. The printing paths also affect the fiber orientation and stress transfer of the material. The optimized printing path can form a structure that is conducive to stress dispersion, hindering the initiation and expansion of cracks, and improve the crack resistance of the material. In addition, under compression loading, the friction between the specimen and the loading surface has a constraint effect on the specimen. The printing path will affect the direction and magnitude of this friction effect, thereby affecting the actual stress condition and test results of the specimen.

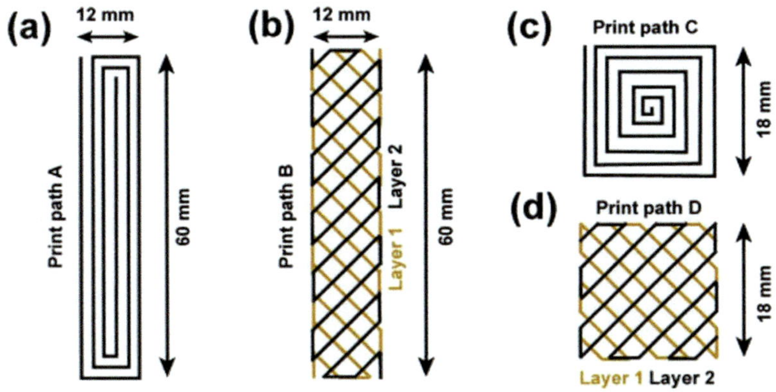

FIGURE 2.13 (a) and (b) Printing process of the different paths for the three-point bending test; (c) and (d) uniaxial compressive strength tests (Hambach and Volkmer 2019).

Studies have shown that a suitable printing path can improve the compressive strength of 3DPC, enhance its flexural properties, and improve interlayer bonding. For example, Hambach and Volkmer (2019) found that a spiral path can significantly improve the compressive strength of fiber-reinforced cement-based materials. Figure 2.13 shows the printing paths for three-point bending tests (Figure 2.13a,b) and uniaxial compressive strength tests (Figure 2.13c,d) in the study, and the results show that printing paths A and C (parallel printing) outperform in flexural and compressive strength tests, while staggered printing (printing paths B and D) increases the porosity of the sample, resulting in decreased mechanical properties. Hambach and Volkmer (2019) revealed that different printing paths lead to changes in the internal microstructure of the material, thereby affecting the compressive strength and elastic modulus. Wolfs et al. (2018) found that the printing path affects the bonding properties of the interlayer interface and affects the development of early compressive strength; Le et al. (2012) optimized the printing path to reduce interlayer defects and enhance interlayer adhesion.

2.3.4 Influence of Printing Interval Time

An appropriate printing interval is crucial for ensuring that the previous layer of concrete retains optimal moisture and plasticity to form a strong bond with the subsequent layer. If the interval is too short, the previous layer may lack sufficient strength, potentially leading to deformation or poor bonding between layers. Conversely, if the interval is too long, the surface of the previous layer may become excessively dry or hardened, reducing interlayer bonding and increasing interface defects. Determining the optimal printing interval has long been a challenging parameter, as it requires balancing the mechanical properties of 3DPC with printing efficiency.

Previous studies have extensively examined this parameter to enhance the mechanical performance of 3DPC. For instance, Le et al. (2012) investigated tensile bond strength using 11 sets of specimens with varying time intervals between successive

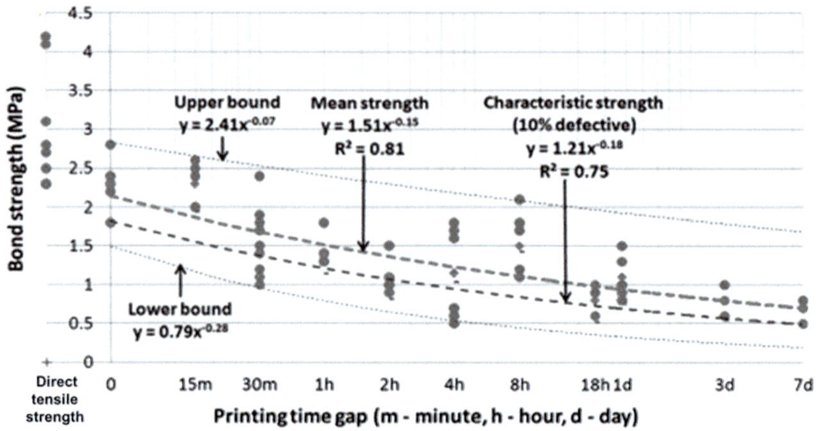

FIGURE 2.14 Variation in bond strength with printing gap and comparison with direct tensile strength (Le et al. 2012).

layers. Their results, as presented in Figure 2.14, showed a gradual decrease in bond strength as the interval time increased. Similarly, Sanjayan et al. (2018) examined the impact of surface moisture on interlayer strength and observed that longer printing intervals led to a decrease in surface humidity, which significantly weakened interlayer bond strength. This finding highlights the critical role of surface moisture in improving interlaminar strength. Tay et al. (2019) explored the effect of time intervals on the bond strength of 3DPC and found that as the interval extended, the stiffness of the previous layer increased, limiting its integration with subsequent layers and thereby reducing bond strength. Nerella et al. (2019) reported that prolonged intervals could result in interface defects such as micropores, which negatively impact mechanical properties. They emphasized the importance of maintaining continuity during the printing process to preserve mechanical integrity. Zhang et al. (2022) focused on the relationship between moisture transport behavior and interlayer voids in 3DPC. Their findings revealed that longer time intervals affected the moisture gradient between layers, increasing air retention at the interface. This led to the formation of pores and a reduction in interlayer bond strength. Sanjayan et al. (2018) studied the influence of printing intervals on early mechanical properties and observed that within a time interval of 20–30 minutes, the interlayer bond strength is improved due to increased surface bleeding. However, as the interval lengthened further, the bond strength began to decline.

2.3.5 Influence of Nozzle Size and Lifting Height

The effects of nozzle size and lifting height on mechanical properties are closely related. Huang et al. (2022) investigated the influence of different layer-to-diameter ratios (defined as the ratio of printing height to nozzle inner diameter, *D*) on printing performance using nozzles with inner diameters of 10 mm and 20 mm. Their results demonstrated that when *D* was small, the extruded material appeared flatter and wider. As *D* increased, the printing pattern transitioned from linear to corrugated, thereby impacting both printing quality and mechanical properties, as illustrated in Figure 2.15. This study highlighted that nozzle size dictates the lifting height, which subsequently has a direct impact on the mechanical properties of the printed structure.

Generally, the inner diameter of the nozzle determines both the extrusion speed and the volume of material extruded. A larger nozzle size increases the material output, resulting in thicker layers. However, this may compromise interlayer bonding, increase the porosity of the printed component, and ultimately reduce its mechanical properties. Conversely, a smaller nozzle size may lead to insufficient material extrusion, resulting in printing defects and reduced structural integrity. Moreover, nozzle size also influences the morphology and precision of the printed object. An appropriately sized nozzle ensures print stability, promotes even material deposition across layers, and prevents the formation of ripples or irregular shapes.

The lifting height, defined as the vertical movement of the nozzle between layers during the printing process, also plays a critical role. If the lifting height is too small, excessive pressure may be exerted on the previously printed layer, leading to over-extrusion, increased interlayer density, or even deformation due to material compression. On the other hand, if the lifting height is too large, the material may spend excessive time in the air, increasing the possibility of deformation under gravity. This can result in poor interlayer bonding and a subsequent reduction in structural stability.

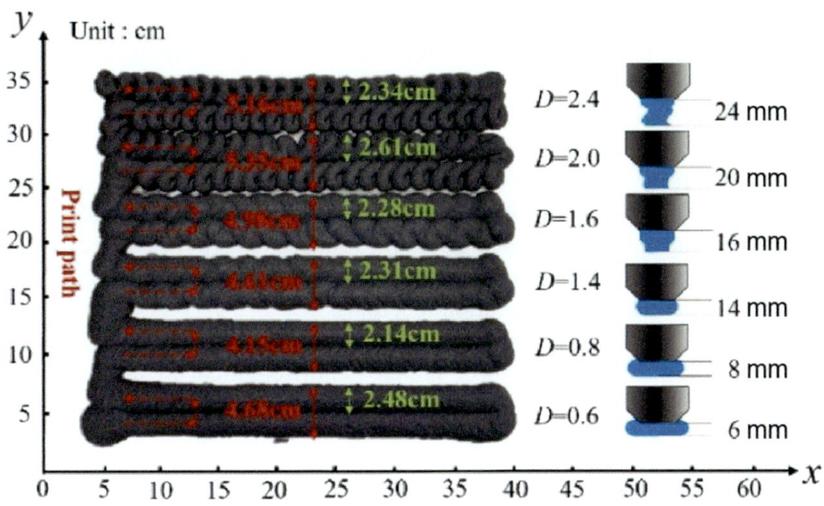

FIGURE 2.15 Influence of nozzle lifting height on printing effect (Huang et al. 2022).

2.3.6 Influence of Printing Environmental Conditions

There are significant differences in the construction processes and physical properties of 3DPC and traditional poured concrete, which result in distinct responses to environmental conditions, particularly in terms of wind, temperature, and humidity. For poured concrete, the use of formwork provides protection from direct environmental exposure, resulting in lower rates of water evaporation. The formwork not only minimizes water loss but also creates a relatively stable internal environment, promoting the full hydration of cement and ensuring the proper development of mechanical properties. Additionally, conventional curing measures, such as watering and covering, help maintain a high-humidity environment, stabilizing internal temperature and humidity. This stability mitigates the impact of temperature fluctuations during the curing process and ensures continued hydration, ultimately leading to denser pore structures, fewer internal defects, and more stable performance. Formwork and protective measures further shield the material from wind, preserving a suitable hydration environment and supporting optimal mechanical property development.

In contrast, 3DPC is exposed directly to the environment without formwork during and after the printing process, increasing the surface area susceptible to environmental effects. This exposure causes internal water to evaporate more easily, especially under conditions of high temperature and low humidity. Rapid water loss inhibits the hydration reaction of cement, reducing the production of hydration products, increasing porosity, and subsequently compromising mechanical properties. High temperatures exacerbate water evaporation and further hinder hydration. Additionally, elevated temperatures may accelerate the hydration rate, leading to premature hardening of the internal structure. This premature hardening can create shrinkage stress and microcracks, compromising long-term performance.

Researchers have conducted extensive studies on environmental factors. Ma et al. (2022) investigated the compressive, flexural, and bonding strengths of 3DPC under three different environmental conditions: Standard (S, relative humidity (RH) > 95%, 20°C ± 5°C), drying (D, RH = 60% ± 5%, 20°C ± 5°C), and drying and wind (W, RH = 60% ± 5%, 20°C ± 5°C, wind speed: 3 m/s). Their results revealed that under dry conditions (RH = 60% ± 5%), the compressive and flexural strengths of 3DPC were reduced by 20%–30% compared to standard curing conditions. When wind was introduced (wind speed: 3 m/s), the strength reductions exceeded 30%, as shown in Figure 2.16. This strength loss was attributed to increased water evaporation, which led to insufficient hydration reactions.

Figure 2.17a illustrates the splitting tensile test configuration, while Figure 2.17b shows the splitting tensile strength of samples at different curing ages. The results revealed that regardless of curing conditions, the splitting tensile strength was nearly identical when loading was applied along the horizontal or vertical interface. Sanjayan et al. (2018) examined the effect of surface humidity on the interlayer bonding performance of 3DPC. They found that under dry conditions (23°C ± 3°C), moisture loss within the concrete resulted in a 47.7% reduction in interlayer bond strength.

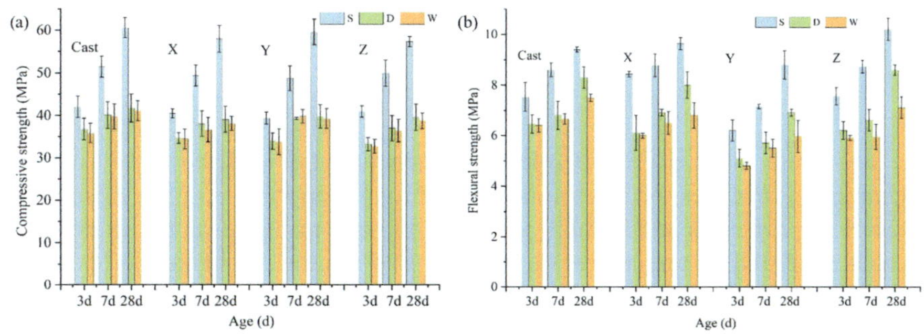

FIGURE 2.16 Effect of conditions on (a) compressive strength; (b) flexural strength of the printed sample at different directions (*X*, *Y*, and *Z*) and of cast samples (Ma et al. 2022).

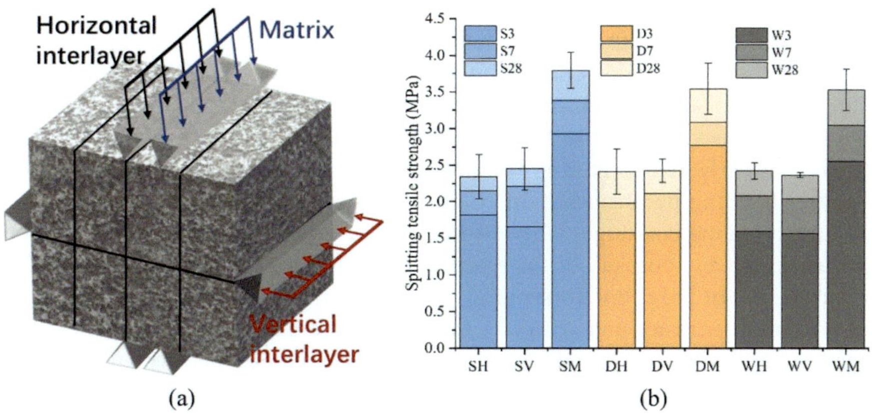

FIGURE 2.17 Splitting tensile strength at different curing conditions: (a) loading positions: horizontal interlayer (*H*), vertical interlayer (*V*), and matrix (*M*); (b) splitting tensile strength under S, D, and W curing conditions at 3, 7, and 28 days (M et al. 2022).

This highlights that the combined effects of temperature, humidity, and wind speed significantly influence the interlayer bonding performance and overall mechanical properties of 3DPC. Alchaar et al. (2021) investigated the mechanical properties of 3D-printed fiber-reinforced concrete under various temperature conditions. They developed a novel testing method to evaluate interlayer bond shear strength at different printing time intervals and also measured compressive and flexural strengths. Their results indicated that high temperatures accelerated water evaporation and surface dehydration, leading to a 56% reduction in compressive strength. However, under high-temperature printing conditions, the flexural strength increased by 21% and 18% compared to the control group and samples printed at normal temperatures, respectively.

2.4 EFFECT OF MIXTURE COMPONENTS ON MECHANICAL PROPERTIES

2.4.1 Fine Aggregate

Fine aggregate plays a critical role in influencing the workability, density, and mechanical properties of cast concrete. Through appropriate mix design, the mechanical properties and durability of concrete can be significantly improved. In the context of 3DPC, fine aggregate not only impacts the mechanical properties of the material but also has a profound effect on its printability, extrusion performance, and interlayer bond strength.

The particle size, gradation, and shape of fine aggregate directly influence the rheological properties of 3DPC. Properly selected fine aggregate can enhance the thixotropy and cohesion of the concrete, enabling smooth extrusion and maintaining the shape of the printed material during the printing process. This is crucial for ensuring robust interlayer bonding and the stability of the overall structure.

The interface of 3DPC is a potential weakness. The characteristics of fine aggregate affect the fluidity and permeability of the mortar during printing, which in turn influences the interlayer bond strength. For instance, the particle size of sand has a significant impact on moisture retention. Finer sand, due to its larger specific surface area, is more effective to retain moisture and form a water film, thereby enhancing interlayer bonding performance. In contrast, coarser sand may lead to faster moisture loss, reducing bonding quality. Moreover, the type of fine aggregate affects the microstructure and porosity of the interface, which directly influences the mechanical properties of the printed structure. A well-designed fine aggregate composition can minimize porosity and optimize the microstructure of the interlayer, thereby improving both interlayer bonding and the overall mechanical performance of 3DPC.

Different types of sand (such as natural river sand, machine-made sand, and recycled sand) have different effects on the performance of 3DPC. For instance, recycled sand may contain fine particles and impurities, which affect the rheological properties and interlayer bonding properties. However, through appropriate treatment and mix design, recycled sand can be successfully used in 3D printing to enhance sustainability. Ding, Xiao, Zou et al. (2020) studied the mechanical properties of 3DPC containing recycled sand and found that recycled sand affected the compressive strength, splitting tensile strength, and flexural strength of printed concrete. The results revealed that the compressive strength of printed concrete had obvious anisotropy, while the replacement of recycled sand had limited effect on the anisotropy. The variation in the effect of recycled sand on the splitting tensile strength does not follow a clear pattern; however, the anisotropy observed in 3D-printed samples incorporating recycled sand is consistent with that of samples without recycled sand, as shown in Figure 2.18. Zou et al. (2021) explored the printability and advantages of 100% recycled sand for 3D printing mortar and pointed out that through appropriate mix design, recycled sand can be successfully used in 3D printing. Additionally, environmental conditions (such as humidity) affect the water content and surface properties of sand, thereby affecting the setting and hardening of concrete and the interlayer bonding properties. Appropriate humidity maintains the workability of

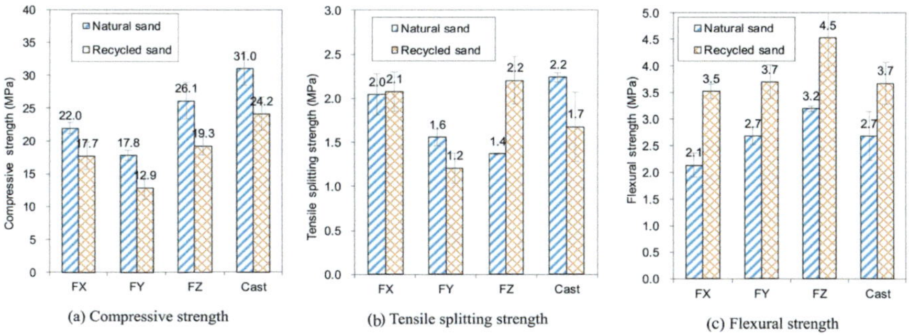

FIGURE 2.18 Mechanical properties of 3D-printed concrete and cast concrete containing recycled sand and natural sand (Ding, Xiao, Zou et al. 2020).

the sand and promotes the bonding between the new and old concrete layers. If the surface of the sand is excessively dry, the interlayer bonding performance of 3DPC is significantly reduced. Since the characteristics of fine aggregate directly influence the extrusion performance and stacking stability of concrete during the printing process, the printing equipment must satisfy specific requirements. For instance, incorporating coarse-grained fine aggregate increases the viscosity of the concrete, which places greater demands on the conveying capacity of the equipment and nozzle diameter. To maintain printing quality and mechanical properties, it is essential to prevent nozzle clogging and ensure the accuracy of the printing path.

2.4.2 Coarse Aggregates

In traditional cast concrete, coarse aggregate serves as the primary skeletal material, playing a critical role in filling and reinforcing the structure. The inclusion of coarse aggregate enhances the compressive and flexural strength of concrete, as the aggregate itself possesses high strength and stiffness, enabling it to resist external loads. Additionally, coarse aggregate limits the shrinkage and creep of concrete, increases its elastic modulus, and minimizes deformation. The interface transition zone (ITZ) between coarse aggregate and cement paste is a crucial factor influencing the overall performance of concrete. Surface properties of coarse aggregate, such as roughness and water absorption, significantly affect the strength and bonding characteristics of the ITZ.

In 3DPC, the inclusion of coarse aggregate presents numerous challenges. The size and shape of coarse aggregate particles can lead to nozzle clogging during the printing process, reducing the extrudability of the material and ultimately compromising printing quality (Ding et al. 2021). Additionally, the presence of coarse aggregate may result in incomplete interlayer contact, increased interlayer porosity, and reduced interlayer bonding, all of which negatively impact the mechanical properties of the printed structure. The orientation of coarse aggregate during the printing process can also contribute to material anisotropy, leading to variations in mechanical properties along different directions (Chen, Zhang et al. 2021). Furthermore, coarse aggregate may introduce pores and defects during printing, reducing the density and strength of the hardened concrete. It is important to note that cast concrete and 3DPC have different requirements for coarse aggregate due to the distinct nature of their production processes. While cast concrete is highly adaptable to a wide range of aggregate sizes and shapes, printed concrete demands smaller aggregate sizes and more regular shapes to ensure a smooth extrusion, prevent clogging, and maintain the quality of the printed structure.

Currently, extensive research has been conducted on the mechanical properties of 3DPC incorporating coarse aggregate. Ding et al. (2021) investigated the effects of coarse aggregate on the printability and mechanical properties of 3DPC, reporting that the inclusion of coarse aggregate increased the complexity of the printing process and reduced interlayer adhesion. Similarly, Rahul et al. (2022) explored the rheological and mechanical properties, as well as the shrinkage behavior, of 3DPC containing natural and recycled coarse aggregates. Their findings highlighted that the water absorption rate and surface characteristics of coarse aggregate significantly influence printing quality. Liu, Liu, Bai et al. (2022) analyzed the impact of pore defects on the hardened properties of 3DPC containing coarse aggregate, revealing that pores formed during the printing process reduced both the compressive strength and durability of the concrete. Wang et al. (2022) successfully developed 3DPC with coarse aggregate by optimizing the mix design and printing process, resulting in improved printability and mechanical properties. Their study further examined the influence of coarse aggregate content on compressive, flexural, and splitting tensile strength. As shown in Figure 2.19a, compressive strength achieves optimal performance when the volume fraction of coarse aggregate is approximately 60%. Additionally, Figure 2.19b indicates that increasing the amount of coarse aggregate enhances the crack resistance and fracture behavior of concrete. The trend for splitting tensile strength aligns with that of compressive strength, as illustrated in Figure 2.19c. Chen, Zhang et al. (2021) studied extruded 3DPC containing coarse aggregate and found that the inclusion of coarse aggregate introduced directional dependence in the mechanical properties of the material, with a reduction in strength observed along the printing direction.

2.4.3 Fiber

In conventional cast concrete, the incorporation of fibers primarily serves to enhance the crack resistance and toughness of the material. Fibers create a bridging effect within the concrete matrix, which inhibits the initiation and propagation of cracks. This mechanism mitigates cracking induced by plastic shrinkage and drying shrinkage, thereby improving the overall durability of the material. Furthermore, the inclusion of fibers enhances the capacity to bear dynamic loads, improves the impact and fatigue resistance, and ultimately extends the service life of the structure.

In 3DPC, the incorporation of fibers can enhance interlayer bonding performance, mitigate the adverse effects of interface weakness, and improve the structural integrity of the printed components (Nerella et al. 2019). Due to the inherent layer-by-layer deposition process in 3D printing, the structures often exhibit anisotropic behavior. The orientation and distribution of fibers play a critical role in optimizing the

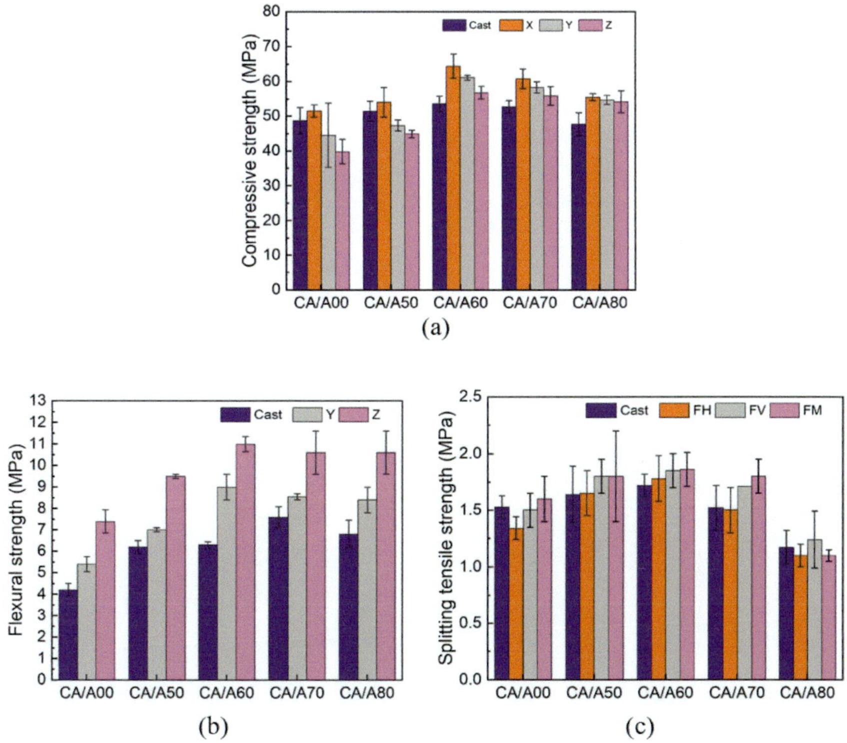

FIGURE 2.19 Effect of coarse aggregate on the mechanical performance of 3DPC: (a) Compressive strength; (b) Flexural strength; (c) Splitting tensile strength (Wang et al. 2022).

mechanical properties of the material across different directions. However, this fiber alignment can also lead to more pronounced anisotropy, which may present additional challenges (Ma et al. 2024).

Figure 2.20 illustrates the microstructure of polyacrylonitrile (PAN) fiber-reinforced concrete, highlighting the alignment of fibers along the printing direction and their role in addressing anisotropy issues. Fibers not only enhance the viscosity of concrete in its fresh state but also improve its yield stress and thixotropy, preventing collapse and deformation during the printing process. This contributes to improved printing quality and enhanced mechanical properties after hardening.

Currently, a wide variety of fibers are utilized in 3DPC, and extensive research has been conducted on their influence on mechanical performance. Pham et al. (2020) investigated the effects of steel fibers with varying sizes and contents on the mechanical properties of 3DPC. Their findings revealed that longer fibers and higher fiber contents significantly improved the flexural strength and toughness of the material, although the degree of improvement varied with direction, as shown in Figure 2.21. Similarly, Bos et al. (2019) examined the impact of short straight steel fibers on the ductility and crack resistance of 3DPC, reporting that the inclusion of fibers enhanced the material's ductility. Sun et al. (2022) developed a polyvinyl

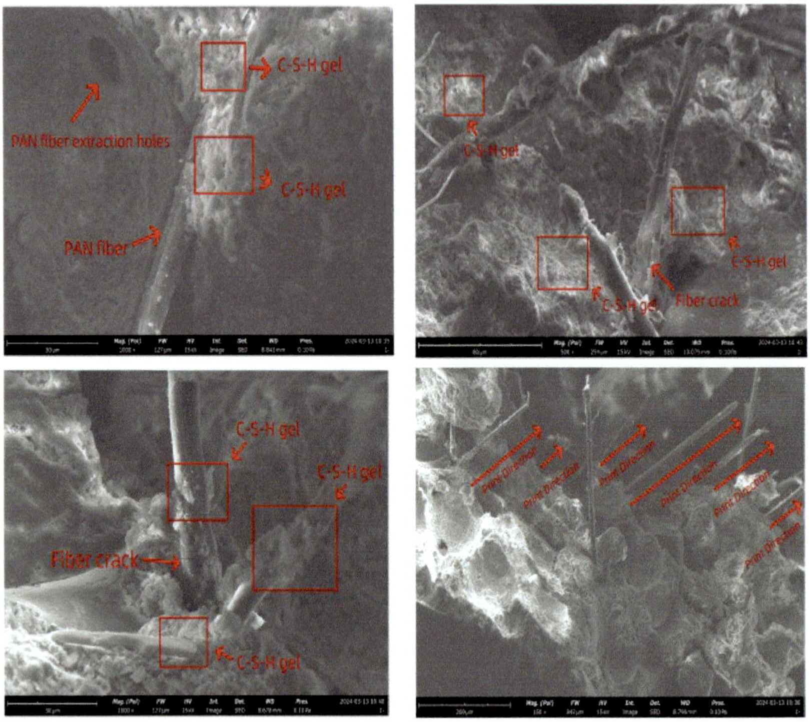

FIGURE 2.20 Microstructure of PAN fiber concrete (Ma et al. 2024).

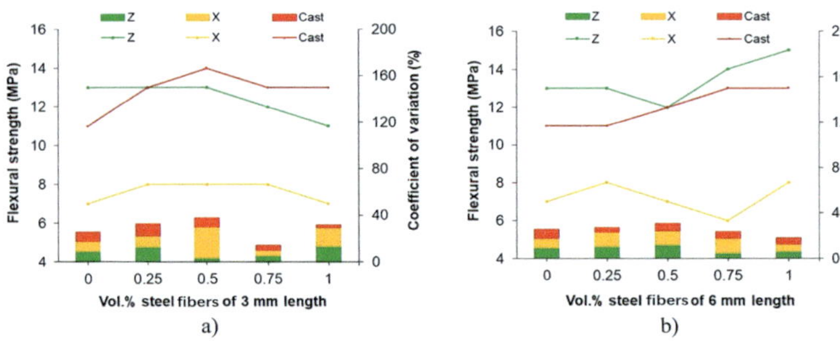

FIGURE 2.21 Effects of (a) 3-mm-long steel fiber and (b) 6-mm-long steel fiber on the mechanical properties of 3D-printed concrete (Pham et al. 2020).

alcohol (PVA) fiber-reinforced high-strength cementitious composite specifically for 3D printing, which demonstrated significant improvements in both mechanical properties and durability. Ye et al. (2021a) studied the effect of polythene (PE) fiber content on the workability and mechanical anisotropy of 3D-printed ultra-high ductility concrete. Their research indicated that an optimal fiber content could enhance isotropy and improve overall mechanical performance. Hambach and Volkmer (2019) investigated directional carbon fiber-reinforced 3D-printed material and achieved a flexural strength exceeding 100 MPa. Chu et al. (2021) developed an extrudable high-strength fiber-reinforced concrete incorporating nano-calcium carbonate and glass fibers, which not only enhances the mechanical properties but also ensures the suitability for 3D printing applications. Sun et al. (2021) examined the bonding performance between basalt fiber-reinforced polymer (BFRP) bars and printed concrete, emphasizing its crucial role in improving structural performance. These studies offer valuable insights into the design and optimization of printed concrete, providing practical guidance for enhancing its mechanical properties and structural capabilities.

2.4.4 Supplementary Cementitious Materials

Common supplementary cementitious materials (SCMs) include fly ash, silica fume, ground granulated blast-furnace slag (GGBFS) (Zhao et al. 2023), steel slag (SS) (Yu et al. 2023), natural pozzolans, and metakaolin. These materials are widely used in 3DPC to enhance its performance. SCMs play a critical role in influencing the mechanical properties of 3DPC through several mechanisms: (1) SCMs such as fly ash and silica fume improve the rheological properties of concrete, enhancing material printability and ensuring the stability of the printing process. (2) The fine particles in SCMs fill the pores within the cement matrix, increasing the density of the concrete and thereby improving both compressive and flexural strength. (3) Certain SCMs exhibit pozzolanic reactivity, which contributes to reducing shrinkage, lowering the risk of cracking, and enhancing the durability of the concrete.

SCMs actively participate in the hydration reaction, producing additional calcium-silicate-hydrate (C-S-H) gel, which improves the ITZ and strengthens the bond between aggregate and cement paste. Furthermore, the utilization of industrial by-products as SCMs, such as fly ash and slag, decreases reliance on traditional cement, reduces the carbon footprint, and aligns with the principles of sustainable development.

Many researchers have adopted SCMs to optimize the mechanical properties of 3DPC. Lu, Qian et al. (2019) developed a sprayable 3D-printed cement-based material using fly ash beads and air-entraining agents, aiming to improve the printability and mechanical properties of the material while reducing the material density. Ye et al. (2023) developed a 3D-printed engineering cement-based composite (ECC) containing incineration bottom ash (IBA), aiming to achieve sustainable and digital construction. The study showed that the addition of IBA had a positive effect on the mechanical properties of the material. Ye et al. (2021b) studied the performance of 3D-printed ultra-high ductility concrete containing rubber particles. The results showed that the combined use of rubber particles and SCMs (fly ash and silica fume) can improve the fresh mix performance and anisotropic mechanical behavior of the material. Zhao et al. (2023) investigated the macroscopic properties and pore structure of a low-carbon 3D printing mortar developed using recycled GGBFS and clay brick powder (CBP). Their study confirmed the feasibility of 3D printing utilizing low-carbon cementitious materials such as GGBFS and CBP. Yu et al. (2023) studied the effect of SS content on the anisotropy of 3DPC at different ages, as shown in Figure 2.22. Their findings revealed that increasing SS content resulted in reductions

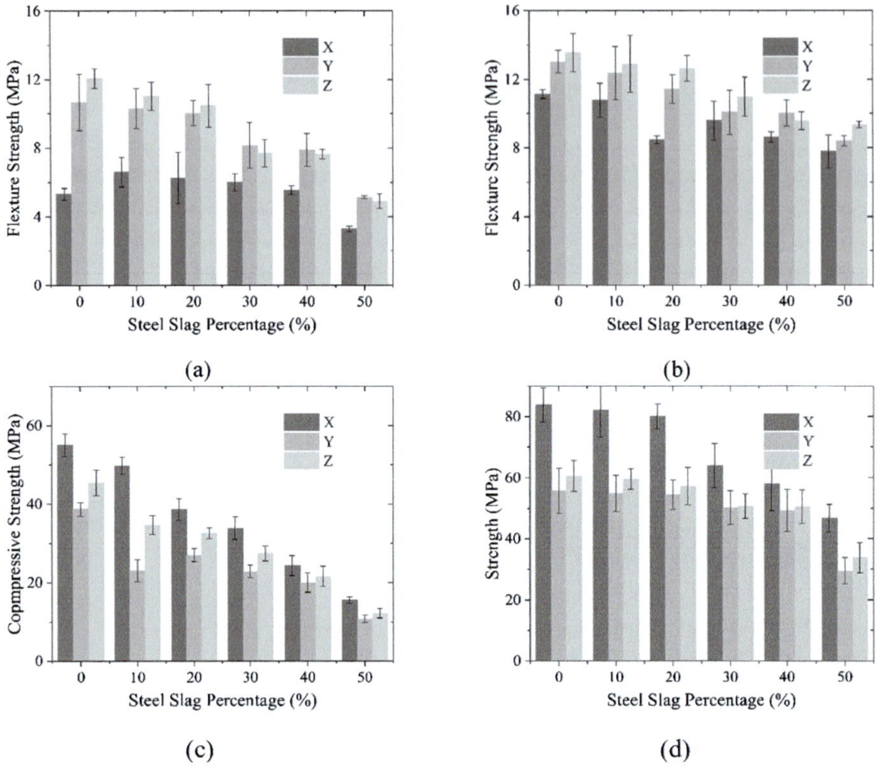

FIGURE 2.22 Test results of flexural strength at (a) 7 days and (b) 28 days; compressive strength at (c) 7 days and (d) 28 days (Yu et al. 2023).

in compressive and flexural strengths. However, neither SS content nor solidification time significantly influenced bending mechanical anisotropy. Across all groups, the highest strength was observed in the Z direction, while the lowest strength occurred in the X direction.

2.5 CONCLUSIONS

This chapter systematically investigated various factors affecting the mechanical properties of 3DPC. First, the mechanical performance testing methods of 3DPC were summarized, including the compressive, flexural, and interlayer bonding tests. The impact of anisotropy caused by the layered construction process on the mechanical properties in different loading directions is also emphasized in this chapter.

Second, the study analyzed how printing process parameters affect the performance of 3DPC. Parameters such as interlayer time intervals, printing speed, and layer height were proven to play critical roles in the mechanical properties of 3DPC. Optimizing these parameters can enhance the overall mechanical performance of the material, ensuring print quality and structural stability.

Finally, the mechanical properties of 3DPC are also heavily influenced by material composition and mixture design. The incorporation of suitable SCMs, such as fly ash and silica fume, along with appropriate admixtures, can significantly enhance the rheological and mechanical properties of the material. A rational material design not only improves printability but also strengthens mechanical performance and durability, thereby providing a robust foundation for the practical application of 3DPC in engineering projects.

REFERENCES

Al-Qutaifi, Sarah, Ali Nazari, and Ali Bagheri. 2018. "Mechanical Properties of Layered Geopolymer Structures Applicable in Concrete 3D-Printing." *Construction and Building Materials* 176. 690–99. https://doi.org/10.1016/j.conbuildmat.2018.04.195.

Alchaar, Aktham S., and Adil K. Al-Tamimi. 2021. "Mechanical Properties of 3D Printed Concrete in Hot Temperatures." *Construction and Building Materials* 266. 120991. https://doi.org/10.1016/j.conbuildmat.2020.120991.

Arunothayan, Arun R., Behzad Nematollahi, Ravi Ranade, Shin Hau Bong, and Jay Sanjayan. 2020. "Development of 3D-Printable Ultra-High Performance Fiber-Reinforced Concrete for Digital Construction." *Construction and Building Materials* 257. 119546. https://doi.org/10.1016/j.conbuildmat.2020.119546.

Arunothayan, Arun R., Behzad Nematollahi, Ravi Ranade, Kamal H. Khayat, and Jay G. Sanjayan. 2022. "Digital Fabrication of Eco-Friendly Ultra-High Performance Fiber-Reinforced Concrete." *Cement and Concrete Composites* 125. 104281. https://doi.org/10.1016/j.cemconcomp.2021.104281.

Bos, F. P., E. Bosco, and T. A. M. Salet. 2019. "Ductility of 3D Printed Concrete Reinforced with Short Straight Steel Fibers." *Virtual and Physical Prototyping* 14 (2). 160–74. https://doi.org/10.1080/17452759.2018.1548069.

Bos, Freek, Rob Wolfs, Zeeshan Ahmed, and Theo Salet. 2016. "Additive Manufacturing of Concrete in Construction: Potentials and Challenges of 3D Concrete Printing." *Virtual and Physical Prototyping* 11 (3). 209–25. https://doi.org/10.1080/17452759.2016.1209867.

Chen, Yidong, Yunsheng Zhang, Bo Pang, Zhiyong Liu, and Guojian Liu. 2021. "Extrusion-Based 3D Printing Concrete with Coarse Aggregate: Printability and Direction-Dependent Mechanical Performance." *Construction and Building Materials* 296. 123624. https://doi.org/10.1016/j.conbuildmat.2021.123624.

Chen, Yu, Oğuzhan Çopuroğlu, Claudia Romero Rodriguez, Fernando F.de Mendonca Filho, and Erik Schlangen. 2021. "Characterization of Air-Void Systems in 3D Printed Cementitious Materials Using Optical Image Scanning and X-Ray Computed Tomography." *Materials Characterization* 173. https://doi.org/10.1016/j.matchar.2021.110948.

Chen, Yu, Koen Jansen, Hongzhi Zhang, Claudia Romero Rodriguez, Yidong Gan, Oğuzhan Çopuroğlu, and Erik Schlangen. 2020. "Effect of Printing Parameters on Interlayer Bond Strength of 3D Printed Limestone-Calcined Clay-Based Cementitious Materials: An Experimental and Numerical Study." *Construction and Building Materials* 262. 120094. https://doi.org/10.1016/j.conbuildmat.2020.120094.

Chu, S. H., L. G. Li, and A. K. H. Kwan. 2021. "Development of Extrudable High Strength Fiber Reinforced Concrete Incorporating Nano Calcium Carbonate." *Additive Manufacturing* 37. https://doi.org/10.1016/j.addma.2020.101617.

Ding, Tao, Jianzhuang Xiao, and Viktor Mechtcherine. 2023. "Microstructure and Mechanical Properties of Interlayer Regions in Extrusion-Based 3D Printed Concrete: A Critical Review." *Cement and Concrete Composites* 141. 105154. https://doi.org/10.1016/j.cemconcomp.2023.105154.

Ding, Tao, Jianzhuang Xiao, Fei Qin, and Zhenhua Duan. 2020. "Mechanical Behavior of 3D Printed Mortar with Recycled Sand at Early Ages." *Construction and Building Materials* 248. 118654. https://doi.org/10.1016/j.conbuildmat.2020.118654.

Ding, Tao, Jianzhuang Xiao, Shuai Zou, and Yu Wang. 2020. "Hardened Properties of Layered 3D Printed Concrete with Recycled Sand." *Cement and Concrete Composites* 113. 103724. https://doi.org/10.1016/j.cemconcomp.2020.103724.

Ding, Tao, Jianzhuang Xiao, Shuai Zou, and Jiangtao Yu. 2021. "Flexural Properties of 3D Printed Fibre-Reinforced Concrete with Recycled Sand." *Construction and Building Materials* 288. 123077. https://doi.org/10.1016/j.conbuildmat.2021.123077.

Ding, Tao, Jianzhuang Xiao, Shuai Zou, and Xinji Zhou. 2020. "Anisotropic Behavior in Bending of 3D Printed Concrete Reinforced with Fibers." *Composite Structures* 254. 112808. https://doi.org/10.1016/j.compstruct.2020.112808.

Geng, Zifan, Wei She, Wenqiang Zuo, Kai Lyu, Hao Pan, Yunsheng Zhang, and Changwen Miao. 2020. "Layer-Interface Properties in 3D Printed Concrete: Dual Hierarchical Structure and Micromechanical Characterization." *Cement and Concrete Research* 138. 106220. https://doi.org/10.1016/j.cemconres.2020.106220.

Gomaa, Shady, Elmer M. Irizarry, Ayesha Ahmed, Raul Marrero Rosa, Hassan Ahmed, Jedadiah Burroughs, Eric Kreiger, Jiaqi Liu, Matthew Troemner, and Gianluca Cusatis. 2024. "3D Printing of Ultra-High-Performance Concrete: Shape Stability for Various Printing Systems." *Construction and Building Materials* 456. 139039. https://doi.org/10.1016/j.conbuildmat.2024.139039.

Hambach, Manuel, and Dirk Volkmer. 2019. "Properties of 3D-Printed Fiber-Reinforced Portland Cement Paste." *Cement and Concrete Composites* 79. 62–70. https://doi.org/10.1016/j.cemconcomp.2017.02.001.

He, Lewei, Jolyn Ze Mei Tan, Wai Tuck Chow, Hua Li, and Jiahui Pan. 2021. "Design of Novel Nozzles for Higher Interlayer Strength of 3D Printed Cement Paste." *Additive Manufacturing* 48. 102452. https://doi.org/10.1016/j.addma.2021.102452.

Heever, Marchant van den, Anton du Plessis, Jacques Kruger, and Gideon van Zijl. 2022. "Evaluating the Effects of Porosity on the Mechanical Properties of Extrusion-Based 3D Printed Concrete." *Cement and Concrete Research* 153. 106695. https://doi.org/10.1016/j.cemconres.2021.106695.

Huang, Xin, Weihao Yang, Fangnian Song, and Jiuqun Zou. 2022. "Study on the Mechanical Properties of 3D Printing Concrete Layers and the Mechanism of Influence of Printing Parameters." *Construction and Building Materials* 335. 127496. https://doi.org/10.1016/j.conbuildmat.2022.127496.

Ji, Guangchao, Jianzhuang Xiao, Peng Zhi, Yu Ching Wu, and Nv Han. 2022. "Effects of Extrusion Parameters on Properties of 3D Printing Concrete with Coarse Aggregates." *Construction and Building Materials* 325. 126740. https://doi.org/10.1016/j.conbuildmat.2022.126740.

Le, T. T., S. A. Austin, S. Lim, R. A. Buswell, R. Law, A. G. F. Gibb, and T. Thorpe. 2012. "Hardened Properties of High-Performance Printing Concrete." *Cement and Concrete Research* 42 (3). 558–66. https://doi.org/10.1016/J.CEMCONRES.2011.12.003.

Ler, Kee Hong, Chau Khun Ma, Chee Loong Chin, Izni Syahrizal Ibrahim, Khairul Hazman Padil, Mohd Aminul Izmeer Ab Ghafar, and Alva Ami Lenya. 2024. "Porosity and Durability Tests on 3D Printing Concrete: A Review." *Construction and Building Materials* 446. 137973. https://doi.org/10.1016/j.conbuildmat.2024.137973.

Liu, Huawei, Chao Liu, Guoliang Bai, Yiwen Wu, Chunhui He, Rongfei Zhang, and Youqiang Wang. 2022. "Influence of Pore Defects on the Hardened Properties of 3D Printed Concrete with Coarse Aggregate." *Additive Manufacturing* 55. 102843. https://doi.org/10.1016/j.addma.2022.102843.

Liu, Huawei, Chao Liu, Yiwen Wu, Guoliang Bai, Chunhui He, Rongfei Zhang, and Youqiang Wang. 2022. "Hardened Properties of 3D Printed Concrete with Recycled Coarse

Aggregate." *Cement and Concrete Research* 159. 106868. https://doi.org/10.1016/j.cemconres.2022.106868.

Lu, Bing, Ye Qian, Mingyang Li, Yiwei Weng, Kah Fai Leong, Ming Jen Tan, and Shunzhi Qian. 2019. "Designing Spray-Based 3D Printable Cementitious Materials with Fly Ash Cenosphere and Air Entraining Agent." *Construction and Building Materials* 211. 1073–84. https://doi.org/10.1016/j.conbuildmat.2019.03.186.

Lu, Bing, Yiwei Weng, Mingyang Li, Ye Qian, Kah Fai Leong, Ming Jen Tan, and Shunzhi Qian. 2019. "A Systematical Review of 3D Printable Cementitious Materials." *Construction and Building Materials* 207. 477–90. https://doi.org/10.1016/j.conbuildmat.2019.02.144.

Ma, Lei, Qing Zhang, Zijian Jia, Chao Liu, Zhicong Deng, and Yamei Zhang. 2022. "Effect of Drying Environment on Mechanical Properties, Internal RH and Pore Structure of 3D Printed Concrete." *Construction and Building Materials* 315. 125731. https://doi.org/10.1016/j.conbuildmat.2021.125731.

Ma, Wei, Guosheng Wang, Yaya Zhou, Qinghu Xu, and Yuntong Dai. 2024. "Polyacrylonitrile Fiber Reinforced 3D Printed Concrete: Effects of Fiber Length and Content." *Journal of Building Engineering* 97. 110869. https://doi.org/10.1016/j.jobe.2024.110869.

Nair, Sooraj A. O., Avinaya Tripati, and Narayanan Neithalath. 2021. "Examining Layer Height Effects on the Flexural and Fracture Response of Plain and Fiber-Reinforced 3D-Printed Beams." *Cement and Concrete Composites* 124. 104254. https://doi.org/10.1016/j.cemconcomp.2021.104254.

Napolitano, Rosanna, Daniele Forni, Costantino Menna, Domenico Asprone, and Ezio Cadoni. 2021. "Dynamic Characterization of the Layer-Interface Properties of 3D-Printed Concrete Elements." *Case Studies in Construction Materials* 15. e00780. https://doi.org/10.1016/j.cscm.2021.e00780.

Nerella, Venkatesh Naidu, Simone Hempel, and Viktor Mechtcherine. 2019. "Effects of Layer-Interface Properties on Mechanical Performance of Concrete Elements Produced by Extrusion-Based 3D-Printing." *Construction and Building Materials* 205. 586–601. https://doi.org/10.1016/j.conbuildmat.2019.01.235.

Nicolas, Roussel, Buswell Richard, Ducoulombier Nicolas, Ivanova Irina, Kolawole John Temitope, Lowke Dirk, Mechtcherine Viktor, et al. 2022. "Assessing the Fresh Properties of Printable Cement-Based Materials: High Potential Tests for Quality Control." *Cement and Concrete Research* 158. 106836. https://doi.org/10.1016/j.cemconres.2022.106836.

Pan, Tinghong, Yaqing Jiang, Hui He, Yu Wang, and Kangting Yin. 2021. "Effect of Structural Build-up on Interlayer Bond Strength of 3D Printed Cement Mortars." *Materials* 14 (2). 1–17. https://doi.org/10.3390/ma14020236.

Pan, Zuanfeng, Doudou Si, Jinghong Tao, and Jianzhuang Xiao. 2023. "Compressive Behavior of 3D Printed Concrete with Different Printing Paths and Concrete Ages." *Case Studies in Construction Materials* 18. e01949. https://doi.org/10.1016/j.cscm.2023.e01949.

Panda, Biranchi, Suvash Chandra Paul, Nisar Ahamed Noor Mohamed, Yi Wei Daniel Tay, and Ming Jen Tan. 2018. "Measurement of Tensile Bond Strength of 3D Printed Geopolymer Mortar." *Measurement* 113. 108–16. https://doi.org/10.1016/j.measurement.2017.08.051.

Panda, Biranchi, Suvash Chandra Paul, and Ming Jen Tan. 2017. "Anisotropic Mechanical Performance of 3D Printed Fiber Reinforced Sustainable Construction Material." *Materials Letters* 209. 146–49. https://doi.org/10.1016/j.matlet.2017.07.123.

Paul, Suvash Chandra, Yi Wei Daniel Tay, Biranchi Panda, and Ming Jen Tan. 2018. "Fresh and Hardened Properties of 3D Printable Cementitious Materials for Building and Construction." *Archives of Civil and Mechanical Engineering* 18 (1). 311–19. https://doi.org/10.1016/j.acme.2017.02.008.

Pham, Luong, Phuong Tran, and Jay Sanjayan. 2020. "Steel Fibres Reinforced 3D Printed Concrete: Influence of Fibre Sizes on Mechanical Performance." *Construction and Building Materials* 250. 118785. https://doi.org/10.1016/j.conbuildmat.2020.118785.

Rahul, A. V., Manu K. Mohan, Geert De Schutter, and Kim Van Tittelboom. 2022. "3D Printable Concrete with Natural and Recycled Coarse Aggregates: Rheological, Mechanical and Shrinkage Behaviour." *Cement and Concrete Composites* 125. 104311. https://doi.org/10.1016/j.cemconcomp.2021.104311.

Sanjayan, Jay G., Behzad Nematollahi, Ming Xia, and Taylor Marchment. 2018. "Effect of Surface Moisture on Inter-Layer Strength of 3D Printed Concrete." *Construction and Building Materials* 172. 468–75. https://doi.org/10.1016/j.conbuildmat.2018.03.232.

Silfwerbrand, J. 2003. "Shear Bond Strength in Repaired Concrete Structures." *Materials and Structures/Materiaux et Constructions* 36 (260). 419–24. https://doi.org/10.1617/13859.

Singh, Amardeep, Qiong Liu, Jianzhuang Xiao, and Qifeng Lyu. 2022. "Mechanical and Macrostructural Properties of 3D Printed Concrete Dosed with Steel Fibers under Different Loading Direction." *Construction and Building Materials* 323. 126616. https://doi.org/10.1016/j.conbuildmat.2022.126616.

Sun, Xiaoyan, Chao Gao, and Hailong Wang. 2021. "Bond Performance between BFRP Bars and 3D Printed Concrete." *Construction and Building Materials* 269. 121325. https://doi.org/10.1016/j.conbuildmat.2020.121325.

Sun, Xiaoyan, Jiawei Zhou, Qun Wang, Jiangpeng Shi, and Hailong Wang. 2022. "PVA Fibre Reinforced High-Strength Cementitious Composite for 3D Printing: Mechanical Properties and Durability." *Additive Manufacturing* 49. 102500. https://doi.org/10.1016/j.addma.2021.102500.

Tay, Yi Wei Daniel, Guan Heng Andrew Ting, Ye Qian, Biranchi Panda, Lewei He, and Ming Jen Tan. 2019. "Time Gap Effect on Bond Strength of 3D-Printed Concrete." *Virtual and Physical Prototyping* 14 (1). 104–13. https://doi.org/10.1080/17452759.2018.1500420.

Wang, Chaofan, Bing Chen, Thanh Liem Vo, and Mohammad Rezania. 2023. "Mechanical Anisotropy, Rheology and Carbon Footprint of 3D Printable Concrete: A Review." *Journal of Building Engineering* 76. 107309. https://doi.org/10.1016/j.jobe.2023.107309.

Wang, Xianggang, Lutao Jia, Zijian Jia, Chao Zhang, Yuning Chen, Lei Ma, Zhibin Wang, Zhicong Deng, Nemkumar Banthia, and Yamei Zhang. 2022. "Optimization of 3D Printing Concrete with Coarse Aggregate via Proper Mix Design and Printing Process." *Journal of Building Engineering* 56. 104745. https://doi.org/10.1016/j.jobe.2022.104745.

Wolfs, R. J. M., F. P. Bos, and T. A. M. Salet. 2018. "Early Age Mechanical Behaviour of 3D Printed Concrete: Numerical Modelling and Experimental Testing." *Cement and Concrete Research* 106. 103–16. https://doi.org/10.1016/J.CEMCONRES.2018.02.001.

Wolfs, R. J. M., F. P. Bos, and T. A. M. Salet. 2019. "Hardened Properties of 3D Printed Concrete: The Influence of Process Parameters on Interlayer Adhesion." *Cement and Concrete Research*, 119. 132–40. https://doi.org/10.1016/J.CEMCONRES.2019.02.017.

Ye, Junhong, Can Cui, Jiangtao Yu, Kequan Yu, and Fangyuan Dong. 2021a. "Effect of Polyethylene Fiber Content on Workability and Mechanical-Anisotropic Properties of 3D Printed Ultra-High Ductile Concrete." *Construction and Building Materials* 281. 122586. https://doi.org/10.1016/j.conbuildmat.2021.122586.

Ye, Junhong, Can Cui, Jiangtao Yu, Kequan Yu, and Jianzhuang Xiao. 2021b. "Fresh and Anisotropic-Mechanical Properties of 3D Printable Ultra-High Ductile Concrete with Crumb Rubber." *Composites Part B: Engineering* 211. 108639. https://doi.org/10.1016/j.compositesb.2021.108639.

Ye, Junhong, Fei Teng, Jie Yu, Shiwei Yu, Hongjian Du, Dong Zhang, Shaoqin Ruan, and Yiwei Weng. 2023. "Development of 3D Printable Engineered Cementitious Composites with Incineration Bottom Ash (IBA) for Sustainable and Digital Construction." *Journal of Cleaner Production* 422. 138639. https://doi.org/10.1016/j.jclepro.2023.138639.

Yu, Qian, Binrong Zhu, Xuesen Li, Lingqi Meng, Jingming Cai, Yamei Zhang, and Jinlong Pan. 2023. "Investigation of the Rheological and Mechanical Properties of 3D Printed Eco-Friendly Concrete with Steel Slag." *Journal of Building Engineering* 72. 106621. https://doi.org/10.1016/j.jobe.2023.106621.

Zhang, Kaijian, Wenqiang Lin, Qingtian Zhang, Dehui Wang, and Surong Luo. 2024. "Evaluation of Anisotropy and Statistical Parameters of Compressive Strength for 3D Printed Concrete." *Construction and Building Materials* 440. 137417. https://doi.org/10.1016/j.conbuildmat.2024.137417.

Zhang, Yu, Hongxia Qiao, Rusheng Qian, Cuizhen Xue, Qiong Feng, Li Su, Yunsheng Zhang, Guojian Liu, and Hongjian Du. 2022. "Relationship between Water Transport Behaviour and Interlayer Voids of 3D Printed Concrete." *Construction and Building Materials* 326. https://doi.org/10.1016/j.conbuildmat.2022.126731.

Zhao, Yasong, Yangyunzhi Gao, Gaofeng Chen, Shujun Li, Amardeep Singh, Xu Luo, Cheng Liu, Jianming Gao, and Hongjian Du. 2023. "Development of Low-Carbon Materials from GGBS and Clay Brick Powder for 3D Concrete Printing." *Construction and Building Materials* 383. 131232. https://doi.org/10.1016/j.conbuildmat.2023.131232.

Zou, Shuai, Jianzhuang Xiao, Tao Ding, Zhenhua Duan, and Qingtian Zhang. 2021. "Printability and Advantages of 3D Printing Mortar with 100% Recycled Sand." *Construction and Building Materials* 273. 121699. https://doi.org/10.1016/j.conbuildmat.2020.121699.

3 Fracture Mechanics of Layered and Additively Manufactured Materials

Theoretical Frameworks and Experimental Approaches

Shashank Gupta, Aimane Najmeddine, and Reza Moini

3.1 INTRODUCTION

Fracture mechanics has remained largely underutilized in additively manufactured materials and structures, despite the need for understanding the role of heterogeneous interfaces in fracture response and in the design and engineering across scales (R. Moini 2023). The advent of additive manufacturing has allowed for the design of more advanced concrete infrastructure components, allowing engineering of the macroscopic response through control over the local compositions, properties, and overall morphology (materials' architecture). The discretization of materials into filaments leads to the formation of horizontal and vertical interfaces that can significantly alter the size of the fracture process zone (FPZ). Moreover, the crack path can be significantly altered in layered and additively manufactured materials due to the presence of heterogeneity, thereby dominating the mechanics of failure and fracture, either in the absence or presence of a pre-existing notch (Gupta et al. 2023; Prihar, Garlock, et al. 2024). The presence of layered defects, also referred to as *weak* interfaces, stems from a large range of capillary pores or larger macroscopic pores than those within the filaments (Moini, Olek, Magee, et al. 2018; Moini, Baghaie, et al. 2021; Gupta et al. 2026; Najmeddine and Moini 2025; Najmeddine, Gupta, and Moini 2025). These layered microstructural heterogeneities are absent in the cast counterparts and, therefore, can change the fracture properties and fracture behavior associated with the softening law. Many similarities exist among additively manufactured cement paste, mortar, and concrete with their cast counterparts in terms of crack initiation and initial fracture energy. However, the total fracture energy required for propagation and the interaction of the crack during propagation with the weak interfaces substantially differ from those of the cast counterparts. Such a key

DOI: 10.1201/9781003434863-4

distinction presents the need for a better understanding of fracture in layered cement-based materials. It also offers a new opportunity for engineering the arrangement of the materials and properties of the interfaces, as a means to control the fracture behavior, structural response, and overall fracture properties and damage resistance of these materials and structures. The efficacy and impact of the underlying damage mechanisms in layered materials, nevertheless, remain highly dependent on the size (i.e., size of the FPZ relative to the size of the structure) and scale (i.e., size effect), which will determine the appropriate fracture mechanics theory to understand the fracture response.

Moreover, the mechanisms of energy dissipation in these layered and 3D-printed materials – including microcracking at the interfaces ahead of the crack tip, interface and aggregate bridging, friction between the interfaces behind the crack tip, and crack–microstructure (i.e., interface) interactions – can promote the spread of damage over larger volumes compared to cast counterparts, effectively enhancing damage dissipation and/or enlarging the FPZ. Hence, exploitation of such mechanisms can allow for new approaches to enhance the intrinsic fracture energy of cement-based materials, by exploring interfacial damage phenomena and increasing the fracture toughness by design. Such an undertaking has previously only been viable through new compositions (e.g., addition of fibers), whereas the manufacturing-enabled design approach has recently promoted the development of architected cement-based materials through a wide range of additive processes (Gupta and Moini 2024; Gupta, Esmaeeli, and Moini 2024; Prihar, Gupta, et al. 2024; Rodriguez et al. 2024; Daneshvar et al. 2024; Prihar, Garlock, et al. 2024; Moini and Makinen 2023; Ralston, Gupta, and Moini 2024; Li et al. 2025; Zhou et al. 2023; Xie, He, et al. 2025; Zhou, Xu, et al. 2025; Zhou, Bol, et al. 2025) and other digital fabrication technologies across scales (Lowke et al. 2024; Tomholt, Meggers, and Moini 2024; Dobrzanski et al. 2025; Dobrzanski et al. 2022; Rupp et al. 2023). In recent years, the use of purposeful arrangements of material to engineer fracture and other mechanical properties has gained significant attention. By engineering the crack–microstructure (crack–interface) interactions under both quasi-static and impact loading conditions (Liu et al. 2022; Pan et al. 2024; Ramesh et al. 2024; Zhang et al. 2025; Xie, Xu, et al. 2025), new degrees of freedom for tailoring the fracture response of cement-based materials have emerged. These studies have demonstrated significant improvement in damage delocalization and damage dissipation, for instance, through crack deflection along the interfaces in contrast to penetration into the bulk filaments (Moini, Olek, Youngblood, et al. 2018; Savija 2020; Chang et al. 2022; Najmeddine, Gupta, and Moini 2025; Najmeddine and Moini 2025), not achievable in conventional monolithic counterparts fabricated through casting. Nevertheless, utilizing the appropriate fracture mechanics methods is necessary for both the layered and more complex architected classes of additively manufactured cement-based materials.

Theoretical studies have also demonstrated the potential for improvement in safety with respect to the tail of strength for some architected arrangements compared to monolithic cast materials based on fishnet statistics (Luo and Bažant 2017; Bažant et al. 2021). The field of fracture mechanics of layered, additively manufactured, and digitally fabricated architected materials is in its infancy, aiming to answer abundant fundamental questions on fracture in these materials.

Similarly, long-term damage associated with durability characteristics of additively manufactured cement-based materials, such as freeze–thaw and transport, is currently not well-understood, especially with regard to the presence of the interface (Ghantous et al. 2023; Das et al. 2022; Surehali et al. 2023; Van Der Putten et al. 2022; Dong et al. 2025; Ghourchian et al. 2021; Gupta et al. 2023). The damage due to physical and chemical degradation in layered and 3D-printed cement-based materials may require pairing with a fracture mechanics approach to the problem, given that degradation can take place in the bulk filaments (i.e., within the interfacial transition zone (ITZ) between the paste and the aggregates) or in the interfacial zone (IZ) between the filaments. The resulting mechanical degradation and long-term durability performance can therefore be highly determined by the local cohesive properties and fracture characteristics of these IZs, in addition to the bulk ITZ in the microstructure of the filaments. The fracture toughness and cohesive strength of the interfaces and filaments can be altered due to the physical and chemical degradation. In either case, the coupling between the physicochemical degradation and fracture properties in the presence of interfaces has not been well-understood or examined.

Utilizing the linear and nonlinear elastic fracture mechanics developed for brittle and quasi-brittle materials (Bažant and Planas 2019; Bažant et al. 2021) in conjunction with the fracture mechanics of layered materials (as discussed in Chapter 12 in this book) (Salviato et al. 2019; Bažant 2002) is necessary to better understand the fracture and damage in additively manufactured cementitious materials. The foundational models can help provide the advancement in the understanding of fracture across all classes of cementitious materials whose behavior can be dominated by the presence of weak interfaces (i.e., fiber-reinforced concrete, repair, and cold joints).

In recent years, the design and fabrication of additively manufactured structures have rapidly grown (Ma et al. 2022; Bos et al. 2022). The large-scale deployments in the building and renewable energy sectors have necessitated the formal design of concrete structures using additive technology, with or without steel reinforcement (Wang et al. 2025; Heever et al. 2022; Gebhard et al. 2021, 2022). These structures are often fabricated as additively manufactured thin walls integrated with the casting of reinforced concrete for gravity and lateral load-bearing systems (Gebhard et al. 2022). These structures are versatile in design and often customized, given the flexibility provided by additive manufacturing. In the absence of a design code, they must be structurally analyzed through extensive experimental and theoretical approaches. One major challenge in the experimental approach to analyze the structural performance of these layered structures, unlike conventionally cast structures, is to experimentally account for the presence and mechanical characteristics of the interface and the associated heterogeneous local properties. Such characteristics, present due to the layered interfaces, persist and potentially differ across all processing techniques, such as extrusion-based (Rabiei and Moini 2025; Rodriguez et al. 2021; Moini et al. 2022; M. Moini 2020; Moini, Olek, Zavattieri, et al. 2021) and powder-bed techniques (Lowke et al. 2021; Lowke et al. 2018), further requiring the experimental characterization of the local mechanical properties of the interface (cohesive strength and toughness) in modes I and II, for incorporation into appropriate numerical methods for structural analysis.

In brief, the design and engineering of additively manufactured materials and structures across laboratory and large scales, whether carried out with or without reinforcement, need to address the fracture mechanics questions including the following:

i. What are the size-dependent aspects of fracture and failure in layered additively manufactured materials in the presence of highly variable filament size relative to the size of the structure?
ii. What are the relevant linear and nonlinear foundational models in fracture mechanics of brittle and quasi-brittle materials that are appropriate for studying additively manufactured materials and structures across sizes and scales?
iii. What are the appropriate approaches to experimentally characterize the fracture properties of layered and additively manufactured materials and the local interfacial properties?

The principles of mechanics of layered materials, while used in layered composites and layered rocks, have not yet been widely applied to additively manufactured concrete as a layered material or as a structure, except for limited studies (Gupta et al. 2023; Nair, Tripathi, and Neithalath 2021; Prihar, Garlock, et al. 2024; Tang et al. 2024; Najmeddine, Gupta, and Moini 2025; Najmeddine and Moini 2025; Prihar, Gupta, et al. 2024). The developments in the additively manufactured concrete have also not directly dealt with the foundational methods and models available in the layered and interfacial fracture mechanics literature (e.g., in brittle and composite materials) (Begley and Hutchinson 2017). For instance, the three-point-bend test (i.e., un-notched specimen) has often been used to characterize the interlayer bond strength in additively manufactured cement-based materials, which may or may not lead to failure initiation or propagation at the interface, as it can initiate at or propagate through the bulk filaments. Furthermore, the strength and fracture toughness of additively manufactured cement-based materials, similar to their conventional counterparts, are dependent on size. Size effect law (SEL; types 1 and 2) (Bažant 1999, 2000) can be used to reveal a first-hand perspective on how layered structures with interfaces differ from their counterparts and how their properties depend on scale. This is of practical importance, especially given the wide range for sizes used in the fabrication of 3D-printed materials and structures from laboratory (cm) to large renewable energy (tens of meters) scale today. From a standardization perspective, the quantification of "material" properties (i.e., strength and fracture toughness) for the fabricated structures may benefit from visiting the SEL, as these properties remain size-dependent. As mentioned earlier, the size of the FPZ (including interfacial and bulk flaws) relative to the size of the structures can also determine appropriate linear or nonlinear elastic fracture mechanics approaches for characterizing the fracture properties and fracture behavior of additively manufactured cement-based materials and structures.

This chapter revisits the fundamentals of linear and nonlinear elastic fracture mechanics and discusses the foundational theoretical and experimental methods and models to describe, examine, and understand fracture in brittle and quasi-brittle materials. Due to limited space, only selected classical approaches have been discussed. The use of these models for characterizing interfaces or examining the fracture properties and fracture response of bulk layered compositions with interfaces is discussed.

Lastly, the current gaps and future directions in experimental approaches for better understanding the fracture mechanics of additively manufactured and architected materials and structures are discussed. This chapter walks the reader through basic linear and more advanced nonlinear elastic fracture mechanics approaches and discusses the potential application of these methods in layered and more complex architected cement-based materials.

3.2 THEORETICAL FOUNDATION FOR BRITTLE AND QUASI-BRITTLE MATERIALS

3.2.1 Linear Elastic Fracture Mechanics

3.2.1.1 Critical Stress Intensity Factor Approach

For certain cracked geometries subjected to external loading, it is possible to derive closed-form expressions for the stresses within the body, assuming isotropic linear elastic material behavior (Anderson and Anderson 2005). Early contributions to such analytical solutions were made by Westergaard (1939), Irwin (1957), and Sneddon (1946).

When a polar coordinate system is defined with its origin at the crack tip, the stress field in any linear elastic cracked body can be expressed as

$$\sigma_{ij} = \frac{k}{\sqrt{r}} f_{ij}(\theta) + \text{HOT}, \tag{3.1}$$

where σ_{ij} is the stress tensor; r and θ are the radial and angular coordinates measured from the crack tip, respectively; k is a proportionality constant; and $f_{ij}(\theta)$ is a dimensionless function of θ. The higher-order terms (HOT) depend on the geometry of the cracked body and remain finite or vanish as $r \to 0$, while the leading term, proportional to $\frac{1}{\sqrt{r}}$, becomes unbounded. This singularity indicates that the stress magnitude increases sharply as the distance to the crack tip decreases (as shown in Figure 3.1), regardless of the overall geometry.

Cracks can be loaded in three fundamental modes:

- **Mode I** (opening mode): Principal load acts normal to the crack plane, opening the crack faces.
- **Mode II** (sliding mode): In-plane shear loading causes one crack face to slide relative to the other.
- **Mode III** (tearing mode): Out-of-plane shear loading results in tearing of the crack surfaces.

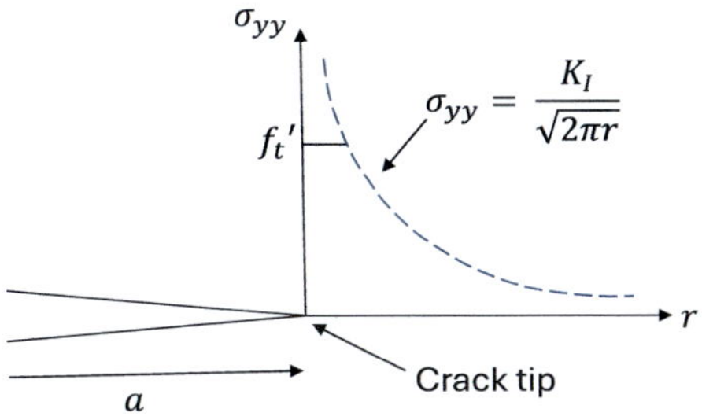

FIGURE 3.1 Stress field in the vicinity of the crack tip.

Each mode produces the $\frac{1}{\sqrt{r}}$ stress singularity at the crack tip, but the proportionality constant k and the angular function $f_{ij}(\theta)$ differ by mode. It is common to replace k with the *stress intensity factor* K, where $K = k\sqrt{2\pi}$ (Anderson and Anderson 2005). A subscript is used to indicate the mode of loading, resulting in K_I, K_{II}, and K_{III}. The stress field ahead of the crack tip in each mode can then be written as

$$\lim_{r\to 0} \sigma_{ij}^{(I)} = \frac{K_I}{\sqrt{2\pi r}} f_{ij}^{(I)}(\theta), \tag{3.2}$$

$$\lim_{r\to 0} \sigma_{ij}^{(II)} = \frac{K_{II}}{\sqrt{2\pi r}} f_{ij}^{(II)}(\theta), \tag{3.3}$$

$$\lim_{r\to 0} \sigma_{ij}^{(III)} = \frac{K_{III}}{\sqrt{2\pi r}} f_{ij}^{(III)}(\theta), \tag{3.4}$$

where K_I, K_{II}, and K_{III} quantify the intensity of the stress field for each loading mode.

The stress intensity factor quantifies the amplitude of the crack tip singularity such that stresses in the vicinity of the crack tip increase proportionally with K (Anderson and Anderson 2005). It fully characterizes the crack tip conditions, meaning that if K is known, all components of stress, strain, and displacement can be determined as functions of r and θ in the linear elastic fracture mechanics (LEFM) framework (Ritchie and Liu 2021). In front of the crack tip $(\theta = 0)$, the normal stress component σ_{yy} at the crack tip can be expressed using the singular solution of LEFM as $\sigma_{yy} = \frac{K_I}{\sqrt{2\pi r}}$, as shown in Figure 3.1. However, in LEFM, the normal stress ahead of the crack tip (σ_{yy}) is physically limited by the tensile strength of the material (f_t'), as illustrated in Figure 3.1.

While the stress intensity factor is derived from an idealized, singular elastic solution, it remains central to fracture mechanics because the inelastic deformation near the crack tip is typically confined to a small zone (depending on the material and the size of the structure), often known as the FPZ. As Erdogan (1983) emphasizes, the stress intensity factor scales a universal near-tip field that is largely insensitive to global geometry or boundary conditions, making it a powerful and transferable parameter for predicting crack initiation and growth (propagation) across diverse materials and structures, provided the assumptions of LEFM with a small FPZ relative to specimen size are satisfied.

Crack initiation in brittle and quasi-brittle materials such as cementitious materials is commonly assessed using the Mode I fracture toughness K_{IC}, defined as the critical value of the stress intensity factor at which a crack begins to propagate under tensile loading (Zhu and Joyce 2012). When the applied K_I reaches K_{IC}, the energy available for crack initiation equals the material's resistance to fracture, leading to crack growth. This criterion forms the basis of LEFM and is widely used to characterize the crack resistance of materials that have minimal inelastic deformation ahead of the crack tip (i.e., small FPZ) (Erdogan 1983).

The Mode I stress intensity factor K_I describes the stress field near the crack tip under pure tension and is given by (Anderson and Anderson 2005)

$$K_I = \sigma\sqrt{\pi a} \cdot Y, \tag{3.5}$$

where σ is the nominal stress applied at a characteristic distance from the crack tip, a is the crack length, and Y is a dimensionless geometry factor that depends on the crack length and the specimen geometry.

To experimentally determine the critical stress intensity factor K_{IC}, many fracture tests have been proposed for cementitious materials over the past several decades. The single-edge notch bend (SENB) test (in either three-point or four-point configurations) (Baldie and Pratt 1985; Karihaloo and Nallathambi 1991; Wang et al. 2016; Ye et al. 2024), wedge-splitting test (Brühwiler and Wittmann 1990; Xiao et al. 2004; Ishiguro 2007; Banthia and Nandakumar 2003), and Brazil-nut (or notched Brazilian) (Gupta et al. 2023; Venu et al. 2024; Ojdrovic and Petroski 1985) tests are among some of the most commonly used. These tests have been extensively adopted at various scales and for a variety of cementitious systems, such as cement paste (Baldie and Pratt 1985), mortar (Ishiguro 2007), and conventional concrete (Karihaloo and Nallathambi 1991; Ojdrovic and Petroski 1985), and a wide range of fiber-reinforced concrete (Venu et al. 2024; Ye et al. 2024; Xiao et al. 2004). More recently, they have been used in additively manufactured cementitious materials for cement paste (Gupta et al. 2023), mortar (Prihar, Garlock, et al. 2024; Prihar, Gupta, et al. 2024), and concrete (Subramaniam et al. 2024; Thakur et al. 2023).

The stress intensity factor K_I for the SENB test, as shown in Figure 3.2a, can be defined as (Jenq and Shah 1985a; ASTM E1820 2025)

$$K_I = \left[\frac{PS}{BW^{3/2}}\right] f\left(\frac{a}{W}\right), \tag{3.6}$$

where B is the specimen thickness, W is the specimen depth, a is the total crack length, P is the load, and S is the span length. The geometry factor $f(a/W)$ for $S = 4W$ is defined as

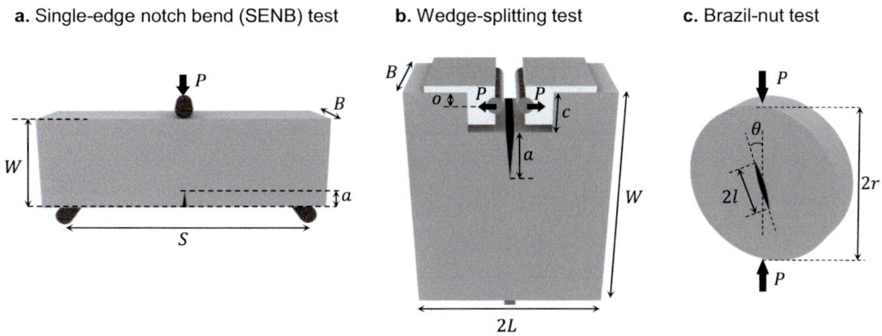

FIGURE 3.2 Schematic of the (a) SENB test, (b) Wedge-splitting test, and (c) Brazil-nut test to determine stress intensity factor (K_I) in cementitious materials.

$$f(a/W) = \frac{3\left(\frac{a}{W}\right)^{1/2}\left[1.99-\left(\frac{a}{W}\right)\left(1-\frac{a}{W}\right)\left(2.15-3.93\left(\frac{a}{W}\right)+2.7\left(\frac{a}{W}\right)^{2}\right)\right]}{2\left(1+\frac{2a}{W}\right)\left(1-\frac{a}{W}\right)^{3/2}}. \quad (3.7)$$

For the wedge-splitting test, the stress intensity factor K_I was evaluated using the finite-element method (Guinea, Elices, and Planas 1996). This formulation provides asymptotically correct behavior for both short and long cracks. The geometry of the wedge-splitting test is illustrated in Figure 3.2b. The stress intensity factor K_I for the wedge-splitting test, as shown in Figure 3.2b, can be defined as (Guinea, Elices, and Planas 1996)

$$K_I = \frac{P}{B\sqrt{c}}\frac{\alpha^{1/2}\left\{g_2+\left[g_1\left(1+A_1+A_2\right)-g_2-A_3\right]\alpha+A_3\alpha^{3/2}\right\}}{\left(1-\alpha\right)^{3/2}\left(1+A_1\alpha^{1/2}+A_2\alpha^{3/2}\right)}, \quad (3.8)$$

where P is the applied force, B is the specimen thickness, c is the height of the groove, $\alpha = \frac{a}{W-c}$, a is the notch length, W is the total specimen height, and A_1, A_2, A_3 are coefficients depending on the specimen shape. The functions g_1 and g_2 are defined as

$$g_1 = 3.975\left(\frac{W}{c}-\frac{o}{c}\right)\left(\frac{W}{c}-1\right)^{-3/2}, \quad (3.9)$$

$$g_2 = 1.122\pi^{1/2}\left(\frac{W}{c}-1\right)^{1/2} f\left(\frac{W}{c}\right). \quad (3.10)$$

The dimensionless function $f(W/c)$ depends on the ratio L/W and is given by

$$f\left(\frac{W}{c}\right)=\begin{cases} 0.845+22.91\left(\frac{W}{c}\right)^{-2}, \text{for } \frac{L}{W}=0.5 \\ 0.690+21.65\left(\frac{W}{c}\right)^{-2}, \text{for } \frac{L}{W}=1.0 \text{ or } 2.0. \end{cases} \quad (3.11)$$

Here, L is the half-width of the specimen and o is the distance of application of load from the top of the specimen (Figure 3.2b).

For the SENB and wedge-splitting tests, the stress intensity factors are evaluated under Mode I loading. Thus, the setup of the Brazil-nut test, as shown in Figure 3.2c, enables the evaluation of not only Mode I (tensile) fracture toughness but also Mode II (shear) and mixed-mode fracture by varying the notch orientation relative to the applied load (Atkinson, Smelser, and Sanchez 1982; Gupta et al. 2023; Tong, Wong, and Lupton 2007). The fracture mode is governed by the diametral compression angle θ, which is defined as the angle between the loading direction and the plane of the notch. At $\theta=0$, the specimen fails in pure Mode I, while at $\theta=27.2$ (for $l/r=0.3$) it fails in pure Mode II; intermediate angles correspond to mixed-mode conditions (Atkinson, Smelser, and Sanchez 1982).

The Mode I and Mode II stress intensity factors for the Brazil-nut test, as shown in Figure 3.2c, can be defined as (Atkinson, Smelser, and Sanchez 1982)

$$K_I = f_I \frac{P}{\pi r D}\sqrt{\pi l}, \quad (3.12)$$

$$K_{II} = f_{II} \frac{P}{\pi r D}\sqrt{\pi l}. \quad (3.13)$$

In these expressions, P is the critical load at fracture, r is the radius of the cylindrical specimen, D is the specimen thickness, and l is the half-notch length. The dimensionless coefficients f_I and f_{II} depend on the ratio of half-notch length (l), radius (r), and compression angle θ.

3.2.1.2 Limitations and Sensitivities

An LEFM approach using K_{IC} to characterize the fracture properties (i.e., fracture toughness) of additively manufactured concrete components enables the evaluation of both bulk and interfacial characteristics. It also provide a pathway for understanding how processing controls properties, e.g., how factors such as filament orientation and design of the materials architecture (Moini, Olek, Youngblood, et al. 2018; Moini et al. 2024; Prihar, Garlock, et al. 2024; Prihar, Gupta, et al. 2024; Gupta et al. 2023; Ralston, Gupta, and Moini 2024; Tomholt, Meggers, and Moini 2024; Daneshvar et al. 2024) control the fracture response. Strength tests can provide insight into the bulk vs. interfacial properties (i.e., "bond" strength); however, they can fall short in capturing the mechanical response of the interface and bulk due to the lack of a pre-existing crack (notch) that helps in understanding the local crack–microstructure (crack-interface) interaction. In other words, it is challenging to assess the interfacial properties through strength without a notched specimen with the tip located sharply at the interface. In the absence of a pre-existing notch, the initiation and propagation may not take place at the interfaces.

Moreover, in additively manufactured cementitious materials, the determination of fracture toughness of the bulk filaments and interfaces using K_{IC} becomes particularly sensitive to notch location (relative to the interface) and to the presence of directional interfacial defects. These interfacial defects can also alter the size of the FPZ ahead of the tip and the applicability of the LEFM for a given size and material (cement paste, mortar, and concrete). Unlike conventional cementitious materials, the processing-induced interfacial features (R. Moini 2023) can alter energy dissipation pathways ahead (e.g., crack deflection into the interface) (Prihar, Gupta, et al. 2024; Gupta et al. 2023) and behind the crack tip (e.g., friction and interlocking at the interface) (Prihar, Gupta, et al. 2024). These energy dissipation pathways may therefore require nonlinear elastic fracture mechanics principles. As illustrated in Figure 3.3, the fracture toughness in 3D-printed cementitious materials (Figure 3.3a) varies depending on the location of the notch relative to the filament orientation, whether placed within a printed filament (Figure 3.3b) or along an IZ (Figure 3.3c,d). The former characterizes the fracture toughness of the bulk 3D-printed cementitious materials, and the latter characterizes that of the interfaces in each orientation. Such considerations must be investigated for the application of LEFM for the assessment of the fracture mechanics and fracture properties of 3D-printed cementitious materials using scalar fracture toughness quantities (K_{IC} and K_{IIC}).

In concrete and other quasi-brittle materials, fracture mechanics often departs from the assumptions of LEFM due to the presence of a sizable FPZ ahead of the crack tip. These FPZs can potentially be further enlarged due to the presence of interfacial heterogeneities that are absent at the meso-scale in conventional counterparts

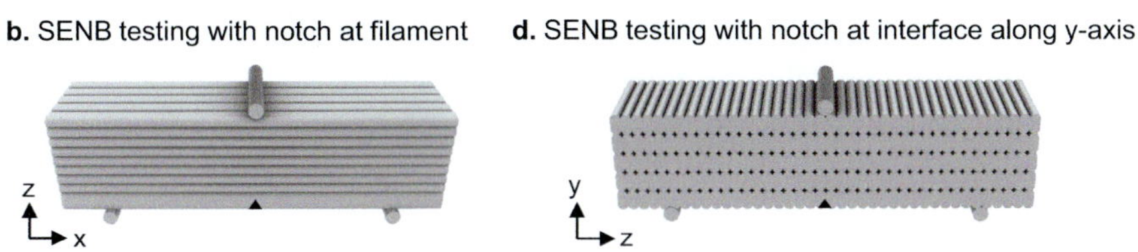

FIGURE 3.3 Schematic of (a) 3D-printed cementitious material, (b) SENB specimen with notch at the filament, (c) interface along the z-axis (build direction), and (d) interface along the y-axis (orthogonal to print and build directions).

beyond the ITZ (Figure 3.4). In LEFM, the FPZ is assumed to be negligibly small compared to the structure size, allowing the crack tip to be treated as a singularity. In contrast, for quasi-brittle materials such as concrete, rock, coarse-grained ceramics, and various fiber-reinforced composites, the FPZ can occupy a substantial portion of the nonlinear zone, with softening damage (microcracking as schematically shown in Figure 3.4 in both the ITZ in concrete and layered interfaces in 3D-printed counterparts) determining the fracture properties and behavior (Bažant 2002). This enlarged FPZ, due to the presence of interfaces, can play a central role in governing the fracture across sizes and dimensions of structures relative to the spacing of the filaments and can dictate whether and to what extent LEFM assumptions remain valid.

The size of the FPZ, denoted as l_{FPZ}, is a key parameter in determining the applicability of different fracture mechanics approaches. When l_{FPZ} is small compared to the structural dimensions, LEFM assumptions are often valid. However, in many materials of interest – particularly those exhibiting quasi-brittle behavior – the FPZ can represent a significant fraction of the sample size in some cases (i.e., thin walls with a single filament), and in some cases, it may extend across the entire cross section.

Hence, l_{FPZ} plays a central role in governing fracture behavior. For example, in normal-strength concrete or coarse-grained rock, $l_{FPZ} \approx 0.5\,m$; in dam concrete with very large aggregates, $l_{FPZ} \approx 3\,m$; in grouted soil masses, $l_{FPZ} \approx 10\,m$; and in jointed rock masses, l_{FPZ} can reach approximately 50 m. On a smaller scale, in fine-grained silicon oxide ceramics, $l_{FPZ} \approx 0.1\,\text{mm}$, and in silicon wafers, $l_{FPZ} \approx 10-100\,\text{nm}$. Depending on the structural size D, defined as the characteristic dimension of the cross section, different fracture models are applicable (Bažant 2002):

$$\frac{D}{l_{\text{FPZ}}} \geq 100;\ \text{LEFM} \tag{3.14}$$

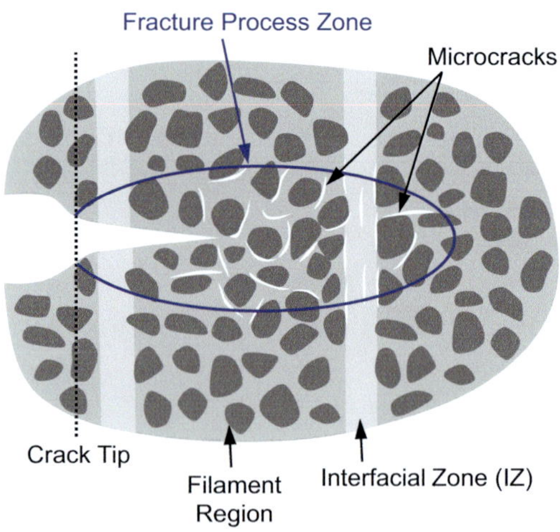

FIGURE 3.4 Schematic of a representative volume element (RVE) showing the FPZ in a 3D-printed cementitious material, showing microcracking in the potential interfacial zone between filaments and within the filament.

$$5 \leq \frac{D}{l_{\text{FPZ}}} < 100;\ \text{nonlinear quasi-brittle fracture mechanics} \tag{3.15}$$

$$\frac{D}{l_{\text{FPZ}}} < 5;\ \text{non-local damage, discrete element models, plasticity} \tag{3.16}$$

When $D / l_{\text{FPZ}} \geq 100$, LEFM assumptions hold because the FPZ is much smaller than the structural size. For intermediate ratios $5 \leq D / l_{\text{FPZ}} < 100$, nonlinear quasi-brittle fracture mechanics is required to account for significant inelastic deformation within the FPZ. For small ratios $D / l_{\text{FPZ}} < 5$, the FPZ governs the entire fracture process, and size-independent plastic limit or discrete models become more suitable.

The applicability of LEFM to 3D-printed cementitious materials depends critically on the ratio D / l_{FPZ}.
Due to printing-induced heterogeneities, such as layered interfaces and pore alignment, the FPZ length in 3D-printed materials may vary compared to that in cast counterparts of the same constituent material. As illustrated in Figure 3.4, the presence of a weak IZ between filaments may introduce additional microcracking, which can increase the size of the FPZ (see Figure 3.4). If l_{FPZ} is small relative to *D*, LEFM assumptions may remain valid; however, if l_{FPZ} constitutes a substantial fraction of the sample size, nonlinear quasi-brittle fracture models are more appropriate. In cases where l_{FPZ} approaches *D*, nonlinear models can be applied to understand the fracture behavior of the material.

3.2.2 Nonlinear Elastic Fracture Mechanics

The stress intensity factor K_I is a robust parameter for characterizing fracture in brittle, linear elastic materials such as cement paste, where the FPZ is small and the assumptions of LEFM hold (Bažant and Planas 2019). However, for quasi-brittle materials such as mortar and concrete, several studies have shown that significant nonlinear deformation precedes crack propagation due to toughening mechanisms such as microcracking, aggregate–matrix interface breakage, and frictional slip (Bažant and Planas 2019). These mechanisms contribute to a larger FPZ, invalidating the LEFM assumption of a negligible nonlinear zone and rendering K_{IC} insufficient as a standalone fracture parameter. In contrast to LEFM, nonlinear fracture mechanics approaches often characterize fracture response with more than one parameter. For instance, to study fracture in quasi-brittle materials, several parameters, including fracture energy (G_F), tensile strength $\left(f_t'\right)$, and critical crack opening distance (w_c), are typically used. To study *crack* propagation and nonlinear phenomena during damage in cementitious composites, parameters including fracture toughness (K_R), crack extension (Δa), crack mouth opening distance (CMOD), and crack tip opening distance (CTOD) and their relationships are used.

The nonlinear fracture models have been primarily developed and validated for quasi-brittle, homogeneous materials such as conventional concrete. In concrete, the presence of coarse aggregates contributes to significant nonlinearities ahead of the crack tip, resulting in a relatively large FPZ (Bažant and Planas 2019). In contrast, cement paste and mortar, commonly used in 3D-printed cementitious materials, exhibit much smaller nonlinear zones due to the absence of coarse aggregates (Bažant and Planas 2019). Therefore, the assumptions and formulations underlying these nonlinear models, developed for concrete, may differ for 3D-printed materials, which contain additional layered heterogeneity and often contain smaller to no aggregates (i.e., small FPZ, otherwise) (Moini, Olek, Youngblood, et al. 2018; M. Moini 2020; Moini, Baghaie, et al. 2021; R. Moini 2023; Ghantous et al. 2023; Gupta et al. 2023, 2026). The layered nature of 3D-printed cementitious materials introduces new sources of heterogeneity, in particular, IZs between adjacent filaments with anisotropic strength and fracture toughness along different orientations. These interfaces can act as weak planes or localized defects, theoretically giving rise to additional nonlinearities in the FPZ, despite the otherwise more brittle matrix behavior. Here, the nonlinear fracture mechanics models that prescribe the characterization of the fracture response of concrete are discussed. Considerations for the applicability of these models to 3D-printed cementitious materials are discussed for each model.

The direction of the notch relative to that of the filaments and interfaces in three orthogonal directions can lead to a local microstructure and properties along the crack path that may be captured by the nonlinear fracture response. For instance, for a smaller layer height for a given sample size, the higher is the number of interfaces ahead of the crack, which, in the case of a weak interface, can increase the size of the FPZ, further complicating the characterization of the fracture behavior using only LEFM. In other words, 3D-printed cementitious materials, depending on the design, can have various lengths of FPZs (relative to the sample size), which determines the fracture properties and alters whether LEFM is sufficient or nonlinear fracture mechanics approaches should be used. The nonlinear fracture mechanics methods, thus, can be carefully used to study 3D-printed materials. Here, we have provided some of the key nonlinear fracture mechanics methods that capture the characteristics of fracture using two or three parameters, such as the cohesive crack model, the two-parameter fracture model (TPFM), and resistance curve (*R*-curve) methods. These methods can provide a framework for analyzing the nonlinear aspects of fracture and crack propagation in layered, anisotropic, and heterogeneous cementitious systems.

More importantly, size effect models (types I and II) are discussed as they can be used for 3D-printed materials to provide a high-level understanding of the limits of strength as a function of increasing the size of the structure (for Type I) or the domain in which the linear vs. nonlinear approaches are applicable (for Type II). The definition of a 3D-printed system as a material or a structure, and how it determines its fracture toughness and strength, is particularly the subject of the size effect models. Future studies may employ the nonlinear fracture methods to quantify the interaction between interfacial defects and nonlinear fracture processes and use the size effect to understand how they change as a function of the size of the 3D-printed materials and structures.

3.2.2.1 Size Effect Law

Size effect refers to the phenomenon in which the nominal strength of geometrically similar specimens decreases with increasing structural size, despite identical boundary conditions and loading types (Bažant and Planas 2019; Bažant et al. 2021). Bažant classifies size effects in quasi-brittle structures into two fundamental types: type 1 applies to structures that fail at the onset of macrocrack initiation, whereas type 2 pertains to structures in which a large, stable crack develops before reaching the peak load (Bažant et al. 2021).

Type 1 size effect is observed when failure is caused by the initiation of a macrocrack at the weakest location in the structure and is influenced by both energetic and statistical effects due to material randomness (Bažant et al. 2021). In such cases, the size effect arises from a combination of the finite size of the FPZ and the randomness of local material strength, giving it both energetic and statistical (Weibull-type) components (Figure 3.5a). As the structural size increases, the probability of encountering a critically weak location grows, leading to a statistically driven reduction in nominal strength. For sufficiently large structures, the influence of the FPZ becomes negligible, and the size effect transitions to being purely statistical, in agreement with Weibull weakest-link chain theory (Bažant et al. 2021). This theory assumes that the structure fails as soon as a macrocrack initiates in the weakest RVE, which is statistically equivalent to a chain failing at its weakest link, leading to the Weibull statistical distribution.

To capture the type 1 size effect, Bažant et al. (2021) proposed the following expression for nominal strength:

$$\sigma_N = f_\infty \left(1 + \frac{rD_b}{D}\right)^{1/r}, \tag{3.17}$$

where each term has a specific physical and mathematical definition:

- f_∞ is the nominal strength of very large structures and is defined as

$$f_\infty = \sqrt{\frac{E' G_f}{g_0' c_f}}, \tag{3.18}$$

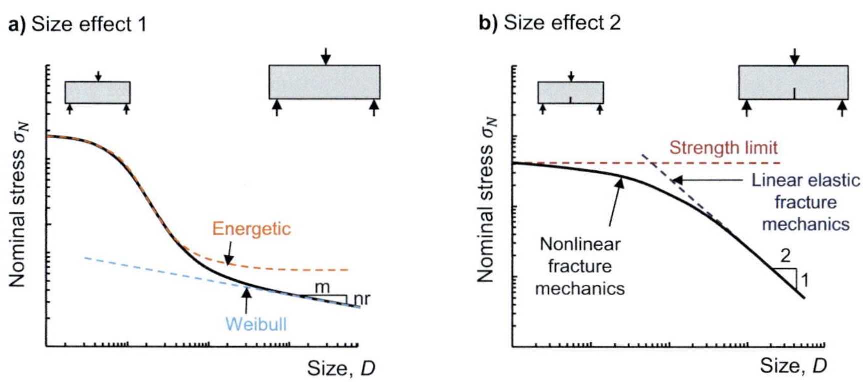

FIGURE 3.5 Schematic of (a) type 1 and (b) type 2 size effect curves. Adapted from Bažant et al. (2021).

where $E' = E$ is Young's modulus for plane stress, and $E' = \frac{E}{1-\nu^2}$ for plane strain (ν is Poisson's ratio), G_f is the fracture energy of the material, g_0' is the first derivative of the dimensionless energy release rate function $g(\alpha)$, and c_f is the effective size of the FPZ.

- r is a positive fitting parameter determined from optimal fitting of the experimentally measured size effect curve.
- D is the characteristic size of the specimen.
- D_b is defined as

$$D_b = \left\langle \frac{-g_0''}{4g_0'} \right\rangle c_f , \tag{3.19}$$

where g_0'' is the second derivative of the dimensionless energy release rate function, and the angular brackets $\langle \cdot \rangle$ denote the Macaulay brackets, representing the positive part of the argument. These brackets are used because the ratio g_0''/g_0' can sometimes be positive, in which case there is no size effect (i.e., D_b will vanish).

The quantities g_0' and g_0'' are obtained from the Taylor series expansion of the dimensionless energy release rate function $g(\alpha)$, which is expressed as

$$g(\alpha) = g_0 + g_0'\theta + \frac{1}{2} g_0''\theta^2 + \ldots \tag{3.20}$$

Here, $\alpha_0 = a_0 / D$, $g_0 = g(\alpha_0)$, $g_0' = \frac{dg(\alpha)}{d\alpha}|_{\alpha=\alpha_0}$, and $\theta = c_f / D$. For the type 1 size effect, we have

$a_0 = 0 \Rightarrow \alpha_0 = 0$, so $g_0 = 0$; however, the second g_0' and third g_0'' order terms are used in defining the nominal strength (σ_N) in the type 1 size effect.

Size effect type 2 occurs when failure follows the stable growth of a dominant crack (Figure 3.5b). In this case, the strain energy release from the undamaged portion of the structure into the crack front governs the fracture process. As the structure size increases, the strain energy release scales quadratically with size, while the energy dissipated at the crack front scales only linearly. This mismatch leads to an energetic size effect on the nominal structural strength. This principle forms the basis of a simple yet general SEL for nominal structural strength in type 2 failures (Bažant et al. 2021).

Bažant and Planas (2019) proposed the type 2 size effect method as a rigorous approach to evaluate fracture properties in quasi-brittle materials. Unlike the single-parameter K_{IC}-based approach of LEFM to characterize fracture, which assumes a negligible FPZ, the type 2 size effect method explicitly accounts for the finite FPZ and the associated nonlinear damage zone ahead of the crack through an equivalent LEFM (Bažant and Planas 2019). The equivalent LEFM approach approximates the nonlinear behavior of a finite FPZ by replacing it with an equivalent elastic crack such that the far-field stress, strain, and displacement fields are identical up to second-order terms of $(R_c / D)^2$, where R_c is the characteristic size of the FPZ and D is the characteristic size of the structure (Bažant and Planas 2019).

The type 2 SEL equation is expressed as

$$\sigma_N = Bf_t\left(1+\frac{D}{D_0}\right)^{-1/2}, \tag{3.21}$$

where σ_N is the nominal strength, f_t is the material tensile strength, D is the characteristic size of the structure, D_0 is the transitional size that governs deviation from the LEFM scaling limit, and B is a parameter depending on structural geometry.

The expressions for D_0 and B are given by

$$D_0 = c_f\frac{g_0'}{g_0}, \quad B = \frac{1}{f_t}\sqrt{\frac{E'G_f}{c_f g_0'}}, \tag{3.22}$$

Here, c_f is the effective size of the FPZ; g_0 and g_0' are the dimensionless energy release rate and its first derivative, respectively, evaluated at the initial relative crack length α_0; and E' is the effective modulus, where $E' = E$ for plane stress and $E' = \dfrac{E}{1-\nu^2}$ for plane strain (ν is the Poisson ratio).

This expression, derived by truncating the Taylor series expansion of $g(\alpha)$ at the second term, captures not only the effect of structure size but also the influence of geometry through $g(\alpha)$. Therefore, this law is sometimes referred to as a "size–shape" effect law.

By analyzing how nominal strength decreases with increasing specimen size, the type 2 SEL captures the transition from strength-based failure in small specimens to fracture-based failure in large specimens (Bažant and Planas 2019). This transition thereby overcomes a central limitation of LEFM, which assumes that the FPZ is negligibly small and cannot account for nonlinear fracture behavior in quasi-brittle materials.

A key outcome of the type 2 size effect method is the estimation of an effective crack length (c_f), which accounts for both the initial crack length and the additional crack extension caused by microcracking within the FPZ. The method involves fitting experimental data from geometrically scaled specimens (Figure 3.6a) to a type 2 SEL using regression analysis, commonly performed with linear and bi-logarithmic (log–log) forms (Bažant and Planas 2019).

The type 2 SEL can be expressed in a *linear regression form*, as proposed by Bažant and Planas (2019). In this form, the SEL is written in a linear expression:

$$Y = AX + C, \text{where } Y = \frac{1}{\sigma_N^2}, \ X = D. \tag{3.23}$$

Here, D is the characteristic size of the specimen (e.g., beam depth), and σ_N is the nominal stress at failure, calculated from the peak load. By plotting $1/\sigma_N^2$ against

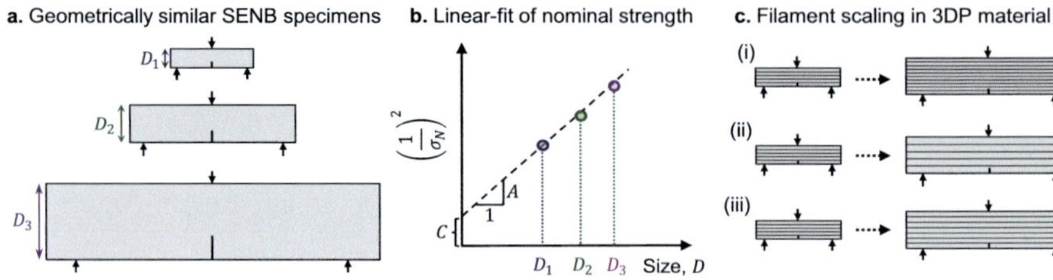

FIGURE 3.6 (a) Geometrically scaled SENB specimens with varying depths D_1, D_2, and D_3; (b) linear regression of nominal strength σ_N vs. size D used in the type 2 size effect method to estimate fracture toughness; (c) illustration of different scaling strategies for 3D-printed materials, including increasing the (i) number of filaments (constant filament thickness), (ii) filament thickness (constant number of layers), or (iii) both number of filaments and their thickness, each introducing varying distributions of IZs that may influence fracture behavior.

D, the slope A and intercept C can be obtained through linear regression (see Figure 3.6b). These regression constants are then used to determine key fracture parameters, including the *fracture toughness* (K_{IC}), *fracture energy* (G_f), and *effective crack extension* (c_f), using the following expressions:

$$K_{IC} = k_0 \cdot \frac{1}{\sqrt{A}}, \quad G_f = \frac{k_0^2}{E'} \cdot \frac{1}{A}, \quad c_f = \frac{k_0}{2k_0{}'} \cdot \frac{C}{A}. \tag{3.24}$$

Here, k_0 introduces the effect of the geometry of the specimen.

Recent work by Pathirage et al. (2025) examines the regression statistics of the size effect method. The authors present new insight into the widely adopted linear regression formulation based on nominal strength (σ_N) vs. specimen size (D), in disagreement with the RILEM recommendation (Shah 1990a). The authors note that the RILEM-recommended formulation can deviate from key regression assumptions, such as the requirement that the error term be independent of the expectation function and be independently distributed. They further point out that this can lead to asymptotic bias in the estimated fracture energy (G_f) and the effective FPZ length (c_f).

In contrast to the transformed regression space ($1/\sigma_N^2 - D$ space) recommended by RILEM, performing nonlinear least squares regression directly in the untransformed space of peak load (P) vs. specimen size (D) offers notable statistical advantages. In this untransformed approach, the estimators correspond directly to the material fracture properties in Bažant's SEL, namely, the fracture energy (G_f) and the effective crack extension (c_f). Pathirage et al. (2025) demonstrated that when regression is performed in the P–D space, the estimators of G_f and c_f are *asymptotically unbiased*, meaning their expected values converge to the true material properties as the number of specimens tested increases.

The estimators are also *mean square consistent*, which implies that the mean squared error between the estimated and true values of G_f and c_f approaches zero as the dataset size increases. Moreover, they possess the *minimum variance* property, making them statistically efficient. In other words, among all unbiased estimators, those obtained from the *P–D* regression have the smallest possible spread in their sampling distribution. This is in contrast to the transformed regression space recommended by RILEM, where the SEL is linearized and fitted in the $1/\sigma_N^2 - D$ space. Although this transformation simplifies the fitting procedure (with linear regression), it alters the error structure of the regression and can lead to biased and less efficient estimates of fracture parameters. The findings underscore the importance of selecting a regression framework that preserves the statistical integrity of the estimators to ensure both accuracy and efficiency in fracture parameter identification.

While the SEL provides a powerful tool to extract fracture parameters in quasi-brittle materials, its application to layered 3D-printed cementitious systems should consider additional complexity due to the presence of additional defects arranged at the layered interfaces. While the presence of aggregates in normal concrete creates ITZs as weak interfaces and alters the FPZ length depending on the size of the aggregates, in 3D-printed cementitious materials, additional layered interfaces can further change the FPZ depending on the layer height, filament width, and processing and in situ environmental parameters that locally change the water content, degree of hydration, and pore phase content and morphology. As a result, the size of the designed features and the process can determine the size of the defects, which in turn can change the size of the FPZ, relative to the specimen size. Therefore, in studying the scaling law in layered systems, one can begin to probe the size effect in multiple ways: (i) by increasing the number of printed filaments while maintaining filament size (Figure 3.6c (i)), (ii) by increasing the size of each filament while keeping the number of filaments constant (Figure 3.6c (ii)), or (iii) by simultaneously increasing both the number and the size of filaments (Figure 3.6c (iii)). Each approach alters the FPZ ahead of the crack tip differently within the specimen, which can significantly affect the fracture characteristics across scales. Understanding how the variables in the design of the spatial arrangement of the materials and defects and the corresponding local properties control the fracture behavior presents a promising direction for material development and engineering of the bulk properties using the geometry, properties, and size.

3.2.2.2 Cohesive Crack Model

The cohesive crack model, also called the fictitious crack model, originally introduced by Hillerborg, Modéer, and Petersson (1976), is widely used to describe fracture in quasi-brittle materials, such as concrete, under uniaxial tensile loading (Bažant 2002). Rather than assuming a sharp crack that propagates instantaneously, the model postulates a *fictitious crack zone* ahead of the real crack tip (as shown in Figure 3.7a) where tensile stresses are transferred across the crack surfaces through cohesive traction (Bažant and Planas 2019; Elices, Rocco, and Roselló 2009). Fracture propagation is governed by a stress–crack opening relation $\sigma = f(w)$, known as the *softening curve* (Figure 3.7a) (Bažant 2002). In concrete-like materials, the softening function $f(w)$ is often approximated using a bi-linear law for simplicity (Figure 3.7b) (Elices, Rocco, and Roselló 2009). As illustrated in Figure 3.7b, this bi-linear curve captures two essential mechanisms: large-scale ITZ debonding or fracture of aggregates in its

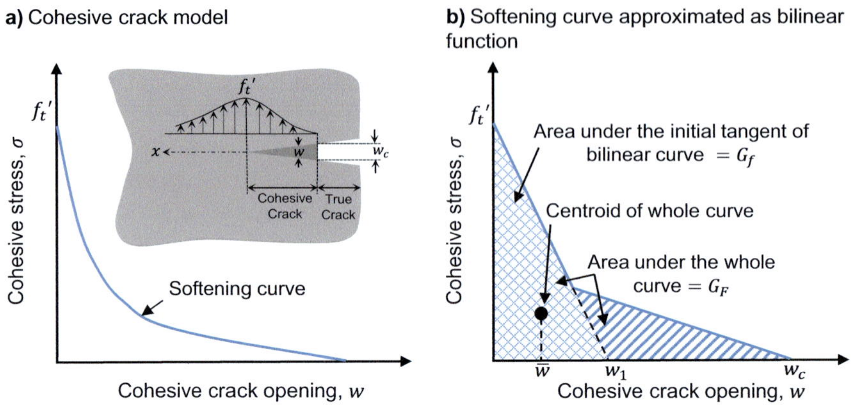

FIGURE 3.7 (a) Schematic of the cohesive crack model and (b) its bi-linear softening approximation, where $f_t^{'}$ is the tensile strength, G_F is the total fracture energy corresponding to the area under the entire softening curve, G_f is the initial fracture energy corresponding to the area under the initial slope of the softening curve, $\overline{w}$ is the centroid of the softening curve, and w_1 is the intercept of the initial tangent with the crack opening axis.

steep initial segment, followed by a shallow tail associated with frictional aggregate pull-out and crack face bridging due to aggregates (Elices, Rocco, and Roselló 2009).

A few primary parameters are typically used to define the general softening behavior of the cohesive crack model, depending on the type and quality of the testing, with more than three parameters being generally challenging to identify in normal testing (Bažant 2002). In many normal testing situations, only two parameters can be reliably identified: the local *tensile strength* $f_t^{'}$, which corresponds to the peak cohesive stress at the onset of cracking $(w = 0)$, as shown in Figure 3.7b, and the *fracture energy* G_F, which represents the energy required to fully open the cohesive zone and is defined as the area under the softening curve (Figure 3.7b):

$$G_F = \int_0^{w_c} f(w)\, dw, \tag{3.25}$$

where w_c is the critical crack opening displacement at which the cohesive stress becomes zero.

In addition to f_t' and G_F, the third parameter, the *initial fracture energy* G_f, is defined as the area under the initial tangent of the softening curve extended to the crack opening axis w:

$$G_f = \frac{f_t^{'2}}{2\sigma_0^{'}} = \frac{w_1^2\sigma_0^{'}}{2} = \frac{1}{2} f_t^{'} w_1, \tag{3.26}$$

where $\sigma_0^{'} = \frac{df(w)}{dw}|_{w=0}$ is the initial tangent of the slope of the bi-linear softening curve, and w_1 is the horizontal intercept of this tangent with the crack opening axis

(Figure 3.7b). Within a realistic size range, G_f predominantly governs the maximum load capacity and the size effect in quasi-brittle fracture (Planas, Guinea, and Elices 1999; Bažant 2002).

To reduce the softening curve to a dimensionless form (Bažant and Planas 2019), the characteristic crack opening displacement is defined as

$$w_{ch} = \frac{G_F}{f_t'} . \tag{3.27}$$

The dimensionless formulation is then written as

$$\hat{\sigma} = \frac{\sigma}{f_t'}, \hat{w} = \frac{w}{w_{ch}}, \hat{\sigma} = \hat{f}\left(\hat{w}\right), \tag{3.28}$$

with

$$\hat{f}\left(0\right) = 1, \int_0^{\infty} \hat{f}\left(\hat{w}\right) d\hat{w} = 1. \tag{3.29}$$

To construct the bi-linear softening curve shown in Figure 3.7b, four parameters are required: (i) the tensile strength f_t', (ii) the fracture energy G_F, (iii) the abscissa of the centroid $\overline{w}$, and (iv) the horizontal intercept w_1 of the initial slope. Parameters G_F and f_t' are obtained experimentally, while the abscissa of the centroid $\overline{w}$ can be expressed as a function of G_F and f_t'. The horizontal intercept w_1 can be evaluated as a function of f_t', the effective Young's modulus E', and the nominal strength σ_{Nu} of the material. The detailed determination of each parameter is explained as follows.

The tensile strength f_t' is typically obtained from splitting tensile (Brazilian) tests (Rocco et al. 2001; ASTM C496 2017):

$$f_t' = \frac{2P_u}{\pi b D}, \tag{3.30}$$

where P_u is the peak load, b is the specimen thickness, and D is its diameter.

The fracture energy G_F can be measured from notched beam or wedge-splitting tests (Bažant and Planas 2019):

$$G_F = \frac{W_F}{b\left(W - a_0\right)}, \tag{3.31}$$

where W_F is the total work-of-fracture (the entire area under the curve of the load-displacement plot), a_0 is the notch length, and W and b are the beam depth and thickness, respectively.

The abscissa of the centroid $\overline{w}$ of the softening curve is computed as

$$\overline{w} = \alpha \frac{G_F}{f_t'} , \tag{3.32}$$

where α is a dimensionless shape factor. For bi-linear softening, $\alpha = 0.987$ (Petersson 1981).

The horizontal intercept of the initial linear tangent of the bi-linear softening curve w_1, shown in Figure 3.7b, which governs the initial slope of the bi-linear curve, is given by

$$w_1 = \frac{2f_t' l_1}{E'}, \tag{3.33}$$

where l_1 is a characteristic length and E' is the effective Young's modulus (equal to E for plane stress and $E/\left(1-\nu^2\right)$ for plane strain).

The characteristic length l_1 can be estimated based on the splitting tensile strength f_t' and nominal strength $\left(\sigma_{Nu}\right)$ obtained from the notched three-point bend test (Planas, Guinea, and Elices 1999; Bažant and Planas 2019). Assuming the tensile strength f_t' is known from splitting tests, the ratio σ_{Nu} / f_t' can be expressed as a function of the dimensionless parameter D / l_1, such that

$$\frac{\sigma_{Nu}}{f_t'} = \phi\left(\frac{D}{l_1}\right), \tag{3.34}$$

where $\phi(\cdot)$ is a numerically computed function specific to the testing geometry (Bažant and Planas 2019).

This function can be inverted numerically or graphically to give

$$\frac{D}{l_1} = \chi\left(\frac{\sigma_{Nu}}{f_t'}\right), \tag{3.35}$$

where $\chi(\cdot)$ denotes the inverse of $\phi(\cdot)$. Hence, the characteristic length is computed as

$$l_1 = D \cdot \left[\chi\left(\frac{\sigma_{Nu}}{f_t'}\right)\right]^{-1}, \tag{3.36}$$

The characteristic length l_1 is determined using the approximation (Planas, Guinea, and Elices 1999)

$$l_1 = \kappa D\left[\frac{13.11}{(x-1)^2} + \frac{2.68}{x}\right], \tag{3.37}$$

where

$$x = \left(\eta_0 \frac{f_t'}{\sigma_{Nu}}\right)^2 \tag{3.38}$$

and the constants κ and η_0 are given by

$$\kappa = 1-\alpha_0^{1.45}, \eta_0 = 3(1-\alpha_0)^2, \tag{3.39}$$

with $\alpha_0 = a_0 / D$ as the relative notch length.

The nominal strength σ_{Nu} for notched beams in three-point bending is

$$\sigma_{Nu} = \frac{3P_u S}{2BW^2}, \tag{3.40}$$

where P_u is the peak load, S is the span, B is the width, and W is the depth of the beam.

The characteristic length l_1 discussed above can also be expressed in terms of the initial fracture energy G_f as

$$l_1 = \frac{E'G_f}{f_t'^2}, \tag{3.41}$$

where E' is the effective Young's modulus of elasticity.

An alternative fracture characteristic length, denoted as l_{ch}, is defined based on the total fracture energy G_F as

$$l_{ch} = \frac{E'G_F}{f_t'^2}. \tag{3.42}$$

This expression for l_{ch}, interpreted as a measure of FPZ length, was first introduced by Irwin in 1958 in the context of LEFM (Bažant 2002). Its systematic application to concrete fracture mechanics was initiated by Hillerborg, Modéer, and Petersson 1976 within the framework of the fictitious crack model.

These two characteristic lengths are directly related through the ratio of the total and initial fracture energies:

$$\frac{l_{ch}}{l_1} = \frac{G_F}{G_f}. \tag{3.43}$$

Experimental studies typically show that $G_F \approx 2.5 G_f$, leading to $l_{ch} \approx 2.5\, l_1$ (Bažant and Planas 2019).

If the stress at the slope change of the bi-linear softening curve occurs at a fraction $\psi f_t'$ of the tensile strength, the critical crack opening displacement w_c can be expressed as

$$w_c = \frac{2}{\psi f_t'}\left[G_F - (1-\psi) G_f\right]. \tag{3.44}$$

The slope of the initial tangent of the bi-linear softening curve is given by

$$\sigma_0^{'} = \frac{df(w)}{dw}\Big|_{w=0} = \frac{f_t^{'2}}{2G_f}. \tag{3.45}$$

For example, using $\psi = 0.25$ and assuming $G_F / G_f = 2.5$ (Bažant 2002), we can obtain

$$w_c \approx \frac{14G_f}{f_t^{'}} \text{ or } w_c \approx \frac{5.6G_F}{f_t^{'}}. \tag{3.46}$$

Hence, the intercept of the initial slope with the crack opening axis, denoted by w_1, is related to w_c by

$$w_1 = \frac{1}{7} w_c. \tag{3.47}$$

Brühwiler, Wittmann, and Rokugo (1987) studied the influence of the loading rate on the fracture energy and the bi-linear softening curve. They found that the total fracture energy (G_F) is significantly influenced by the loading rate. At a loading rate 5,000 times higher than that in the quasi-static case, G_F increases by about 70% compared to the quasi-static value. The quasi-static loading condition, as referred to in the RILEM recommendation (RILEM 1985), is defined as a loading duration of 30–60 seconds from the start of the test until the peak load is reached, corresponding to a strain rate of about $2.5 \times 10^{-6} s^{-1}$ at the initial crack tip, as estimated approximately.

To facilitate comparison across different rates, the bi-linear strain-softening curves were normalized in both axes:

$$X = \frac{w}{G_F / f_t^{'}}, \tag{3.48}$$

$$Y = \frac{\sigma}{f_t^{'}}. \tag{3.49}$$

where σ is the cohesive stress, $f_t^{'}$ is the tensile strength, and w is the crack opening displacement. The shape of the normalized strain-softening curve (in the *X–Y* space) was found to be largely independent of the loading rate. Therefore, Brühwiler, Wittmann, and Rokugo (1987) claim that only the rate-dependent values for $f_t^{'}$ and G_F need to be known to model rate effects in fracture mechanics for concrete.

This formulation enables an accurate and efficient representation of cohesive fracture behavior using parameters derived from standard mechanical tests. In the context of 3D-printed cement paste or mortar, the FPZ may be smaller than that in concrete due to the absence of coarse aggregates. Moreover, the 3D-printed systems can have a larger FPZ compared to their cast counterparts due to the presence of additional interfaces ahead of the crack. As a result, the suitability and characterization of the fracture response using the bi-linear softening curve and its parameters must be carefully carried out with regard to microstructural damage dissipation in layered systems. Moreover, energy dissipation mechanisms in layered and 3D-printed cement-based materials are diverse and include those behind the crack tip (e.g., friction and interlocking of the fractured surfaces/filaments), beyond those ahead of the crack tip in the FPZ. These aspects require significant new considerations to account for the impact of the fracture parameters and the methods that are developed for cohesive fracture in conventional brittle and quasi-brittle materials.

3.2.2.3 Crack Band Model

To address mesh size sensitivity and energy dissipation inconsistency in modeling localized fracture in finite element (FE) analysis, Zdeněk P. Bažant and Byung H. Oh (1983a) introduced the crack band model for quasi-brittle materials. Unlike the line crack approach (Figure 3.8a), the crack band model simulates fracture as a smeared strain localization occurring over a finite band of elements (Figure 3.8b). This approach is particularly effective for quasi-brittle materials like concrete, where the

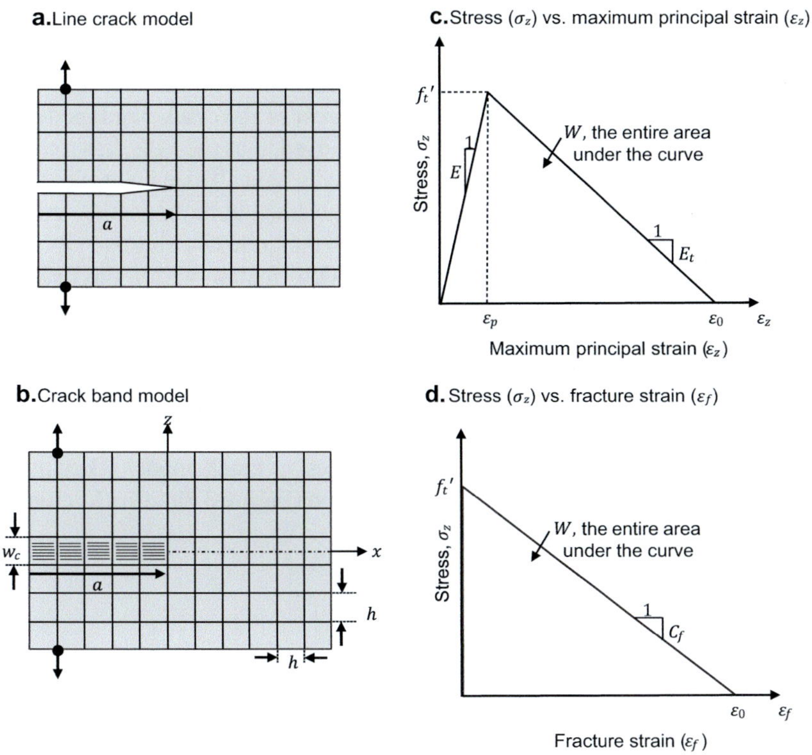

FIGURE 3.8 Comparison of (a) the line crack model and (b) the crack band model with band width w_c and mesh h. (c) Stress–strain relation in terms of maximum principal strain ε_z, showing elastic limit ε_0, strain corresponding to peak stress ε_p, and tangent strain-softening modulus E_t. (d) Stress–fracture strain curve used in the crack band model with slope C_f. Adapted from Zdeněk P. Bažant and Byung H. Oh (1983a).

FPZ is not a sharp discontinuity but a distributed region of microcracking. Zdeněk P. Bažant and Byung H. Oh (1983a) demonstrated through FE simulations that replacing an idealized single-line crack with a finite band of localized cracks produces equivalent results for fracture propagation, as the energy flux into the FPZ is a global property of the structure and is largely insensitive to the local stress and strain variations near the crack front. The fundamental idea is to ensure that the area under the stress–strain curve, scaled by the element (mesh) width h, matches the material's fracture energy G_F, thereby preserving energy dissipation regardless of the mesh size.

The formulation begins with the assumption of uniformly distributed microcracks normal to the loading direction z, as shown in Figure 3.8b. Under constant stress conditions, the strain in the direction of the loading (ε_z) increases due to the opening of the microcracks, while the strains parallel to the microcracks (ε_x and ε_y) remain unaffected. The principal strain–stress relationship in the presence of microcracking is written as

$$\begin{bmatrix} \varepsilon_x \\ \varepsilon_y \\ \varepsilon_z \end{bmatrix} = \frac{1}{E} \begin{bmatrix} 1 & -v & -v \\ -v & 1 & -v \\ -v & -v & 1 \end{bmatrix} \begin{bmatrix} \sigma_x \\ \sigma_y \\ \sigma_z \end{bmatrix} + \begin{bmatrix} 0 \\ 0 \\ \varepsilon_f \end{bmatrix}, \tag{3.50}$$

where ε_f is the fracture strain representing the additional strain due to microcrack opening. ε_f can be further understood in the context of the crack band model.

The front of an advancing crack band, also referred to as the FPZ, has a characteristic width w_c (Figure 3.8b). For plain concrete, w_c is typically several times the maximum aggregate size (Bažant and Oh 1983a). Experimental findings have shown that w_c is 1.5–4 times the maximum aggregate size in concrete (Bažant and Oh 1983a).

The fracture strain ε_f represents the average deformation within the FPZ caused by microcracking, defined as $\varepsilon_f = \delta_f / w_c$, where $\delta_f = \sum_i \delta_f^i$ is the total opening displacement of all individual microcracks (Figure 3.8b). Neglecting rate effects, ε_f can be considered a function of the normal stress σ_z, i.e., $\varepsilon_f = f(\sigma_z)$, and this function must be monotonically decreasing. Zdeněk P. Bažant and Byung H. Oh (1983a) assumed a linear softening law for this relationship, which differs from the bi-linear approximation typically adopted in the cohesive crack model (Bažant 2002). Assuming a linear softening law in uniaxial tension (Figure 3.8d), this fracture strain (ε_f) can be defined in terms of stress σ_z as

$$\varepsilon_f = \frac{1}{C_f}\left(f_t' - \sigma_z\right), \tag{3.51}$$

where C_f is the slope of the $\sigma_z - \varepsilon_f$ curve as shown in Figure 3.8d. Substituting Eq. (3.51) into Eq. (3.50) leads to the total strain vector in this regime as follows:

$$\begin{bmatrix} \varepsilon_x \\ \varepsilon_y \\ \varepsilon_z \end{bmatrix} = \begin{bmatrix} E^{-1} & -vE^{-1} & -vE^{-1} \\ -vE^{-1} & E^{-1} & -vE^{-1} \\ -vE^{-1} & -vE^{-1} & E_t^{-1} \end{bmatrix} \begin{bmatrix} \sigma_x \\ \sigma_y \\ \sigma_z \end{bmatrix} + \begin{bmatrix} 0 \\ 0 \\ \varepsilon_0 \end{bmatrix}, \tag{3.52}$$

where E_t is the tangent softening modulus of the strain-softening segment of the uniaxial stress–strain plot in the z direction (Figure 3.8c) and can be defined in terms of Young's modulus (E) and slope of the $\sigma_z - \varepsilon_f$ curve $\left(C_f\right)$ using Eq. (3.53):

$$\frac{1}{E_t} = \frac{1}{E} - \frac{1}{C_f}. \tag{3.53}$$

$\varepsilon_0 = f_t' / C_f$ is the strain at the onset of softening (Figure 3.8c,d).

For a general stress–strain relation (e.g., in anisotropic materials), the constitutive law can be expressed as

$$\varepsilon = \boldsymbol{C}\left(\mu\right)\sigma, \tag{3.54}$$

where $\varepsilon = (\varepsilon_x, \varepsilon_y, \varepsilon_z)^T$ is the strain vector, $\sigma = (\sigma_x, \sigma_y, \sigma_z)^T$ is the stress vector, and $\boldsymbol{C}\left(\mu\right)$ is the 3 × 3 compliance matrix, which is expressed as

$$\begin{bmatrix} \varepsilon_x \\ \varepsilon_y \\ \varepsilon_z \end{bmatrix} = \begin{bmatrix} C_{11} & C_{12} & C_{13} \\ C_{12} & C_{22} & C_{23} \\ C_{13} & C_{23} & C_{33}\mu^{-1} \end{bmatrix} \begin{bmatrix} \sigma_x \\ \sigma_y \\ \sigma_z \end{bmatrix}. \tag{3.55}$$

Here, $C_{11}, C_{12}, \ldots, C_{33}$ are the initial compliances before cracking, and μ is the *cracking parameter*.

This parameter varies within a range $0 < \mu < 1$ and depends on the stress and strain states. The value of $\mu = 1$ corresponds to an *uncracked material*, while $\mu \to 0$ represents the *state of continuous cracking*. The effect of cracking in the z direction is modeled by increasing the corresponding compliance term C_{33} by a factor μ^{-1}, reflecting stiffness degradation normal to the crack plane (z direction) while maintaining the other components unchanged.

Using Eq. (3.55), the strain in the z direction can be expressed as

$$\varepsilon_z = C_{13}\sigma_x + C_{23}\sigma_y + C_{33}\mu^{-1}\sigma_z. \tag{3.56}$$

For an elastic body, referring to the compliance matrix in Eq. (3.50), Eq. (3.56) becomes

$$\varepsilon_z = -\frac{v}{E}\sigma_x - \frac{v}{E}\sigma_y + \frac{1}{\mu E}\sigma_z. \tag{3.57}$$

Now, based on Eq. (3.52), the strain in the z direction can also be written as

$$\varepsilon_z = -\frac{v}{E}\sigma_x - \frac{v}{E}\sigma_y + \frac{1}{E_t}\sigma_z + \varepsilon_0. \tag{3.58}$$

Equating Eqs. (3.57) and (3.58) yields

$$\frac{1}{\mu} = E\left(\frac{\varepsilon_0}{\sigma_z} + \frac{1}{E_t}\right). \tag{3.59}$$

Furthermore, substituting Eq. (3.53) into Eq. (3.59), we can express μ as

$$\frac{1}{\mu} = E\left(\frac{\varepsilon_0}{\sigma_z} + \frac{1}{E_t}\right) = \frac{E}{-E_t} \cdot \frac{\varepsilon_z}{\varepsilon_0 - \varepsilon_z}. \tag{3.60}$$

When cracking starts at $\sigma_z = f_t' = C_f \varepsilon_0$, substituting into Eq. (3.60) and using Eq. (3.53) yields $\mu = 1$. As the state of continuous cracking is approached, the stress in the z direction (σ_z) approaches zero, and consequently, μ also approaches zero. In other words, the strain in the z direction approaches ε_0 in the case of continuous cracking (see Figure 3.8c), and μ approaches zero as shown by Eq. (3.60).

The fracture energy G_f is defined as the energy required to open all microcracks over the width of the FPZ, given by

$$G_f = w_c \int_{f_t'}^{0} \sigma_z d\varepsilon_f. \tag{3.61}$$

Based on the linear softening law (Figure 3.8d), the fracture energy G_f simplifies to

$$G_f = \frac{f_t'^2}{2C_f} w_c. \tag{3.62}$$

From this, the material parameters are obtained:

$$C_f = \frac{f_t'^2 w_c}{2G_f}, \varepsilon_0 = \frac{f_t'^2}{C_f} = \frac{2G_f}{f_t' w_c}. \tag{3.63}$$

The energy dissipated per unit area under the complete stress–strain curve can also be expressed as

$$G_f = \frac{1}{2}\left(f_t' \varepsilon_p + f_t' \left(\varepsilon_0 - \varepsilon_p\right)\right) w_c = W w_c, \tag{3.64}$$

where ε_p is the plastic strain, and $W = \int \sigma_z d\varepsilon_z$ is the area under the nominal uniaxial stress–strain curve (Figure 3.8c).

Rearranging gives a direct expression for the crack band characteristic width:

$$w_c = \frac{2G_f}{f_t'^2}\left(\frac{1}{E} - \frac{1}{E_t}\right)^{-1}. \tag{3.65}$$

This formulation provides a rigorous link between fracture energy $\left(G_f\right)$, tensile strength $\left(f_t^{'}\right)$, post-peak tangent modulus $\left(E_t\right)$, and the effective width $\left(w_c\right)$ of the FPZ. The crack band model also involves important computational considerations to ensure objectivity. In a step-by-step incremental solution, where unloading is not involved, switching from elastic behavior to softening and fully cracked states requires well-defined criteria. Zdeněk P. Bažant and Byung H. Oh (1983a) proposed two criteria to determine the transition:

1. When the maximum principal strain ε_z in an element reaches the strain ε_p, the softening process is initiated, and the tangential compliance matrix (Eq. (3.52)) is used. The local coordinate system is then aligned with the principal strain directions to maintain consistent crack orientation.
2. When the normal strain ε_z in the previous increment exceeds ε_0, the element is considered fully cracked, and the following stiffness matrix for fully cracked concrete is applied (Bažant and Oh 1983a; Suidan and Schnobrich 1973):

$$\begin{Bmatrix} \sigma_x \\ \sigma_y \\ \sigma_z \end{Bmatrix} = \begin{bmatrix} D_{11} - D_{13}^2 D_{33}^{-1} & D_{12} - D_{13} D_{23} D_{33}^{-1} & 0 \\ \text{sym} & D_{22} - D_{23}^2 D_{33}^{-1} & 0 \\ \text{sym} & \text{sym} & 0 \end{bmatrix} \begin{Bmatrix} \varepsilon_x \\ \varepsilon_y \\ \varepsilon_z \end{Bmatrix}, \tag{3.66}$$

where $D_{\alpha\beta}\left(\alpha, \beta = 1,2,3\right)$ are components of the initial stiffness matrix $\tilde{D}$ for uncracked concrete.

To consider the effect of mesh orientation, Zdeněk P. Bažant and Byung H. Oh (1983a) addressed the issue of crack band shape when not aligned with the mesh. For square meshes, the effective crack band width w_c is related to the mesh size h and crack angle α by

$$w_c = \frac{h}{\cos\alpha}. \tag{3.67}$$

An essential requirement in numerical modeling is *objectivity*, meaning that the results should not depend on the chosen mesh size (aside from numerical errors that vanish with refinement) (Bažant and Oh 1983a). Earlier works on linear fracture analysis (Bažant and Oh 1983a; Bažant and Cedolin 1984; Cedolin and Bažant 1980) have shown that objectivity can be achieved if the strength limit is not kept constant but is allowed to vary with the element size, ensuring the correct energy release rate. In the crack band model introduced by Zdeněk P. Bažant and Byung H. Oh (1983a), in addition to fracture energy, there are additional material parameters – the strength limit $f_t^{'}$ and the strain-softening modulus E_t. It is important to ensure objectivity of the results with respect to these two material parameters ($f_t^{'}$ and E_t).

The strain ε_z within the FPZ is not, by itself, a critical quantity. A true continuum with strain-softening is unstable. The main role of ε_z is that, together with the width

w_c of the FPZ, it defines the relative displacement $\varepsilon_z w_c$, which determines the work done by the surrounding structure on the FPZ. This work represents the main part of the fracture energy. Therefore, it makes no difference if an element-wide crack band of width h (equal to the mesh size) is used in place of the actual width w_c, as long as the correct value of fracture energy is preserved. This can be done by adjusting the softening modulus E_t while keeping G_f constant. By replacing w_c with h in the governing equations, the necessary adjustments when h is larger than w_c can be determined. Following these adjustments, the parameters C_f (slope of the $\sigma_z - \varepsilon_f$ curve), ε_0 (strain at the onset of softening), and E_t (tangent softening modulus of the strain-softening segment in the $\sigma_z - \varepsilon_f$ curve) can be calculated as follows:

$$C_f = \frac{f_t'^2 h}{2G_f}, \varepsilon_0 = \frac{2G_f}{f_t' h}, E_t = \left(\frac{1}{E} - \frac{2G_f}{f_t'^2 h} \right)^{-1}. \tag{3.68}$$

To maintain numerical stability and ensure $E_t < 0$, the element size must satisfy the condition

$$h < \frac{2G_f E}{f_t'^2}. \tag{3.69}$$

In practice, the FE size should be less than approximately one-half of the limit given in Eq. (3.69) (Bažant and Oh 1983a; Bažant 2002). The limiting case $h = 2G_f E / f_t'^2$ corresponds to $E_t^{-1} = 0$ or $E_t \to -\infty$, which results in a sudden (or vertical) stress drop.

It is worth noting that, aside from the factor of 2, the expression in Eq. (3.69) has the same form as Irwin's expression for the size of the yielding zone in ductile fracture (Knott 1973; Bažant and Oh 1983a; Bažant and Planas 2019).

This condition ensures a gradual strain-softening behavior rather than an instantaneous vertical stress drop. For larger structures or coarse meshes, where this condition is violated, an equivalent strength f_{eq} may be used to preserve the fracture energy.

$$f_{\text{eq}} = \left(\frac{2EG_f}{w_c} \right)^{1/2} \tag{3.70}$$

These computational refinements guarantee that the fracture energy dissipation remains independent of element size. Without these adjustments, coarse mesh results deviate significantly and may incorrectly predict peak load and failure behavior (Bažant and Oh 1983a).

The characteristic width of the FPZ, w_c, cannot be uniquely determined from standard fracture tests. This is because such tests can often be matched equally well using a range of FE sizes (h), and experimental observations indicate that the density of damage decreases gradually from the crack centerline, making w_c inherently ambiguous. Bažant and Planas 2019 proposed several approaches for estimating w_c.

One method is based on determining the minimum spacing of parallel cracks, which may be measured experimentally during early cooling or drying (with limitations due to rapid localization), inferred from theoretical spacing calculations in pavements using fracture mechanics principles, or obtained through hydraulic fracturing tests in special cases. Another approach involves restraint tests, where localization is prevented in a tensioned concrete layer restrained by bonded steel plates (Bažant et al. 2021).

Finally, calibration from structural tests can be carried out by selecting an element size (h_{ref}) that matches the energy dissipation (W_{ref}) in the fracture process, particularly when extrapolating from small-scale laboratory tests to large-scale structures (Bažant and Planas 2019). This is typically achieved through a multistep fitting procedure to minimize variations in results across different test types. In practice, the calibration of w_c is performed by selecting a reference element size (h_{ref}) and adjusting the strength and softening law such that the resulting fracture energy G_f matches the values obtained from laboratory-scale tests. This h_{ref} is then applied to larger structures or different geometries, with adjustments to the softening modulus ensuring fracture energy preservation. The success of this calibration critically depends on adopting a realistic tensorial damage constitutive law for the crack band model, ensuring that the chosen w_c or h_{ref} produces accurate and mesh-objective numerical predictions (Bažant and Planas 2019).

A key advantage of the crack band model over the cohesive crack model is its compatibility with complex tensorial damage constitutive laws. Unlike scalar cohesive laws, tensorial formulations (Bažant and Planas 2019) can capture the influence of all stress components. This is particularly important for normal compressive stresses, which can promote fracture or cause splitting, and for shear components, which can alter the crack growth direction (Bažant and Oh 1983a).

For realistic but complex constitutive laws, identifying parameter adjustments to regularize energy dissipation in the softening regime is challenging. An external adjustment is typically applied, in which the material properties remain unchanged, while the crack band is coupled with fictitious springs (or elastic elements) in the principal stress directions (Bažant 2002; Červenka, Bažant, and Wierer 2005). These springs retain the elastic properties of the material but have an effective length larger than the crack band width. During fracture, the elastic energy stored in these springs is transferred to the crack band and dissipated, enabling the use of complex constitutive laws in commercial FE codes such as ATENA (Bažant 2002).

The crack band approach works best when the mesh is aligned with the crack propagation direction. If the direction is unknown, misalignment can cause zig-zag propagation and shear stress development parallel to the crack band, reducing accuracy. Remedies include using triangular meshes with multiple orientations for better approximation of arbitrary crack paths (Bažant and Planas 2019) or applying empirical correction factors based on the angle between the mesh lines and the crack band direction (Bažant and Oh 1985). The crack band model is an embedded automatic feature in commercial software such as SBETA, ATENA, DIANA, and ABAQUS, as well as in open-source codes like object-oriented FEM (OOFEM) (Patzák 2012). For 3D-printed cement paste and mortar, the absence of coarse aggregates may result in a narrower FPZ compared to conventional concrete. However, the presence of interfaces in 3D-printed concrete, mortar, or paste can lead to a larger FPZ compared to the native counterparts (Figure 3.4). Consequently, the effective crack band width (w_c) along with the softening law (C_f, ε_0, and E_t) and equivalent strength $\left(f_{eq}\right)$ used in simulations may need to be adjusted accordingly to accurately capture the fracture behavior in these systems. Moreover, the crack path in 3D-printed cement-based materials often involves deflection (into the interface) in addition to propagation (in the bulk filaments). The latter can be an inclined path due to deflection or the presence of shear, away from a centerline straight crack. These characteristics of crack in 3D-printed cement-based materials can require more accurate and refined meshes, or the use of specific meshes, to properly capture the often tortuous and complex fracture path in these classes of concrete materials, for which the crack band model is a good contender. Nevertheless, the representation of the weak interfaces with regard to the FPZ requires significant considerations. To ensure reliable simulation of fracture in 3D-printed cement-based materials, the crack band model must account for the stiffness, stress, and strain fields both inside and outside the FPZ. While weak interfacial layers can change the size of FPZs, they also introduce preferential crack paths along interfaces, which may not be explicitly represented in traditional crack band implementations. Without capturing these interface-driven propagation mechanisms, even a well-calibrated softening law may yield inaccurate predictions of crack trajectory and failure behavior.

3.2.2.4 Microplane (M7) Model

The microplane model originated from the pioneering work of G.I. Taylor in 1938, in which the plasticity model was formulated based on vectors of stress and strain acting on arbitrarily oriented planes within the material (Bažant and Prat 1988a). Batdorf and Budiansky (1949) developed this into the slip theory of plasticity, in which stress vectors on plastic slip planes are projections of the continuum stress tensor, and the plastic strain vectors are superposed to form the continuum plastic strain tensor. This formulation led to Taylor models, which later evolved into more general microplane models. Zdeněk P. Bažant, in 1984, introduced the term "microplane" to better account for inelastic phenomena such as cracking and damage, where the stress–strain relationship is defined on planes of all orientations. Key advancements necessary for modeling damage included (i) replacing the static constraint with a kinematic constraint, (ii) calculating the continuum stress tensor from the microplane stress vectors using the principle of virtual work, and (iii) incorporating the elastic strain into the microplane constitutive law itself, rather than relying solely on the macro-level elasticity tensor. The principle of virtual work integrates stress tensor components over a unit hemisphere of microplane orientations. To reduce numerical errors in far post-peak response, at least 21, and preferably 37 or more, microplanes are recommended. The microplane model is categorized as a semi-multiscale model, capturing interactions among microscale inelastic deformations of different orientations without representing distance interactions explicitly.

Since 1983, a series of progressively refined microplane models labeled M0 through M7 have been developed (Bažant and Oh 1983b, 1985; Bažant and Prat 1988a, 1988b; Bažant, Xiang, and Prat 1996; Bažant et al. 2000; Caner and Bažant 2000; Bažant and Caner 2005a, 2005b; Caner and Bažant 2011, 2013a, 2013b). The M7 model overcomes the problems of the previous model and exhibits better and broader modeling capability than the previous microplane (M0–M6) models (Caner and Bažant 2013a).

It should be emphasized that the microplane model defines only the constitutive (or stress–strain) behavior. Due to post-peak softening, it needs to be paired with a crack band model in practical applications. In the crack band model, M7 provides the stress–separation relation for a cohesive crack, incorporating the influence of shear and longitudinal compression along the crack (Caner and Bažant 2013a).

3.2.2.4.1 Basic Relations for Kinematically Constrained Microplanes

The microplane formulation departs from traditional tensorial models by defining stress–strain behavior on planes of arbitrary orientation, known as microplanes. For kinematically constrained models, the central assumption is that the microplane strain vector is a projection of the macroscopic strain tensor $\left(\varepsilon_{ij}\right)$ onto the microplane (Figure 3.9a) (Caner and Bažant 2013a; Bažant 1984).

The normal strain ε_N on a generic microplane is given by

$$\varepsilon_N = N_{ij}\varepsilon_{ij} = n_i n_j \varepsilon_{ij}, \tag{3.71}$$

where n_i are the components of the unit normal vector n to the microplane (Figure 3.9a), and $i,\ j = 1, 2, 3$ denote the Cartesian coordinate directions.

The in-plane shear strain vector on each microplane is expressed in terms of two orthogonal components along the in-plane unit vectors m and l, which are generated randomly for each microplane. The corresponding shear strains are defined as

$$\varepsilon_M = M_{ij}\varepsilon_{ij} = \frac{1}{2}\left(m_i n_j + n_i m_j\right)\varepsilon_{ij}, \quad \varepsilon_L = L_{ij}\varepsilon_{ij} = \frac{1}{2}\left(l_i n_j + n_i l_j\right)\varepsilon_{ij}. \tag{3.72}$$

The total in-plane shear strain resultant ε_T on the microplane is then given by

$$\varepsilon_T = \sqrt{\varepsilon_L^2 + \varepsilon_M^2}\,. \tag{3.73}$$

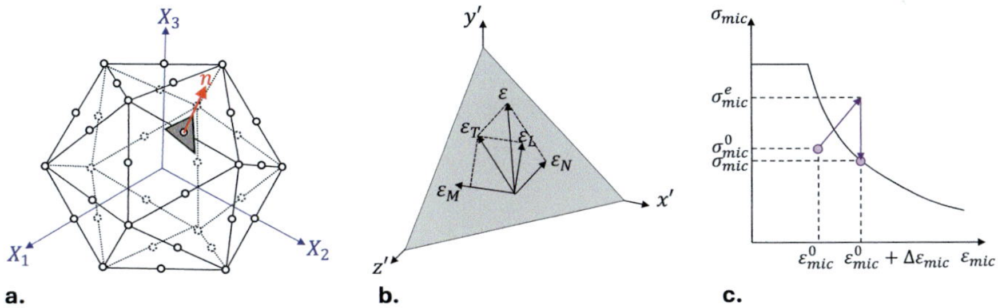

FIGURE 3.9 (a) Microplane strain vector and its component; (b) microplane strain components; (c) stress–strain boundary at constant strain. Adapted from Caner and Bažant (2013a).

In Eqs. (3.71)–(3.73), the subscript N refers to the normal direction n_i, whereas the subscripts L and M refer to the shear directions l_i and m_i, respectively. The microplane strain components in the N, L, and M directions are illustrated in Figure 3.9b.

To describe compression-related softening, especially under confinement, the deviatoric normal strain is introduced by subtracting the volumetric strain ε_V (defined as one-third the trace of ε_{ij}):

$$\varepsilon_D = \varepsilon_N - \varepsilon_V, \varepsilon_V = \frac{1}{3}\varepsilon_{kk}. \tag{3.74}$$

Because the stress and strain vectors cannot both be direct projections on the microplanes, the equilibrium between them must be enforced. This is achieved via the principle of virtual work, leading to the following variational condition:

$$\frac{2\pi}{3}\sigma_{ij}\delta\varepsilon_{ij} = \int_\Omega \left(\sigma_N \delta\varepsilon_N + \sigma_L \delta\varepsilon_L + \sigma_M \delta\varepsilon_M\right) d\Omega. \tag{3.75}$$

In Eq. (3.75), Ω represents the surface of a unit hemisphere centered at the material point, and the left-hand side involves the macroscopic (continuum) stress tensor. Substituting the strain projections from Eqs. (3.71) and (3.72) in Eq. (3.75) and factoring out $\delta\varepsilon_{ij}$ yield the macroscopic stress tensor as an integral of the microplane contributions:

$$\sigma_{ij} = \frac{3}{2\pi}\int_\Omega s_{ij} d\Omega \approx 6\sum_{\mu=1}^{N_m} w_\mu s_{ij}^{(\mu)}, \tag{3.76}$$

where N_m is the number of total microplanes. The microplane stress contribution s_{ij} is constructed from the individual stress components:

$$s_{ij} = \sigma_N N_{ij} + \sigma_L L_{ij} + \sigma_M M_{ij}. \tag{3.77}$$

In numerical applications, the integral in Eq. (3.76) is approximated using Gaussian quadrature over a discrete set of microplane orientations $\boldsymbol{n}_\mu$ with associated weights w_μ, satisfying $\sum_\mu w_\mu = \frac{1}{2}$.

The microplane model prescribes constitutive laws directly on each microplane. The normal and shear stress components are functions of the microplane strain history:

$$\sigma_N(t) = F\,|_{\tau=0}^{t}\left[\varepsilon_N(\tau), \varepsilon_T(\tau)\right], \sigma_L(t) = G\,|_{\tau=0}^{t}\left[\varepsilon_N(\tau), \varepsilon_T(\tau)\right], \sigma_M(t) = H\,|_{\tau=0}^{t}\left[\varepsilon_N(\tau), \varepsilon_T(\tau)\right] \tag{3.78}$$

Here, F, G, and H are functionals representing the material's constitutive response based on the strain history, with ε_T denoting the shear strain resultant defined in Eq. (3.73).

3.2.2.4.2 Elastic Behavior and Material Damage for Unloading

When the normal microplane strains ε_N are not decomposed into volumetric and deviatoric parts, the normal and shear stiffnesses on the microplanes, E_N and E_T, can be defined using macrolevel constants under a kinematic constraint (Caner and Bažant 2013a). The corresponding expressions are

$$E_N = \frac{E}{1-2v}, E_T = E_N\left(\frac{1-4v}{1+v}\right), \tag{3.79}$$

where E is the Young's modulus and v is the Poisson ratio. To ensure E_N and E_T remain positive, v must lie within the range $v \in [-1, 0.25]$.

This range is adequate for concrete, which typically has $v \approx 0.18$, but is not sufficient for materials like polymers, metals, or ceramics with v in the range 0.25–0.5. To address this limitation, a remedy proposed by Bažant and Byung H. Oh (1985) involves coupling the microplane model with an isotropic volumetric element subjected to the same strain. This parallel element contributes a bulk modulus K' and assumes a shear modulus $G' = 0$. The combined bulk and shear moduli become

$$\tilde{K} = K + K', \tilde{G} = G + G' = G,$$

and the effective Poisson ratio $\tilde{v}$ is given by Ferhun C. Caner and Zdeněk P. Bažant (2013a):

$$\tilde{v} = \frac{3K(1+G/G')-2G}{6K(1+G/G')+2G}, \tag{3.80}$$

which tends to 0.5 as $G'/G \to 0$. This allows larger values of v to be reproduced without introducing a volumetric-deviatoric strain split.

In modeling of concrete failure in Ferhun C. Caner and Zdeněk P. Bažant (2013a), such coupling with parallel elements is generally not used, as Poisson's ratios around 0.18 can still be achieved by calibrating E_N and E_T without coupling. However, for materials with v close to 0.25, this approach may need further investigation.

3.2.2.4.3 Microplane Stress–Strain Boundaries and Numerical Algorithm

Damage in a microplane model is governed by stress boundaries: the compressive volumetric stress boundary (σ_V^b), the compressive deviatoric stress boundary (σ_D^b), and the tensile normal stress boundary (σ_N^b). If the stress in a microplane exceeds its boundary in a load or time increment, the stress is reduced vertically back to the boundary while maintaining the current strain value, as shown in Figure 3.9c.

The numerical algorithm is implemented at each integration point of each FE, looping through all microplanes (Caner and Bažant 2013a, 2013b). Here, the superscript "o" refers to values at the beginning of the current step, and the subscript "0" denotes the initial (undamaged) values. The computation proceeds as follows:

1. Using Eqs. (3.71) and (3.72), microplane strain increments $\Delta\varepsilon_N$ (normal), $\Delta\varepsilon_M$ (shear in M direction), and $\Delta\varepsilon_L$ (shear in L direction) are calculated as

$$\Delta\varepsilon_N = N_{ij}\Delta\varepsilon_{ij}, \Delta\varepsilon_M = M_{ij}\Delta\varepsilon_{ij}, \Delta\varepsilon_L = L_{ij}\Delta\varepsilon_{ij}. \tag{3.81}$$

2. Calculate volumetric strain at the current step ε_V (average normal strain), using

$$\varepsilon_V^o = \frac{\varepsilon_{kk}}{3}, \Delta\varepsilon_V = \frac{\Delta\varepsilon_{kk}}{3}, \varepsilon_V = \varepsilon_V^o + \Delta\varepsilon_V. \tag{3.82}$$

3. Compute the ε_e defined as

$$\varepsilon_e = \langle -\sigma_V^o / E_{N0} \rangle, \tag{3.83}$$

where σ_V^o is the volumetric stress from the previous step, E_{N0} is the initial normal stiffness constant of the microplane, and $\langle \cdot \rangle$ is the Macaulay bracket. Then evaluate the compressive volumetric stress–strain boundary:

$$\sigma_V^b = -Ek_1k_3e^{-\varepsilon_V / k_1\alpha}, \tag{3.84}$$

where E is the Young's modulus, and k_1 and k_3 are adjustable empirical parameters. The exponent α is

$$\alpha = \frac{k_5}{1 + \min\left(\langle -\sigma_V^0 \rangle, c_{21}\right) / E_{N0}} \left(\frac{\varepsilon_I^0 - \varepsilon_{III}^0}{k_1}\right)^{c_{20}} + k_4, \tag{3.85}$$

where ε_I^o and ε_{III}^o are the maximum and minimum principal strains, respectively, at the beginning of the step; k_1, k_4, and k_5 are adjustable empirical parameters; and c_{20} is a fixed empirical parameter.

4. Compute deviatoric strain increment $(\Delta\varepsilon_D)$ and total deviatoric strain (ε_D) (difference between normal and volumetric strains):

$$\Delta\varepsilon_D = \Delta\varepsilon_N - \Delta\varepsilon_V, \varepsilon_D^o = \varepsilon_N - \varepsilon_V^o, \varepsilon_D = \varepsilon_D^o + \Delta\varepsilon_D. \tag{3.86}$$

   Then, compute the compressive deviatoric boundary:

$$\sigma_D^b = \frac{-Ek_1\beta_3}{1 + [\langle -\varepsilon_D \rangle / (k_1\beta_2)]^2}, \tag{3.87}$$

where $\beta_2 = c_5\gamma_1 + c_7$, $\beta_3 = c_6\gamma_1 + c_8$, and

$$\gamma_0 = \frac{f'_{c0}}{E_0} - \frac{f'_c}{E}, \gamma_1 = e^{\gamma_0} \tan h\left(\frac{c_9 \langle -\varepsilon_V \rangle}{k_1} \right), \tag{3.88}$$

where f'_{c0} is the initial (undamaged) compressive strength and f'_c is the compressive strength at the current step.

5. Update the normal microplane strain:

$$\varepsilon_N = \varepsilon_V + \varepsilon_D \tag{3.89}$$

and retrieve ε_N^{0+} and ε_N^{0-} (positive and negative maximum history variables, respectively). Then compute the damaged modulus:

$$E_N = E_{N0} e^{-c_{13}\varepsilon_N^{0+}} f(\zeta) \qquad \text{if } \sigma_N^o \geq 0, \tag{3.90}$$

$$E_N = E_{N0} \text{ if } \sigma_N^o > E_{N0}\epsilon_N \text{ and } \sigma_N^o \Delta\epsilon_N < 0, \tag{3.91}$$

$$E_N = E_{N0} \left(e^{-c_{14}\left|\varepsilon_N^{0-}\right| / (1 + c_{15}\varepsilon_e)} + c_{16}\varepsilon_e \right) \text{if } \sigma_N^o < 0, \tag{3.92}$$

where $f(\zeta) = (1 + a\zeta^2)^{-1}$ in which $\zeta = \int \langle d\epsilon_V \rangle$ and $a = 0.1$.

6. Calculate elastic normal microplane stress:

$$\sigma_N^e = \sigma_N^o + E_N \Delta\varepsilon_N. \tag{3.93}$$

7. Compute the tensile normal boundary:

$$\sigma_N^b = E k_1 \beta_1 e^{-\langle \varepsilon_N - \beta_1 c_2 k_1 \rangle / [c_4 \varepsilon_e + k_1 c_3]}, \tag{3.94}$$

where

$$\beta_1 = -c_1 + c_{17} e^{-c_{19}\langle -\sigma_V^0 - c_{18} \rangle / E_{N0}}, \tag{3.95}$$

8. Enforce a vertical drop in the stress, at constant strain, onto the stress–strain boundary in the normal stress direction (as shown in Figure 3.9c):

$$\sigma_N = \max\left[\min\left(\sigma_N^e, \sigma_N^b \right), \sigma_V^b + \sigma_D^b \right]. \tag{3.96}$$

9. If the normal stress–strain boundary is exceeded, the overall maximum values of the normal microplane strain are stored as ε_N^{0+} and ε_N^{0-}.
10. Compute an approximate average value of normal microplane stresses (σ_N) as volumetric stress:

$$\sigma_V \approx \frac{1}{2\pi} \sum_{\mu=1}^{N_m} w_\mu \sigma_N. \tag{3.97}$$

11. Retrieve shear stress history $\left(\sigma_L^o, \sigma_M^o\right)$ and compute:

$$\hat{\sigma}_N^o = E_T \left\langle k_1 c_{11} - c_{12} \left\langle \varepsilon_V \right\rangle \right\rangle, \tag{3.98}$$

Then evaluate the shear boundary:

$$\sigma_\tau^b = [(c_{10} \left\langle \hat{\sigma}_N^o - \sigma_N \right\rangle)^{-1} + (E_T k_1 k_2)^{-1}]^{-1}, \text{ if } \sigma_N \leq 0. \tag{3.99}$$

$$\sigma_\tau^b = [(c_{10} \hat{\sigma}_N^o)^{-1} + (E_T k_1 k_2)^{-1}]^{-1}, \text{ if } \sigma_N > 0. \tag{3.100}$$

12. Compute shear stress:

$$\sigma_\tau^e = \sqrt{(\sigma_L^o + E_T \Delta\varepsilon_L)^2 + (\sigma_M^o + E_T \Delta\varepsilon_M)^2}, \tag{3.101}$$

$$\sigma_\tau = \min(\sigma_\tau^b, |\sigma_\tau^e|), \tag{3.102}$$

$$\sigma_L = \left(\sigma_L^o + E_T \Delta\varepsilon_L\right) \frac{\sigma_\tau}{\sigma_\tau^e}, \sigma_M = \left(\sigma_M^o + E_T \Delta\varepsilon_M\right) \frac{\sigma_\tau}{\sigma_\tau^e}. \tag{3.103}$$

Finally, the stress tensor σ_{ij} is obtained using a sum over all the microplanes in Eqs. (3.76) and (3.77), with all internal variables updated for the next step. Parameters $k_1, \ldots, k_5$ are adjustable, and $c_1, \ldots, c_{21}$ are fixed. Delaying σ_V by one step leads to first-order accuracy, which can be improved via second-order iteration using the averaged value of σ_V in the first iteration.

3.2.2.5 Two-Parameter Fracture Model

To account for the FPZ and its influence on fracture toughness in concrete, Jenq and Shah (1985a) proposed the Two-parameter fracture model (TPFM), which incorporates both the critical stress intensity factor K_{IC}^S and the critical crack tip opening displacement $CTOD_C$. In the TPFM framework, the critical crack tip opening displacement $CTOD_C$ is introduced as an additional parameter to complement the critical stress intensity factor K_{IC}^S, providing a more complete description of the fracture by accounting for the effective crack extension caused by the formation of a larger FPZ ahead of the crack.

Jenq and Shah conceptualized the fracture process in three stages (as illustrated in Figure 3.10a): an initial *linear elastic* region, a *nonlinear softening* region driven by FPZ development, and a *critical peak point* corresponding to the onset of unstable crack propagation. The procedure for evaluating fracture parameters from a standard SENB test involves the following steps:

1. Calculation of Young's Modulus, E: The initial compliance $\left(C_i\right)$ of the linear part of the load-CMOD plot for an SENB specimen can be defined using Eq. (3.104) (Steenkamp, Read, and Reed 1988).

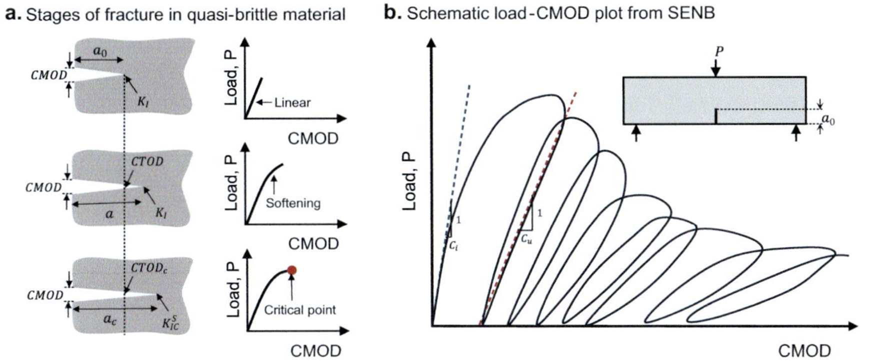

FIGURE 3.10 (a) Stages of fracture in a quasi-brittle material; (b) schematic load–CMOD response obtained from cyclic SENB. Adapted from Surendra P. Shah (1990).

$$C_i = \frac{6Sa_0V_1(\alpha_0)}{EBW^2}, \tag{3.104}$$

where S, W, and B are the span length, depth, and width of the beam, respectively; a_0 is the initial notch length; and $V_1(\alpha_0)$ is a geometry function defined as

$$V_1(\alpha) = 0.76 - 2.28\alpha + 3.87\alpha^2 - 2.04\alpha^3 + \frac{0.66}{(1-\alpha)^2}, \tag{3.105}$$

where $\alpha_0 = a_0 / W$ is the normalized initial notch length.

As the initial compliance (C_i) can be obtained from the load–CMOD plot (as shown in Figure 3.10b), and beam geometry $(S, W, \text{and } B)$ and initial notch length (a_0) are known, the Young's modulus (E) can be obtained by rearranging Eq. (3.104) as follows:

$$E = \frac{6Sa_0V_1(\alpha_0)}{C_iBW^2}. \tag{3.106}$$

2. Calculation of Critical Effective Crack Length, a_c:

   In the TPFM procedure, the loading–unloading–reloading cycle is introduced to enable the calculation of unloading compliance (C_u) near the peak load, which is critical for determining the effective crack length (a_c). As specified in the testing protocol (Shah 1990b), unloading begins when the applied load reaches the peak load and then drops to approximately 95% of the maximum load. The load is then reduced to zero and re-applied, completing the cycle. The unloading–reloading rate does not need to match the initial loading rate, and the entire cycle should be completed within a short time period (1 minute is recommended by Shah 1990b). This process, schematically illustrated in Figure 3.10b, enables accurate estimation of the unloading compliance C_u, which is essential for evaluating the critical effective crack length (a_c).

$$a_c = a_0 + \Delta a, \tag{3.107}$$

where Δa is the stable crack growth at the peak load.

The unloading compliance at 95% of the peak load (C_u), as schematically shown in Figure 3.10b, is used to calculate the critical effective crack length (a_c) using Eq. (3.108). In this equation, both C_u and the Young modulus E (determined from Eq. (3.106)) are known, while a_c is the unknown. Since the geometry function $V_1(\alpha_c)$ is a cubic function of the normalized crack length $\alpha_c = a_c / W$, the equation is nonlinear in a_c. An iterative numerical method, such as the Newton–Raphson method or fixed-point iteration, can be employed to compute a_c.

$$E = \frac{6Sa_c V_1(\alpha_c)}{C_u BW^2}, \tag{3.108}$$

where $\alpha_c = a_c / W$.

3. Calculation of Critical Stress Intensity Factor, K_{IC}^S:

The critical stress intensity factor (K_{IC}^S) is calculated using the maximum peak load (P_{max}), critical effective crack length (a_c), and beam geometry $(S, W, \text{and } B)$.

$$K_{IC}^S = \frac{P_{max} S}{BW^{\frac{3}{2}}} F(\alpha_c), \tag{3.109}$$

with

$$F(\alpha_c) = \frac{3(\alpha_c)^{\frac{1}{2}}}{2} \frac{\left(1.99 - \alpha_c(1-\alpha_c)\left(2.15 - 3.93\alpha_c + 2.7\alpha_c^2\right)\right)}{(1+2\alpha_c)(1-\alpha_c)^{3/2}}. \tag{3.110}$$

4. Calculation of Critical Crack Tip Opening Displacement, $CTOD_C$:

The critical crack tip opening distance $(CTOD_C)$ corresponding to critical effective crack length a_c is defined as follows:

$$CTOD_C = \frac{6P_{max} S\alpha_c V_1(\alpha_c)}{EWB}\left[(1-\beta_c)^2 + (1.081 - 1.149\alpha_c)(\beta_c - \beta_c^2)\right]^{1/2}, \tag{3.111}$$

where $\beta_c = a_0 / a_c$.

Jenq and Shah also proposed a size-independent parameter Q to further generalize the fracture behavior:

$$Q = \left(\frac{E \cdot CTOD_C}{K_{IC}^S} \right)^2, \tag{3.112}$$

This framework allows fracture resistance to be characterized with consideration of both the elastic energy release and the inelastic deformation at the crack tip, making it suitable for quasi-brittle materials. In 3D-printed cement paste and mortar, the absence of coarse aggregates reduces the extent of nonlinearity near the crack tip compared to concrete. However, the presence of interfaces can increase the size of FPZ compared to the monolithic cement paste, mortar, and concrete counterparts, all of which can alter both the stress intensity factor K_{IC}^S and the critical crack tip opening displacement CTOD_C. Furthermore, due to the layered architecture, the two fracture parameters may vary with the orientation of the notch relative to the printed filaments and IZs.

3.2.2.6 *R*-Curve

The resistance curve, or R -curve, describes how crack growth resistance R changes as a crack extends by an effective length Δa (Bažant and Planas 2019; Zhao, Liu, and Xu 2021):

$$R = R(\Delta a). \tag{3.113}$$

As the crack advances, R typically increases from an initial value (possibly zero) toward a plateau corresponding to the fracture energy G_f predicted by LEFM. This plateau is reached when $\Delta a = c_f$, the equivalent crack extension. Introducing a dimensionless form $\bar{R}(\cdot)$ that equals unity at $\Delta a = c_f$, the resistance can be expressed as

$$R = G_f \bar{R}\left(\frac{\Delta a}{c_f} \right). \tag{3.114}$$

This evolving fracture resistance throughout crack propagation in quasi-brittle materials arises from toughening mechanisms such as microcracking, crack deflection, aggregate bridging, and aggregate interlocking (Shah and Ouyang 1994) that exist in conventional concrete. Two widely used methods for *R*-curve evaluation are the size effect method and the compliance-based method. The size effect approach estimates an initiation-based *R*-curve using peak load measurements from multiple geometrically scaled specimens, without requiring direct monitoring of crack growth for each specimen. In contrast, the compliance-based method utilizes the load–CMOD response from a single specimen to continuously estimate crack extension and fracture resistance. In the following subsections, we present both *R*-curves based on size effect and based on compliance methods.

3.2.2.6.1 R-Curve Determined Using Size Effect Analysis

An R-curve describes the relationship between a material's crack growth resistance and the crack extension Δa (Bažant and Planas 2019; Bažant and Jirásek 1993). The R-curve model is based on the concept of the equivalent elastic crack but also describes the process of complete crack growth (Bažant and Planas 2019). The concept of the R-curve maintains the fundamental framework of LEFM; however, instead of adopting the fracture criterion $G = G_f$ with a constant G_f, the crack is assumed to advance when $G = R$, where R represents a variable crack growth resistance.

In the size effect-based approach, a series of geometrically similar specimens with the same a_0 / W ratio are tested (as illustrated in Figure 3.11a). Each specimen, due to its size, fails at a different nominal stress σ_{Nu}, and the corresponding energy release rate at peak load is calculated as (Bažant and Planas 2019)

$$G\left(\frac{a}{W}\right) = \frac{1}{E'}\sigma_{Nu}^2 D\left[k\left(\frac{a}{W}\right)\right]^2, \tag{3.115}$$

where E' is the effective elastic modulus (equal to E for plane stress and $E/\left(1-\nu^2\right)$ for plane strain), W is the specimen depth, D is the characteristic dimension, and $k\left(a/W\right)$ is the dimensionless stress intensity factor function. The total crack length at failure is

$$a = a_0 + \Delta a. \tag{3.116}$$

By plotting G vs. Δa for each specimen (blue curves in Figure 3.11b), a set of $G - \Delta a$ curves is obtained. The R-curve is defined as the envelope of these curves (black curve in Figure 3.11b) and represents the material's intrinsic resistance to crack growth as a function of crack extension.

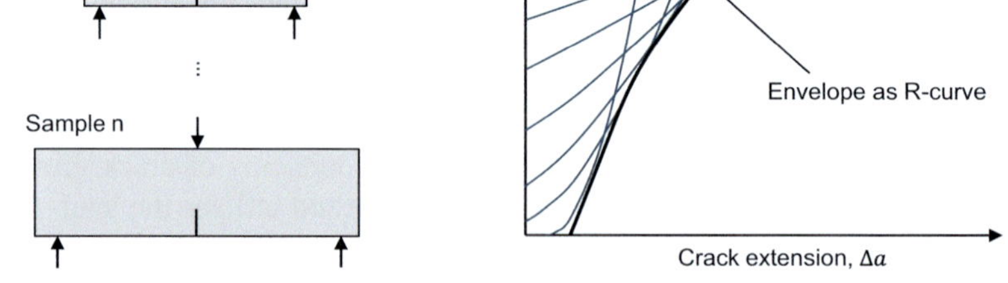

FIGURE 3.11 (a) Geometrically similar specimens used for fracture testing; (b) R-curve constructed as the envelope of energy release rate G vs. crack extension Δa for different specimen sizes.

In this approach, the extraction of fracture toughness relies on analyzing geometrically scaled specimens and provides valuable insight into the resistance behavior across scales and the identification of the plateau. It allows probing size-dependent behavior across various microstructures of 3D-printed cementitious systems enabled by design and fabrication. As discussed in Section 3.2.1 and illustrated in Figure 3.6c, scaling layered 3D-printed cementitious materials requires further understanding using either the SEL to capture the fracture parameters or the *R*-curve to capture how resistance scales with increasing size.

3.2.2.6.2 R-Curve Determined Using Compliance

The compliance-based method enables the construction of *R*-curves by evaluating the evolution of crack resistance (in terms of stress intensity factor K_R) as a function of effective crack extension during a single notched specimen test (Shah and Ouyang 1994; Banthia and Sheng 1996; Banthia and Nandakumar 2003; Gupta, Esmaeeli, and Moini 2024; Gupta and Moini 2024; Prihar, Gupta, et al. 2024). Unlike size effect-based methods, which rely on peak load data from geometrically scaled specimens, this approach allows for continuous monitoring of crack growth using load–CMOD measurements. This approach is used in quasi-brittle materials and often in fiber-reinforced concrete, where fracture resistance increases with crack propagation due to inelastic processes such as microcracking and aggregate or fiber bridging.

The effective crack length $\left(a_{\text{eff}}\right)$ can be determined from the compliance calculated using load–CMOD data obtained during the SENB test (Jenq and Shah 1985b; Mobasher 2011). The CMOD for each increment of load in SENB is given by

$$\text{CMOD} = \frac{6P\alpha_{\text{eff}}S}{WBE'}V_1\left(\alpha_{\text{eff}}\right), \tag{3.117}$$

where P is the applied load at the current increment, S is the span, W is the specimen depth, B is the specimen thickness, $\alpha_{\text{eff}} = a_{\text{eff}}/W$, and E' is the effective Young's modulus, with $E' = E/\left(1-\nu^2\right)$ for plane strain and $E' = E$ for plane stress.

The geometry function $\left(V_1\right)$ for SENB is defined as (Jenq and Shah 1985b; Mobasher 2011)

$$V_1\left(\alpha\right) = 0.76 - 2.28\alpha + 3.87\alpha^2 - 2.04\alpha^3 + \frac{0.66}{(1-\alpha)^2}. \tag{3.118}$$

Rearranging Eq. (3.117), the effective Young's modulus is expressed as

$$E' = \frac{6\alpha_{\text{eff}}SP}{WB\text{CMOD}}V_1\left(\alpha_{\text{eff}}\right). \tag{3.119}$$

The compliance at each increment of load is defined as

$$C\left(\alpha\right) = \frac{\text{CMOD}}{P}. \tag{3.120}$$

Substituting Eq. (3.120) into Eq. (3.119) gives

$$E' = \frac{6\alpha_{\text{eff}} S}{WB} \frac{1}{C(\alpha_{\text{eff}})} V_1(\alpha_{\text{eff}}). \tag{3.121}$$

Since E' is a material property and remains constant during the test, the following relationship holds for both α_{eff} and the initial $\alpha_0 = a_0 / W$:

$$\frac{\alpha_{eff}}{C(\alpha_{eff})} V_1(\alpha_{\text{eff}}) = \frac{\alpha_0}{C(\alpha_0)} V_1(\alpha_0), \tag{3.122}$$

where $C(\alpha_0)$ is the initial compliance obtained from the slope of the elastic portion of the load–CMOD curve (as shown in Figure 3.12a).

The effective crack length is then given by

$$a_{eff} = a_0 \frac{C(\alpha_{eff})}{C(\alpha_0)} \frac{V_1(\alpha_0)}{V_1(\alpha_{\text{eff}})}, \tag{3.123}$$

and the corresponding crack extension (Δa) is defined as

$$\Delta a = a_{eff} - a_0. \tag{3.124}$$

It is important to note that Eq. (3.123) is nonlinear, as a_{eff} appears on both sides of the equation, requiring iterative or numerical methods for its solution, such as fixed-point iteration or the Newton–Raphson method.

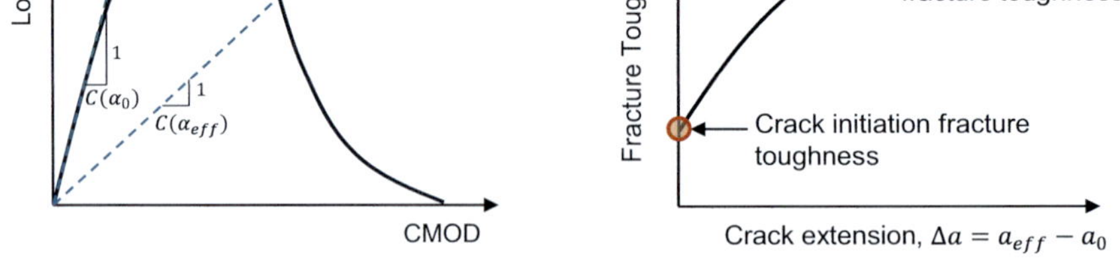

FIGURE 3.12 (a) Schematic load–CMOD curve from a notched beam test; (b) *R*-curve obtained using the compliance method, showing the evolution of fracture toughness with crack extension.

Once the effective crack length is known, the fracture toughness (K_R) is calculated for each increment of load using (Mobasher 2011)

$$K_R = \left[\frac{PS}{BW^{3/2}}\right] f(\alpha_{eff}), \tag{3.125}$$

where

$$f(\alpha) = \frac{3\alpha^{1/2}\left[1.99 - \alpha(1-\alpha)(2.15 - 3.93\alpha + 2.7\alpha^2)\right]}{2(1+2\alpha)(1-\alpha)^{3/2}}. \tag{3.126}$$

For the notched four-point bend setup, the same formulation given in Eq. (3.123) can be used to evaluate a_{eff} but with a modified geometry function $V_1(\alpha)$ defined as (Banthia and Sheng 1996)

$$V_1(\alpha) = 0.8 - 1.7\alpha + 2.4\alpha^2 + \frac{0.66}{1-\alpha^2}. \tag{3.127}$$

In the notched four-point bend setup, the fracture toughness (K_R) is computed as (Banthia and Sheng 1996)

$$K_R = \frac{PS}{BW^2} \sqrt{\pi a_{\text{eff}}} \cdot f(\alpha_{\text{eff}}), \tag{3.128}$$

where

$$f(\alpha) = 1.122 - 1.40\alpha + 7.33\alpha^2 - 13.08\alpha^3 + 14.0\alpha^4. \tag{3.129}$$

By computing K_R for each increment of crack extension $\Delta a = a_{\text{eff}} - a_0$, an *R*-curve can be obtained, as schematically illustrated in Figure 3.12b. This curve reflects the material's increasing resistance to crack growth (or fracture resistance or fracture toughness). It provides valuable insight into the fracture behavior beyond initiation and how fracture toughness evolves as the crack propagates through and interacts with the materials in a given specimen size. It also provides insight into the stability and rate of crack growth. Considerations regarding the size of the sample should be made in relation to the length of the FPZ for the examined materials, such as 3D-printed systems.

A similar compliance-based method has also been successfully applied to other fracture geometries, such as the contoured double cantilever beam (CDCB) test (Banthia and Nandakumar 2003; Sorelli, Banthia, Plizzari, et al. 2014), to evaluate crack growth resistance in fiber-reinforced cementitious materials.

In the compliance-based *R*-curve approach, fracture resistance is inferred from the evolution of specimen compliance with crack growth. Although cement paste and mortar are generally more brittle than conventional concrete, exhibiting a smaller FPZ and limited intrinsic toughening, the presence of weak interfacial layers in layered 3D-printed systems can introduce additional toughening mechanisms. One such mechanism is crack deflection along interfaces or potential microcracking along the layered IZs, which increases the energy required for crack propagation and enhances damage resistance, as schematically illustrated in Figure 3.12b. These mechanisms can be further engineered to proactively alter the length of the FPZ and toughen the materials by controlling the geometry (through design of spatial arrangement of materials architecture) and the chemistry of the interfacial vs. bulk filaments in the layered 3D-printed systems. The compliance approach, similar to those used in fiber-reinforced concrete with more nuanced toughening mechanisms, can be used in 3D-printed cementitious materials that contain more complex, architected designs at the micro and meso-scales, to better understand the energy dissipation resulting from crack–microstructure interactions both ahead (e.g., microcracking and crack deflection in FPZ) and behind the crack tip (e.g., interlocking due to rough fracture surfaces).

While these analytical frameworks have been essential in providing theoretical insights into the fracture behavior at or due to the presence of interfaces, they have been limited by their assumptions of linear elasticity and simplified (bilayer) geometries. Real layered and composite materials often exhibit complex microstructures, nonlinear material behavior, and three-dimensional fracture and crack evolution that cannot be fully captured by classical LEFM approaches. Furthermore, the probing and prediction of crack initiation, propagation, complex interaction with the microstructure through multiple deflections or branching, and the associated stress and displacement fields during crack propagation all require more sophisticated computational tools. These limitations have recently motivated the development of advanced numerical methods capable of modeling the complexity of fracture in heterogeneous materials (containing interfaces and various bulk constitutive properties) while preserving the fundamental physics established by classical fracture mechanics theory.

3.3 CONCLUDING REMARKS

Several linear and nonlinear elastic approaches to characterize fracture in cement-based materials and structures are discussed. The applicability of the existing approaches to layered additively manufactured materials and structures, however, is rather nuanced. That is, the applicability of LEFM and non-LEFM models discussed in this chapter to additively manufactured materials and structures depends on the size of the FPZ (depending on whether paste, mortar, or concrete are used) relative to the size of the structure.

Interfacial properties can be characterized using the LEFM approach through critical stress intensity factor (K_{IC}). For a crack impinging on an interface, utilizing LEFM by experimentally characterizing the interfacial and bulk properties ($K_{IC,interface} / K_{IC,bulk}$) can inform us of the anticipated propagation phenomena (as discussed in chapter 12 in this book) and assist in engineering the interfacial and bulk material properties. This approach can help infer potential additional energy dissipation and toughening mechanisms that may reside in additively manufactured structures not present macroscopically in cast counterparts in the absence of abundant interfaces in the structure.

Nevertheless, real layered composite materials often have complex microstructures and exhibit nonlinear behavior, three-dimensional fracture phenomena, and crack evolution that may not be fully captured by classical LEFM approaches. Numerous approaches for characterizing fracture properties of quasi-brittle material are discussed and should be used if the constituent materials used in additively manufactured structures are concrete containing aggregates, especially coarse aggregates. However, considerations to include interfacial properties must be incorporated in the non-LEFM models.

Determination of the correct and sufficient number of parameters to accurately characterize fracture in additively manufactured materials can also vary considerably depending on the problem, material, size of the structures (scale), and boundary conditions. Scale and size govern almost everything in fracture mechanics, and that includes additively manufactured cement-based materials and structures. The aspects discussed in this chapter only provide a high-level framework for thinking about solving fracture mechanics problems across scales and sizes. The SEL provides a fundamental framework to understand the failure and fracture in a broad range of sizes of structures explored in the additive manufacturing community. The use of this law can help understand the scaling laws involved in the fabrication of structures whose sizes several orders of magnitude different. It can be used to expand laboratory studies to determine the scaling laws in terms of energy dissipation and toughening mechanisms from laboratory to large scales.

Moreover, approaches that aim to characterize fracture initiation can differ from those that aim to characterize the crack propagation (or crack–microstructure interaction) throughout the overall fracture response. These approaches must be selected carefully for additively manufactured structures and according to the problem of interest.

Nonetheless, as mentioned earlier, if the interfaces are not explicitly represented in a given conventional fracture model developed for cement-based materials, as they often are not, even with a well-experimentally calibrated softening law, one may not yield accurate predictions of crack propagation path, fracture properties, and fracture response pertaining to ignoring or underestimating the dissipative contribution of energy from interfacial damage phenomena in layered materials.

3.4 ACKNOWLEDGMENTS

The authors acknowledge funding from the National Science Foundation, CAREER program (No. 2238992), which partially supported Moini Lab's knowledge-sharing and dissemination efforts leading to this work, to establish the relevant fundamental knowledge for the growing additive technology and to accelerate the translation of knowledge for industry adoption and research community use. The content development in this book chapter is also supported by the educational efforts throughout a new course offered at the Civil and Environmental Engineering Department at Princeton University: CEE 374 – Autonomous Fabrication and Robotics.

REFERENCES

Anderson, Ted L. and Anderson, T. L. 2005. *Fracture Mechanics: Fundamentals and Applications*. CRC Press.

ASTM C496. 2017. *Standard Test Method for Splitting Tensile Strength of Cylindrical Concrete Specimens*. Technical report C496/C496M–17. ASTM C496/C496M–17. West Conshohocken, PA: ASTM International. https://doi.org/10.1520/C0496-96. https://www.astm.org/c0496_c0496m-17.html.

ASTM E1820. 2025. *ASTM E1820–25a: Standard Test Method for Measurement of Fracture Toughness*. West Conshohocken, PA: ASTM International. https://www.astm.org/e1820-18.html.

Atkinson, C., Smelser, R. E. and Sanchez, Jo. 1982. "Combined mode fracture via the cracked Brazilian disk test." *International Journal of Fracture* 18: 279–291.

Baldie, Keith David and Pratt, P. L. 1985. "Crack growth in hardened cement paste." *MRS Online Proceedings Library (OPL)* 64: 47.

Banthia, N. and Sheng, J. 1996. "Fracture toughness of micro-fiber reinforced cement composites." *Cement and Concrete Composites* 18 (4): 251–269.

Banthia, Nandakumar and Nandakumar, N. 2003. "Crack growth resistance of hybrid fiber reinforced cement composites." *Cement and Concrete Composites* 25 (1): 3–9.

Batdorf, S. B. and Budiansky, B. 1949. *A Mathematical Theory of Plasticity Based on the Concept of Slip*. Technical Report Technical Note 1871. Washington, DC: National Advisory Committee for Aeronautics.

Bažant, Zdeněk P. 1984. "Microplane model for strain-controlled inelastic behavior." Chapter 3 in *Mechanics of Engineering Materials*, edited by Desai, C. S. and Gallagher, R. H., 45–59. London: Wiley.

Bažant, Zdeněk P. 1999. "Size effect on structural strength: A review." *Archive of Applied Mechanics* 69 (9): 703–725.

Bažant, Zdeněk P. 2000. "Size effect." *International Journal of Solids and Structures* 37 (1–2): 69–80.

Bažant, Zdeněk P. 2002. "Concrete fracture models: Testing and practice." *Engineering Fracture Mechanics* 69 (2): 165–205.

Bažant, Zdeněk P. and Caner, F. C. 2005a. "Microplane model M5 with kinematic and static constraints for concrete fracture and anelasticity. I. Theory." *Journal of Engineering Mechanics* 131 (1): 31–40. https://doi.org/10.1061/(ASCE)0733-9399(2005)131:1(31).

Bažant, Zdeněk P. and Caner, F. C. 2005b. "Microplane model M5 with kinematic and static constraints for concrete fracture and anelasticity. II. Computation." *Journal of Engineering Mechanics* 131 (1): 41–47. https://doi.org/10.1061/(ASCE)0733-9399(2005)131:1(41).

Bažant, Zdeněk P., Caner, F. C., Carol, I., Adley, M. D. and Akers, S. A. 2000. "Microplane model M4 for concrete: I. Formulation with work-conjugate deviatoric stress." *Journal of Engineering Mechanics* 126 (9): 944–953.

Bažant, Zdeněk P. and Cedolin, Luigi. 1984. "Finite element modeling of crack band propagation." *Journal of Structural Engineering (United States)* 110 (3): 660–662.

Bažant, Zdeněk P. and Jirásek, Milan. 1993. "R-curve modeling of rate and size effects in quasi-brittle fracture." *International Journal of Fracture* 62: 355–373.

Bažant, Zdeněk P., Le, Jia-Liang, Salviato, Marco, et al. 2021. *Quasibrittle Fracture Mechanics and Size Effect: A First Course*. Oxford University Press.

Bažant, Zdeněk P. and Oh, Byung-Ho. 1983a. "Crack band theory for fracture of concrete." *Matériaux et construction* 16: 155–177.

Bažant, Zdeněk P. and Oh, Byung-Ho. 1983b. "Microplane model for fracture analysis of concrete structures." In *Proceedings of the Symposium on the Interaction of Non-Nuclear Munitions with Structures*, 49–53. U.S. Air Force Academy, Colorado Springs, CO.

Bažant, Zdeněk P. and Oh, Byung-Ho. 1985. "Microplane model for progressive fracture of concrete and rock." *Journal of Engineering Mechanics* 111 (4): 559–582.

Bažant, Zdeněk P. and Planas, Jaime. 2019. *Fracture and Size Effect in Concrete and Other Quasi-Brittle Materials*. Routledge.

Bažant, Zdeněk P. and Prat, Pere C. 1988a. "Microplane model for brittle-plastic material: I. Theory." *Journal of Engineering Mechanics* 114 (10): 1672–1688.

Bažant, Zdeněk P. and Prat, Pere C. 1988b. "Microplane model for brittle-plastic material: II. Verification." *Journal of Engineering Mechanics* 114 (10): 1689–1702.

Bažant, Zdeněk P., Xiang, Y. and Prat, P. C. 1996. "Microplane model for concrete. I. Stress–strain boundaries and finite strain." *Journal of Engineering Mechanics* 122 (3): 245–254.

Begley, Matthew R. and Hutchinson, John W. 2017. *The Mechanics and Reliability of Films, Multilayers and Coatings*. Cambridge University Press.

Bos, Freek P., Menna, Costantino, Pradena, Mauricio, et al. 2022. "The realities of additively manufactured concrete structures in practice." *Cement and Concrete Research* 156: 106746.

Brühwiler, E., Wittmann, F. H. and Rokugo, K. 1987. "Influence of rate of loading on fracture energy and strain softening of concrete." In *Transactions of the 9th International Conference on Structural Mechanics in Reactor Technology (SMiRT 9)*, Lausanne, Switzerland, 25–33. SMiRT Conference Organization.

Brühwiler, E. and Wittmann, F. H. 1990. "The wedge splitting test, a new method of performing stable fracture mechanics tests." *Engineering Fracture Mechanics* 35 (1–3): 117–125.

Caner, Ferhun C. and Bažant, Zdeněk P. 2000. "Microplane model M4 for concrete. II: Algorithm and calibration." *Journal of Engineering Mechanics* 126 (9): 954–961.

Caner, Ferhun C. and Bažant, Zdeněk P. 2011. "Microplane model M6f for fiber reinforced concrete." In *Proceedings of the XI International Conference on Computational Plasticity: Fundamentals and Applications (COMPLAS 2011)*, edited by Oñate, E., Owen, R. J. and Suárez, B., 796–807. Barcelona, Spain: Centro Internacional de Métodos Numéricos en Ingeniería, Universitat Politècnica de Catalunya.

Caner, Ferhun C. and Bažant, Zdeněk P. 2013a. "Microplane model M7 for plain concrete. I: Formulation." *Journal of Engineering Mechanics* 139 (12): 1714–1723.

Caner, Ferhun C. and Bažant, Zdeněk P. 2013b. "Microplane model M7 for plain concrete. II: Calibration and verification." *Journal of Engineering Mechanics* 139 (12): 1724–1735.

Cedolin, Luigi and Bažant, Zdeněk P. 1980. "Effect of finite element choice in blunt crack band analysis." *Computer Methods in Applied Mechanics and Engineering* 24 (3): 305–316.

Červenka, Jan, Bažant, Zdeněk P. and Wierer, Martin. 2005. "Equivalent localization element for crack band approach to mesh-sensitivity in microplane model." *International Journal for Numerical Methods in Engineering* 62 (5): 700–726.

Chang, Ze, Liang, Minfei, Xu, Yading, Schlangen, Erik and Šavija, Branko. 2022. "3D concrete printing: Lattice modeling of structural failure considering damage and deformed geometry." *Cement and Concrete Composites* 133: 104719.

Daneshvar, Dana, Rabiei, Mahsa, Gupta, Shashank, Najmeddine, Aimane, Prihar, Arjun and Moini, Reza. 2024. "Geometric fidelity of interlocking bodies in two-component robotic additive manufacturing." In *RILEM International Conference on Concrete and Digital Fabrication*, 134–141. Springer.

Das, Arnesh, Sanchez, Asel Maria Aguilar, Wangler, Timothy and Flatt, Robert J. 2022. "Freeze-thaw performance of 3D printed concrete: Influence of interfaces." In *RILEM International Conference on Concrete and Digital Fabrication*, 200–205. Springer.

Dobrzanski, James, Buswell, Richard, Cavalaro, Sergio, Kinnell, Peter, Wang, Weiqiang, Xu, Jie and Kolawole, John. 2022. "Milling a cement-based 3D printable mortar in its green state using a ball-nosed cutter." *Cement and Concrete Composites* 125: 104266.

Dobrzanski, James, Xu, Jerry, Bartek, Rasti, et al. 2025. "From digital crafting to digital manufacturing: Automated production using hybrid 3D concrete printing." *Journal of Building Engineering* 108: 112640.

Dong, Liang, Wu, Chengqing, Liu, Zhongxian, Wu, Pengtao, Shao, Ruizhe, Ren, Quanchang and Liu, Jiaojian. 2025. "Chloride transport anisotropy and interfacial degradation in 3D-printed ultra-high performance concrete: Multi-scale evaluation and engineering implications." *Construction and Building Materials* 491: 142722.

Elices, M., Rocco, C. and Roselló, C. 2009. "Cohesive crack modelling of a simple concrete: Experimental and numerical results." *Engineering Fracture Mechanics* 76 (10): 1398–1410.

Erdogan, F. 1983. "Stress intensity factors." *Journal of Applied Mechanics*, 992–1002.

Gebhard, Lukas, Burger, Joris, Mata-Falcón, Jaime, Lloret Fritschi, Ena, Gramazio, Fabio, Kohler, Matthias and Kaufmann, Walter. 2022. "Towards efficient concrete structures with ultra-thin 3D printed formwork: Exploring reinforcement strategies and optimisation." *Virtual and Physical Prototyping* 17 (3): 599–616.

Gebhard, Lukas, Mata-Falcón, Jaime, Anton, Ana, Dillenburger, Benjamin and Kaufmann, Walter. 2021. "Structural behaviour of 3D printed concrete beams with various reinforcement strategies." *Engineering Structures* 240: 112380.

Ghantous, Rita Maria, Evseeva, Anastasiia, Dickey, Brandon, Gupta, Shashank, Prihar, Arjun, Esmaeeli, Hadi Shagerdi, Moini, Reza and Weiss, W Jason. 2023. "Examining effect of printing directionality on freezing-and-thawing response of three-dimensional-printed cement paste." *ACI Materials Journal* 120 (4).

Ghourchian, Sadegh, Butler, Marko, Krüger, Markus and Mechtcherine, Viktor. 2021. "Modelling the development of capillary pressure in freshly 3D-printed concrete elements." *Cement and Concrete Research* 145: 106457.

Guinea, G. V., Elices, M. and Planas, J. 1996. "Stress intensity factors for wedge-splitting geometry." *International Journal of Fracture* 81: 113–124.

Gupta, Shashank, Esmaeeli, Hadi S. and Moini, Reza. 2024. "Tough and ductile architected nacre-like cementitious composites." *Advanced Functional Materials* 34 (39): 2313516.

Gupta, Shashank, Esmaeeli, Hadi S., Prihar, Arjun, Ghantous, Rita M., Weiss, W. Jason and Moini, Reza. 2023. "Fracture and transport analysis of heterogeneous 3D-printed lamellar cementitious materials." *Cement and Concrete Composites* 140: 105034.

Gupta, Shashank and Moini, Reza. 2024. "Tough cortical bone-inspired tubular architected cement-based material with disorder." *Advanced Materials* 36 (52): 2313904.

Gupta, Shashank, Weiss, W. Jason, Ghantous, Rita M. and Moini, Reza. 2026. "Pore connectivity and arrangement in additively manufactured cementitious material." *Cement and Concrete Research* 204: 108206.

Heever, Marchant van den, Bester, Frederick, Kruger, Jacques and Zijl, Gideon van. 2022. "Numerical modelling strategies for reinforced 3D concrete printed elements." *Additive Manufacturing* 50: 102569.

Hillerborg, Arne, Modéer, Mats and Petersson, P-E. 1976. "Analysis of crack formation and crack growth in concrete by means of fracture mechanics and finite elements." *Cement and Concrete Research* 6 (6): 773–781.

Irwin, George R. 1957. "Analysis of stresses and strains near the end of a crack traversing a plate." *Journal of Applied Mechanics* 24 (3): 361–364. https://doi.org/10.1115/1.4011547.

Ishiguro, S. 2007. "Experiments and analyses of fracture properties of grouting mortars." In *Proceedings of the 6th International Conference on Fracture Mechanics of Concrete and Concrete Structures (FraMCoS-6)*, 47–57. Catania, Italy, June.

Jenq, Yeoushang and Shah, Surendra P. 1985a. "Two parameter fracture model for concrete." *Journal of Engineering Mechanics* 111 (10): 1227–1241.

Jenq, Yeoushang and Shah, Surendra P. 1985b. "A fracture toughness criterion for concrete." *Engineering Fracture Mechanics* 21 (5): 1055–1069.

Karihaloo, B. L. and Nallathambi, P. 1991. *Notched Beam Test: Mode I Fracture Toughness*. Vol. 89. London: Chapman & Hall.

Knott, John Frederick. 1973. *Fundamentals of Fracture Mechanics*. Gruppo Italiano Frattura.

Li, Shuai, Liu, Junli, Cheng, Chi-Tsun, Xuan, Hung Nguyen and Tran, Phuong. 2025. "Design and performance of 3D printed Bouligand steel fiber-reinforced cementitious composite curved beams." *Cement and Concrete Composites* 163: 106188.

Liu, Junli, Li, Shuai, Fox, Kate and Tran, Phuong. 2022. "3D concrete printing of bioinspired Bouligand structure: A study on impact resistance." *Additive Manufacturing* 50: 102544.

Lowke, Dirk, Dini, Enrico, Perrot, Arnaud, Weger, Daniel, Gehlen, Christoph and Dillenburger, Benjamin. 2018. "Particle-bed 3D printing in concrete construction – Possibilities and challenges." *Cement and Concrete Research* 112: 50–65.

Lowke, Dirk, Freund, Niklas, Böhler, David and Herding, Friedrich. 2024. *Fourth RILEM International Conference on Concrete and Digital Fabrication: Digital Concrete 2024*. Vol. 53. Springer Nature.

Lowke, Dirk, Vandenberg, Aileen, Pierre, Alexandre, Thomas, Amaury, Kloft, Harald and Hack, Norman. 2021. "Injection 3D concrete printing in a carrier liquid – Underlying physics and applications to lightweight space frame structures." *Cement and Concrete Composites* 124: 104169.

Luo, Wen and Bažant, Zdeněk P. 2017. "Fishnet statistics for probabilistic strength and scaling of nacreous imbricated lamellar materials." *Journal of the Mechanics and Physics of Solids* 109: 264–287.

Ma, Guowei, Buswell, Richard, Silva, Wilson Ricardo Leal da, Wang, Li, Xu, Jie and Jones, Scott Z. 2022. "Technology readiness: A global snapshot of 3D concrete printing and the frontiers for development." *Cement and Concrete Research* 156: 106774.

Mobasher, Barzin. 2011. *Mechanics of Fiber and Textile Reinforced Cement Composites*. CRC Press.

Moini, Mohamadreza. 2020. "Buildability and mechanical performance of architectured cement-based materials fabricated using a direct-ink-writing process." PhD diss., Purdue University.

Moini, Mohamadreza, Olek, Jan, Magee, Bryan, Zavattieri, Pablo and Youngblood, Jeffrey. 2018. "Additive manufacturing and characterization of architectured cement-based materials via X-ray micro-computed tomography." In *RILEM International Conference on Concrete and Digital Fabrication*, 176–189. Springer.

Moini, Mohamadreza, Olek, Jan, Youngblood, Jeffrey P., Magee, Bryan and Zavattieri, Pablo D. 2018. "Additive manufacturing and performance of architectured cement-based materials." *Advanced Materials* 30 (43): 1802123.

Moini, Reza. 2023. "Perspectives in architected infrastructure materials." *RILEM Technical Letters* 8: 125–140.

Moini, Reza, Baghaie, Ahmadreza, Rodriguez, Fabian B., Zavattieri, Pablo D., Youngblood, Jeffrey P. and Olek, Jan. 2021. "Quantitative microstructural investigation of 3D-printed and cast cement pastes using micro-computed tomography and image analysis." *Cement and Concrete Research* 147: 106493.

Moini, Reza and Makinen, William. 2023. *Multi-material Concrete 3D Printing with Thermoplastic and Elastomeric Polymers*. US Patent App. 18/122,432, September.

Moini, Reza, Olek, Jan, Zavattieri, Pablo D. and Youngblood, Jeffrey P. 2021. "Open-span printing method for assessment of early-age deformations of additively manufactured cement-based materials using an isosceles triangle." In *ASTM Symposium on Standards Development for Cement and Concrete Used in Additive Construction*, edited by Jones, S. Z. and Kreiger, E. L., 1–12. West Conshohocken, PA: ASTM International. https://doi.org/10.1520/STP163620200089.

Moini, Reza, Olek, Jan, Zavattieri, Pablo D. and Youngblood, Jeffrey P. 2022. "Early-age buildability – Rheological properties relationship in additively manufactured cement paste hollow cylinders." *Cement and Concrete Composites* 131: 104538.

Moini, Reza, Rodriguez, Fabian, Olek, Jan, Youngblood, Jeffrey P. and Zavattieri, Pablo D. 2024. "Mechanical properties and fracture phenomena in 3D-printed helical cementitious architected materials under compression." *Materials and Structures* 57 (7): 170.

Nair, Sooraj A. O., Tripathi, Avinaya and Neithalath, Narayanan. 2021. "Examining layer height effects on the flexural and fracture response of plain and fiber-reinforced 3D-printed beams." *Cement and Concrete Composites* 124: 104254.

Najmeddine, Aimane, Gupta, Shashank and Moini, Reza. 2025. "Coupled large deformation phase-field and cohesive zone model for crack propagation in hard-soft multi-materials." *Journal of the Mechanics and Physics of Solids* 196: 106016.

Najmeddine, Aimane and Moini, Reza. 2025. "Role of the weak interface on toughening in layered brittle materials: A coupled phase field-cohesive zone model approach to fracture." In *12th International Conference on Fracture Mechanics of Concrete and Concrete Structures, FraMCoS-12*, 1–12. Vienna, Austria: IA-FraMCoS. https://doi.org/10.21012/FC12.1164.

Ojdrovic, Rasko and Petroski, Henry J. 1985. "The cracked Brazilian specimen for fracture toughness testing of concrete." *International Journal of Fracture* 27: R75–R80.

Pan, Jinlong, Ping, Pengxin, Ding, Boyin, Zhu, Binrong, Lin, Yuanzheng, Ukrainczyk, Neven, Zhang, Hong and Cai, Jingming. 2024. "Impact behaviour of 3D printed fiber reinforced cementitious composite beams." *Composites Part A: Applied Science and Manufacturing* 182: 108175.

Pathirage, Madura, Pijaudier-Cabot, Gilles, Grégoire, David and Cusatis, Gianluca. 2025. "Transformation and regression statistics of the size-effect method for determining fracture energy and process zone size in quasi-brittle materials." *Materials and Structures* 58 (1): 39.

Patzák, Bořek. 2012. "OOFEM – An object-oriented simulation tool for advanced modeling of materials and structures." *Acta Polytechnica* 52 (6).

Petersson, Per-Erik. 1981. Crack growth and development of fracture zones in plain concrete and similar materials. Doctoral thesis, Division of Building Materials, Lund Institute of Technology (LTH), Lund University, Lund, Sweden.

Planas, J., Guinea, G. V. and Elices, M. 1999. "Size effect and inverse analysis in concrete fracture." *International Journal of Fracture* 95: 367–378.

Prihar, Arjun, Garlock, Maria E. M., Najmeddine, Aimane and Moini, Reza. 2024. "Mechanical performance of sinusoidally architected concrete enabled by robotic additive manufacturing." *Materials & Design* 238: 112671.

Prihar, Arjun, Gupta, Shashank, Esmaeeli, Hadi S. and Moini, Reza. 2024. "Tough double-Bouligand architected concrete enabled by robotic additive manufacturing." *Nature Communications* 15 (1): 7498.

Rabiei, Mahsa and Moini, Reza. 2025. "Extrusion under material uncertainty with pressure-based closed-loop feedback control in robotic concrete additive manufacturing." *Automation in Construction* 180: 106494.

Ralston, Nadia, Gupta, Shashank and Moini, Reza. 2024. "3D-printing of architected calcium silicate binders with enhanced and in-situ carbonation." *Virtual and Physical Prototyping* 19 (1): e2350768.

Ramesh, Akilesh, Rajeev, Pathmanathan, Xu, Shanqing, Sanjayan, Jay and Lu, Guoxing. 2024. "Impact response of textile-reinforced 3D printed concrete panels." *Engineering Structures* 315: 118489.

RILEM. 1985. "Determination of the fracture energy of mortar and concrete by means of three-point bend tests on notched beams." *Materials and Structures* 18 (106): 285–290.

Ritchie, Robert O. and Liu, Dong. 2021. *Introduction to Fracture Mechanics*. Elsevier.

Rocco, C., Guinea, G. V., Planas, J. and Elices, M. 2001. "Review of the splitting-test standards from a fracture mechanics point of view." *Cement and Concrete Research* 31 (1): 73–82.

Rodriguez, Fabian B., Moini, Reza, Agrawal, Shubham, Williams, Christopher S., Zavattieri, Pablo D., Olek, Jan, Youngblood, Jeffrey P. and Varma, Amit H. 2024. "Mechanical response of small-scale 3D-printed steel-mortar composite beams." *Cement and Concrete Composites* 154: 105795.

Rodriguez, Fabian B., Olek, Jan, Moini, Reza, Zavattieri, Pablo D. and Youngblood, Jeffrey P. 2021. "Linking solids content and flow properties of mortars to their three-dimensional printing characteristics." *ACI Materials Journal* 118 (6): 371–382.

Rupp, Marco, Hayashi, Shuichiro, Dashe, Claire, Gupta, Shashank, Moini, Reza and Arnold, Craig B. 2023. "Continuous wave laser grooving of cementitious materials." *Applied Physics A* 129 (2): 160.

Salviato, Marco, Kirane, Kedar, Bažant, Zdeněk P. and Cusatis, Gianluca. 2019. "Mode I and II interlaminar fracture in laminated composites: A size effect study." *Journal of Applied Mechanics* 86 (9): 091008.

Savija, Branko. 2020. "Use of 3D printing to create multifunctional cementitious composites: Review, challenges and opportunities." *RILEM Technical Letters* 5: 16–25.

Shah, S. P. 1990a. "Size-effect method for determining fracture energy and process zone size of concrete." *Materials and Structures* 23 (1): 461–465.

Shah, S. P. 1990b. "Determination of fracture parameters (KIcs and CTODc) of plain concrete using three-point bend tests." *Materials and Structures* 23 (6): 457–460.

Shah, S. P. and Ouyang, C. 1994. "Fracture mechanics for failure of concrete." *Annual Review of Materials Science* 24 (1): 293–320.

Sneddon, Ian Naismith. 1946. "The distribution of stress in the neighbourhood of a crack in an elastic solid." *Proceedings of the Royal Society of London. Series A. Mathematical and Physical Sciences* 187 (1009): 229–260.

Sorelli, Luca, Banthia, Nemkumar, Plizzari, Giovanni A., et al. 2014. "Crack growth resistance of thin mortar layers with hybrid fiber reinforcement." *American Concrete Institute: Farmington Hills, MI, USA. Spec. Publ* 224: 161–178.

Steenkamp, P. A. J. M., Read, D. T. and Reed, R. P. 1988. *JR-Curve Testing of Three-Point Bend Specimens by the Unloading Compliance Method.* ASTM International.

Subramaniam, Kolluru VL, Paritala, Spandana, Kulkarni, Omkar and Thakur, Manideep S. 2024. "Fracture in 3D-printed concrete beams: Deflection and penetration of impinging cracks at layer interfaces." *Journal of Engineering Mechanics* 150 (12): 04024093.

Suidan, Mubbada and Schnobrich, William C. 1973. "Finite element analysis of reinforced concrete." *Journal of the Structural Division* 99 (10): 2109–2122.

Surehali, Sahil, Tripathi, Avinaya, Nimbalkar, Atharwa Samir and Neithalath, Narayanan.2023. "Anisotropic chloride transport in 3D printed concrete and its dependence on layer height and interface types." *Additive Manufacturing* 62: 103405.

Tang, Yuxiang, Xiao, Jianzhuang, Ding, Tao, Liu, Haoran, Zhang, Mingzhong and Zhang, Junhui. 2024. "Trans-layer and inter-layer fracture behavior of extrusion-based 3D printed concrete under three-point bending." *Engineering Fracture Mechanics* 296: 109836.

Thakur, Manideep Singh, Kulkarni, Omkar, Kamakshi, Tippabhotla A., Paritala, Spandana and Subramaniam, Kolluru V. L. 2023. "Influence of cold joint on fracture behaviour of 3D printed concrete." In *Interdisciplinary Symposium on Smart & Sustainable Infrastructures*, 194–202. Springer.

Tomholt, Lara, Meggers, Forrest and Moini, Reza. 2024. "3D-printing channel networks with cement." In *Fourth RILEM International Conference on Concrete and Digital Fabrication: Digital Concrete 2024*, 53:74. Springer Nature.

Tong, J., Wong, K. Y. and Lupton, C. 2007. "Determination of interfacial fracture toughness of bone – Cement interface using sandwich Brazilian disks." *Engineering Fracture Mechanics* 74 (12): 1904–1916.

Van Der Putten, Jolien, De Volder, Melissa, Van den Heede, Philip, Deprez, Maxim, Cnudde, Veerle, De Schutter, Geert and Van Tittelboom, Kim. 2022. "Transport properties of 3D printed cementitious materials with prolonged time gap between successive layers." *Cement and Concrete Research* 155: 106777.

Venu, Harish, Murali, G., Ramkumar, V. R., Kiong, Tiong Sieh and Razali, Nur Mazlini. 2024. "Assessment of fracture toughness in fibrous concrete via Brazilian test: Experimental study with low-clinker cementitious binder." *Theoretical and Applied Fracture Mechanics* 133: 104641.

Wang, Qiang, Yang, Wenwei, Wang, Li, Bai, Gang and Ma, Guowei. 2025. "Reinforcement design and structural performance for the topology optimized 3D printed concrete truss beams." *Engineering Structures* 332: 120064.

Wang, Yusuo, Hu, Xiaozhi, Liang, Li and Zhu, Wancheng. 2016. "Determination of tensile strength and fracture toughness of concrete using notched 3-pb specimens." *Engineering Fracture Mechanics* 160: 67–77.

Westergaard, Harold M. 1939. "Bearing pressures and cracks: Bearing pressures through a slightly waved surface or through a nearly flat part of a cylinder, and related problems of cracks." *Journal of Applied Mechanics* 6 (2): A49–A53. https://doi.org/10.1115/1.4008919.

Xiao, Jianzhuang, Schneider, Holger, Dönnecke, Cindy and König, Gert. 2004. "Wedge splitting test on fracture behaviour of ultra high strength concrete." *Construction and Building Materials* 18 (6): 359–365.

Xie, Jinbao, He, Shan, Xu, Yading, Meng, Zhaozheng, Zhou, Wen, Schlangen, Erik and Šavija, Branko. 2025. "Enhanced elastomer-like auxetic cementitious materials through strain-hardening cementitious composites (SHCC) with extended softening properties." *Cement and Concrete Composites* 161: 106069.

Xie, Jinbao, Xu, Yading, Meng, Zhaozheng, Liang, Minfei, Zhou, Yubao and Šavija, Branko. 2025. "Impact behavior of auxetic cementitious cellular composites (ACCCs) architected through additive manufacturing (AM) assisted casting: Experiment and modelling." *Construction and Building Materials* 471: 140692.

Ye, Junhong, Yang, Minxin, Yu, Jiangtao, Dai, Yecheng, Yin, B. B. and Weng, Yiwei. 2024. "Size effect on flexural and fracture behaviors of 3D printed engineered cementitious composites: Experimental and numerical studies." *Engineering Structures* 298: 117062.

Zhang, Chao, Ren, Juanjuan, Zhang, Shuhao, Guo, Yipu, Li, Ning, Li, Wengui and Qu, Fulin. 2025. "Advanced impact resistance design through 3D-printed concrete technology: Unleashing the potential of additive manufacturing for protective structures." *Journal of Building Engineering* 111: 113533.

Zhao, Yanhua, Liu, Yang and Xu, Bohan. 2021. "Determination of crack extension resistance of concrete from boundary effect model." *Engineering Fracture Mechanics* 244: 107562.

Zhou, Wen, Bol, Rowin JM, Zhou, Yubao, Meng, Zhaozheng, Xu, Yading, Xie, Jinbao and Šavija, Branko. 2025. "Reinforcing mechanism of lattice-reinforced cementitious composites: Insight into flexural performance and material interactions." *Materials & Design* 256: 114332.

Zhou, Wen, Mcgee, Wes, Gökçe, H. Süleyman and Li, Victor C. 2023. "A bio-inspired solution to alleviate anisotropy of 3D printed engineered cementitious composites (3DP-ECC): Knitting/tilting filaments." *Automation in Construction* 155: 105051.

Zhou, Wen, Xu, Yading, Meng, Zhaozheng, Xie, Jinbao, Zhou, Yubao, Schlangen, Erik and Šavija, Branko. 2025. "Filament stitching: An architected printing strategy to mitigate anisotropy in 3D-printed engineered cementitious composites (ECC)." *Cement and Concrete Composites* 160: 106044.

Zhu, Xian-Kui and Joyce, James A. 2012. "Review of fracture toughness (G, K, J, CTOD, CTOA) testing and standardization." *Engineering Fracture Mechanics* 85: 1–46.

4 Durability of 3D-Printed Concrete

Mohammed Alnaggar and Tathagata Bhaduri

4.1 INTRODUCTION

Durability is one of the key aspects for quality control in the construction of concrete structures. As 3D printing of concrete (3DPC) is emerging as a transformative construction method, it is important to predict possible durability issues for the sustainability of 3D-printed built infrastructure.

Typically, 3DPC includes concrete with fine aggregates and chemical admixtures, excluding coarse aggregate particles for the rheological needs of the printing process. Thus, the material's pore structure is different from that of traditional concrete. The printing process involving the extrusion and deposition of concrete by layers can also create porous interfaces, which are a significant challenge from a durability perspective since they can reduce local material resistance to environmental vulnerabilities. Furthermore, in the absence of formwork, shrinkage can impact the development of interlayer bonding and overall mechanical properties.

While research for developing printable concrete mix and 3D-printing methods has gained momentum, the investigation of long-term durability is still limited. The current literature base primarily focuses on the use of supplementary cementitious materials and additives like fibers and nanoparticles, or adjusting printing variables to mitigate identified durability challenges. This chapter will review such durability challenges and discuss the existing and future potential solutions to provide foundations for the sustainability of 3DPC in construction.

4.2 DURABILITY ASPECTS OF 3DPC

This section reviews the fundamentals of durability considerations in traditional concrete and extrapolates the idea of 3DPC. The American Concrete Institute (ACI) defines durability as "the ability of a material to resist weathering action, chemical attack, abrasion, and other conditions of service" ("Durability of Concrete Topic" 2022). Based on this definition, durability can be broadly subdivided into chemical durability and physical durability.

Chemical durability refers to the ability of concrete to resist various chemical attacks, such as alkali–silica reaction (ASR), acid attack, sulfate attack, and carbonation. In general, quality control of concrete to various chemical durability aspects is done at the material mix design level. The primary difference for 3DPC is the presence of interlayer (and sometimes intra-layer) voids, which increase the permeability of the material and thus reduce its chemical durability.

DOI: 10.1201/9781003434863-5

Physical durability provides a measure of the ability of concrete to resist environmental conditions, such as temperature and relative humidity changes, which induce volumetric changes that can lead to cracking if they are restrained. So, physical durability is linked with the resistance to freeze–thaw, wetting-drying, and other weathering effects like abrasion resistance. Similar to chemical durability, most of the quality control can be done at the mix design level and is not different for 3DPC. However, the extrusion process in 3DPC can result in varying the void structure and content, which may introduce large pores with interlayer air voids (Yu et al. 2021), as shown in Figure 4.1a. Also, compared to conventional casting, 3DPC exhibits higher porosity with increasing pore diameter size (Yu et al. 2021), as shown in Figure 4.1b. Similar issues with extrusion-based 3DPC were reported in other literature (Nerella, Hempel, and Mechtcherine 2019; Skibicki et al. 2022; Van Der Putten et al. 2019; Sikora et al. 2022; Nodehi et al. 2022). Another major issue here is the surface texture of the 3D-printed structure and its effects on stress distribution. As in many other 3DPC applications, stress concentrations can reduce fatigue life.

Based on the above discussion, predicting and controlling the durability features of 3DPC is more complex than that of traditional concrete structures due to the unique manufacturing process and structural topology. A major source of complexity is the interplay of chemical and physical processes with mechanical properties during and after 3D printing and their dependence on 3D printing process parameters. Another important aspect for 3DPC is the fact that it is loaded before solidification, and that raises fundamental questions and quality control challenges because fluid deformations do not typically affect long-term deformation, while solid deformations do. A key challenge here is to separate fluid behavior (plastic flow) from solid behavior (creep and damage) and understand their transition, which is still an open research subject today. If creep deformations are neglected, 3DPC load-induced cracking can be controlled by maintaining principal tensile and shear stresses within their time-dependent increasing tensile and shear strengths, respectively. However, creep deformations add an extra layer of complexity due to the possibility of load redistribution leading to possible creep buckling and/or nonlinear creep damage. Since

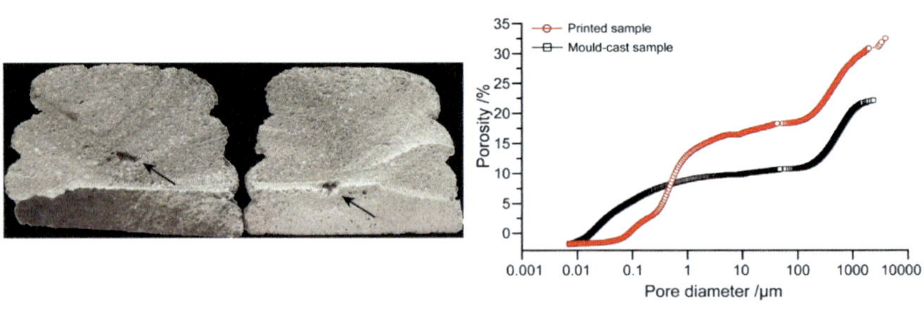

a) b)

FIGURE 4.1 (a) Air voids at the interlayer zones of 3DPC; (b) porosity vs. pore diameter/μm plot for printed and mold-cast samples (Yu et al. 2021).

creep in concrete is a function of its composition (chemistry), the surrounding environmental conditions (physics), and the state of stresses induced by the service load conditions (mechanics), creep can be seen as one of the chemo-physical durability problems.

Given these considerations, the following section reviews three major durability aspects influenced by the 3DPC process and structural topology: shrinkage, creep, and permeability.

4.3 SHRINKAGE CONSIDERATIONS IN 3DPC

4.3.1 Types of Shrinkage

Shrinkage in general is a result of two major phenomena. The first is chemical, and the second is physical. Chemical shrinkage refers to the fact that the volume of hydration products is less than the volume of reactants. Therefore, depending on the hydration degree, the water and cementitious materials that are involved in the reaction produce C-S-H gel that is smaller in volume. Simultaneously, a physical shrinkage phenomenon occurs at various length scales due to the formation of surface tension forces within the material's internal structure whenever a fully saturated pore becomes partially saturated. The driving mechanism of physical shrinkage is any consumed water due to hydration, as well as any water that can be lost due to exposure to the environment through evaporation. Since perfect separation between these phenomena is difficult, shrinkage types are instead defined based on material age and boundary conditions, as discussed below.

Plastic shrinkage is the total shrinkage experienced during the plastic material stage (before hardening). If the material is sealed for moisture loss, the plastic shrinkage is almost equal to the chemical shrinkage. If moisture is lost in unsealed conditions, plastic shrinkage includes both chemical shrinkage and volumetric loss from evaporation. For 3DPC, this evaporation process is more significant than in conventional construction due to the large, exposed surface area at an early age.

After solidification, two other types of shrinkage are defined. Autogenous shrinkage is defined as the shrinkage occurring in the absence of moisture exchange (ACI Committee 209 2008). This is the macroscopically measured volume change occurring due to chemical shrinkage and water loss in the pores due to consumption in the hydration process. For autogenous shrinkage, the material must not lose or gain any moisture, so it must be sealed. Drying shrinkage, then, could be defined as shrinkage occurring in a specimen that is allowed to dry (ACI Committee 209 2008). Here, it includes both the chemical and physical shrinkage phenomena described above. However, some researchers refer to this total shrinkage minus autogenous shrinkage. Readers need to carefully look at the definition used in the literature before making conclusions. In 3DPC, drying shrinkage is expected to be higher due to the larger exposed area, too, unless watering is done periodically to reduce it. It is also important to remember that since water loss induces self-equilibrated internal stresses, creep is also triggered, and therefore, creep and shrinkage are coupled. This coupling becomes significant in drying conditions.

4.3.2 Shrinkage in 3DPC

Different studies have evaluated shrinkage in 3DPC. All studies agree that shrinkage in 3DPC is higher than that of regular concrete. To understand the relation between the microstructure characteristics and observed shrinkage in 3D-printable pastes, a study compared a control paste with no additives with two pastes that had increasing amounts of microsilica and superplasticizer (Slavcheva 2019). In this study, samples were first cured in the molds for 24 hours and then exposed to different relative humidity (RH) values between 20% and 70%. The samples' weight loss and shrinkage strains were measured over time, along with their pore structure characteristics. In general, microsilica improves the ability of the paste to retain water as it increases pores smaller than 20 nm and generally results in a lower total pore volume. This is because the finer the pores, the harder it is to evaporate the water contained in them. Mass loss measurements confirmed this relation and showed that at any RH, the finer the pores (more microsilica content), the lower the mass loss. For example, at 20% RH, the control sample lost 12.3% of its mass, while MB-30 (microsilica-to-cement ratio of 0.3) lost only 5.1%. At 70% RH, the control lost 5.6% of its mass, while MB-30 lost only 1.1%. Nevertheless, microsilica increases the shrinkage coefficient considerably. The study showed that the fitted linear shrinkage coefficient for MB-30 is almost twice as high as compared to the control sample (0.036 compared to 0.017). As a result, the reduction in total shrinkage is not as efficient as the reduction in water loss, and still, considerable shrinkage is present. It is important to note here that the initial plastic shrinkage during the first 24 hours was not recorded, and as will be discussed later, this amount of shrinkage is very significant compared to the later one.

In another study, the effects of different temperatures and RH conditions on the shrinkage behavior of 3D printing mortars were assessed by comparing control samples initially sealed for 24 hours and then cured at 24°C at 50% RH with other free-formed unsealed samples that were exposed to different conditions from the onset of casting (24–35°C and 45%–85% RH) (Shahmirzadi et al. 2021). Different sand-to-cement ratios and different surface-to-volume ratios were considered in the study, but in all cases, only one layer was made, so the interlayer effects were not observed. For the first 24 hours, the sealed curing had a considerable impact on increasing both the compressive and flexural strength of the sealed samples compared to that of all unsealed samples. At the same temperature and RH conditions, unsealed samples showed 61% lower compressive strength and 55% lower flexural strength compared to sealed ones. These reductions were attributed to the water loss, which caused the exposed free-formed samples to have a lower degree of hydration (21.3%–23%) compared to the reference ones (34.7%–35%). Also, water loss increased microporosity and the presence of microcracking. An important observation in this study is the significant amount of mass loss during the first day for the free-formed samples, which was followed by mass gains afterward as the hydration process continued. At the same RH conditions, the total mass loss of free-formed samples became similar to that of the reference ones at around 168 days. This study demonstrates the effect of exposure conditions on hydration and moisture evaporation, which affect shrinkage and deviate from typical concrete performance. Sealed curing for the first 24 hours enables sufficient hydration to reduce evaporation rates and strengthen the material, allowing it to better withstand tensile stresses from subsequent drying shrinkage. Such importance of the first 24-hour period was further highlighted by another novel study to monitor the shrinkage of 3D-printed filaments from the onset of printing onward (Federowicz et al. 2020). The method depends on using laser gauge measurements of the displacement of reference points added to the ends of the filament immediately after printing it. This novel method was used to evaluate the shrinkage of 3D-printable mortars with and without shrinkage-reducing agents (SRAs). Two reference filaments and two other filaments with two different doses of an SRA were tested. One of the reference filaments was covered with foil to prevent surface evaporation so the effects of exposure could be evaluated. Results show that for all samples, a significant shrinkage was recorded within the first day. Exposed samples developed 95% of their 7-day shrinkage only within the first 24 hours, while the foil-covered sample developed only 59% of its 7-day shrinkage within the first 24 hours. Also, shrinkage magnitudes varied significantly, with the reference filament reaching 5,178 μm/m compared to 4,795 and 3,976 μm/m for the two filaments with the SRA. These numbers are four to five times higher than the total shrinkage of the foil-covered filament at 7 days, which reached 1,031 μm/m, although the foil-covered one did not contain any SRA. To evaluate the effect of restraints and multiple layers' interaction, another study investigated 3D-printed mortar walls with and without restraints (Moelich, Kruger, and Combrinck 2020). In this study, digital image correlation (DIC) was used to develop the displacement maps of the walls made from multiple layers. Also, free shrinkage was compared to restrained shrinkage using the ASTM C1579 plastic shrinkage cracking test and the ring test. This study highlighted the occurrence of early-age interlayer slip in the shrinkage transfer from one layer to the other, which is unique to 3DPC. This interlayer slip (observed during demolding, as shown in Figure 4.2) can contribute to reduced interlayer bond strength and durability; hence, more research is needed in this area.

FIGURE 4.2 Cracking along interlayer planes due to differential displacement under a restraint combination (Moelich, Kruger, and Combrinck 2020).

4.3.3 Creep Considerations in 3DPC

Creep is defined as a property of solids that behave in a viscoelastic/viscoplastic manner under sustained loading. However, 3DPC undergoes time-dependent deformation before solidification during the printing process, creating unique challenges. Currently, no fundamental unifying theory exists that describes the stress-induced time-dependent deformation of a solidifying 3DPC material. Therefore, existing studies primarily provide phenomenological observations and empirical quantifications rather than unified physical models applicable across the fresh to hardened concrete transition. For example, Beersaerts, Lucas, and Pontikes (2020) compared commercial 3DPC mixes with Fe-rich slag-based mortars under various conditions. While they reported elastic modulus evolution within the first 24 hours using ultrasonic measurements, creep comparisons were performed on solid samples that were cured for the first 8 days and then loaded. Results showed that Fe-rich slag-based mortars had lower creep coefficients than commercial mixes after 8 days of curing. However, since testing was performed on fully hardened samples, these findings do not completely address the critical early-age creep behavior during printing that affects buildability and stability. This study suggests a typical solid concrete creep test that is only relevant to 3DPC as a guide to qualitatively identify which material creeps more. Another study evaluated the creep behavior of cured samples from 3D-printed bridge segments at both 7 and 28 days from printing (Kinomura et al. 2020). The tests are done to determine creep coefficients only, and as expected, the creep coefficient for the 7-day loading was higher than that of the 28-hour loading. Also, this is an indicative study of the quality of the 3D-printed material and is not intended for 3D printing process analysis. Similarly, other researchers have evaluated the creep performance of hardened 3DPC at different loading ages after printing (Blaakmeer and Lobo 2020; Salet et al. 2018).

For early-age creep deformations of a 3D-printable mortar, one of the notable works was performed by Casagrande et al. (2020). This work determined the time-dependent evolution of compressive strength, elastic modulus, and creep strain over 60 minutes. Compression tests were conducted under displacement control up to a vertical strain of 12%, and the elastic modulus was determined as the secant modulus between 0% and 2% strains. Note here that this modulus is not purely elastic and is called an apparent elastic modulus from a creep point of view. This is because, theoretically, the modulus will always be a function of the loading rate, and that is why ultrasonic measurements typically show a higher elastic modulus compared to the one measured in compression tests. Also, for the early-age creep tests, the authors evaluated creep deformations at different ages (t = 0, 15, 30, and 60 minutes) by imposing 8 N for 900 seconds (including a short ramp time). As expected, strength and apparent elastic modulus increased over time, while creep strains decreased. Nevertheless, it is obvious from the shown curves that the observed deformation is more like a viscous deformation (relevant to fluid behavior) with a clear plateau rather than a continuous creep deformation (relevant to solid behavior). However, an interesting observation is made by the authors, which may indicate a transition from fluid to solid behavior. In the creep test, they determined the strain at the end of the loading ramp and termed it the instantaneous strain, then determined the creep strain as the remaining part, and then found the ratio. At $t = 0$ minutes, the ratio was only 5%, while at 60 minutes, it reached 12.57%. This means that a solid-like creep behavior is starting to show over time, especially because at 60 minutes, the material is obviously stronger, and thus, its elastic deformation under loading should be smaller, but that reduction was not the same in creep. Nevertheless, the results in this work should also be taken as qualitative, given the variations in ramp times between different ages and the fact that the results will depend on the mortar type, its strength evolution, and the loading level and rate. Another interesting study (Chen et al. 2020) used a shear stress-controlled protocol in their rheological measurements to evaluate creep deformations at an early age by imposing a first stage of constant shear stress and measuring the shear strain. The study was mainly performed to evaluate the beneficial effects of adding bentonite to calcium sulfoaluminate cement composites. Results showed a reduction in creep deformations with bentonite additions.

Despite experimental efforts to provide valuable insights, direct measurement of deformation during full-scale printing remains challenging. Therefore, numerical modeling approaches have also been developed to understand the early-age time-dependent deformation process during printing. Andersen et al. (2020) developed and presented preliminary results from a creep model based on the multi-Maxwell chain model by Bažant et al. (1997) to simulate time-dependent deformation during printing. This study extended its application phenomenologically to the fresh state by calibrating the model parameters on fresh 3DPC deformation captured using digital image correlation (DIC). While it is not a unifying model, the procedure presented is promising for deducing theoretical creep models by comparing model predictions to 3D printing deformation. Nevertheless, the limitations of finite element simulations forced the researchers to represent the printing process through the addition of segments rather than accurately capturing the full printing process. Another interesting numerical study by Chang et al. (2023) explored the influence of early-age creep on 3DPC buildability with different failure modes (shown in Figure 4.3).

It was shown that with increasing creep compliance (C_0) into the 3DPC numerical model, the critical printed height decreases due to increased deformation, which leads to structural failure (shown in Figure 4.4). For structures prone to elastic buckling, buildability remains largely unaffected, as system stiffness outweighs the

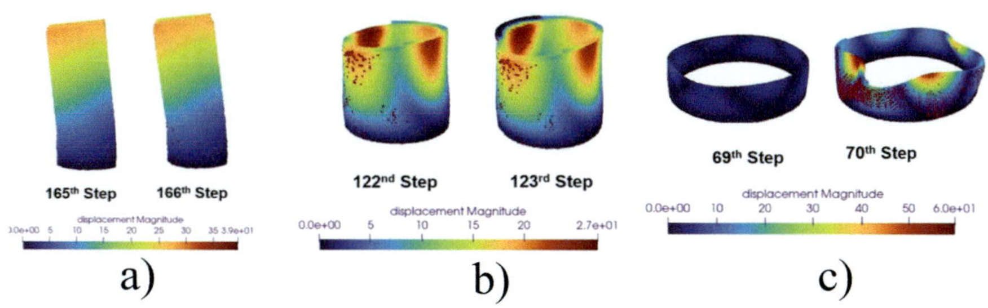

FIGURE 4.3 Different failure modes in 3DPC: (a) plastic collapse; (b) combined failure; (c) elastic buckling (Chang et al. 2023).

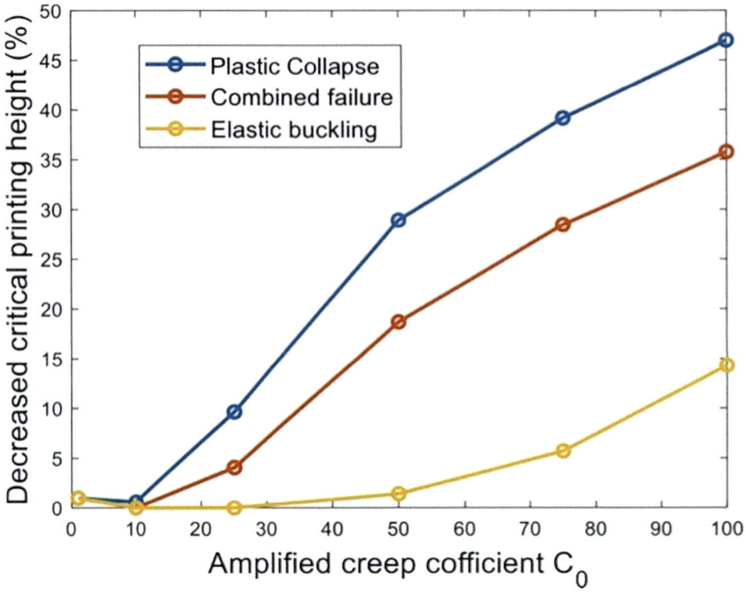

FIGURE 4.4 Decreased critical printing height vs. creep compliance (C_0) for different failure modes (Chang et al. 2023).

impact of additional creep-induced damage. In contrast, structures susceptible to plastic collapse show reduced buildability and a shift in failure mode. Overall, this work quantitatively linked the creep coefficient with critical printing heights and final failure modes.

Thus, from the literature collected here, one can directly conclude that creep deformation during 3D printing is still an open area for researchers to find suitable testing methods and, most importantly, a unifying theory for deformation under loading that can transition from fluid to solid states.

4.3.4 Permeability Considerations in 3DPC

Permeability will be discussed from two aspects here. First, we will discuss the effects of the 3DPC process on the formation and/or changes to the permeability of concrete. Then, we will discuss in general the impact of permeability on durability as it pertains to 3DPC.

4.3.4.1 Effects of the 3DPC process on permeability

As mentioned before, permeability in 3DPC comes from two sources. The first is the intrinsic void structure of the 3DPC material itself, which can be affected to some extent by the pumping/extrusion process. The other source is the geometric shape produced during the printing process, which contains extra voids between filaments and layers. Thus, many researchers have evaluated the permeability and pore structure of various 3DPC materials and methods (Rehman and Kim 2021)Click or tap here to enter text.. An interesting study by Mazhoud et al. (2019) controlled fresh state permeability to design 3D-printable mortars for underwater 3DPC. The study shows the relation between fresh state permeability and washout under water. Another study correlated process parameters to the flexural strength and gas permeability of 3D-printed sand molds (Sivarupan et al. 2020). While the study pertains to sand 3D printing, not concrete, recoating speed (speed at re-coater spreads pre-mixed sand), which correlates to printing speed, and higher resolution printing, which correlates to filament size, are process parameters that affect permeability. In the study, it was shown that slow recoating and high-resolution printing reduce permeability and improve packing density.

Another study investigated the mechanical performance and air-void structure of 3D-printed samples and compared the results with those of cast samples (Che et al. 2021). The study showed that the air-void morphology had a substantial influence on the mechanical performance of hardened 3DPC. In the study, anti-foaming agents (AFAs) were introduced, and their effects were also evaluated. It was shown that the introduction of AFAs improved all mechanical properties of both the fresh and hardened states. Similarly, but for 3DPC, a study investigated the effects of layer height, pass time, and printing speed on the alteration of the microstructure (Kruger, du Plessis, and van Zijl 2021). It was shown that air voids changed from a spherical shape in cast samples to an elongated, ellipsoidal shape in the 3D-printed ones, which may worsen the stress concentration. However, the study is limited to porosity observations, and no mechanical tests were done. Another interesting finding was that vertical and horizontal interlayer porosity in this study were similar, which may contradict the common assumption that horizontal porosity would be lower than vertical porosity. Note that this observation is significantly dependent on the printing parameters and printed geometry, though. Similar investigations were also performed for fiber-reinforced 3DPC by van den Heever et al. (2022). In this study, total defect content; 3D void topology; pore spatial, size, and compactness distributions; and interconnectivity were quantitatively investigated. Figure 4.5 shows the formation of interconnected pores at interlayer locations, which generally negatively affects both strength and durability.

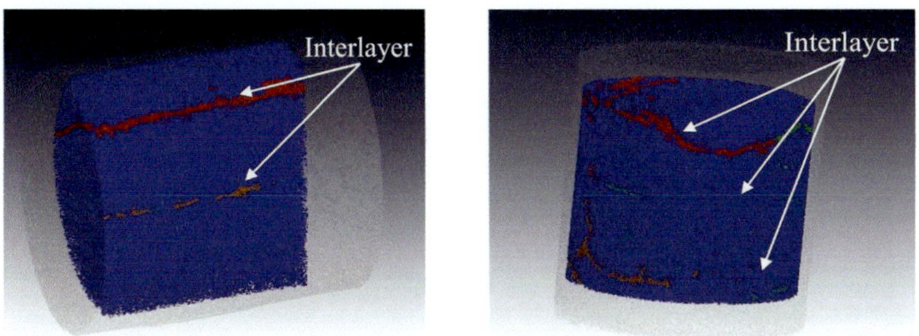

FIGURE 4.5 Interconnected pores from extracted samples from 3DPC (van den Heever et al. 2022).

Also, results show the significant effects of these parameters on the mechanical behavior of the printed samples. To improve the pore size distribution, researchers investigated the addition of nano-calcium carbonate to 3DPC (Yang, Li, and Che 2020). Results show enhancement in various characteristics of the 3D printing process parameters and show that the volume fraction of large pores (> 2 mm) is reduced, while the medium pore volume fraction (1–2 mm) is increased. As can be seen from the presented literature, the pore structure of 3DPC is different from that of cast concrete counterparts. The microstructure within filaments can be optimized through material design and extrusion parameters, while interlayer porosity is primarily controlled by printing process parameters that affect layer-to-layer bonding. Additives can also reduce it, and this improves both mechanical and permeability characteristics. Nevertheless, the lack of standard testing methods is also obvious here, given the lack of a defined shape of an element. This must be considered when future regulations are made.

4.3.4.2 Permeability Effects on the Durability of 3DPC Structures

Permeability issues are studied on 3DPC structures from a durability perspective, particularly for chloride ingress and acid/sulfate attack. These aspects are briefly covered in this section.

The first aspect discussed is the effect of chloride ingress on 3DPC durability. In general, it has been observed that 3DPC structures have less resistance to chloride ingress compared to conventionally cast concrete (Surehali et al. 2023; Zhang et al. 2021; Du et al. 2023), as chloride transport is accelerated via voids created between 3DPC layers (shown in Figure 4.7).

In addition, Surehali et al. (2023) performed a non-steady-state migration test with 2N NaCl solution for 3DPC at different print layer heights (6, 13, and 20 mm). It was shown that the higher the layer height, the greater the chloride penetration along the print direction, as inter-filament voids are more prevalent for higher layer heights. Overall, this study highlighted the importance of reducing inter-filament interfaces with wide print layers to improve chloride ingress resistance. Du et al. (2023) showed that perpendicular to the printing direction (along the layers) had a higher non-steady-state migration coefficient (low chloride resistance) compared to the printing direction and filament direction. Another study by Sun et al. (2022) demonstrated that interlayer chloride resistance is lower than that intra-layer, irrespective of direction.

a) b)

FIGURE 4.6 Chloride penetration depth in (a) cast sample and (b) 3DPC (Du et al. 2023).

Collectively, all studies agree that the chloride ingress is primarily dependent on concrete pore structure, which is influenced by concrete mix design, printing parameters, and testing conditions.

Another important aspect of durability is resistance to sulfate or acid attack. Limited research exists on the durability of materials against acid and sulfate attack. One notable study by Baz et al. (2021) used 1% and 3% sulfuric acid solutions for 56 days, which showed similar levels of deterioration in conventionally cast concrete and 3DPC specimens. Interestingly, mass loss was greater in non-printed specimens compared to 3DPC specimens. The authors argued this was due to the greater loss of accessible pores in the non-printed samples, meaning more surface area was vulnerable to acid attack. Additionally, the total porosity level was the same for both printed and non-printed samples, but pore size distribution varied. Scanning electron microscopy analysis indicated that printed samples had more uniform pore sizes, with pore sizes within a certain range. Overall, it was demonstrated that printed samples exhibited similar durability regarding permeability. However, the internal structure of printed specimens showed better resistance against acid and sulfate attack. Another study by Inaty, Baz, and Aouad (2023) with 0.5% sulfuric acid solution for ~140 days had a similar conclusion on a similar level of mass loss in cast and printed samples. However, for the initial 84 days, printed samples had better chemical resistance. The authors highlighted that large pores in 3DPC specimens take a longer filling time by gypsum and ettringite, which can slow acid infiltration; thus, in the initial days, 3DPC had better resistance. Zhang et al. (2021) conducted a wet–dry cycling sulfate attack test on 3DPC using a 5% sodium sulfate solution for 150 cycles, where each cycle corresponds to 24 hours. It was shown that 3DPC had better sulfate resistance via the measurement of compressive strength at different sulfate exposure cycles. The authors attributed this to the possible existence of connected or even continuous pore structure, which gets filled by ettringite, thereby providing better sulfate attack resistance compared to regular cast concrete with a discontinuous pore structure. A similar study was done by Rui et al. (2023) on 3D-printed fiber-reinforced sulfoaluminate cement-based composites. It was shown that wet–dry cycles contributed to increased mass gain. However, the mass gain gradually declined with possible filling of gypsum and ettringite in the pores. Overall, a more comprehensive study with a standardized 3DPC technique and concrete material with different microstructural observations is required to explain microstructural evolution under sulfate attack, linking the durability properties of 3DPC.

4.4 SUMMARY AND CONCLUSIONS

This chapter examined the durability of 3DPC structures, focusing on shrinkage, creep, and permeability aspects. The following insights and mitigation strategies can be extracted from the discussion above.

Shrinkage in 3DPC mainly occurs within the first 24 hours due to rapid moisture loss from exposed surfaces. Early-age curing, surface sealing, and controlled environmental conditions effectively reduce shrinkage-related cracking and strength loss. Adding shrinkage-reducing admixtures or internal curing agents further improves dimensional stability.

Creep influences buildability and structural integrity during printing, as lower layers deform under sustained loads. Mitigation involves optimizing printing speed, layer height, and interval timing, and controlling rheology with accelerators or viscosity modifiers. Real-time monitoring and adaptive control systems may provide additional ways to manage time-dependent deformation.

Permeability concerns arise from interconnected voids at interlayer interfaces, which promote fluid and ion ingress. These issues can be mitigated by refining the pore structure through mix adjustments with fine fillers or supplementary cementitious materials, enhancing interlayer bonding, and applying protective surface treatments or post-processing methods.

Ensuring the long-term durability of 3DPC requires a comprehensive approach that integrates material design, process control, and environmental management. Future research should focus on developing predictive numerical models for early-age behavior, standardized durability testing, and extensive field studies. Addressing these areas will help create robust design guidelines and support the wider adoption of 3DPC as a durable, sustainable construction technology.

REFERENCES

ACI Committee 209. 2008. *ACI PRC−209.2–08 Guide for Modeling and Calculating Shrinkage and Creep in Hardened Concrete*. Detroit.

Andersen, Sebastian, Wilson Ricardo Leal da Silva, Ieva Paegle, and Jens Henrik Nielsen. 2020. "Numerical Model Describing the Early Age Behavior of 3D Printed Concrete – Work in Progress." *RILEM Bookseries* 28. Springer: 175–84. https://doi.org/10.1007/978-3-030-49916-7_18/FIGURES/8.

Baz, Bilal, Georges Aouad, Joelle Kleib, David Bulteel, and Sébastien Remond. 2021. "Durability Assessment and Microstructural Analysis of 3D Printed Concrete Exposed to Sulfuric Acid Environments." *Construction and Building Materials* 290. https://doi.org/10.1016/j.conbuildmat.2021.123220.

Bažant, Zdeněk P., Anders Boe Hauggaard, Sandeep Baweja, and Franz-Josef Ulm. 1997. "Microprestress-Solidification Theory for Concrete Creep. I: Aging and Drying Effects." *Journal of Engineering Mechanics* 123 (11). American Society of Civil Engineers (ASCE): 1188–94. https://doi.org/10.1061/(ASCE)0733-9399(1997)123:11(1188).

Beersaerts, Glenn, Sandra S. Lucas, and Yiannis Pontikes. 2020. "An Fe-Rich Slag-Based Mortar for 3D Printing." *RILEM Bookseries* 28. Springer: 3–12. https://doi.org/10.1007/978-3-030-49916-7_1/FIGURES/4.

Blaakmeer, J., and B. Lobo. 2020. "A Robust Mortar and Printing System." *RILEM Bookseries* 28. Springer: 1091–103. https://doi.org/10.1007/978-3-030-49916-7_104/FIGURES/7.

Casagrande, Lorenzo, Laura Esposito, Costantino Menna, Domenico Asprone, and Ferdinando Auricchio. 2020. "Mechanical Characterization of Cement-Based Mortar Used in 3DCP Including Early-Age Creep Effects." *RILEM Bookseries* 28. Springer: 407–16. https://doi.org/10.1007/978-3-030-49916-7_42/FIGURES/7.

Chang, Ze, Minfei Liang, Yu Chen, Erik Schlangen, and Branko Šavija. 2023. "Does Early Age Creep Influence Buildability of 3D Printed Concrete? Insights from Numerical Simulations." *Additive Manufacturing* 77 (September). Elsevier: 103788. https://doi.org/10.1016/J.ADDMA.2023.103788.

Che, Yujun, Shengwen Tang, Huashan Yang, Weiwei Li, and Mengyuan Shi. 2021. "Influences of Air-Voids on the Performance of 3D Printing Cementitious Materials." *Materials* 14 (16). https://doi.org/10.3390/ma14164438.

Chen, Mingxu, Bo Liu, Laibo Li, Lidong Cao, Yongbo Huang, Shoude Wang, Piqi Zhao, Lingchao Lu, and Xin Cheng. 2020. "Rheological Parameters, Thixotropy and Creep of 3D-Printed Calcium Sulfoaluminate Cement Composites Modified by Bentonite." *Composites Part B: Engineering* 186 (January). Elsevier Ltd: 107821. https://doi.org/10.1016/j.compositesb.2020.107821.

Du, Longyu, Jiehang Zhou, Jianzhong Lai, Kai Wu, Xuexiang Yin, and Yuanyuan He. 2023. "Effect of Pore Structure on Durability and Mechanical Performance of 3D Printed Concrete." *Construction and Building Materials* 400. https://doi.org/10.1016/j.conbuildmat.2023.132581.

"Durability of Concrete Topic." 2022. Accessed March 9. https://www.concrete.org/topicsinconcrete/topicdetail/durabilityofconcrete?search=durability of concrete.

Federowicz, Karol, Maria Kaszyńska, Adam Zieliński, and Marcin Hoffmann. 2020. "Effect of Curing Methods on Shrinkage Development in 3D-Printed Concrete." *Materials* 13 (11). https://doi.org/10.3390/ma13112590.

Heever, Marchant van den, Anton du Plessis, Jacques Kruger, and Gideon van Zijl. 2022. "Evaluating the Effects of Porosity on the Mechanical Properties of Extrusion-Based 3D Printed Concrete." *Cement and Concrete Research* 153 (December 2021). Elsevier Ltd: 106695. https://doi.org/10.1016/j.cemconres.2021.106695.

Inaty, François El, Bilal Baz, and Georges Aouad. 2023. "Long-Term Durability Assessment of 3D Printed Concrete." *Journal of Adhesion Science and Technology* 37 (12). https://doi.org/10.1080/01694243.2022.2102717.

Kinomura, Koji, Satoshi Murata, Yujin Yamamoto, Hirotoshi Obi, and Akihito Hata. 2020. "Application of 3D Printed Segments Designed by Topology Optimization Analysis to a Practical Scale Prestressed Pedestrian Bridge." *RILEM Bookseries* 28. Springer: 658–68. https://doi.org/10.1007/978-3-030-49916-7_66/FIGURES/11.

Kruger, Jacques, Anton du Plessis, and Gideon van Zijl. 2021. "An Investigation into the Porosity of Extrusion-Based 3D Printed Concrete." *Additive Manufacturing* 37 (December 2020). Elsevier B.V.: 101740. https://doi.org/10.1016/j.addma.2020.101740.

Mazhoud, Brahim, Arnaud Perrot, Vincent Picandet, Damien Rangeard, and Eric Courteille. 2019. "Underwater 3D Printing of Cement-Based Mortar." *Construction and Building Materials* 214 (July). Elsevier: 458–67. https://doi.org/10.1016/J.CONBUILDMAT.2019.04.134.

Moelich, Gerrit M., Jacques Kruger, and Riaan Combrinck. 2020. "Plastic Shrinkage Cracking in 3D Printed Concrete." *Composites Part B: Engineering* 200 (July). Elsevier Ltd: 108313. https://doi.org/10.1016/j.compositesb.2020.108313.

Nerella, Venkatesh Naidu, Simone Hempel, and Viktor Mechtcherine. 2019. "Effects of Layer-Interface Properties on Mechanical Performance of Concrete Elements Produced by Extrusion-Based 3D-Printing." *Construction and Building Materials* 205 (April). Elsevier Ltd: 586–601. https://doi.org/10.1016/j.conbuildmat.2019.01.235.

Nodehi, Mehrab, Federico Aguayo, Shahab Edin Nodehi, Aliakbar Gholampour, Togay Ozbakkaloglu, and Osman Gencel. 2022. "Durability Properties of 3D Printed Concrete (3DPC)." *Automation in Construction*. Elsevier B.V. https://doi.org/10.1016/j.autcon.2022.104479.

Putten, Jolien Van Der, Maxim Deprez, Veerle Cnudde, Geert De Schutter, and Kim Van Tittelboom. 2019. "Microstructural Characterization of 3D Printed Cementitious Materials." *Materials* 12 (18). MDPI AG. https://doi.org/10.3390/ma12182993.

Rehman, Atta Ur, and Jung Hoon Kim. 2021. "3d Concrete Printing: A Systematic Review of Rheology, Mix Designs, Mechanical, Microstructural, and Durability Characteristics." *Materials*. MDPI AG. https://doi.org/10.3390/ma14143800.

Rui, Aoyu, Li Wang, Wenyu Lin, and Guowei Ma. 2023. "Experimental Study on Damage Anisotropy of 3D-Printed Concrete Exposed to Sulfate Attack." *Construction and Building Materials* 407. https://doi.org/10.1016/j.conbuildmat.2023.133590.

Salet, Theo A. M., Zeeshan Y. Ahmed, Freek P. Bos, and Hans L. M. Laagland. 2018. "Design of a 3D Printed Concrete Bridge by Testing*." *Virtual and Physical Prototyping* 13 (3): 222–36. https://doi.org/10.1080/17452759.2018.1476064.

Shahmirzadi, Mohsen Rezaei, Aliakbar Gholampour, Alireza Kashani, and Tuan D. Ngo. 2021. "Shrinkage Behavior of Cementitious 3D Printing Materials: Effect of Temperature and Relative Humidity." *Cement and Concrete Composites* 124 (February). Elsevier Ltd: 104238. https://doi.org/10.1016/j.cemconcomp.2021.104238.

Sikora, Pawel, Mateusz Techman, Karol Federowicz, Ahmed M. El-Khayatt, H. A. Saudi, Mohamed Abd Elrahman, Marcin Hoffmann, Dietmar Stephan, and Sang Yeop Chung. 2022. "Insight into the Microstructural and Durability Characteristics of 3D Printed Concrete: Cast versus Printed Specimens." *Case Studies in Construction Materials* 17 (December). Elsevier Ltd. https://doi.org/10.1016/j.cscm.2022.e01320.

Sivarupan, Tharmalingam, Mohamed El Mansori, Nicolas Coniglio, and Matthew Dargusch. 2020. "Effect of Process Parameters on Flexure Strength and Gas Permeability of 3D Printed Sand Molds." *Journal of Manufacturing Processes* 54: 420–37. https://doi.org/10.1016/j.jmapro.2020.02.043.

Skibicki, Szymon, Monika Pułtorak, Maria Kaszyńska, Marcin Hoffmann, Ewa Ekiert, and Daniel Sibera. 2022. "The Effect of Using Recycled PET Aggregates on Mechanical and Durability Properties of 3D Printed Mortar." *Construction and Building Materials* 335 (June). Elsevier Ltd. https://doi.org/10.1016/j.conbuildmat.2022.127443.

Slavcheva, G. S. 2019. "Drying and Shrinkage of Cement Paste for 3D Printable Concrete." *IOP Conference Series: Materials Science and Engineering* 481 (1). https://doi.org/10.1088/1757-899X/481/1/012043.

Sun, Xiaoyan, Jiawei Zhou, Qun Wang, Jiangpeng Shi, and Hailong Wang. 2022. "PVA Fibre Reinforced High-Strength Cementitious Composite for 3D Printing: Mechanical Properties and Durability." *Additive Manufacturing* 49. https://doi.org/10.1016/j.addma.2021.102500.

Surehali, Sahil, Avinaya Tripathi, Atharwa Samir Nimbalkar, and Narayanan Neithalath. 2023. "Anisotropic Chloride Transport in 3D Printed Concrete and Its Dependence on Layer Height and Interface Types." *Additive Manufacturing* 62. https://doi.org/10.1016/j.addma.2023.103405.

Yang, Huashan, Weiwei Li, and Yujun Che. 2020. "3D Printing Cementitious Materials Containing Nano-CaCO3: Workability, Strength, and Microstructure." *Frontiers in Materials* 7 (August): 1–15. https://doi.org/10.3389/fmats.2020.00260.

Yu, Shiwei, Ming Xia, Jay Sanjayan, Lin Yang, Jianzhuang Xiao, and Hongjian Du. 2021. "Microstructural Characterization of 3D Printed Concrete." *Journal of Building Engineering* 44 (December). Elsevier: 102948. https://doi.org/10.1016/J.JOBE.2021.102948.

Zhang, Yu, Yunsheng Zhang, Lin Yang, Guojian Liu, Yidong Chen, Shiwei Yu, and Hongjian Du. 2021. "Hardened Properties and Durability of Large-Scale 3D Printed Cement-Based Materials." *Materials and Structures/Materiaux et Constructions* 54 (1). https://doi.org/10.1617/s11527-021-01632-x.

5 Reinforcement Strategies for 3D-Printed Concrete

Zhengyu Wu, Ali M. Memari, and José P. Duarte

5.1 INTRODUCTION

5.1.1 Role of Reinforcement in 3DPC

Reinforcement plays a fundamental role in the structural performance of concrete, which is strong in compression but weak in tension. By resisting tensile stresses, reinforcement significantly enhances the flexural strength, crack resistance, fatigue performance, and serviceability of concrete structures. In 3D-printed concrete (3DPC), the importance of reinforcement becomes even more pronounced due to the unique layer-by-layer construction. This method introduces anisotropic mechanical properties, with the interfaces between layers being notably weaker than the monolithic cast concrete. These vulnerable interfaces make reinforcement essential, not only for improving durability but also for ensuring the structural integrity of 3DPC elements.

In conventional cast concrete, reinforcement techniques such as steel rebar and welded wire mesh are well-established and widely used. However, adopting these strategies for 3DPC presents unique challenges. These include integrating reinforcement installation into the automated printing process while minimizing manual labor, ensuring strong bonding between the reinforcement and the printed concrete without compaction and formwork, achieving continuity in reinforcement through multiple layers, and accommodating the complex geometries enabled by the 3DPC. Although these challenges are considerable, they have also driven innovation in reinforcement strategies in 3DPC.

5.1.2 Chapter Structure and Focus

The aim of this chapter is to present a detailed review of the latest reinforcement techniques for 3DPC. It explores both the practical methods currently implemented in projects around the world and the innovative approaches introduced in recent research. The chapter evaluates these techniques based on their ability to address the limitations of traditional methods, with a focus on structural performance, ease of application, and integration with automated processes.

This chapter is divided into three main sections. Section 5.2 reviews practical reinforcement techniques currently used in 3DPC projects, focusing on how they are

DOI: 10.1201/9781003434863-6

incorporated into the printing process and their effectiveness in meeting structural demands. Section 5.3 explores innovative reinforcement methods discussed in recent research, analyzing their advantages and disadvantages. Section 5.3.3 evaluates the potential for these methods to achieve widespread industry adoption and outlines future research and development directions. While reinforcement is essential for various structural components, this chapter specifically focuses on the reinforcement of wall elements, which play a key role in most 3DPC applications.

This chapter contributes to the ongoing discussions on advancing additive manufacturing in construction by addressing the challenges and opportunities for reinforcement in 3DPC. It provides a comprehensive understanding of current practices, identifies existing knowledge gaps, and highlights innovative approaches that hold potential for future developments in the field.

5.2 REINFORCEMENT METHODS ADOPTED IN 3DPC PROJECTS

5.2.1 Overview of 3DPC Projects around the World

The adoption of 3DPC has been expanding rapidly worldwide, with a variety of 3DPC projects carried out across various regions. These projects, varying in scale and application, are spearheaded by companies from different countries, driving the advancement of 3DPC. In the United States, several companies are pioneering advancements in 3D printing construction. The pioneer in the field of 3DPC, Contour Crafting Corp., based in California, has developed innovative construction techniques aimed at emergency housing and commercial projects. The company claims their strategies are able to construct the walls of a 2,000-square-foot house in less than 24 hours (Figure 5.1a), emphasizing cost reduction and accelerated construction timelines. Branch Technology, headquartered in Tennessee, has completed architectural projects such as the OneC1TY Pavilion at Nashville in 2018 (Figure 5.1b), showcasing its expertise in lightweight structures using cellular fabrication. ICON, based in Texas, focuses on sustainable and affordable housing solutions. The company developed and expanded the world's first 3D-printed hotel, El Cosmico, in Marfa, Texas, transforming over 60 acres into a community with housing and hospitality amenities, breaking ground in 2024 (Figure 5.1c). Black Buffalo 3D, located in Pennsylvania, specializes in CC-ES AC509 (ICC Evaluation Service 2022) compliant 3D construction printers (Figure 5.1d) capable of building structures up to three stories tall. One of its printers was utilized in 2023 to construct two 3D-printed homes in Newport News, Virginia, as part of Habitat for Humanity Peninsula and Greater Williamsburg's latest housing initiative (Black Buffalo 3D 2025).

In China, WinSun, headquartered in Shanghai, has been at the forefront of large-scale 3DPC developments. Notably, in 2016, the company constructed the world's first additively manufactured office building (Figure 5.2). The modular structure was printed in China, transported, and assembled in Dubai. WinSun's projects demonstrate the potential of 3DPC for urban construction, showcasing its versatility and scalability.

FIGURE 5.1 3DPC projects constructed by US companies: (a) Contour Crafting Corporation (Contour Crafting Corporation 2025); (b) Branch Technology (Drebezgova et al. 2019); (c) ICON (ICON 2024a); (d) Black Buffalo 3D (Bello and Memari 2022). Used with permission.

FIGURE 5.2 Additively manufactured office building in Dubai by WinSun (Buchanan and Gardner 2019). Used with permission.

In the Netherlands, TU Eindhoven completed a large-scale 3D-printed bridge of 29 m length in 2021, a landmark project using additive manufacturing for infrastructure (Figure 5.3a). CyBe Construction, based in Oss, has also undertaken various projects, including the world's first 3D-printed laboratory in Dubai (Figure 5.3b). Completed in 2017, this project involved printing the walls of a 168- m^2 structure on-site in just 46 hours. It highlights CyBe's capability to deliver fast, large-scale 3D-printed buildings with minimal construction time.

In Germany, PERI, based in Weissenhorn, has completed several 3DPC projects in Europe, including Germany's first 3D-printed single-family house (Figure 5.4). This two-story house, finished in 2021, required just 100 hours of printing time and a total construction period of 8 months. This project highlights the company's primary focus on residential construction.

In the United States, Apis Cor, headquartered in Melbourne, Florida, initiated the construction of the first 3D-printed home in Brevard County, located in Melbourne, in September 2023. Their work showcases the use of mobile 3D printers (Figure 5.5) to create affordable housing solutions.

In Italy, WASP, headquartered in Massa Lombarda, utilized its Crane WASP 3D printer to construct a lightweight building with local soil and natural materials

a) b)

FIGURE 5.3 3DPC projects constructed in the Netherlands by (a) TU Eindhoven (Ahmed et al. 2022) and (b) CyBe (Bello and Memari 2022). Used with permission.

FIGURE 5.4 Germany's first printed home by PERI (Weger et al. 2021). Used with permission.

FIGURE 5.5 3D mobile printer, named Frank, from Apis Cor (Staritz, Miller, and McClurg 2020). Used with permission.

(Figure 5.6). Completed in July 2024, this project "TOVA" is located in Collserola Natural Park, Spain. It underscores WASP's commitment to sustainable construction by employing locally sourced materials.

In France, XtreeE, based in Paris, has completed various projects, including a biomimetic underwater structure in 2024 (Figure 5.7a) and a printed table base in 2021

(Figure 5.7b). The company focuses on utilizing robotics and advanced materials to develop innovative construction solutions, demonstrating the versatile applications of 3DPC across various fields.

These projects illustrate the trends toward global adoption of 3DPC, with companies employing unique printing and reinforcement strategies to tackle diverse construction challenges. Each case demonstrated the potential of 3D printing to revolutionize construction practices.

FIGURE 5.6 3D-printed lightweight building using local soil and natural materials by WASP (El-Mahdy et al. 2025). Used with permission.

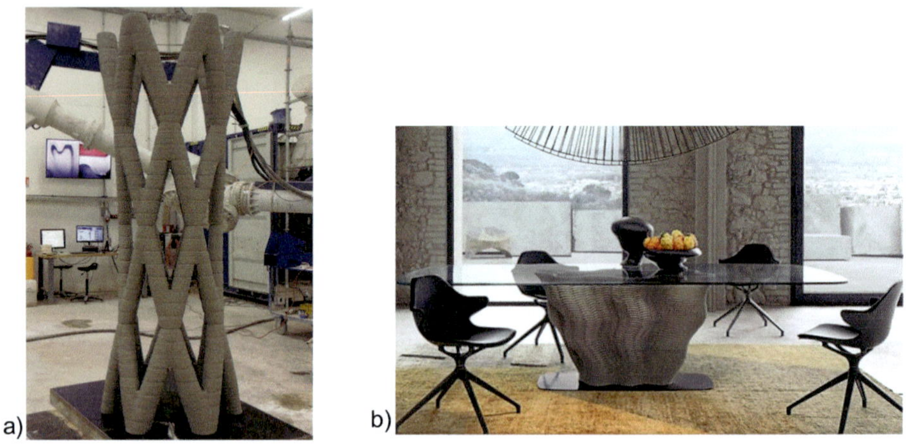

FIGURE 5.7 3DPC projects constructed by XtreeE: (a) biomimetic underwater structure (XtreeE 2025a), courtesy of XtreeE; (b) table base (XtreeE 2025b), Corail Table for Roche Bobois, designed by Fritsch Durisotti and manufactured by XtreeE.

5.2.2 Case Studies of Reinforcement Techniques

Integrating reinforcement in 3DPC projects presents several challenges, such as incorporating reinforcement into the automated printing process, ensuring strong bonding between reinforcement and concrete without compaction or formwork, and adapting to the complex geometry unique to 3DPC. This section explores how various companies address these challenges through practical reinforcement strategies in their 3DPC projects. The selected companies highlight diverse approaches and techniques currently applied in existing 3DPC projects worldwide. ICON places vertical rebars prior to printing for columns and wall elements. For straight walls, horizontal ladder mesh and wire ties are manually applied to printed layers during the printing process, connecting the inner and outer portions of wall segments (Figure 5.8a). In curved wall elements, vertical rebars are tied to horizontal curved rebars to form a mesh structure before printing begins (Figure 5.8b). Arched structures are reinforced with post-tensioned cables, with tension applied after the printed concrete has cured (Figure 5.8c). The process involves manually positioning the rebar system before printing and adding reinforcements as layers are built. Once the shell is complete, concrete is poured into the formwork, creating a solid structure by bonding the printed shell and reinforcement. Key challenges include maintaining vertical rebar alignment and achieving strong bonding between the poured concrete and the reinforcement.

FIGURE 5.8 Reinforcement strategies adopted by ICON in (a) straight walls (ICON 2025, 2024b), (b) curved walls (Wu, Memari, and Duarte 2022), and (c) arched structures (ICON Build 2023). Used with permission.

PERI integrates C-shaped form ties during wall printing to stabilize thin printed wall shells throughout the printing process (Figure 5.9). Instead of fresh concrete, the company fills the cavity within the printed formwork with insulation materials, such as spray foam, to enhance thermal performance. For beams printed on the top of columns, a long-span steel plate is placed on the columns before printing begins. This stay-in-place plate provides temporary support (as a lintel) during printing and improves the flexural capacity and ductility of the beam after curing. Reinforcement is manually incorporated during the printing process, with horizontal ties positioned within the printed layers.

Apis Cor places concentrated reinforcement, including longitudinal rebar and stirrups similar to those used in conventional construction, in the printed column, wall, and wall foundation elements (Figure 5.10a). Unlike other companies, Apis Cor employs inclined lateral rebars in their wall elements (Figure 5.10b). This inclined rebars better constrain the vertical rebars compared to rebars placed perpendicular to the printing direction.

XtreeE mixes fiber reinforcement directly into the concrete mixture, allowing the fibers to be extruded through the nozzle along with the fresh concrete and embedded into the printed layers (Figure 5.11). The layer-by-layer printing process often creates weakened interfaces between layers in printed elements, but the inclusion of fibers helps mitigate localized damage and increases cracking resistance of printed structures.

WinSun employs reinforcement techniques in their 3DPC projects similar to those used by other 3DPC companies for structures under 100 m. These techniques include inserting steel cages and ladder meshes into the printed formwork

FIGURE 5.9 Construction of a 3DPC project by PERI (Weger et al. 2021). Used with permission.

a) b)

FIGURE 5.10 Reinforcement strategies adopted by Apis Cor, including (a) concentrated rebars and (b) inclined lateral rebars (Wu, Memari, and Duarte 2022). Used with permission.

FIGURE 5.11 Fiber reinforcement adopted by XtreeE (XtreeE 2015). Courtesy of XtreeE.

for columns and walls (Figure 5.12a). For larger-scale structures exceeding 100 m, WinSun integrates precast concrete columns with reinforcement into the printed wall shells, achieving a capacity comparable to that of conventional shear walls. For tall walls and columns, the construction process alternates between inserting

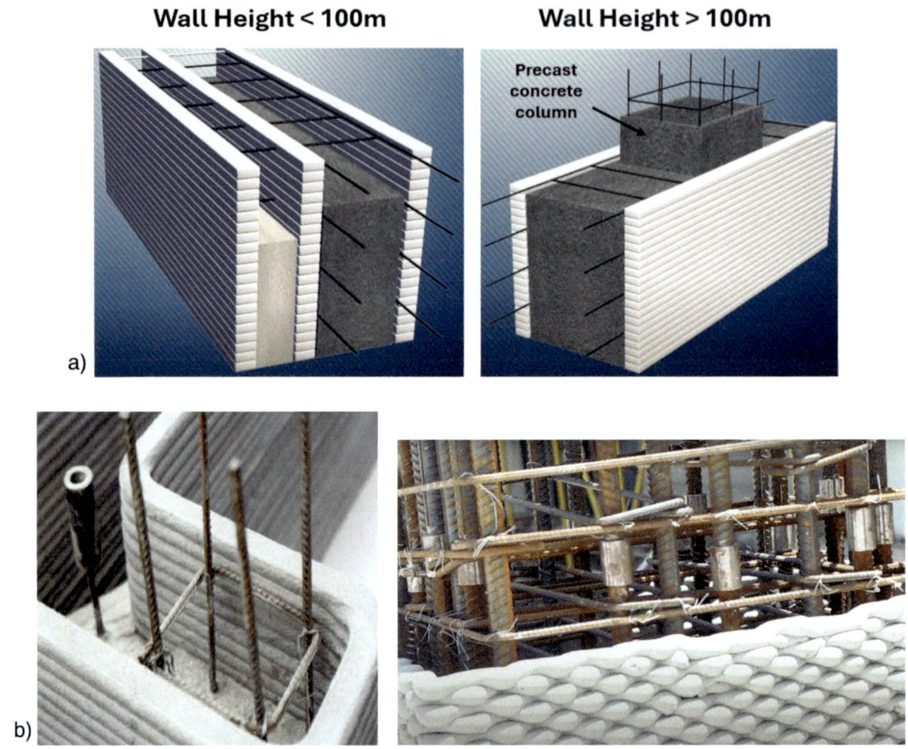

FIGURE 5.12 Reinforcement strategies adopted by WinSun in (a) walls (image adapted from Brandon 2015) and (b) tall columns (Asprone et al. 2018; Wu, Memari, and Duarte 2022). Used with permission.

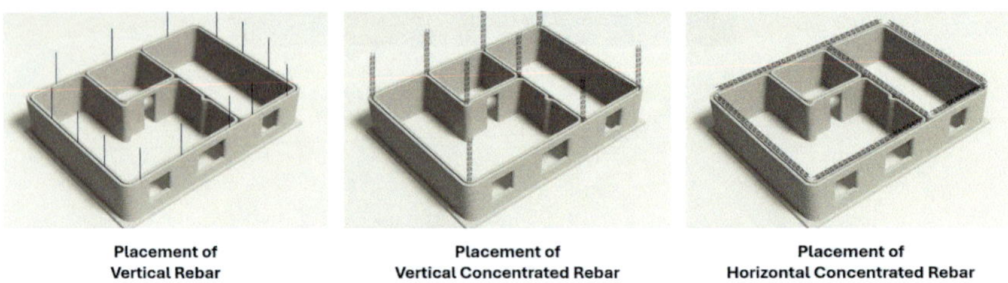

FIGURE 5.13 Construction steps in CyBe's 3DPC project. Image adapted from Dosary (2018).

vertical rebars and printing layers (Figure 5.12b). Vertical rebars are installed first, followed by printing up to the height of the rebars. Once this height is reached, additional rebars are connected on top using rebar couplers, and the printing continues. This step-by-step approach simplifies construction while maintaining structural continuity throughout the column.

CyBe Construction follows a systematic process for their 3DPC projects (Figure 5.13). First, concrete shells for walls and columns are printed directly onto the foundation. Vertical rebars are then inserted into the wall shells, while prefabricated vertical concentrated reinforcements are placed for the column shells. Horizontal concentrated reinforcements are subsequently added on top of the vertical reinforcements. Finally, the walls and columns are cast to complete the structure.

5.2.3 Comparative Analysis of Reinforcement Techniques

Various reinforcement strategies have been implemented in existing 3DPC projects, categorized into traditional rebar systems, ties and mesh reinforcements, fiber reinforcements, and post-tensioning cables. These methods are adopted from conventional cast concrete, as they have been proven reliable over decades of use. This section evaluates these reinforcement strategies, focusing on their effectiveness, implementation feasibility, scalability, and adaptability. Traditional rebar systems, including vertical, horizontal, and concentrated rebars, remain among the most effective reinforcement strategies for load-bearing elements. Vertical rebars can be placed either before or after printing (Figures 5.8a, 5.12). When horizontal rebars are needed, they are tied to vertical rebars and placed into printed cavities, providing additional out-of-plane bending resistance (Figure 5.8b). Concentrated rebars are primarily used for reinforcing columns (Figure 5.10a). For tall structures requiring multiple printing stages, additional vertical rebars are connected to existing ones using couplers followed by the printing of subsequent layers (Figure 5.12b). However, these methods rely heavily on manual labor, which can slow the construction process and require precise alignment, particularly in tall columns or curved walls. The step-by-step process of inserting rebars, printing layers, and connecting reinforcements further increases complexity and construction time.

Form ties and mesh reinforcements, such as ladder meshes, are commonly used to stabilize printed wall shells while the material is fresh and to improve bonding between adjacent printed layers (Figures 5.8a, 5.9, 5.12a). These materials are placed to connect the inner and outer wall shells during the printing process. While relatively easy to apply in straight or simple geometries (e.g., straight walls), this strategy faces challenges adapting to curved or complex shapes due to limited flexibility. The manual placement also limits the scalability and slows down construction.

Fiber reinforcement, added directly to the concrete mixture, increases resistance to crack formation and limits crack widths, while mitigating localized damage at weak layer interfaces (Figure 5.11). This method integrates smoothly with the printing process, requiring no manual labor or interruptions. However, its structural capacity is limited compared to traditional rebar systems, making it more suitable for non-loadbearing structures or as a supplementary reinforcement strategy.

Post-tensioning cables are effective for reinforcing arched or curved structures, significantly enhancing their load-bearing capacity by applying tension after the printed concrete has cured (Figure 5.8c). This process ensures the structure can handle additional stress and maintain stability. However, it requires precise installation and multiple post-processing steps, which may increase construction complexity and demand skilled labor. While this method provides strong structural reinforcement, the added effort can limit its scalability for larger projects.

The evaluation of reinforcement strategies currently adopted in 3DPC projects highlights their strengths and limitations. Traditional rebar systems and post-tensioning cables, widely used due to their proven effectiveness in cast concrete, provide strong structural performance and reliability for large-scale, load-bearing applications. Mesh and fiber reinforcement methods are easier to apply but may require supplemental reinforcement strategies to handle heavy loads. However, these methods heavily rely on manual placement, requiring significant labor and time during construction. This slows down construction speed and reduces the efficiency expected from 3DPC. Additionally, these methods are not suitable for complex geometries, as adapting rigid reinforcements to curved shapes requires further manual adjustments, adding to the overall effort and cost. Despite their reliability, these traditional approaches limit the scalability and automation potential of 3DPC technology, highlighting the need for innovative strategies to reinforce printed structures.

5.3 CHALLENGES AND ONGOING RESEARCH

5.3.1 Key Challenges

The most significant challenge in 3DPC reinforcement is that it still relies on the manual placement of materials. Traditional methods, such as rebar and mesh, are incompatible with the automated, layer-by-layer nature of 3D printing. Manual tasks such as laying and tying rebars, placing meshes, and inserting form ties disrupt automation, slow construction, and limit the scalability of 3DPC technology.

Another challenge is achieving vertical continuity in reinforcement. Current methods require manually connecting rebars between printing stages, which may result in alignment issues and weak points, particularly in tall or multilayered structures.

Material limitations further restrict 3DPC applications. While fibers and hybrid materials are easier to integrate into the mixture, they often lack the tensile strength needed for load-bearing elements, limiting their use in large-scale projects.

Finally, incorporating reinforcement into a complex geometry remains difficult. While the 3DPC technology enables designs for complex shapes, traditional reinforcements are hard to apply in curved or irregular shapes without significant adjustments, increasing labor and material costs.

These challenges underscore the need for innovative reinforcement strategies that support automation, improve structural performance, and align with the capabilities of 3DPC.

5.3.2 Review of Innovative Methods in Literature

Recent advancements in reinforcement techniques for 3DPC have focused on automating the placement of reinforcing materials to overcome the limitations of manual methods. Several innovative approaches have been proposed, including continuous and discontinuous reinforcement, mesh mold, staple reinforcement, vertical rollable steel welded wire mesh, in-process welding, steel printing, and prefabricated steel elements. These methods are designed to improve structural performance and seamlessly integrate reinforcement placement into the printing process. For more details about the classification and characterization of these methods, including the advantages

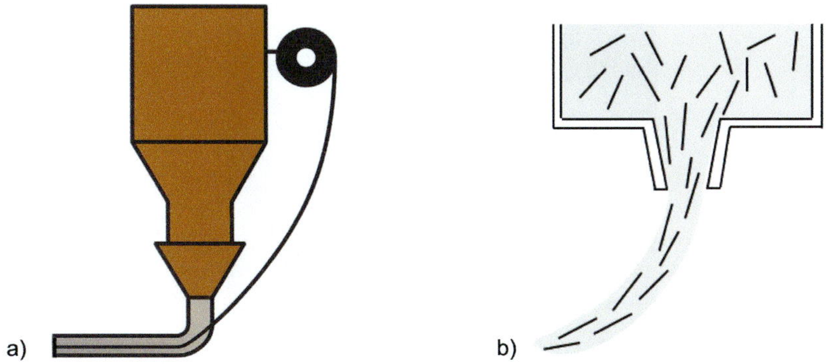

FIGURE 5.14 (a) Continuous and (b) discontinuous reinforcement.

and limitations, refer to the literature by Asprone et al. (2018), Mechtcherine et al. (2021), and Wu, Memari, and Duarte (2022).

Continuous reinforcement involves embedding long and continuous materials, such as steel wires or cables, directly into the printed concrete during the printing process (Figure 5.14a). A spool of flexible steel wire feeds through a specialized nozzle, which places the reinforcement precisely beneath the extruded concrete as layers are built. This simultaneous placement ensures structural continuity and automation (Ma et al. 2019). This method enhances the tensile strength and ductility of the reinforced concrete elements. Discontinuous reinforcement, on the other hand, incorporates short materials like steel fibers into the concrete mixture (Figure 5.14b), as also demonstrated in XtreeE's projects shown in Figure 5.11. These fibers are extruded together with concrete, eliminating the need for separate placement and seamlessly integrating with automated workflows. Fiber materials can enhance tensile performance and reduce weak interlayer bonding, which is a common issue in 3DPC. However, this method remains more effective for non-load-bearing components due to its limited tensile strength compared to traditional reinforcement.

The mesh mold technique offers a unique solution by combining reinforcement and formwork into a single prefabricated structure. Prefabricated steel or polymer meshes act as both stay-in-place formwork and tensile reinforcement, with concrete sprayed or extruded into the mesh during the printing process (Figure 5.15a). This approach is particularly effective for creating lightweight and complex geometries while minimizing material waste. However, current applications are generally limited to lightly loaded or non-structural components due to the lower strength. Staple reinforcement introduces small, steel staple-like elements into the printed layers, enhancing interlayer bonding and providing resistance to localized damage (Figure 5.15b). These staples are automatically inserted during the printing process, effectively addressing the weak interfaces caused by layer-by-layer construction. While this method improves the strength of printed concrete, its performance is currently limited to smaller-scale applications.

a)

b)

FIGURE 5.15 (a) Prefabricated mesh (Hack and Lauer 2014); (b) staple reinforcement (Geneidy 2019). Used with permission.

Vertical rollable steel welded wire mesh provides another innovative solution for reinforcing wall and column elements (Figure 5.16a). In this approach, the flexible mesh is stored in a spool attached to the printing head. The spool feeds the mesh through a long vertical notch in the nozzle head at the same speed as the nozzle moves. Concrete filaments on both sides support the continuous mesh after it comes out from the nozzle, maintaining its vertical orientation. The portion of the mesh extending above the printed filament surface interacts with the upper layer mesh, providing interlocking force to the concrete structure when the next layer is printed. Its prefabricated and rollable nature facilitates transport, handling, and integration into automated workflows, reducing reliance on manual placement. In-process welding involves welding steel rebar segments in horizontal and vertical directions to form

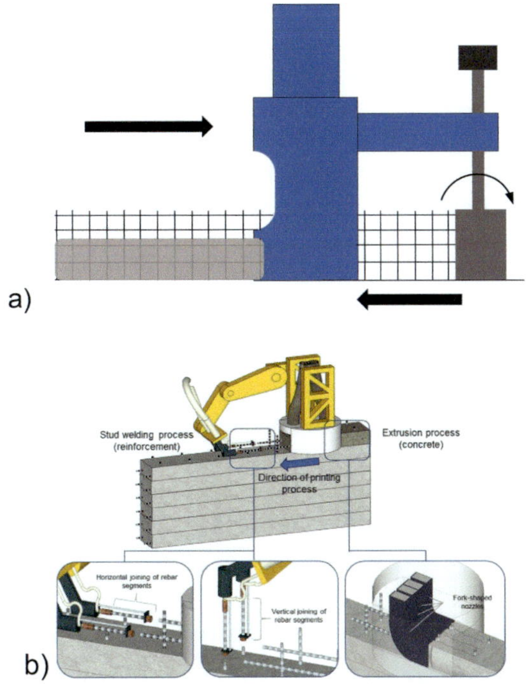

FIGURE 5.16 (a) Vertical rollable steel welded wire mesh; (b) in-process welding (Classen, Ungermann, and Sharma 2020). Used with permission.

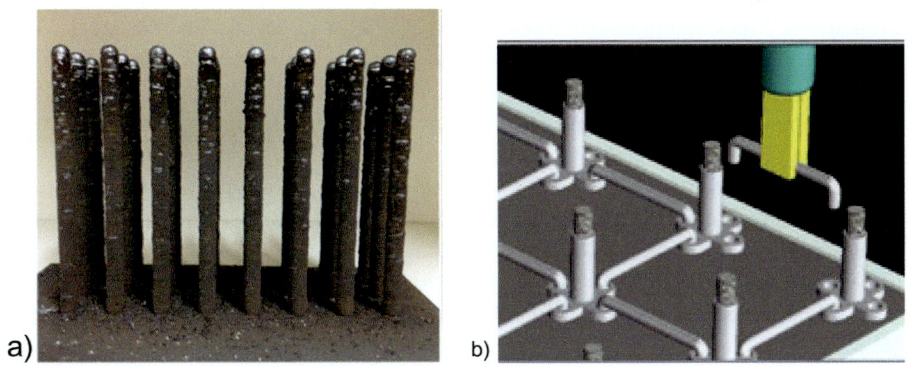

FIGURE 5.17 (a) Steel printing (Mechtcherine et al. 2018); (b) assembling steel element methods (Khoshnevis 2004). Used with permission.

a three-dimensional reinforcement cage ahead of the printing path to integrate reinforcing and printing processes (Figure 5.16b). This method integrates the reinforcing and printing processes, eliminating the need for rebar placement after printing. It also enhances bonding with the printed concrete, as the extruded material envelops the prefabricated mesh.

Steel printing uses a gas metal arc welding process (shape welding) to print steel beads, which can form long horizontal filaments or vertical rebars (Figure 5.17a). This method aims to print steel and concrete through different mechanisms and nozzles simultaneously. Another method relies on prefabricated small steel elements to reinforce concrete members (Figure 5.17b). In this process, a concrete shell (much like formwork) is first printed, then the prefabricated steel elements are assembled to form a cage inside the shell cavity, after which concrete is poured into the cavity. Both methods are still in the conceptual stage and will require extensive testing and resolution of technical challenges before practical implementation.

These innovative methods address a major limitation in current 3DPC projects: the reliance on manual placement of reinforcement materials. By integrating reinforcement directly during the printing process, these methods eliminate the need for manual placement and significantly accelerate construction speed. These reinforcement strategies not only enhance structural performance but also improve overall construction efficiency. While some methods remain in the conceptual stage, they represent an important step toward achieving automated, scalable, and reliable reinforcement solutions for 3DPC structures.

5.3.3 Evaluation of Potential for Industry Adoption and Future Directions

The reinforcement methods reported in the literature aim to reduce manual labor while integrating reinforcement during the printing process. Continuous and discontinuous reinforcement (Figure 5.14a,b), staple reinforcement (Figure 5.15b), and vertical rollable steel welded wire mesh (Figure 5.16a) are relatively simple to implement. These methods use flexible, thin, or short materials, such as steel wires, fibers, or pre-rolled mesh, which can be easily applied in the printed filaments during the printing process. While they improve interlayer bonding and provide localized cracking resistance, the relatively low structural capacity limits their applications to small-scale or non-load-bearing elements.

In contrast, steel printing (Figure 5.17a) and prefabricated steel elements (Figure 5.17b) focus on customization and are suitable for use in complex geometries. However, both remain at the conceptual stage, requiring extensive testing and improvements. Their complexity and limited efficiency make them impractical for large-scale use at this stage, serving more as innovative demonstrations than feasible solutions.

Mesh Mould (Figure 5.15a) combines reinforcement and formwork fabrication, offering a unique approach by automating the production of a mesh structure. However, since the mesh is fabricated before concrete placement, it requires additional labor work and time in the entire workflow.

Among these techniques, in-process welding (Figure 5.16b) seems to have the greatest potential for future construction applications. By welding short rebars concurrently with printing, it achieves strong concrete to steel interface bonding and aligns well with automated workflows. The high strength of the rebar also makes it suitable to reinforce large-scale, load-bearing structures. However, challenges such as welding precision, material consistency, and equipment durability must be addressed before it can be widely used.

The methods reviewed demonstrate a trade-off between ease of implementation and structural capacity. Methods such as continuous reinforcement, vertical rollable steel welded wire mesh, and staple reinforcement are easy to apply due to the used lightweight or flexible materials. However, these same characteristics result in low load-bearing capacity, limiting their application to non-load-bearing elements or as supplementary reinforcement.

Conversely, methods such as in-process welding, steel printing, and prefabricated steel elements use strong materials capable of resisting heavy loads, making them suitable for large-scale and load-bearing applications. However, their complexity and the need for advanced equipment and precise control limit their practicality and scalability. These approaches remain promising but require further refinement to overcome the current challenges.

Future developments should aim to balance these trade-offs by combining the advantages of various methods. For example, hybrid approaches that integrate lightweight reinforcement for ease of implementation with stronger materials in critical load-bearing areas could provide an effective solution. Additionally, advancements in robotic automation, material innovation, and real-time monitoring systems will be essential for improving these methods into scalable and practical solutions. By addressing these challenges, automated reinforcement methods can move beyond conceptual stages and play a crucial role in modern construction.

5.4 CONCLUSION

In summary, the reinforcement methods currently employed in 3DPC projects are largely adopted from conventional cast concrete practices, such as the use of steel rebars. While these methods have proven reliable for decades and provide strong structural performance, they compromise one of the key advantages of 3DPC: automation. Their reliance on manual placement and labor-intensive processes slows down the construction process and limits the scalability and efficiency that 3DPC is designed to deliver.

To address these limitations, innovative reinforcement techniques specifically designed for 3DPC have been proposed in the literature. These methods aim to integrate reinforcement directly into the printing process, leveraging automation to speed up workflows and reduce manual labor. Methods such as continuous reinforcement, in-process welding, and vertical rollable steel welded wire mesh provide solutions to the inefficiencies of traditional methods while enhancing interlayer bonding and structural performance. However, despite their potential, many of these innovative approaches remain in the conceptual or experimental stages. They face significant challenges, such as material limitations, equipment precision, and the need for extensive testing to ensure reliability and compliance with industry standards.

The continued reliance on conventional methods in current 3DPC projects highlights the practical challenges that innovative techniques must overcome before achieving widespread adoption. Overcoming these challenges will require interdisciplinary collaboration among mechanical engineering, material science, and structural engineering to develop solutions that combine automation with high load-bearing capacities. By addressing these issues, 3DPC can realize its full potential as a scalable, efficient, and transformative technology for modern construction.

5.5 CONFLICTS OF INTEREST

One of the co-authors, José P. Duarte, and The Pennsylvania State University own equity in X-Hab 3D, whose activity is related to concrete printing. Their ownership in this company has been reviewed by the University's Individual and Institutional Conflict of Interest Committees and is currently being managed by the University.

REFERENCES

Ahmed, Zeeshan Y., Rob J. M. Wolfs, Freek P. Bos, and Theo A. M. Salet. 2022. "A Framework for Large-Scale Structural Applications of 3D Printed Concrete: The Case of a 29-m Bridge in the Netherlands." *Open Conference Proceedings* 1: 5–19. https://doi.org/10.52825/ocp.v1i.74.

Asprone, Domenico, Costantino Menna, Freek P. Bos, Theo A. M. Salet, Jaime Mata-Falcón, and Walter Kaufmann. 2018. "Rethinking Reinforcement for Digital Fabrication with Concrete." *Cement and Concrete Research* 112: 111–21. https://doi.org/10.1016/j.cemconres.2018.05.020.

Bello, Nicholas D., and Ali M. Memari. 2022. "Comparative Review of the Technology and Case Studies of 3D Concrete Printing of Buildings by Several Companies." *Buildings* 13 (1): 106. https://doi.org/10.3390/buildings13010106.

Black Buffalo 3D. 2025. "Black Buffalo 3D Printed Homes Have New Owners Just in Time for the Holidays." Accessed June 8, 2025. https://bb3d.io/black-buffalo-3d-printed-homes-have-new-owners-just-in-time-for-the-holidays/.

Brandon, J. F. 2015. "Don't Believe the Hype: WinSun Unveils Bland Multi-Story 3D Printed Buildings." *SolidSmack*, February 3. https://www.solidsmack.com/fabrication/dont-believe-hype-winsun-unveils-bland-multi-story-3d-printed-buildings/.

Buchanan, Craig, and Leroy Gardner. 2019. "Metal 3D Printing in Construction: A Review of Methods, Research, Applications, Opportunities and Challenges." *Engineering Structures* 180: 332–48. https://doi.org/10.1016/j.engstruct.2018.11.045.

Classen, Martin, Jan Ungermann, and Rahul Sharma. 2020. "Additive Manufacturing of Reinforced Concrete – Development of a 3D Printing Technology for Cementitious Composites with Metallic Reinforcement." *Applied Sciences* 10 (11): 3791. https://doi.org/10.3390/app10113791.

Contour Crafting Corporation. 2025. "Building Automated Construction." Accessed June 8, 2025. https://www.contourcrafting.com/building-construction.

Dosary, Wajdi Waleed. 2018. "Consolidated Contractors Company's 3D Concrete Printed House Using CyBe Robotic Arm." *YouTube*, video, November 4. https://www.youtube.com/watch?v=Dn7mxNK2CD8.

Drebezgova, M. Yu, N. V. Chernysheva, V. A. Klimenko, and A. V. Olisov. 2019. "Architectural Design of Residential Modules Using Additive Technologies." *IOP Conference Series: Materials Science and Engineering* 698 (3): 033007. https://doi.org/10.1088/1757-899X/698/3/033007.

El-Mahdy, Deena, Eugene Marais, Marwa Abdel Rahim, Alexandre Dubor, and Edouard Cabay. 2025. "Life Cycle Assessment of Earth-Based Residential Unit 'TOVA': A 3D Printed On-Site Load-Bearing Structure." *Case Studies in Construction Materials*: e04925. https://doi.org/10.1016/j.cscm.2025.e04925.

Geneidy, O. 2019. "Incorporating a Simultaneous Layer Stapling Process While 3D Printing Structures so as to Complement Their Structural Properties." *IAAC Blog*, March 20. http://www.iaacblog.com/programs/incorporating-simultaneous-layer-stapling-process-3d-printing-structures-complement-structural-properties/.

Hack, Norman, and Willi Viktor Lauer. 2014. "Mesh-Mould: Robotically Fabricated Spatial Meshes as Reinforced Concrete Formwork." *Architectural Design* 84 (3): 44–53. https://doi.org/10.1002/ad.1762.

ICC Evaluation Service. 2022. *Acceptance Criteria for 3-D Automated Construction Technology for 3-D Concrete Walls (AC509)*. Whittier, CA: ICC Evaluation Service, LLC.

ICON. 2024a. "World's First 3D-Printed Hotel and Development I El Cosmico in Marfa, TX." *YouTube*, video, September 12. https://www.youtube.com/watch?v=CHlRew73mMk.

ICON. 2024b. "ICON's Technology Showcase I DOMUS EX MACHINA at SXSW 2024." *YouTube*, video, March 28. https://www.youtube.com/watch?v=wUjvmYv3_-I.

ICON. 2025. "3D-Printed Homes in Wimberley TX I ICON." *YouTube*, video, January 27. https://www.youtube.com/watch?v=-N6noQc-GRE.

ICON Build. 2023. "United States Marine Corps and ICON 3D Print Structure at Camp Pendleton." *ICON*, May 17. https://www.iconbuild.com/projects/usmc-vehicle-hide-structure-at-camp-pendleton.

Khoshnevis, Behrokh. 2004. "Automated Construction by Contour Crafting-Related Robotics and Information Technologies." *Automation in Construction* 13 (1): 5–19. https://doi.org/10.1016/j.autcon.2003.08.012.

Ma, Guowei, Zhijian Li, Li Wang, and Gang Bai. 2019. "Micro-Cable Reinforced Geopolymer Composite for Extrusion-Based 3D Printing." *Materials Letters* 235: 144–47. https://doi.org/10.1016/j.matlet.2018.09.175.

Mechtcherine, Viktor, Richard Buswell, Harald Kloft, Freek P. Bos, Norman Hack, Rob Wolfs, Jay Sanjayan, Behzad Nematollahi, Egor Ivaniuk, and Tobias Neef. 2021a. "Integrating Reinforcement in Digital Fabrication with Concrete: A Review and Classification Framework." *Cement and Concrete Composites* 119: 103964. https://doi.org/10.1016/j.cemconcomp.2021.103964.

Mechtcherine, Viktor, Jasmin Grafe, Venkatesh N. Nerella, Erik Spaniol, Martin Hertel, and Uwe Füssel. 2018. "3D-Printed Steel Reinforcement for Digital Concrete Construction – Manufacture, Mechanical Properties and Bond Behaviour." *Construction and Building Materials* 179: 125–37. https://doi.org/10.1016/j.conbuildmat.2018.05.237.

Staritz, Peter, Caleb Miller, and Josiah McClurg. 2020. "The Mobile Gantry: A Robotic Architecture for 3D Printing Structures on Mars." https://doi.org/10.31224/osf.io/5dy9z.

Weger, Daniel, Christoph Gehlen, Waldemar Korte, Fabian Meyer-Brötz, Jennifer Scheydt, and Thorsten Stengel. 2021. "Building Rethought–3D Concrete Printing in Building Practice." *Construction Robotics* 5 (3): 203–10. https://doi.org/10.1007/s41693-021-00057-5.

Wu, Zhengyu, Ali M. Memari, and Jose P. Duarte. 2022. "State of the Art Review of Reinforcement Strategies and Technologies for 3D Printing of Concrete." *Energies* 15 (1): 360. https://doi.org/10.3390/en15010360.

XtreeE. 2015. "The Large-Scale 3D." *YouTube*, video, September 15. https://www.youtube.com/watch?v=PXImY6z1cKY.

XtreeE. 2025a. "KRYPTON 2.0." Accessed June 8, 2025. https://www.xtreee.com/uses-cases/krypton-2.0.

XtreeE. 2025b. "KRYPTON 2.0: A New Chapter for Architectural 3D Printing." *XtreeE News*. Accessed June 8, 2025. https://www.xtreee.com/news/news-title-06.

Part 2

3D-Printable Materials and Mixture Design

6 3D-Printable Geopolymer Concrete

Maryam Hojati and Christopher Vreeland

6.1 INTRODUCTION: THE RISE OF GEOPOLYMERS IN 3D PRINTING

The construction industry is undergoing a transformative shift with the rise of 3D printing, or additive manufacturing (AM), which is redefining traditional building methods by offering greater efficiency, precision, and sustainability (Dilberoglu et al. 2017; Wangler et al. 2016; 2019; Sedghi, Rashidi, and Hojati 2024). This technology enables structures to be fabricated layer by layer directly from digital models, reducing material waste and allowing for unprecedented design flexibility (Wangler et al. 2016; Nazarian et al. 2021; Hojati et al. 2018). Beyond its ability to streamline construction, 3D printing simplifies supply chains, reduces transportation costs, and minimizes reliance on labor-intensive methods (Ayyagari, Chen, and García de Soto 2023). However, one of the most pressing challenges in construction today is the environmental impact associated with conventional cement-based materials, which account for nearly 8% of global CO emissions (Tabassum and Ahmad Mir 2023; Batikha et al. 2022; Ahmed 2023; Khan, Koç, and Al-Ghamdi 2021). Addressing this challenge has led to the increasing focus on geopolymers, sustainable, cement-free binders that offer an environmentally friendly alternative to ordinary portland cement (OPC) (Ziejewska et al. 2022; Barve, Bahrami, and Shah 2023).

Geopolymers present a compelling solution due to their reduced carbon footprint and ability to maintain acceptable mechanical and durability properties. Unlike traditional cement, which requires high-temperature calcination of limestone, geopolymers are synthesized from aluminosilicate sources such as industrial by-products like fly ash (FA), blast furnace slag, and calcined clay (Imtiaz et al. 2020; Z. Zhang et al. 2014; Hojati, Rajabipour, and Radlińska 2022). This reduces energy consumption and promotes the recycling of waste materials, contributing to a circular economy in construction. The sustainability benefits of geopolymers are substantial; studies and research have shown that depending on the composition and processing method, CO emissions from geopolymer material production can be reduced by anywhere from 26% to 45% compared to traditional concrete (Ricciotti et al. 2023). The use of industrial waste materials such as slag and FA in geopolymer production further amplifies these environmental benefits, potentially achieving CO

DOI: 10.1201/9781003434863-8

emission reductions of up to 80% compared to OPC (Ma, Zhang, and Liu 2022; Zhao et al. 2021; Ricciotti et al. 2023). In fact, the synthesis process of geopolymers generates just 20% of the emissions associated with OPC production, with overall energy consumption reduced to only 25% of that of the OPC industry (Ma, Zhang, and Liu 2022; Zhao et al. 2021; Ricciotti et al. 2023). Alkali-activated slag and FA geopolymer concrete have demonstrated the ability to reduce production costs and carbon emissions by 15.7% and 73.8%, respectively (Y. Li et al. 2023), compared to traditional concrete. Moreover, 3D-printable one-part geopolymer mixes, which utilize solid activators, can cut embodied energy and carbon emissions by up to 70% compared to 3D-printed OPC of similar compressive strength (Ricciotti et al. 2023). Even when compared to traditional reinforced concrete, 3D-printing concrete alone has been found to reduce CO emissions by approximately 22% (Ricciotti et al. 2023).

Geopolymer cement/concrete (GPC) has gained significant attention due to its superior environmental performance compared to OPC. Since the calcination of limestone in OPC production is a major source of greenhouse gas emissions, utilizing alternative precursors, such as aluminosilicate-based GPCs, presents a viable strategy for reducing the construction sector's carbon footprint. Simultaneously, 3D printing is advancing as a method to enhance sustainability in construction by optimizing material usage. Beyond their environmental and technical advantages, GPC-based 3D printing offers some economic benefits. The automation of the printing process reduces labor costs and construction time while ensuring consistency in material application. This is particularly beneficial for large-scale projects, where speed and efficiency are crucial factors. Additionally, reduced dependence on imported construction materials makes this technology a viable solution for regions facing material shortages or high transportation costs. Given its potential to reshape the industry, integrating GPC with 3D printing represents a transformative step toward the future of sustainable construction, paving the way for greener, more resilient, and cost-effective building solutions.

This research area has gained interest in recent years, with significant efforts focused on developing GPC formulations compatible with 3D printing techniques. However, integrating GPC into 3D printing presents considerable challenges due to their specific rheological requirements. Unlike OPC pastes, which generally exhibit fresh-state viscosity and strength characteristics that are well-suited for 3D printing, geopolymer systems require more complex adjustments to achieve printability (Tabassum and Ahmad Mir 2023). The diverse chemical compositions of geopolymers result in a wide range of viscosities, setting times, and mechanical properties, making process optimization more intricate (Shilar et al. 2023).

This chapter presents a comprehensive overview of the current state of knowledge on geopolymers as construction materials, with a particular focus on automated manufacturing methods such as 3D printing. It begins with a discussion of the chemical reaction mechanisms and the historical development of geopolymers. This is followed by an introduction of material components and mix design, a discussion of 3D printing technologies and key considerations for their application in construction, and an analysis of the fresh and hardened properties of geopolymer materials based on the existing literature. Finally, it explores technical applications, current research challenges, and potential future directions in the field.

6.2 INTRODUCTION TO GEOPOLYMERS

6.2.1 Geopolymer Definition

There are multiple pathways to enhancing the environmental sustainability of concrete construction, with one of the most promising being the development of alternative cementitious materials that reduce reliance on traditional OPC. These materials utilize more complex chemical interactions to achieve similar performance without the high carbon footprint associated with OPC production. While various terms exist to describe these materials, no universally accepted terminology has been established. Unlike OPC, which relies on calcium-based binding, these alternative pastes solidify through the interactions of silica and alumina compounds. A key aspect of their reaction process is using a catalyst that significantly alters the mix's pH, breaking down aluminosilicate precursors, and facilitating their reformation into hardened structures. Depending on the specific mix design and chemical composition, the final hardened matrix can take the form of precipitate gels or long-chain polymers. The variation in precursor and reagent concentrations contributes to the diversity of properties and applications for these materials, as well as the ongoing debate regarding their classification.

The terminology surrounding alternative cementitious materials remains a topic of debate, particularly regarding "alkali-activated cement" (AAC) and "GPC." While some sources use these terms interchangeably, others consider GPC a subset of the broader AAC category. However, neither term fully encompasses the diverse range of minerals and chemical reactions involved. The primary distinction lies in their chemistry and resulting molecular structures. Additionally, aluminosilicate precursors can be activated by acidic sources, such as phosphate-based geopolymer cement (PGC), further expanding the classification of these materials (Zribi and Baklouti 2021; Yan-Shuai et al. 2017; Zribi and Baklouti 2022). Alkali/acidic activation initiates the geopolymerization process, but geopolymerization entails more extensive chemical reorganization, leading to unique material properties (Davidovits 2017). To minimize confusion arising from the overlapping terminology, in this chapter we will use the term "geopolymer" in general to refer to any aluminosilicate-activated binder that requires a pH-altering component to facilitate the hardening process. Further clarification is needed to differentiate these terms in the future; however, we will use the general terminology mentioned above for this chapter.

6.2.2 Historical Development of Geopolymer

Geopolymerization, a process pioneered by French researcher Professor Joseph Davidovits in the early 1970s, involves the formation of long, highly structured, and stable polymer chains of inorganic aluminosilicate compounds (Davidovits 2016, 2017; Davidovits and Sawyer 1985). A defining characteristic of this reaction is the presence of predictably repeated oligomeric links that form the polymeric structure. This unique chemistry enables the production of materials with exceptional strength and thermal stability, making them suitable for applications ranging from fire-resistant coatings to aerospace components. When applied to cement and concrete, geopolymerization results in an ambient-temperature cured solid, similar to fired ceramics or sialic metamorphic rocks such as feldspar, granite, or basalt (Aleem and Arumairaj 2012). This process closely mirrors the natural geological mechanisms responsible for forming essential rock minerals in the Earth's crust, as reflected in the composition of zeolitic materials. Inspired by this natural phenomenon, Davidovits (Davidovits and Cordi 1979) developed a novel binder using "earth materials" such as kaolinite, limestone, and dolomite. Drawing from their geological origins and the chemical processes involved, Davidovits coined the term "geopolymer" to describe these materials, emphasizing their connection to natural mineral formation and sustainable, earth-derived composition (Davidovits 2016). Davidovits also suggested that this cementitious paste method dates back 2,000 years to Roman civilizations, which used heated ingredients and catalysts like sea salts for high-performance concretes (Davidovits 2016).

The study of AACs began in the 1940s when a Ukrainian scientist discovered that ground granulated blast furnace slag (GGBFS), a by-product of steel production, could be hardened using sodium hydroxide (Feret 1939; Roy 1999). This discovery marked the foundation of alkali activation as a sustainable alternative to OPC. In recent years, AACs have gained widespread attention for their ability to utilize industrial by-products, such as FA, slag, and metakaolin (MK), as precursor materials. These materials, often referred to as supplementary cementitious materials (SCMs), can replace OPC in concrete mixes, significantly reducing the embodied carbon footprint of construction. Unlike geopolymers, which rely on the polymerization of aluminosilicate precursors, AACs harden through mechanisms more similar to OPC hydration chemistry.

The hydration products of cementitious systems are defined by the $MeO–Me_2O_3–SiO_2$ system, where *MeO* represents divalent metal oxides (e.g., CaO and MgO) and Me_2O_3 denotes trivalent metal oxides (e.g., Al_2O_3 and Fe_2O_3). This system produces key phases essential for material strength, like calcium silicate hydrate (C-S-H) and calcium aluminate hydrate (C-A-H). Besides, AACs can form two distinct systems (Krivenko 1994; Roy 1999): (i) $Me_2O–Me_2O_3–SiO_2–H_2O$, where Me_2O refers to alkali metal oxides (e.g., Na_2O and K_2O), resulting in alkaline zeolitic minerals like sodium aluminosilicate hydrate (N-A-S-H); and (ii) $Me_2O–MeO–Me_2O_3–SiO_2–H_2O$, which incorporates alkaline earth oxides (e.g., CaO and MgO) alongside alkali oxides, yielding mixed alkali–alkaline earth zeolites, calcium hydrosilicates (C-S-H), or carbonates.

The AAC process begins with breaking the pre-existing bonds in the precursor materials under highly alkaline conditions. However, if the mixture lacks enough alkali reagents, the chemical process cannot be sustained indefinitely due to the depletion of hydroxide compounds. The primary hydration products of AACs vary depending on the precursor composition and alkali source (Krivenko 1994; Nasir,

Mahmood, and Bahraq 2024; Provis, Palomo, and Shi 2015). In calcium-rich systems like GGBFS, calcium (alumino)silicate hydrate (C-(A)-S-H) gels are formed, resembling the amorphous, unorganized microstructure of OPC pastes. In contrast, sodium or potassium aluminosilicate hydrate (N-A-S-H or K-A-S-H) gels are produced in low-calcium systems, such as those using class-F FA or MK. These gels are the result of the dissolution of aluminosilicate precursors in alkaline solutions, followed by polycondensation into a stable network. Despite their potential, AACs face challenges related to long-term durability. The presence of soluble sodium (Na^+) or potassium (K^+) ions within the chemical structure can lead to leaching over time, causing the migration of alkaline components and potentially compromising material integrity. This issue is particularly critical in environments with high moisture or chemical exposure (Provis 2018; Revathi et al. 2022; Srinivasamurthy et al. 2021). Furthermore, while AACs and geopolymers share some similarities in their initial hardening mechanisms, their final chemical structures and performance characteristics differ significantly. This distinction limits the interchangeability of the terms "AAC" and "GPC," as they represent distinct material systems with unique chemistries and applications.

Beyond GPC and AAC, another category of aluminosilicate-based materials exists: PGC (Occhicone et al. 2024). Although their use traces back to ancient times (Barsoum, Ganguly, and Hug 2006), the concept of "phosphate chemically bonded ceramics" was formally introduced by Argonne National Laboratory (Wagh 2016). Subsequent research by Cao et al. (2005) explored aluminosilicate phosphate geopolymers, and Davidovits officially incorporated phosphate geopolymers into his geopolymer chemistry and applications monograph in 2011 (Davidovits 2011). PGCs are formed by reacting aluminosilicate precursors with acidic activators, such as phosphoric acid or phosphate solutions. Unlike AACs, which harden in a strongly alkaline environment, PGC geopolymerization occurs in an acidic medium, where phosphorus partially or entirely substitutes aluminum or silicate within the network structure (Occhicone et al. 2024). This unique chemistry endows PGCs with exceptional mechanical strength, thermal resistance, and dielectric properties. For instance, PGCs can achieve compressive strengths of up to 146 MPa and withstand temperatures as high as 1,500°C, making them highly suitable for extreme-environment applications (Ma, Zhang, and Liu 2022; Perera et al. 2008).

6.2.3 Geopolymer Formation

Although GPC, AAC, and PGC differ, their fundamental chemistry follows a similar process, which is summarized here. For GPC and AAC, the reaction begins with a significant pH shift, breaking atomic bonds in the precursor material and releasing silicate and aluminate ions (J. Davidovits 2017). These ions then form silicate and aluminate tetrahedrons, which share oxygen molecules, leading to the formation of oligomers and precipitates. If the reaction stops at this stage, the resulting hydration gel resembles the C-S-H phase in OPC hydration (Davidovits 2013), as shown in steps 1–3 in Figure 6.1. However, if the reaction continues, polycondensation occurs, forming a longer polymeric macrostructure characteristic of true GPC, as outlined

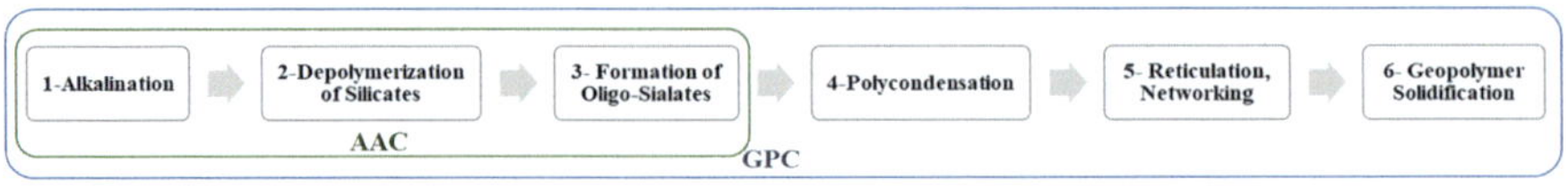

FIGURE 6.1 Six steps of geopolymerization, according to Professor Davidovits.

in steps 4–6. Davidovits defines the six steps of geopolymerization as displayed in Figure 6.1 (Davidovits 2013).

The degree of structural organization in geopolymerization depends on a precise chemical balance between precursor minerals and alkaline components. Factors such as silica, aluminum, and calcium content; the type of aluminum ion; and the reagent's concentration collectively determine the extent of chemical reorganization. While many aluminosilicate-rich powders can be alkalized to form hardened pastes, only a well-balanced mix achieves true polymerization. GGBFS, with its higher calcium content and dense structure, is less reactive than FA, a glassy aluminosilicate that readily dissociates. As a result, slag requires a stronger activator to break down its molecular bonds. In contrast, Davidovits' research on MK, a calcined clay, shows it is highly reactive and easily breaks down with alkalinization (Davidovits 2018). He also highlights that calcination temperature affects the aluminum ion type, influencing polycondensation levels.

Variations in reagents lead to different outcomes, particularly in the alkalinization process, which influences material properties and performance (Ghazy, Elaty, and Mostafa 2022). There are two main approaches: "one-part" (1P) and "two-part" (2P) mixes. The term "part" refers to the number of ingredients before water is added. In 1P mixes, a dry powder alkaline reagent is combined with the binder precursor to form a single dry ingredient, which is then activated by water. In contrast, 2P mixes use a separate chemical activator solution. Research suggests that 1P mixes often produce AACs rather than true GPCs due to differences in reagent interactions (Davidovits 2017). When dry reagents in 1P are mixed with water, they undergo incomplete transformations, leading to variations in silicate structures and uneven spatial distribution. In 2P mixes, the pre-dissolved activator ensures more stable chemical transformations, enabling better control over polymerization and material consistency.

PGC is also called a silico-aluminophosphate binder and follows a mechanism similar to traditional geopolymerization shown in Figure 6.1, with the key distinction being the initial alkalinization step being replaced by acidification (Y.-S. Wang, Alrefaei, and Dai 2019). This process begins with acid-driven depolymerization of aluminosilicate precursors, releasing reactive species (Ma, Zhang, and Liu 2022). Depending on the availability of activation agents, polycondensation occurs among PO_4^{3-}, Al^{3+}, and Si-O units, facilitating the formation of both amorphous and crystalline phases, such as $AlPO_4$. As polycondensation progresses, these units further interlink to develop a complex 3D geopolymeric network. In this structure, crystalline $AlPO_4$ is dispersed within an amorphous geopolymer matrix, ultimately forming a composite material with both crystalline and amorphous characteristics. A fundamental difference between acidic and alkaline activation lies in substituting aluminum or silicate with phosphorus within the polymeric network. The resulting phosphate-based geopolymer units commonly take the form of –Si–O–Al–O–P–, –Si–O–P–O–Al–, or –Al–O–P– (Mathivet et al. 2019; Louati, Baklouti, and Samet 2016; Zribi and Baklouti 2021).

It is important to note that the precise reaction mechanisms and microstructural characteristics of PGC remain a subject of ongoing research, with no universally accepted definition. Early studies proposed that geopolymerization in acidic environments occurs through an adhesion reaction between dissolved aluminum species from aluminosilicate precursors and phosphorus–oxygen tetrahedra. Another perspective suggests that in an acidic medium, the SiO_4^{4-} unit is partially replaced by PO_4^{3-} following the breakdown of Si–O–Al bonds, leading to an amorphous geopolymer network dominated by Al–O–P, Si–O–P, and –Si–O–P–O–Al– linkages (Perera et al. 2008; Zribi and Baklouti 2021).

Figure 6.2 illustrates the activation process of aluminosilicate precursors under acidic and alkaline conditions. By adjusting the pH, the precursor can be activated; however, beyond the differences in pH, the final product exhibits significant structural variations. Notably, the chemical environment of aluminum within the formed gels differs, with aluminum predominantly existing as Al(VI) in

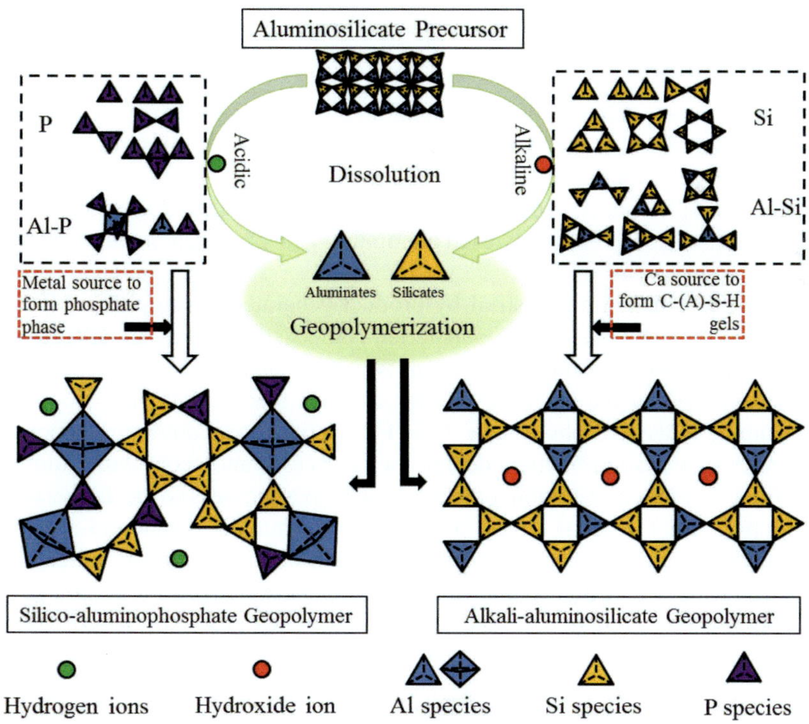

FIGURE 6.2 Geopolymer matrices formed through the activation of aluminosilicate precursors under either acidic or alkaline conditions. Reproduced from Wang, Alrefaei, and Dai 2019 licensed under CC BY 4.0.

silico-aluminophosphate-based geopolymers and as Al(IV) in alkali-aluminosilicate-based geopolymers (Wang, Alrefaei, and Dai 2019).

To lessen the confusion caused by overlapping terminology, this study will consistently use the term "geopolymer" to describe any aluminosilicate-based binder system (e.g., AAC, GPC, and PGC) that relies on a pH-modifying agent to initiate and sustain the hardening reaction.

6.3 MATERIAL COMPONENTS FOR GEOPOLYMER

As previously noted, geopolymer typically combines a powdered binder precursor, chemical reagent, and water to create a paste. Like OPC applications, this paste is then mixed with additional components to function as either mortar or concrete. Both alkaline and acidic activators can be used, with sodium or potassium salts being the most common alkaline initiators for geopolymerization. However, acidic reagents, such as phosphoric acid, have also been employed. Acidic activation breaks molecular bonds, enabling the formation of new structures incorporating phosphate ions, a process studied in PGCs. While both activation methods are viable, this discussion will focus on alkaline reagents, aligning with current research on the 3D printing of this material. The following section will majorly delve into geopolymer chemistry, covering precursors, reagents, fillers, admixtures, and mix design in greater detail.

6.3.1 Precursors

As mentioned, many different powdered aluminum- and silica-rich sources can be used as either partial SCMs or the main ingredient for a geopolymer mixture. Much interest in geopolymer development arose because certain industrial waste products make effective precursors, offering environmental benefits through industrial symbiosis. The following is a list of some commonly used aluminum- and silica-rich precursor materials:

- GGBFS: This is an industrial by-product generated during the production of iron in blast furnaces. This material, composed primarily of silicates and aluminosilicates of molten calcium, is periodically extracted from the furnace as part of the smelting process. Rich in reactive calcium, silica, and alumina in its amorphous state, GGBFS is a highly effective supplementary material for use in blended cementitious systems and geopolymer binders. Notably, among common precursors, GGBFS possesses the highest calcium content, leading to the predominant formation of calcium-alumino-silicate-hydrate (C-(A)-S-H) gel, which exhibits a disordered, amorphous microstructure similar to that found in OPC pastes (Mishra et al. 2024).
- FA: FA, a by-product of coal combustion, is a key precursor in geopolymer production due to its high aluminosilicate content. It is classified into Class C (high calcium, self-cementing) and Class F (low calcium, requiring activators), both of which play a role in geopolymerization. However, Class F FA often requires elevated curing temperatures to set, prompting studies on binary mixes combining FA with calcium-rich compounds to promote the chemical reaction at room temperature. Alkali activation of FA forms

sodium or potassium aluminosilicate hydrate (N-A-S-H or K-A-S-H) gels (Hojati, Rajabipour, and Radlińska 2016; Hojati and Radlińska 2017; Z. Li et al. 2020; Hojati, Rajabipour, and Radlińska 2019). Despite its widespread use as a primary precursor in geopolymer research, the shift toward greener energy sources is expected to reduce coal combustion, leading to a decline in FA availability (Assi et al. 2020). While some view FA as an abundant resource, the anticipated scarcity highlights the need to explore alternative binder precursors to meet future concrete demands.

- Red Mud (RM): RM is a highly alkaline by-product generated during alumina extraction from bauxite via the Bayer process. Composed of iron oxides, aluminum, and silica, it holds potential as a geopolymer precursor due to its aluminosilicate content. However, its high alkalinity and the presence of heavy metals, classified as hazardous and radioactive waste, require careful handling and processing. Consequently, research has predominantly focused on blending RM with other precursors, such as FA or GGBFS, rather than using it independently. This approach enhances stability, as the residual geopolymers and oxidizable heavy metal ions, initially present in the binder as weak acid-soluble and reducible fractions, mitigate environmental risks and improve material performance (Alam, Das, and Rao 2019; X. Li et al. 2024). RM's high NaOH content also reduces the need for expensive alkali activator solutions, lowering overall production costs (Mounika et al. 2023).
- Palm Oil Fuel Ash (POFA): POFA is an agro-industrial biomass waste generated in large quantities by the palm oil industry, particularly in Southeast Asian countries. Often considered a disposal challenge, this by-product can be repurposed through alkali activation, transforming it into a valuable resource. POFA can function as a pozzolan and a precursor for geopolymer, contributing to its strength and durability. Alkali-activated POFA also enhances soil strength and stability, making it suitable for construction and other engineering applications (Santhosh, Subhani, and Bahurudeen 2022; Zain et al. 2017).
- Kaolinite Clay and MK: Kaolinite is the most widely used clay mineral for alkali activation and as an SCM in blended cements (Hollanders et al. 2016; Biel et al. 2020). When heated to temperatures between 650 and 750°C, kaolinite undergoes complete dehydroxylation, resulting in MK, a highly reactive material (Tironi et al. 2013; Ilic, Mitrovic, and Milicic 2010). MK, rich in aluminum and silica, is particularly effective for geopolymers due to its strong polymerization potential, especially when calcined. While kaolinite can be used in its raw form, calcined kaolinite (i.e., MK) is preferred for its enhanced reactivity. MK can be used in its pure form for geopolymer or, in some cases, as part of a broader category of calcined clays with some level of impurity. Like FA, MK forms sodium or potassium aluminosilicate hydrate (N-A-S-H or K-A-S-H) gels during the alkali activation process, making it a valuable precursor for geopolymer production (Khalifa et al. 2020).
- Rice Husk Ash (RHA): RHA, an abundant agricultural waste by-product, is a sustainable and cost-effective alternative to conventional cementitious

materials. Through alkali activation, the silica-rich RHA undergoes dissolution in an alkaline solution, forming a polymerized sodium silicate gel, which contributes to the geopolymerization process (Kumar Das et al. 2022).

- Mine tailings (MT): MT generated from a wide range of global mining operations have emerged as promising precursor materials for alkali-activated binder systems (Akinyemi, Alaba, and Rashedi 2022). In a study by Obenaus-Emler et al. (Obenaus-Emler, Falah, and Illikainen 2020), leaching experiments conducted under varying parameters, such as time, alkaline solution concentration, and temperature, revealed limited dissolution of silica and alumina. Despite the low release levels, the resulting silicon-to-aluminum (Si/Al) ratios were favorable for forming binding gels, similar to those seen in traditional AACs. Additionally, the release of calcium under these conditions can contribute to strength development through the formation of C-S-H gels when sufficient silicon is present. To enhance the mechanical properties of tailings-based binders, researchers have employed strategies such as adding reactive components like GGBFS, particle size reduction through grinding to boost reactivity, and elevated temperature curing (Kiventerä et al. 2020). With the potential for on-site application in mining areas, MT-based binders present a sustainable and viable alternative to traditional cementitious systems.
- Space regolith simulant: Alkali activation of extraterrestrial regolith simulants, particularly those mimicking lunar and Martian soils, presents a compelling pathway for in situ resource utilization (ISRU) to support off-Earth construction. The mineralogical composition of both lunar and Martian regolith simulants has been found to be rich in aluminosilicate phases, chemically analogous to conventional precursors used in geopolymer binder systems (Geng et al. 2023; Yao et al. 2025). This inherent compatibility has driven a growing body of research investigating the feasibility of geopolymerization using these regolith analogs as the primary source material. Several studies have demonstrated the successful synthesis of geopolymer composites from space regolith simulants, meeting critical performance criteria for structural applications. Efforts have been made to adapt these materials for 3D printing processes, which offer scalability, precision, and reduced labor requirements in extraterrestrial environments (Hojati et al. 2018a, 2018b; Mills, Katzarova, and Wagner 2022). The ability to fabricate structural elements and potential habitat components directly from in situ materials reduces payload demands from the Earth and aligns with the long-term goals of sustainable human presence on the Moon and Mars.

The origin and composition of precursor materials are critical determinants in alkali-activated systems, influencing both the required activator concentration and the overall polymerization efficiency. Key chemical parameters, such as reactive silica and alumina availability, the Si/Al ratio, and the presence of calcium, directly impact the geopolymerization process and final binder performance. Notably, Davidovits emphasizes that the mineralogical characteristics and molecular structure of precursors are more influential than bulk oxide composition when designing effective mix formulations (Davidovits 2018). This perspective underscores the importance of detailed structural analysis in precursor selection and is further elaborated in the subsequent Mix Design section.

6.3.2 Chemical Reagents

The development of geopolymers heavily relies on the use of alkaline reagents, with commonly utilized catalysts including highly basic compounds such as potassium hydroxide and sodium metasilicate, as well as milder options like sodium carbonate. While acidic reagents like phosphoric and humic acid have been explored, the focus here remains on alkaline components due to their widespread application and availability. These reagents can be delivered in either dry 1P or solution-based 2P systems. Historically, 2P systems have dominated early research due to their superior geopolymerization consistency; however, they pose significant challenges, such as handling risks, higher transportation costs, and greater environmental and safety concerns. In contrast, 1P systems offer improved practicality, cost-effectiveness, and ease of use, particularly in small-scale or remote construction settings, but currently lack extensive literature and standardization. Despite their logistical complexities, 2P systems may still be more suitable for large-scale applications, provided appropriate safety measures and training are in place. Continued research into 1P geopolymer could significantly advance their reliability and accessibility, promoting broader adoption in the construction industry (Al-Noaimat et al. 2023; Davidovits 2017).

6.3.3 Fillers

Inherent reinforcement sources in geopolymers mirror those used in OPC research, with mix designs tailored to achieve specific properties and performance outcomes. Unlike OPC, geopolymer research has emphasized incorporating recycled or plant-based materials, reflecting its recent development and the growing focus on sustainability. The typically higher compressive strength of geopolymers reduces reliance on carbon-intensive virgin stone aggregates, enabling the exploration of less durable but more renewable alternatives. Research often focuses on sand for mortar paste testing, while traditional aggregates are commonly used in both laboratory and large-scale geopolymer projects to enhance compressive strength. Additionally, nontraditional materials, such as recycled glass and plant fibers, have been investigated for their potential to improve the ecological performance of geopolymers, further advancing sustainable construction practices.

6.3.4 Admixtures

Similar to OPC practices, various admixtures have been incorporated into geopolymer formulations to enhance specific properties. Although the range of admixtures is limited, their role in optimizing geopolymer performance is critical. Due to the often more challenging rheological behavior of geopolymers compared to OPC, specific additives are employed to modify viscosity and thixotropy, which is essential for adapting geopolymers from casting to 3D printing applications. Some of the key additives identified in the literature, particularly for 3D printing, include superplasticizers/high-range water reducers (HRWR), nano clay (NC), seeding crystals (e.g., hydromagnesite), and sodium carboxymethyl starch (CMS) (Qaidi et al. 2022; Panda et al. 2020; Sun et al. 2020). These admixtures help fine-tune workability, flowability, and printability, ensuring optimal processing and performance of geopolymer mixtures. Additionally, water-reducing admixtures, akin to those used in OPC, can further enhance the rheological properties of geopolymers, making them more adaptable to advanced manufacturing techniques like 3D printing.

In addition to traditional additives, several advanced materials are being explored to enhance the performance of geopolymers for specialized applications. For instance, nano-TiO_2 has been incorporated into geopolymer mixtures to impart self-cleaning properties and improve surface reflectivity. Its photocatalytic activity inhibits the growth of biological organisms, such as mold and algae, while enhancing durability (Ambikakumari Sanalkumar and Yang 2021). Another promising additive is graphene oxide, which has shown potential for enabling self-sensing and energy-storing capabilities in geopolymers, although this remains an emerging area of research (Qaidi et al. 2022). Additionally, certain metal elements, such as titanium and copper, are being investigated for their ability to introduce unique characteristics like conductivity and antimicrobial properties. These additives expand the functionality of materials, making them suitable for innovative applications in construction and beyond.

6.3.5 Mix Design

Unlike OPC, which primarily involves the interaction of calcium and water, geopolymer proportioning requires a more nuanced approach due to the variability of raw aluminosilicate precursors and industrial by-products. Davidovits (2011, 2013) categorizes geopolymer cements into four primary types: GGBFS -, rock-, FA-, and ferro-sialate-based, with MK and granite included in the rock subset. To achieve optimal performance and durability, Davidovits emphasizes the importance of ensuring complete geopolymerization rather than gel precipitation, outlining the following steps for effective geopolymer formulation (J. Davidovits 2016):

i. Master the Aluminosilicates: Precursor mineralogy is critical. For instance, FA composition varies significantly based on combustion temperature. At 850°C, many original minerals remain intact, while at 1,200°C, they transform into altered phases, and at 1,800°C, they fully convert to a glassy state. This glassy phase is essential for FA reactivity during alkali activation.
ii. Understand Chemical Reactivity: The molecular structure of *precursors*, such as GGBFS, is as influential as mineralogy. For example, while GGBFS has high CaO content, its behavior differs from free CaO in lime, necessitating higher alkalinity for effective activation.
iii. User-Friendly Systems: To mitigate hazards associated with highly caustic reagents, safer alternatives like soluble silicates with a $SiO_2:Na_2O$ ratio > 1.45 are recommended for on-site applications, while more aggressive reagents are reserved for laboratory or pre-cast use.

iv. Master Soluble Silicates: Alkaline reagents vary in properties due to differences in sodium silicate molecular structures, which influence viscosity and final product performance. Proper preparation and control of these reagents are essential for obtaining consistent results.
v. Replace Na^+ with K^+: Substituting sodium with potassium can double compressive strength, allowing equivalent performance with half the potassium content. Potassium reagents may offset expenses through reduced usage and enhanced performance despite higher costs.
vi. Standardize Mixing Steps: Proper mixing protocols are critical. Unlike OPC, geopolymers require careful preparation of the alkaline solution before combining it with dry ingredients to avoid premature interaction with aggregates and ensure optimal binder formation.

The baseline proportioning of geopolymers depends on the intended application and ingredient properties. Key ratios to regulate include Si/Al, Si/Na, Ca/Si, precursor/reagent, liquid/solid, and aggregate/binder (Al-Noaimat et al. 2023; Farooq et al. 2021; Qaidi et al. 2022; Shilar et al. 2023). Reagent concentrations typically range below 7% for safety, although some studies explore up to 20% NaOH. For 3D printing, admixtures such as superplasticizers, NC, and seeding crystals are often incorporated to optimize rheology and printability.

Davidovits provides an example formulation for cast applications using FA and GGBFS (Davidovits 2016), consisting of 10% potassium silicate solution (($SiO_2 : K_2O > 1.45$), 15–30% GGBFS, 5% water, and 50–85% FA (Class F, CaO < 4%). In this formulation, GGBFS and potassium silicate form the binder, which bonds with the glassy surface of high-temperature FA. Crucially, FA is not included in the initial binder paste to prevent its dissolution, which would reduce binding power and overall strength. This highlights the importance of precise processing steps in geopolymer design to achieve desired mechanical and durability properties. By adhering to these principles, geopolymers can be tailored for diverse applications, including advanced 3D printing, while addressing sustainability and performance challenges inherent in traditional OPC systems.

6.4 3D PRINTING OF GEOPOLYMERS FOR LARGE-SCALE CONSTRUCTION

In light of the substantial benefits associated with replacing OPC with geopolymer binders, particularly in terms of sustainability, durability, and reduced carbon emissions, and considering the construction industry's increasing adoption of advanced technologies, this chapter presents a comprehensive review of the current advancements in the use of geopolymer binders as a primary material for 3D printing in structural applications. Multiple factors, such as the size and complexity of the structure, the desired mechanical performance, and the rheological behavior of the geopolymer mixture, influence the selection of an appropriate 3D printing technique.

According to the existing literature, various 3D printing techniques have been investigated to identify the most suitable methods for utilizing geopolymers in 3D printing. Among these, extrusion-based 3D printing has emerged as the most widely adopted approach, primarily due to its operational simplicity, cost-effectiveness, and ability to produce large-scale components rapidly. This section begins with an overview of the 3D printing technologies applicable to geopolymer-based construction, followed by a detailed examination of the material requirements specific to extrusion-based systems. Emphasis is placed on optimizing the key properties of geopolymer mixtures, such as workability, setting time, buildability, and mechanical strength, to ensure compatibility with the demands of extrusion-based printing (EBP).

6.4.1 3D Printing Techniques for Geopolymer Applications

ASTM 52900 (ISO/ASTM 52900 2015) is an international standard on AM (i.e., 3D printing), which defines 3D printing as creating any physical entity through the continuous addition of material. 3D printing utilizes different techniques to fabricate 3D-printed objects, such as heating to bind the materials, using laser light to fuse them, and using liquid/activating solution to combine particles. 3D printing processes are categorized into seven classes according to ASTM 52900 (ISO/ASTM 52900 2015): binder jetting, directed energy deposition, material extrusion, powder bed fusion, material jetting, sheet lamination, and vat photopolymerization. In construction, three categories of 3D printing have been vastly used in different studies, including EBP, binder jetting, and shotcrete-based printing.

6.4.1.1 Extrusion-Based 3D Printing

EBP is one of the most popular and promising techniques used in the digital-aided construction of large structural parts. "Contour crafting" (Khoshnevis et al. 2006), established at the University of Southern California, and "concrete printing" (Buswell et al. 2007, 2008), proposed at Loughborough University, are two famous EBP examples. In EBP, structural elements are created by layer-wise deposition of fresh cementitious material from a nozzle along a predefined path (Buswell et al. 2018; Perrot, Rangeard, and Pierre 2016; Beek 2020; ANTON et al. 2020). Most of the research teams around the world are currently working on applying 3D printing in construction using the concrete extrusion method of fabrication because of the promising potential of EBP in the construction of mega-scale components with high printing quality, safe construction environment, comparable/superior mechanical performance compared to the conventional construction, and the capability of constructing any geometrically complex parts. The material presented in the following sections of this chapter is also all from the existing literature that used EBP for their manufacturing.

In EBP, geopolymer materials, whether in paste, mortar, or concrete, must exhibit carefully tailored rheological properties to ensure successful printing. Specifically, the mix should possess low dynamic yield stress to allow smooth flow through the nozzle while maintaining high static yield stress and thixotropic behavior to support the deposited layers and retain shape after extrusion. Key process parameters, including printing speed, layer height, and nozzle geometry (shape and size), play a critical role and may vary depending on the desired application. EBP has gained particular attention for geopolymer 3D printing due to its suitability for fabricating structurally sound components with intricate geometries. EBP is the most compatible with geopolymers among various 3D printing methods, especially for large-scale construction applications (Hojati et al. 2018a, 2018b; Singh et al. 2015; Ricciotti et al. 2023; Shilar et al. 2023). In the following sections, a more comprehensive discussion on the application of geopolymers for 3D printing is provided.

6.4.1.2 Binder Jetting 3D Printing

Binder jet printing (BJP) in construction, also known as D-shape printing, developed by Enrico Dini ("D-Shape" n.d.), is an up-scale layer-by-layer fabrication process through jetting (spraying) a liquid, binder, or adhesive agent on powder-based material for solidification and bonding to form stone-like printed objects (Jakupovic 2016; Shakor et al. 2019a, 2022). The powder-based material for large-scale printing in construction mainly consists of fine aggregates (i.e., sand) or a mix of sand and binder (i.e., OPC or geopolymers) that needs to be activated by a proper agent. The adhesive agent is primarily a liquid material such as water, aqueous solutions, or cementitious pastes (i.e., OPC or geopolymers) used as an activator to bond the powder-based materials (Paolini, Kollmannsberger, and Rank 2019; Elsayed et al. 2022). This method could be used for the large-scale construction of the partition walls of gypsum; any concrete members of OPC/geopolymer mortars including the wall, column, or cladding façade of the building, load, and non-load-bearing structural components made of clay; and the creation of artificial sandstone made of quartz-sand. However, there are some challenges and limitations for BJP application in construction, such as the high porosity of printed parts and high dosage of fine powder particles in the job sites, which make the printing site dusty and the cleaning process very hard (Shakor et al. 2019b; Elsayed et al. 2022).

BJP has emerged as a promising 3D printing method for fabricating geopolymer-based components, particularly for applications where complex geometries, high printing resolution, and architectural flexibility are prioritized. BJP offers advantages in effective printable volume, rapid printing speed, and surface finish (Elsayed et al. 2022). However, its adoption for structural-grade geopolymer parts has been limited due to binder formulation and powder-bed preparation challenges. Geopolymer systems, particularly those based on MK, provide superior mechanical strength compared to OPC, which is a critical advantage in offsetting the porosity inherently introduced by BJP. Desamanera company ("Desamanera Brings Large-Scale Stone 3D Printing to the US," n.d.) pioneered the selective jetting of alkaline solutions onto MK-containing powder beds, triggering in situ geopolymerization. This technique, by using MK as the solid phase and the alkaline activator as the jetted liquid, demonstrated the feasibility of printing geopolymer structures with improved strength and dimensional accuracy, critical metrics for architectural and artistic applications.

A key advancement in this process is the development of tailored feedstock powders with optimized flowability and reactivity. Fine particles, such as those below 20 μm, pose significant challenges in spreading due to interparticle forces that lead to agglomeration, macro-defects, and uneven deposition (Averardi et al. 2020; Yang, Zou, and Yu 2003). To address this, a granulation process was introduced, wherein sand particles were coated with up to 30 wt% MK, achieving homogeneous MK distribution and enhancing the interaction with the alkaline solution. This coating not only improved the print resolution and binder activation but also ensured that the jetted liquid effectively wet the reactive surface, enabling rapid viscosity build-up and geopolymer gel formation. Comparative experiments using MK-only powders and MK with a small percentage of rapid hardening OPC (RHC) (Nuaklong et al. 2018; Nath and Sarker 2015; Mehta and Siddique 2017) revealed that the MK-only system could deliver comparable print quality and mechanical performance, affirming the potential of OPC-free BJP for geopolymer-based printing. This highlights the opportunity for sustainable, low-carbon 3D printing using entirely aluminosilicate-based systems without compromising the structural integrity or resolution.

6.4.1.3 Shotcrete-Based 3D Printing

Shotcrete-based printing (SBCP) was first proposed at the Braunschweig University of Technology in 2017 (Heidarnezhad and Zhang 2022). SBCP is a new manufacturing process in concrete that refers to spraying pre-mixed fresh concrete with high-pressure air (like the wet-mix shotcrete process) from the nozzle to create layer-by-layer objects. The previous studies indicated that using high pressure in SBCP improved the compaction of freshly printed materials and led to high interlayer bond and mechanical performance (Lindemann et al. 2018; Ahuja 2017). This method also offers 3D printing of overhangs and vertical surfaces with a fast printing rate. However, some challenges in applying SBCP in large-scale construction include the design of shootable but printable concrete mixtures, material loss because of rebound of aggregates and fibers, and achieving high-quality printed material with high geometry precision (Hack et al. 2021; Heidarnezhad and Zhang 2022; Lindemann et al. 2018; Kloft et al. 2019). This method is the newest construction technique, gaining more popularity for a specific application.

In contrast to geopolymers, the use of OPC-based mortars in shotcrete applications has been extensively studied and is well established. Key factors influencing shotcrete performance include workability, thixotropy, yield stress, plastic viscosity, rebound rate (the proportion of material that bounces off the substrate), and sloughing rate (the downward flow of material on vertical surfaces) (Chen et al. 2022; Yun, Choi, and Yeon 2015). These properties are optimized using admixtures like viscosity modifiers, air-entraining agents, and mineral additives (e.g., FA and silica fume) (Bindiganavile and Banthia 2009; Yun, Choi, and Yeon 2018; Castellanos et al. 2023). Operational parameters, such as nozzle design and spray distance, also impact performance. Wang et al. (X. Wang, Islam, and Zhang 2024) showed that direct material–air contact improves spray quality, while Su et al. (Su, Wu, and Zheng 2022) linked spray distance to rebound behavior influenced by rheological and setting properties. These findings provide a foundation for advancing geopolymer-based shotcrete.

Recent studies have explored the feasibility of using geopolymers for shotcrete applications. Hwalla et al. (2024) investigated FA- and GGBFS-based geopolymer mixtures incorporating dune sand, finding that geopolymers are suitable for shotcrete, with yield stress and plastic viscosity directly affecting flow and buildup thickness. Their results highlighted that mixtures with higher GGBFS content, increased sand-to-binder ratios, elevated sodium hydroxide molarity, and reduced solution content achieved greater buildup thickness due to enhanced viscosity. Similarly,

Zou et al. (2024) developed geopolymer mortars for shotcrete, analyzing the impact of liquid–binder and binder–aggregate ratios on flowability, P-wave velocity, and compressive strength. They identified optimal mix proportions (binder–aggregate ratios of 0.209–0.321) that achieved early-age compressive strengths of 28.4–41.1 MPa (1 day) and 44.2–55.4 MPa (28 days), meeting the requirements for wet shotcrete applications. These studies demonstrate the potential of geopolymers as a viable alternative for shotcrete, offering both mechanical performance and workability.

6.4.2 Key Properties and Performance Factors of Extrusion-Based 3D-Printed Geopolymers

As previously noted, this section focuses on the EBP process for geopolymers. It reviews the fresh and hardened properties of components fabricated using this technique, emphasizing the behavior of 3D-printed geopolymers at both stages. In the fresh state, key properties such as flowability, extrudability, and buildability are introduced, along with their relationship to the rheology of geopolymer mixes. Assessing rheological characteristics is highlighted as a critical factor in determining the fresh properties of geopolymer mixtures. In the hardened state, important mechanical and durability properties, including density, flexural strength, compressive strength, and interlayer bond strength, are reviewed. Additionally, the test methods and key factors influencing these hardened properties are reviewed, providing a comprehensive understanding of the performance of 3D-printed geopolymers.

6.4.2.1 Printability and Rheology

Unlike conventionally mold-cast concrete, 3D-printed concrete is characterized by a layer-by-layer construction process that eliminates the need for formwork. This unique method imposes requirements that extend beyond the conventional standards for concrete as defined by codes or guidelines. Printable concrete mixtures must be sufficiently fresh to enable pumping from the mixer through hoses and extrusion through a nozzle, yet they must rapidly gain enough strength after deposition to support their weight and that of subsequent layers. Consequently, the key process-related properties for successful 3D printing, flowability, extrudability, and buildability, have been widely recognized (C. Zhang et al. 2021; D. Zhang and Wei 2017; Z. Li et al. 2020). Flowability (or pumpability) defines the ability of freshly mixed geopolymer concrete to flow through the printing system, extrudability refers to its capacity to pass smoothly through a nozzle as a continuous filament, and buildability (or green strength) measures the material's ability to maintain its extruded shape under loading from itself and following layers. These properties, which are highly dependent on the rheological behavior of the mixture, have been recently characterized through emerging test methods (Wangler et al. 2022; Marchon et al. 2018).

The fresh-state behavior of geopolymers must be specifically engineered to meet the processing demands of 3D printing technologies. Unlike OPC systems, where hydration-driven mechanisms and water-reducing admixtures dominate workability control, geopolymers exhibit fundamentally different rheological characteristics governed by the dissolution and polycondensation of aluminosilicate precursors under alkaline activation. These systems manifest complex time-dependent viscoelasticity, with rheological evolution closely tied to the kinetics of precursor dissolution and subsequent gelation processes (Shilar et al. 2023; Tanyildizi et al. 2025). Their unique polymeric structure, comprising a dense network of cross-linked silicate and aluminate chains, yields substantially higher viscosity and enhanced elasticity compared to OPC pastes. This elastic framework imparts superior resistance to shear and deformation, along with improved energy dissipation capacity, attributes particularly beneficial for extrusion-based and injection-assisted AM processes, as well as self-leveling applications (Shilar et al. 2023).

Due to this structural complexity, rheological assessment of geopolymer pastes demands rigorous experimental characterization. Tunability of these properties is achievable through strategic incorporation of chemical modifiers such as superplasticizers and setting accelerators, allowing for control over flow behavior and setting kinetics. Beyond admixture optimization, the composition of the alkaline activator exerts a critical influence on fresh properties and buildability (Şahin and Mardani-Aghabaglou 2022).

Early-age strength is essential to limit deformation under self-weight and subsequent loads, driving the need for accelerated setting times. However, a faster setting must be carefully balanced with adequate workability, flowability, and interlayer adhesion, which tend to degrade as the setting accelerates. Therefore, strategies that promote early strength gain without significantly increasing viscosity or decreasing setting time are of high interest. Moreover, printable materials must remain workable for a practical duration, as geopolymers undergo time-dependent chemical reactions and molecular rearrangements that influence their rheological properties throughout the printing process. In this context, continuous or in-line mixing methods are preferred to maintain process control and consistency in material characteristics.

Thixotropy, defined as the ratio of viscosity during agitation to viscosity at rest, has gained particular attention in geopolymer systems due to its critical role in balancing flowability and buildability (Qaidi et al. 2022). Lower thixotropy values indicate a significant viscosity drop under agitation, promoting better workability. This behavior is attributed to the breakdown of molecular networks during movement, which reform at rest to support the deposited layers. Adjustments in water content and base ingredients can fine-tune viscosity, though typically within a narrow range (Chaiyotha et al. 2023). Notably, the Si/Na ratio in the alkaline activator solution is inversely correlated with viscosity and yield stress, with a value of 3.2 reported as optimal for workability in 3D printing applications (Qaidi et al. 2022). Since the viscosity of the reagent strongly influences the overall paste viscosity, proper control of mix chemistry is crucial.

Admixtures offer additional means for tailoring fresh properties. For instance, incorporating NC and hydromagnesite nucleation seeds has enhanced early strength development and thixotropy in geopolymers (Panda et al. 2020). NC promotes flocculation, increasing viscosity, while nucleation seeds facilitate the formation of C-S-H and C-(A)-S-H gels. These changes improve material stability at rest, allowing deflocculation and reducing viscosity under shear. Similarly, CMS has been studied for its effects on buildability, with optimal performance observed at 4–6% CMS by weight. Performance deteriorated above or below this range (Sun et al. 2020).

Yield stress, both static and dynamic, is a key parameter for quantifying early strength and buildability in EBP (Qaidi et al. 2022). Static yield stress represents the force needed to initiate flow, while dynamic yield stress (plastic viscosity) reflects the resistance to ongoing motion. High-yield stress can reduce extrudability and interlayer bonding. For example, the optimal CMS-modified mix showed plastic yield stress in the 32.5–66.7 Pa range, while mixes with >6% CMS displayed excessive yield stresses, compromising performance (Sun et al. 2020).

Flowability has also been closely studied, with target values depending on the application. For instance, rendering mortars typically require 100–110% flowability, while 65–75% is suited for brick joints. In one study, increasing the liquid-to-binder (L/B) and silica-to-FA ratios led to substantial improvements in flowability; a 5% increase in L/B resulted in a 30% increase in flowability, highlighting the sensitivity of mixture design (Chaiyotha et al. 2023).

Setting time, another critical factor, is highly variable and influenced by mix components. For instance, adding 8% CMS extended the final setting time from 66 to 287 minutes, rendering the mix unprintable, thus illustrating the wide range of behaviors achievable through formulation adjustments (Sun et al. 2020; Qaidi et al. 2022). The rearrangement of aluminosilicate networks not only contributes to strength development at rest but also accelerates setting at the edges of mixing or delivery systems, potentially leading to inconsistent performance and efficiency issues (Qaidi et al. 2022).

Interestingly, thermal treatment has been explored to enhance 3D printing performance. Preheating mixes before extrusion significantly improved buildability by accelerating hardening, capitalizing on the chemistry of geopolymerization prior to filament formation (Tramontin Souza et al. 2021). This approach aligns with traditional practices of curing geopolymers at elevated temperatures and offers a strategy to enhance material performance. While the study employed advanced automated mixing systems, such equipment is expected to become standard in large-scale commercial 3D printing applications.

6.4.2.2 Mechanical Strength and Durability

In 3D concrete printing, the absence of vibration for compaction commonly leads to the formation of inter-filament voids between printed layers, resulting in anisotropic mechanical behavior in the hardened stage. These voids can compromise structural performance and have become a key concern in the field. Conversely, the high pump pressure applied during extrusion can reduce intra-filament voids within individual filaments, promoting a denser microstructure. Therefore, the combined influence of both inter- and intra-filament voids plays a crucial role in determining the hardened properties of 3D-printed mixtures (Z. Li et al. 2020). Additionally, the density of printed concrete largely depends on the interaction between pump pressure and printing speed. A well-balanced combination minimizes voids and enhances print quality, often resulting in denser specimens than mold-cast ones (Le et al. 2012). However, under-extrusion, caused by insufficient pump pressure relative to the printing speed, results in higher porosity and reduced density in the printed material. Thus, optimizing these parameters is essential for achieving desirable material performance in 3D printing applications.

The mechanical performance of geopolymer materials is highly variable and depends on several factors, including the initial ingredients (precursors, chemical reagents, fillers, and admixtures), the curing method (room temperature vs. elevated temperature, and duration of curing), and the alignment of the 3D printing direction relative to the loading direction. These variables significantly influence the final mechanical properties of the material. Mechanical strengths, particularly compressive and flexural, are vital for structural applications. While most findings report higher strengths in mold-cast samples (Z. Li et al. 2020), exceptions exist (Nerella and Mechtcherine 2019), likely due to variations in materials, geometry, printing speed, and pump pressure. Ziejewska et al. (2022) developed concrete–geopolymer hybrid materials and demonstrated that 3D-printed hybrid specimens exhibited higher compressive strength than their molded counterparts.

As previously discussed in the material components section, some geopolymer mixes require elevated temperature curing to achieve acceptable strength and a higher degree of precursor activation. This need for high-temperature curing poses a significant limitation for large-scale, in situ 3D printing applications, as it is both energy-intensive and impractical for on-site projects. However, such geopolymer formulations are well-suited for the precast industry, where 3D-printed geopolymer blocks can be manufactured and cured under controlled conditions before being transported to construction sites. Elevated temperature curing enhances the mechanical, structural, and fire-resistant properties of geopolymers, making them an excellent choice for applications demanding high performance in these areas. In addition to elevated temperature curing, some geopolymer concrete mixtures require supplementary curing steps to achieve adequate strength. For instance, researchers working with powder-based 3D printing employed post-processing by immersing printed samples in a sodium metasilicate solution to enhance alkalinity and promote further geopolymerization (Xia and Sanjayan 2016; Maciej Serda 2013). Although this approach led to a notable strength increase, it appears to be specific to the limitations of the "drop-on-demand" printing process, where limited reaction time results in initially low strength. While such a method may offer benefits, it may not yield a fully developed geopolymer cement system. In addition to elevated temperatures, some geopolymer mixtures require additional curing steps to achieve sufficient strength. Researchers working with powder-based 3D printing methods reported soaking their printed samples in a bath of wetted sodium metasilicate to promote further alkalization and enhance strength development (Xia and Sanjayan 2016). Although this approach resulted in notable strength gains, it may address a specific limitation of the "drop-on-demand" printing process, where limited chemical reaction time leads to reduced initial strength. While potentially beneficial, this curing method may not yield a fully developed geopolymer cement.

Due to process-induced anisotropy, numerous studies have compared printed and mold-cast specimens for OPC and geopolymer materials. Compressive strength is typically highest when the load is applied in the filament deposition direction, benefiting from better compaction (Panda, Chandra Paul, and Jen Tan 2017; Ricciotti et al. 2023). Flexural strength, on the other hand, peaks in the perpendicular direction, where tensile stresses are maximized at the specimen's bottom. Lower layers often exhibit better performance due to compaction from upper layers and reduced water-to-binder ratios caused by bleeding (Le et al. 2012; Ziejewska et al. 2022). These anisotropic traits offer opportunities to tailor mechanical properties for targeted functions. Interlayer bonding is another critical issue. The layer-by-layer method can create cold joints with longer time gaps between layer depositions. Recent advances aim to improve bonding through enhanced mechanical interlocking (Zareiyan and Khoshnevis 2017) and engineered interface properties (Marchment, Sanjayan, and Xia 2019).

The setting and hardening behaviors of geopolymers vary significantly from that of OPC mixtures, necessitating careful consideration of cold joint formation, particularly in fast-setting geopolymer systems. Some geopolymer mixes are set within a day, while others can harden in under an hour, posing challenges for maintaining adequate interlayer bonding during printing. Panda et al. (2018) investigated the interlayer bond strength of geopolymers and observed that a time gap of 3 minutes between layers resulted in the strongest bonds and minimal evidence of cold joints while still maintaining printability. In contrast, a 5-minute delay led to smoother fracture surfaces along single planes, indicative of weaker interlayer cohesion, whereas the shorter gap produced jagged failure lines, suggesting better mechanical interlocking between layers. These findings, however, may not directly scale to large-format printing, where process timing and layer deposition rates are more difficult to control.

The reduction in bond strength with increased time gaps has been attributed to surface stiffening and the accumulation of air voids, which inhibit proper fusion between layers. A similar issue arises with increased printing speeds, which have been shown to lower interlayer bond strength (Qaidi et al. 2022), likely due to higher rates of air entrapment and reduced contact time for adhesion. In addition, nozzle standoff distance has emerged as a critical parameter in achieving consistent interlayer bonding. While it is common to set the standoff height equal to the desired filament height, recent studies (Chaiyotha et al. 2023) suggest that reducing the nozzle height to allow physical interaction between the nozzle and previously printed filaments may enhance interfacial consolidation, leading to a more homogeneous structure. However, this approach can introduce dimensional variability and inconsistency across prints, which poses a challenge to repeatability and structural integrity in scaled-up applications.

Durability and long-term performance data for geopolymers remain relatively limited; however, initial studies investigating key indicators such as freeze–thaw resistance, fire resistance, and chemical durability suggest that geopolymers generally demonstrate greater stability compared to OPC systems (Farooq et al. 2021). Within the broader category of geopolymers, AACs have been observed to suffer from reduced mechanical strength and increased water solubility, which can limit their viability in long-term structural applications. In contrast, geopolymer-based formulations tend to perform more favorably in these domains. Notably, geopolymers have shown successful deployment in outdoor environments and exhibit exceptional resistance to acidic conditions (Farooq et al. 2021). These advantageous properties are largely attributed to the highly cross-linked, crystalline, 3D network structure formed during geopolymerization. This structure imparts glass-like characteristics to the hardened material, enabling it to withstand aggressive environmental exposures over extended periods. In fact, archaeological evidence points to the long-term resilience of geopolymer materials, with examples enduring in exposed conditions for millennia (Davidovits 2017).

Shrinkage has received relatively little attention in the 3D printing of construction materials. The lack of formwork increases water evaporation, exacerbating shrinkage. Moreover, printable mixtures typically lack coarse aggregates, reducing internal restraint. Differential shrinkage between newer and older layers can further compromise interfacial bonds (Le et al. 2012). Mitigation strategies such as incorporating fibers, clay minerals, shrinkage-reducing admixtures, and superabsorbent polymers have been proposed (Mohan et al. 2021), although their broader impacts remain underexplored.

Previous studies have shown that using GGBFS as a precursor can increase the likelihood of shrinkage in geopolymers (Zhu et al. 2021; Huang et al. 2021; Hojati and Radlińska 2017). Room-temperature cured geopolymers, particularly those activated with FA or a binary blend of GGBFS and FA, have exhibited significantly greater shrinkage compared to OPC mixes (Hojati, Rajabipour, and Radlińska 2016, 2019; Hojati and Radlińska 2017). Conversely, other studies have reported that specific geopolymer formulations demonstrate excellent volumetric stability, with shrinkage values typically below 1%, due to the complete chemical incorporation of water during aluminosilicate oligomer formation (Davidovits 2017; Farooq et al. 2021). While these geopolymers generally resist shrinkage-induced cracking, they may still develop cracks influenced by the presence of fillers and the composite microstructure, with crack propagation often occurring along interfaces between aggregates, fibers, or reinforcements (Davidovits 2017; Shilar et al. 2023). Hojati et al. (Hojati, Rajabipour, and Radlińska 2022) also noted that geopolymers cured at elevated temperatures exhibit significantly lower creep and shrinkage than OPC, whereas room-temperature cured geopolymers showed substantial drying shrinkage and long-term creep. Since 3D-printed structures are often highly exposed to drying environments, such shrinkage behavior can lead to cracking and performance concerns for in situ 3D printing using shrinkage-prone geopolymers. Therefore, additional studies are required to comprehensively characterize and control the shrinkage performance of these emerging materials for 3D printing applications.

6.5 APPLICATIONS AND CASE STUDIES

While geopolymers have been explored for various construction applications since their development in the late 1970s, their integration with 3D printing technology is a more recent advancement. This section highlights key applications and case studies of the use of geopolymers in 3D printing, showcasing their potential for innovative construction solutions.

The first commercial geopolymer product, Pyrament, was developed in the late 1970s by Lone Star Industries Inc. in collaboration with Professor Joseph Davidovits. Designed as a heat-resistant, fast-setting concrete, Pyrament was ultimately discontinued due to conflicts with traditional cement manufacturing and internal challenges. Despite this setback, geopolymers have seen significant development globally, particularly in Australia. For example, Brisbane Airport utilized over 30,000 m^3 of

GGBFS/FA-based geopolymer for runways, sidewalks, and buildings, demonstrating its durability and performance (Davidovits 2016). In 2022, the University of Queensland's Global Change Institute employed geopolymer in a structural application, using 33 precast panels for three suspended floors, further validating its structural capabilities (Davidovits 2016).

Integrating geopolymers with 3D printing technology is still in its early stages but holds immense promise. One notable example is the work by Geopolymer International (GPI), which has demonstrated the feasibility of 3D-printed GPC structures. GPI printed a 500-ft^2 display home in Las Vegas, NV, using a 2P MK-based GPC mix. The project, completed in 12 hours over three nights to avoid daytime heat, highlights the potential for rapid construction with GPC. GPI has also developed specialized 3D printing equipment, including electronically controlled mixing mechanisms and rotating nozzle heads with side baffles, which improve print precision and enable complex geometries (Shilar et al. 2023).

Another pioneer in 3D printing of GPC is Apis Cor, which has been researching self-healing FA-based geopolymer since 2018. By immobilizing bacteria or fungal spores in biochar within the mix, Apis Cor's geopolymer can autonomously repair cracks through the production of $CaCO_3$ when exposed to water. This innovation addresses durability concerns and extends the lifespan of 3D-printed structures.

6.6 CHALLENGES AND FUTURE DIRECTIONS

Despite promising developments, large-scale applications of 3D-printed geopolymers remain limited. Key limitations include the need for precise control over rheology, curing conditions, print parameters, and issues related to the high alkalinity of geopolymers, which complicates nozzle design and printer maintenance.

Current research is actively addressing these barriers through optimized mix designs, advanced printing equipment, and the incorporation of novel additives like graphene oxide. Leading companies such as GPI and Apis Cor demonstrate the feasibility of geopolymers as a viable alternative to OPC in 3D printing. However, achieving consistent printability remains a major hurdle, particularly due to the geopolymer's complex rheological behavior. Thixotropy and viscosity are critical for extrudability, buildability, and interlayer adhesion, all of which are influenced by precursor materials, activator ratios, and mix design.

Durability is another area of concern. While GPCs generally exhibit superior resistance to freeze–thaw cycles, fire, and acid compared to OPC, distinctions between AACs and true GPCs must be clarified. Although similar in composition, AACs lack the molecular stability of geopolymers and may degrade over time, particularly in freeze–thaw environments, due to soluble sodium and potassium ions. Thus, long-term performance assessments and durability testing are critical. Additionally, there are limited studies on 3D printing of acidic-based geopolymers, i.e., PGC.

Future directions for 3D printing of geopolymers include innovations like multi-filament printing, tailored functionality within a single structure, and AI-assisted mix design for improved performance predictability. To further reduce environmental impact, efforts must continue to develop low-carbon activators and enhance the use of industrial by-products like FA and GGBFS. Standardizing mix designs and performance criteria will ensure consistency, while strong collaboration between academia and industry will help bridge the gap between research and real-world deployment. By tackling these technical and material challenges, 3D printing of geopolymers can emerge as a transformative solution for sustainable, durable, and adaptable construction.

REFERENCES

Ahmed, Ghafur H. 2023. "A Review of '3D Concrete Printing': Materials and Process Characterization, Economic Considerations and Environmental Sustainability." *Journal of Building Engineering* 66 (May): 105863. https://doi.org/10.1016/j.jobe.2023.105863.

Ahuja, Vishwajeet. 2017. "Non-Destructive Approach for Sprayed Concrete Lining Strength Monitoring." University of Warwick. http://wrap.warwick.ac.uk/103087/1/WRAP_Theses_Ahuja_2017.pdf.

Akinyemi, Banjo A., Peter Adeniyi Alaba, and A. Rashedi. 2022. "Selected Performance of Alkali-Activated Mine Tailings as Cementitious Composites: A Review." *Journal of Building Engineering* 50 (June): 104154. https://doi.org/10.1016/j.jobe.2022.104154.

Al-Noaimat, Yazeed A., Seyed Hamidreza Ghaffar, Mehdi Chougan, and Mazen J. Al-Kheetan. 2023. "A Review of 3D Printing Low-Carbon Concrete with One-Part Geopolymer: Engineering, Environmental and Economic Feasibility." *Case Studies in Construction Materials* 18 (July): e01818. https://doi.org/10.1016/j.cscm.2022.e01818.

Alam, Shamshad, Sarat Kumar Das, and B. Hanumantha Rao. 2019. "Strength and Durability Characteristic of Alkali Activated GGBS Stabilized Red Mud as Geo-Material." *Construction and Building Materials* 211 (June): 932–42. https://doi.org/10.1016/j.conbuildmat.2019.03.261.

Aleem, M. I. Abdul, and P. D. Arumairaj. 2012. "Geopolymer Concrete – A Review." *International Journal of Engineering Sciences & Emerging Technologies* 1 (2): 118–22. https://doi.org/10.7323/ijeset/v1_i2_14.

Ambikakumari Sanalkumar, Krishnan U., and En-Hua Yang. 2021. "Self-Cleaning Performance of Nano-TiO2 Modified Metakaolin-Based Geopolymers." *Cement and Concrete Composites* 115 (January): 103847. https://doi.org/10.1016/j.cemconcomp.2020.103847.

Anton, Ana, Patrick Bedarf, Angela Yoo, Benjamin Dillenburger, Lex Reiter, Timothy Wangler, and Robert J. Flatt. 2020. "Concrete Choreography: Prefabrication of 3D-Printed Columns." *Fabricate 2020*, July, 286–93. https://doi.org/10.2307/J.CTV13XPSVW.41.

Assi, Lateef N., Kealy Carter, Edward Deaver, and Paul Ziehl. 2020. "Review of Availability of Source Materials for Geopolymer/Sustainable Concrete." *Journal of Cleaner Production* 263 (August): 121477. https://doi.org/10.1016/j.jclepro.2020.121477.

Averardi, Alessandro, Corrado Cola, Steven Eric Zeltmann, and Nikhil Gupta. 2020. "Effect of Particle Size Distribution on the Packing of Powder Beds: A Critical Discussion Relevant to Additive Manufacturing." *Materials Today Communications* 24 (September): 100964. https://doi.org/10.1016/j.mtcomm.2020.100964.

Ayyagari, Ramani, Qian Chen, and Borja García de Soto. 2023. "Quantifying the Impact of Concrete 3D Printing on the Construction Supply Chain." *Automation in Construction* 155 (November): 105032. https://doi.org/10.1016/j.autcon.2023.105032.

Barsoum, M. W., A. Ganguly, and G. Hug. 2006. "Microstructural Evidence of Reconstituted Limestone Blocks in the Great Pyramids of Egypt." *Journal of the American Ceramic Society* 89 (12): 3788–96. https://doi.org/10.1111/j.1551-2916.2006.01308.x.

Barve, Prasad, Alireza Bahrami, and Santosh Shah. 2023. "Geopolymer 3D Printing: A Comprehensive Review on Rheological and Structural Performance Assessment, Printing Process Parameters, and Microstructure." *Frontiers in Materials* 10 (November). https://doi.org/10.3389/fmats.2023.1241869.

Batikha, Mustafa, Rahul Jotangia, Mohamad Yasser Baaj, and Ibrahim Mousleh. 2022. "3D Concrete Printing for Sustainable and Economical Construction: A Comparative Study." *Automation in Construction* 134 (February): 104087. https://doi.org/10.1016/j.autcon.2021.104087.

Beek, Van. 2020. "Van Beek Screw Conveyor Component Used in New 3D Printer." https://www.powderbulksolids.com/van-beek-screw-conveyor-component-used-new-3d-printer.

Biel, Oliwia, Piotr Rożek, Paulina Florek, Włodzimierz Mozgawa, and Magdalena Król. 2020. "Alkaline Activation of Kaolin Group Minerals." *Crystals* 10 (4): 268. https://doi.org/10.3390/cryst10040268.

Bindiganavile, Vivek, and Nemkumar Banthia. 2009. "Effect of Particle Density on Its Rebound in Dry-Mix Shotcrete." *Journal of Materials in Civil Engineering* 21 (2): 58–64. https://doi.org/10.1061/(ASCE)0899-1561(2009)21:2(58).

Buswell, R. A., W. R. Leal de Silva, S. Z. Jones, and J. Dirrenberger. 2018. "3D Printing Using Concrete Extrusion: A Roadmap for Research." *Cement and Concrete Research* 112 (October): 37–49. https://doi.org/10.1016/J.CEMCONRES.2018.05.006.

Buswell, R. A., R. C. Soar, A. G. F. Gibb, and A. Thorpe. 2007. "Freeform Construction: Mega-Scale Rapid Manufacturing for Construction." *Automation in Construction* 16 (2): 224–31. https://doi.org/10.1016/J.AUTCON.2006.05.002.

Buswell, R. A., A. Thorpe, R. C. Soar, and A. G. F. Gibb. 2008. "Design, Data and Process Issues for Mega-Scale Rapid Manufacturing Machines Used for Construction." *Automation in Construction* 17 (8): 923–29. https://doi.org/10.1016/j.autcon.2008.03.001.

Cao, D., S. U. Dageng, L. U. Bo, and Y. Yang. 2005. "Synthesis and Structure Characterization of Geopolymeric Material Based on Metakaolinite and Phosphoric Acid." *Journal of the Chinese Ceramic Society* 33: 1385–89. https://api.semanticscholar.org/CorpusID:101651887.

Castellanos, Ramiro, Jaime C. Gálvez, Marcos G. Alberti, J. Vera-Agulló, and R. Pina-Zapardiel. 2023. "Rheological Analysis of Wet-Mix Shotcrete with Various Viscosity-Modifiying Admixtures." *Tunnelling and Underground Space Technology* 142 (December): 105269. https://doi.org/10.1016/j.tust.2023.105269.

Chaiyotha, Danai, Watcharapong Kantawong, Panjasila Payakaniti, Supree Pinitsoontorn, and Prinya Chindaprasirt. 2023. "Finding Optimized Conditions for 3D Printed High Calcium Fly Ash Based Alkali-Activated Mortar." *Case Studies in Construction Materials* 18 (July): e01976. https://doi.org/10.1016/j.cscm.2023.e01976.

Chen, Lianjun, Zhenjiao Sun, Guoming Liu, Guanguo Ma, and Xianguo Liu. 2022. "Spraying Characteristics of Mining Wet Shotcrete." *Construction and Building Materials* 316 (January): 125888. https://doi.org/10.1016/j.conbuildmat.2021.125888.

"D-Shape." n.d. Accessed March 1, 2022. https://d-shape.com/.

Davidovits, J. 2011. *Geopolymer Chemistry and Applications*. 5th ed. Edited by Geopolymer Institute. France: Hauts-de-France.

Davidovits, J. 2013. "Geopolymer Cement-a Review." *Geopolymer Science and Technics*. www.geopolymer.org.

Davidovits, J. 2016. "Webinar Spring 2016- Special Focus on 'Geopolymer Cement'." 2016. geopolymer.org/conference/webinar/webinars-videos-collection.

Davidovits, J. 2017. "Geopolymers: Ceramic-Like Inorganic Polymers." *Journal of Ceramic Science and Technology* 8 (3): 335–50. https://doi.org/10.4416/JCST2017-00038.

Davidovits, J., and S. A. Cordi. 1979. "Synthesis of New High Temperature Geo-Polymers for Reinforced Plastics/Composites." *Society of Plastic Engineers, Brookfield Center, USA* 79: 151–54.

Davidovits, Joseph. 2018. "Geopolymers Based on Natural and Synthetic Metakaolin a Critical Review." In *Proceedings of the 41st International Conference on Advanced Ceramics and Composites: Ceramic Engineering and Science Proceedings*, 201–14. https://doi.org/10.1002/9781119474746.ch19.

Davidovits, Joseph, and James L. Sawyer. 1985. "Early High-strength Mineral Polymer, Issued 1985." https://patents.google.com/patent/US4509985A/en.

"Desamanera Brings Large-Scale Stone 3D Printing to the US." n.d. https://3dprintingindustry.com/news/desamanera-brings-large-scale-stone-3d-printing-to-the-us-75472/.

Dilberoglu, Ugur M., Bahar Gharehpapagh, Ulas Yaman, and Melik Dolen. 2017. "The Role of Additive Manufacturing in the Era of Industry 4.0." In *Procedia Manufacturing 11*, 545–54. https://doi.org/10.1016/j.promfg.2017.07.148.

Elsayed, Hamada, Filippo Gobbin, Martiniano Picicco, Antonino Italiano, and Paolo Colombo. 2022. "Additive Manufacturing of Inorganic Components Using a Geopolymer and Binder Jetting." *Additive Manufacturing* 56 (August): 102909. https://doi.org/10.1016/j.addma.2022.102909.

Farooq, Furqan, Xin Jin, Muhammad Faisal Javed, Arslan Akbar, Muhammad Izhar Shah, Fahid Aslam, and Rayed Alyousef. 2021. "Geopolymer Concrete as Sustainable Material: A State of the Art Review." *Construction and Building Materials* 306 (November): 124762. https://doi.org/10.1016/j.conbuildmat.2021.124762.

Feret, Rene. 1939. "Slags for the Manufacture of Cement." *Revue des matériaux de construction et de travaux publics* (July) [cited in: D. M. Roy, "Alkali-Activated Cements: Opportunities and Challenges." Cement and Concrete Research 29 (1999): 249–54].

Geng, Zifan, Lizhi Zhang, Hao Pan, Wei She, Cheng Zhou, Hao Zhou, Zhenglei Yu, and Zhaodong Xu. 2023. "In-Situ Solidification of Alkali-Activated Lunar Regolith: Insights into the Chemical and Physical Origins." *Journal of Cleaner Production* 391 (March): 136147. https://doi.org/10.1016/j.jclepro.2023.136147.

Ghazy, M. F., M. Elaty, and S. M. Mostafa. 2022. "Properties of One-Part versus Two-Part Geopolymers Composites – A Review." *American Journal of Engineering Research* 11 (6): 1–14. www.ajer.org.

Hack, Norman, Mohammad Bahar, Christian Hühne, William Lopez, Stefan Gantner, Noor Khader, and Tom Rothe. 2021. "Development of a Robot-Based Multi-Directional Dynamic Fiber Winding Process for Additive Manufacturing Using Shotcrete 3D Printing." *Fibers* 9 (6): 39. https://doi.org/10.3390/FIB9060039.

Heidarnezhad, Fatemeh, and Qian Zhang. 2022. "Shotcrete Based 3D Concrete Printing: State of Art, Challenges, and Opportunities." *Construction and Building Materials* 323 (March): 126545. https://doi.org/10.1016/J.CONBUILDMAT.2022.126545.

Hojati, M., F. Rajabipour, and A. Radlińska. 2019. "Effect of Activator, Curing and Humidity on Drying Shrinkage of Alkali-Activated Fly Ash." *Green Materials* 7 (2). https://doi.org/10.1680/jgrma.18.00069.

Hojati, Maryam, Shadi Nazarian, Jose P. Duarte, Aleksandra Radlińska, Negar Ashrafi, Flávio Craveiro, and Sven Bilén. 2018. "3D Printing of Concrete: A Continuous Exploration of Mix Design and Printing Process." In *42st International Association for Housing (IAHS)*. Naples, Italy. https://scholar.google.com/citations?view_op=view_citation&hl=en&user=j2cJCSIAAAAJ&citation_for_view=j2cJCSIAAAAJ:Y0pCki6q_DkC.

Hojati, Maryam, and Aleksandra Radlińska. 2017. "Shrinkage and Strength Development of Alkali-Activated Fly Ash-Slag Binary Cements." *Construction and Building Materials* 150 (September): 808–16. https://doi.org/10.1016/j.conbuildmat.2017.06.040.

Hojati, Maryam, Aleksandra Radlińska, Shadi Nazarian, José P. Duarte, Ali M. Memari, and Sven G. Bilén. 2018a. "Design and 3D Printing of Two-Part Geopolymer Mortar." In *1st RILEM International Conference on Concrete and Digital Fabrication*. Zurich, Switzerland.

Hojati, Maryam, Aleksandra Radlińska, Shadi Nazarian, José P. Duarte, Ali M. Memari, and Sven G. Bilén. 2018b. "Synthesis and 3D Printing of One-Part Geopolymer Mortar." In *1st RILEM International Conference on Concrete and Digital Fabrication*. Zurich, Switzerland.

Hojati, Maryam, Farshad Rajabipour, and Aleksandra Radlińska. 2016. "Drying Shrinkage of Alkali Activated Fly Ash: Effect of Activator Composition and Ambient Relative Humidity." In *4th International Conference on Sustainable Construction Materials and Technologies (SCMT4)*, 7–11. Las Vegas, NV, USA. https://d1wqtxts1xzle7.cloudfront.net/55770946/S218.pdf?1518332899=&response-content-disposition=inline%3B+filename%3DDrying_Shrinkage_of_Alkali_Activated_Fly.pdf&Expires=1637006947&Signature=X0lwvdZqsOvXd9iOjf-msfbuXB7akxDLYKJYuuQ70SM0kMXJ5pe~C2Q4ADdVauX.

Hojati, Maryam, Farshad Rajabipour, and Aleksandra Radlińska. 2022. "Creep of Alkali-Activated Cement Mixtures." *Case Studies in Construction Materials* 16 (June): e00954. https://doi.org/10.1016/J.CSCM.2022.E00954.

Hollanders, Sofie, Rieko Adriaens, Jørgen Skibsted, Özlem Cizer, and Jan Elsen. 2016. "Pozzolanic Reactivity of Pure Calcined Clays." *Applied Clay Science* 132–33 (November): 552–60. https://doi.org/10.1016/j.clay.2016.08.003.

Huang, Dunwen, Peng Chen, Hui Peng, Yiwei Yang, Qiaoming Yuan, and Miao Su. 2021. "A Review and Comparison Study on Drying Shrinkage Prediction between Alkali-Activated Fly Ash/Slag and Ordinary Portland Cement." *Construction and Building Materials* 305 (October): 124760. https://doi.org/10.1016/j.conbuildmat.2021.124760.

Hwalla, Joud, Hilal El-Hassan, Abdulkader El-Mir, Joseph J. Assaad, and Tamer El-Maaddawy. 2024. "Development of Geopolymer and Cement-Based Shotcrete Mortar: Impact of Mix Design Parameters and Spraying Process." *Construction and Building Materials* 449 (October): 138457. https://doi.org/10.1016/j.conbuildmat.2024.138457.

Ilic, Biljana, Aleksandra Mitrovic, and Ljiljana Milicic. 2010. "Thermal Treatment of Kaolin Clay to Obtain Metakaolin." *Hemijska Industrija* 64 (4): 351–56. https://doi.org/10.2298/HEMIND100322014I.

Imtiaz, Lahiba, Sardar Kashif Ur Rehman, Shazim Ali Memon, Muhammad Khizar Khan, and Muhammad Faisal Javed. 2020. "A Review of Recent Developments and Advances in Eco-Friendly Geopolymer Concrete." *Applied Sciences* 10 (21): 7838. https://doi.org/10.3390/app10217838.

ISO/ASTM 52900. 2015. "Additive Manufacturing – General Principles – Terminology." In *International Organization for Standardization*. Geneva, Switzerland. www.iso.org-www.astm.org.

Jakupovic, Anes. 2016. "D-SHAPE-Report." 3D Printet Byggeri. http://3dprintetbyggeri.dk/pdf/bes%C3%B8gsrapporter/D-Shape.pdf.

Khalifa, Ahmed Z., Özlem Cizer, Yiannis Pontikes, Andrew Heath, Pascaline Patureau, Susan A. Bernal, and Alastair T. M. Marsh. 2020. "Advances in Alkali-Activation of Clay Minerals." *Cement and Concrete Research* 132 (June): 106050. https://doi.org/10.1016/j.cemconres.2020.106050.

Khan, Shoukat Alim, Muammer Koç, and Sami G. Al-Ghamdi. 2021. "Sustainability Assessment, Potentials and Challenges of 3D Printed Concrete Structures: A Systematic Review for Built Environmental Applications." *Journal of Cleaner Production* 303 (June): 127027. https://doi.org/10.1016/j.jclepro.2021.127027.

Khoshnevis, Behrokh, Dooil Hwang, Ke Thia Yao, and Zhenghao Yeh. 2006. "Mega-Scale Fabrication by Contour Crafting." *International Journal of Industrial and Systems Engineering* 1 (3): 301–20. https://doi.org/10.1504/IJISE.2006.009791.

Kiventerä, Jenni, Priyadharshini Perumal, Juho Yliniemi, and Mirja Illikainen. 2020. "Mine Tailings as a Raw Material in Alkali Activation: A Review." *International Journal of Minerals, Metallurgy and Materials* 27 (8): 1009–20. https://doi.org/10.1007/s12613-020-2129-6.

Kloft, Harald, Martin Empelmann, V. Dettel, and Lukas Ledderose. 2019. "Production of the First Concrete and Reinforced Concrete Columns by Means of 3D Printing with

Concrete." *BFT International* 85 (6): 28–37. https://scholar.google.com/scholar_lookup?title=Production of the first concrete and reinforced concrete columns by means of 3D printing with concrete&publication_year=2019&author=H. Kloft&author=M. Empelmann&author=V. Dettel&author=L. Ledderose.

Krivenko, P. V. 1994. "Alkaline Cements and Concretes." In *Proceedings of the First International Conference Held at the Scientific-Research Institute on Binders and Materials*, 11–130. Kiev, Ukraine. https://catalog.libraries.psu.edu/catalog/19776328.

Kumar Das, Shaswat, Adeolu Adediran, Cyriaque Rodrigue Kaze, Syed Mohammed Mustakim, and Nordine Leklou. 2022. "Production, Characteristics, and Utilization of Rice Husk Ash in Alkali Activated Materials: An Overview of Fresh and Hardened State Properties." *Construction and Building Materials* 345 (August): 128341. https://doi.org/10.1016/j.conbuildmat.2022.128341.

Le, T. T., S. A. Austin, S. Lim, R. A. Buswell, R. Law, A. G. F. Gibb, and T. Thorpe. 2012. "Hardened Properties of High-Performance Printing Concrete." *Cement and Concrete Research* 42 (3): 558–66. https://doi.org/10.1016/j.cemconres.2011.12.003.

Li, Xiaoshuang, Daolin Wang, Qiusong Chen, and Chongchong Qi. 2024. "Alkali Activation of Blast Furnace Slag Using Bayer Red Mud as an Alternative Activator to Prepare Cemented Paste Backfill." *Construction and Building Materials* 453 (November): 139061. https://doi.org/10.1016/j.conbuildmat.2024.139061.

Li, Yue, Jiale Shen, Hui Lin, and Yaqiang Li. 2023. "Optimization Design for Alkali-Activated Slag-Fly Ash Geopolymer Concrete Based on Artificial Intelligence Considering Compressive Strength, Cost, and Carbon Emission." *Journal of Building Engineering* 75 (September): 106929. https://doi.org/10.1016/j.jobe.2023.106929.

Li, Zhanzhao, Maryam Hojati, Zhengyu Wu, Jonathon Piasente, Negar Ashrafi, José P. Duarte, Shadi Nazarian, Sven G. Bilén, Ali M. Memari, and Aleksandra Radlińska. 2020. "Fresh and Hardened Properties of Extrusion-Based 3D-Printed Cementitious Materials: A Review." *Sustainability* 12 (14): 5628. https://doi.org/10.3390/SU12145628.

Lindemann, H., R. Gerbers, S. Ibrahim, F. Dietrich, E. Herrmann, K. Dröder, A. Raatz, and H. Kloft. 2018. "Development of a Shotcrete 3D-Printing (SC3DP) Technology for Additive Manufacturing of Reinforced Freeform Concrete Structures." *RILEM Bookseries* 19 (September): 287–98. https://doi.org/10.1007/978-3-319-99519-9_27.

Louati, S., S. Baklouti, and B. Samet. 2016. "Acid Based Geopolymerization Kinetics: Effect of Clay Particle Size." *Applied Clay Science* 132–33 (November): 571–78. https://doi.org/10.1016/j.clay.2016.08.007.

Ma, Shanliang, Zengqi Zhang, and Xiaoming Liu. 2022. "Comprehensive Understanding of Aluminosilicate Phosphate Geopolymers: A Critical Review." *Materials* 15 (17): 5961. https://doi.org/10.3390/ma15175961.

Maciej Serda. 2013. "Synteza i Aktywność Biologiczna Nowych Analogów Tiosemikarbazonowych Chelatorów Żelaza." Edited by G. Balint, B. Antala, C. Carty, J-M. A. Mabieme, I. B. Amar, and A. Kaplanova. *Uniwersytet Śląski*, 343–54. https://doi.org/10.2/JQUERY.MIN.JS.

Marchment, Taylor, Jay Sanjayan, and Ming Xia. 2019. "Method of Enhancing Interlayer Bond Strength in Construction Scale 3D Printing with Mortar by Effective Bond Area Amplification." *Materials & Design* 169 (May): 107684. https://doi.org/10.1016/J.MATDES.2019.107684.

Marchon, Delphine, Shiho Kawashima, Hela Bessaies-Bey, Sara Mantellato, and Serina Ng. 2018. "Hydration and Rheology Control of Concrete for Digital Fabrication: Potential Admixtures and Cement Chemistry." *Cement and Concrete Research* 112 (December 2017): 96–110. https://doi.org/10.1016/j.cemconres.2018.05.014.

Mathivet, V., J. Jouin, A. Gharzouni, I. Sobrados, H. Celerier, S. Rossignol, and M. Parlier. 2019. "Acid-Based Geopolymers: Understanding of the Structural Evolutions during

Consolidation and after Thermal Treatments." *Journal of Non-Crystalline Solids* 512 (May): 90–97. https://doi.org/10.1016/j.jnoncrysol.2019.02.025.

Mehta, Ankur, and Rafat Siddique. 2017. "Properties of Low-Calcium Fly Ash Based Geopolymer Concrete Incorporating OPC as Partial Replacement of Fly Ash." *Construction and Building Materials* 150 (September): 792–807. https://doi.org/10.1016/j.conbuildmat.2017.06.067.

Mills, Jennifer N., Maria Katzarova, and Norman J. Wagner. 2022. "Comparison of Lunar and Martian Regolith Simulant-Based Geopolymer Cements Formed by Alkali-Activation for in-Situ Resource Utilization." *Advances in Space Research* 69 (1): 761–77. https://doi.org/10.1016/j.asr.2021.10.045.

Mishra, Jyotirmoy, Bharadwaj Nanda, Sanjaya Kumar Patro, and R. S. Krishna. 2024. "A Comprehensive Review on Compressive Strength and Microstructure Properties of GGBS-Based Geopolymer Binder Systems." *Construction and Building Materials* 417 (February): 135242. https://doi.org/10.1016/j.conbuildmat.2024.135242.

Mohan, Manu K., A. V. Rahul, Geert De Schutter, and Kim Van Tittelboom. 2021. "Extrusion-Based Concrete 3D Printing from a Material Perspective: A State-of-the-Art Review." *Cement and Concrete Composites* 115 (August 2020): 103855. https://doi.org/10.1016/j.cemconcomp.2020.103855.

Mounika, Ganta, Uppari Ramakrishna, Gillela Naresh Kumar Reddy, and Kandula Suresh Kumar. 2023. "A Review on Effect of Red Mud on Properties of Alkali Activated Materials (AAMs) and Geopolymers (GPs)." *Materials Today: Proceedings*, April. https://doi.org/10.1016/j.matpr.2023.03.446.

Nasir, Muhammad, Aziz Hasan Mahmood, and Ashraf A. Bahraq. 2024. "History, Recent Progress, and Future Challenges of Alkali-Activated Binders – An Overview." *Construction and Building Materials* 426 (May): 136141. https://doi.org/10.1016/j.conbuildmat.2024.136141.

Nath, Pradip, and Prabir Kumar Sarker. 2015. "Use of OPC to Improve Setting and Early Strength Properties of Low Calcium Fly Ash Geopolymer Concrete Cured at Room Temperature." *Cement and Concrete Composites* 55 (January): 205–14. https://doi.org/10.1016/j.cemconcomp.2014.08.008.

Nazarian, Shadi, José P. Duarte, Sven G. Bilén, Ali Memari, Aleksandra Radlinska, Nicholas A. Meisel, and Maryam Hojati. 2021. "Additive Manufacturing of Architectural Structures: An Interplay Between Materials, Systems, and Design." In *Advances in Science, Technology & Innovation*, 111–19. https://doi.org/10.1007/978-3-030-35533-3_15.

Nerella, Venkatesh Naidu, and Viktor Mechtcherine. 2019. *Studying the Printability of Fresh Concrete for Formwork-Free Concrete Onsite 3D Printing Technology (CONPrint3D). 3D Concrete Printing Technology*. Elsevier Inc. https://doi.org/10.1016/b978-0-12-815481-6.00016-6.

Nuaklong, Peem, Vanchai Sata, Ampol Wongsa, Korb Srinavin, and Prinya Chindaprasirt. 2018. "Recycled Aggregate High Calcium Fly Ash Geopolymer Concrete with Inclusion of OPC and Nano-SiO2." *Construction and Building Materials* 174 (June): 244–52. https://doi.org/10.1016/j.conbuildmat.2018.04.123.

Obenaus-Emler, R., M. Falah, and M. Illikainen. 2020. "Assessment of Mine Tailings as Precursors for Alkali-Activated Materials for on-Site Applications." *Construction and Building Materials* 246 (June): 118470. https://doi.org/10.1016/j.conbuildmat.2020.118470.

Occhicone, Alessio, Stanislao Giorgio Graziuso, Emmanuel De Gregorio, Fabio Montagnaro, Laura Ricciotti, Oreste Tarallo, Giuseppina Roviello, and Claudio Ferone. 2024. "Synthesis and Characterization of New Acid-Activated Red Mud-Metakaolin Geopolymers and Comparison with Their Alkaline Counterparts." *Journal of Cleaner Production* 435 (January): 140492. https://doi.org/10.1016/j.jclepro.2023.140492.

Panda, Biranchi, Suvash Chandra Paul, Nisar Ahamed Noor Mohamed, Yi Wei Daniel Tay, and Ming Jen Tan. 2018. "Measurement of Tensile Bond Strength of 3D Printed Geopolymer Mortar." *Measurement* 113 (January): 108–16. https://doi.org/10.1016/J.MEASUREMENT.2017.08.051.

Panda, Biranchi, Suvash Chandra Paul, and Ming Jen Tan. 2017. "Anisotropic Mechanical Performance of 3D Printed Fiber Reinforced Sustainable Construction Material." *Materials Letters* 209: 146–49. https://doi.org/10.1016/j.matlet.2017.07.123.

Panda, Biranchi, Shaoqin Ruan, Cise Unluer, and Ming Jen Tan. 2020. "Investigation of the Properties of Alkali-Activated Slag Mixes Involving the Use of Nanoclay and Nucleation Seeds for 3D Printing." *Composites Part B: Engineering* 186 (April): 107826. https://doi.org/10.1016/j.compositesb.2020.107826.

Paolini, Alexander, Stefan Kollmannsberger, and Ernst Rank. 2019. "Additive Manufacturing in Construction: A Review on Processes, Applications, and Digital Planning Methods." *Additive Manufacturing* 30 (December). https://doi.org/10.1016/J.ADDMA.2019.100894.

Perera, Dan S., John V. Hanna, Joel Davis, Mark G. Blackford, Bruno A. Latella, Yosuke Sasaki, and Eric R. Vance. 2008. "Relative Strengths of Phosphoric Acid-Reacted and Alkali-Reacted Metakaolin Materials." *Journal of Materials Science* 43 (19): 6562–66. https://doi.org/10.1007/s10853-008-2913-6.

Perrot, A., D. Rangeard, and A. Pierre. 2016. "Structural Built-up of Cement-Based Materials Used for 3D-Printing Extrusion Techniques." *Materials and Structures/Materiaux et Constructions* 49 (4): 1213–20. https://doi.org/10.1617/S11527-015-0571-0/FIGURES/5.

Provis, John L. 2018. "Alkali-Activated Materials." *Cement and Concrete Research* 114 (December): 40–48. https://doi.org/10.1016/j.cemconres.2017.02.009.

Provis, John L., Angel Palomo, and Caijun Shi. 2015. "Advances in Understanding Alkali-Activated Materials." *Cement and Concrete Research* 78 (December): 110–25. https://doi.org/10.1016/j.cemconres.2015.04.013.

Qaidi, S., A. Yahia, B. A. Tayeh, H. Unis, R. Faraj, and A. Mohammed. 2022. "3D Printed Geopolymer Composites: A Review." *Materials Today Sustainability* 20 (December): 100240. https://doi.org/10.1016/j.mtsust.2022.100240.

Revathi, T., N. Vanitha, R. Jeyalakshmi, Baskar Sundararaj, M. Jegan, and P. R. Kannan Rajkumar. 2022. "Adoption of Alkali-Activated Cement-Based Binders (Geopolymers) from Industrial by-Products for Sustainable Construction of Utility Buildings-A Field Demonstration." *Journal of Building Engineering* 52 (July): 104450. https://doi.org/10.1016/j.jobe.2022.104450.

Ricciotti, Laura, Antonio Apicella, Valeria Perrotta, and Raffaella Aversa. 2023. "Geopolymer Materials for Extrusion-Based 3D-Printing: A Review." *Polymers* 15 (24): 4688. https://doi.org/10.3390/polym15244688.

Roy, Della M. 1999. "Alkali-Activated Cements Opportunities and Challenges." *Cement and Concrete Research* 29 (2): 249–54. https://doi.org/10.1016/S0008-8846(98)00093-3.

Şahin, Hatice Gizem, and Ali Mardani-Aghabaglou. 2022. "Assessment of Materials, Design Parameters and Some Properties of 3D Printing Concrete Mixtures; a State-of-the-Art Review." *Construction and Building Materials* 316 (January): 125865. https://doi.org/10.1016/j.conbuildmat.2021.125865.

Santhosh, Kumar Gedela, Sk M. Subhani, and A. Bahurudeen. 2022. "Sustainable Reuse of Palm Oil Fuel Ash in Concrete, Alkali-Activated Binders, Soil Stabilisation, Bricks and Adsorbent: A Waste to Wealth Approach." *Industrial Crops and Products* 183 (September): 114954. https://doi.org/10.1016/j.indcrop.2022.114954.

Sedghi, Reza, Kourosh Rashidi, and Maryam Hojati. 2024. "Large-Scale 3D Wall Printing: From Concept to Reality." *Automation in Construction* 159 (March): 105255. https://doi.org/10.1016/j.autcon.2023.105255.

Shakor, Pshtiwan, S. H. Chu, Anastasiia Puzatova, and Enrico Dini. 2022. "Review of Binder Jetting 3D Printing in the Construction Industry." *Progress in Additive Manufacturing 2021* (January): 1–27. https://doi.org/10.1007/S40964-021-00252-9.

Shakor, Pshtiwan, Shami Nejadi, Gavin Paul, and Sardar Malek. 2019a. "Review of Emerging Additive Manufacturing Technologies in 3D Printing of Cementitious Materials in the Construction Industry." *Frontiers in Built Environment* 4 (January). https://doi.org/10.3389/fbuil.2018.00085.

Shakor, Pshtiwan, Shami Nejadi, Gavin Paul, and Sardar Malek. 2019b. "Review of Emerging Additive Manufacturing Technologies in 3D Printing of Cementitious Materials in the Construction Industry." *Frontiers in Built Environment* 4 (January): 85. https://doi.org/10.3389/FBUIL.2018.00085/BIBTEX.

Shilar, Fatheali A., Sharanabasava V. Ganachari, Veerabhadragouda B. Patil, B. E. Bhojaraja, T. M. Yunus Khan, and Naif Almakayeel. 2023. "A Review of 3D Printing of Geopolymer Composites for Structural and Functional Applications." *Construction and Building Materials* 400 (October): 132869. https://doi.org/10.1016/j.conbuildmat.2023.132869.

Singh, B., G. Ishwarya, M. Gupta, and S. K. Bhattacharyya. 2015. "Geopolymer Concrete: A Review of Some Recent Developments." *Construction and Building Materials* 85 (June): 78–90. https://doi.org/10.1016/J.CONBUILDMAT.2015.03.036.

Srinivasamurthy, Lakshmikanth, Venkata S. Chevali, Zuhua Zhang, and Hao Wang. 2021. "Phase Changes under Efflorescence in Alkali Activated Materials with Mixed Activators." *Construction and Building Materials* 283 (May): 122678. https://doi.org/10.1016/j.conbuildmat.2021.122678.

Su, Chang, Zhuang Wu, and Xiaokui Zheng. 2022. "Analysis of Rebound Rate of Wet Shotcrete Based on Experiment and Discrete Element Method." Edited by Biao Li. *Shock and Vibration* 2022 (August): 1–10. https://doi.org/10.1155/2022/1840580.

Sun, Chenchen, Jichun Xiang, Mengxue Xu, Yan He, Zhangfa Tong, and Xuemin Cui. 2020. "3D Extrusion Free Forming of Geopolymer Composites: Materials Modification and Processing Optimization." *Journal of Cleaner Production* 258 (June): 120986. https://doi.org/10.1016/j.jclepro.2020.120986.

Tabassum, Toiba, and Ajaz Ahmad Mir. 2023. "A Review of 3D Printing Technology-the Future of Sustainable Construction." *Materials Today: Proceedings* 93: 408–14. https://doi.org/10.1016/j.matpr.2023.08.013.

Tanyildizi, Harun, Maksut Seloglu, Mohd Mustafa Al Bakri Abdullah, Rafiza Abdul Razak, and Md Azree Othuman Mydin. 2025. "The Rheological and Mechanical Properties of 3D-Printed Geopolymers: A Review." *Case Studies in Construction Materials* 22 (July): e04679. https://doi.org/10.1016/J.CSCM.2025.E04679.

Tironi, Alejandra, Monica A. Trezza, Alberto N. Scian, and Edgardo F. Irassar. 2013. "Assessment of Pozzolanic Activity of Different Calcined Clays." *Cement and Concrete Composites* 37 (March): 319–27. https://doi.org/10.1016/j.cemconcomp.2013.01.002.

Tramontin Souza, Marcelo, Lisandro Simão, Elisângela Guzi de Moraes, Luciano Senff, José Renato de Castro Pessôa, Manuel J. Ribeiro, and Antonio Pedro Novaes de Oliveira. 2021. "Role of Temperature in 3D Printed Geopolymers: Evaluating Rheology and Buildability." *Materials Letters* 293 (June): 129680. https://doi.org/10.1016/j.matlet.2021.129680.

Wagh, Arun S. 2016. "Raw Materials." In *Chemically Bonded Phosphate Ceramics*, 35–49. Elsevier. https://doi.org/10.1016/B978-0-08-100380-0.00003-8.

Wang, Xiongxin, Md. Mashfiqul Islam, and Qian Zhang. 2024. "Influence of Materials and Nozzle Geometry on Spray and Placement Behavior of Wet-Mix Shotcrete." *Case Studies in Construction Materials* 20 (July): e02852. https://doi.org/10.1016/j.cscm.2024.e02852.

Wang, Yan-Shuai, Yazan Alrefaei, and Jian-Guo Dai. 2019. "Silico-Aluminophosphate and Alkali-Aluminosilicate Geopolymers: A Comparative Review." *Frontiers in Materials* 6 (May). https://doi.org/10.3389/fmats.2019.00106.

Wangler, Timothy, Robert J. Flatt, Nicolas Roussel, Arnaud Perrot, Mohammed Sonebi, Rob Wolfs, Freek Bos, et al. 2022. "Printable Cement-Based Materials: Fresh Properties Measurements and Control." In *RILEM State-of-the-Art Reports*, 36:99–136. Springer International Publishing. https://doi.org/10.1007/978-3-030-90535-4_4.

Wangler, Timothy, Ena Lloret, Lex Reiter, Norman Hack, Fabio Gramazio, Matthias Kohler, Mathias Bernhard, et al. 2016. "Digital Concrete: Opportunities and Challenges." *RILEM Technical Letters* 1 (October): 67. https://doi.org/10.21809/rilemtechlett.2016.16.

Wangler, Timothy, Nicolas Roussel, Freek P. Bos, Theo A. M. Salet, and Robert J. Flatt. 2019. "Digital Concrete: A Review." *Cement and Concrete Research* 123 (March). https://doi.org/10.1016/j.cemconres.2019.105780.

Xia, Ming, and Jay Sanjayan. 2016. "Method of Formulating Geopolymer for 3D Printing for Construction Applications." *Materials & Design* 110 (November): 382–90. https://doi.org/10.1016/j.matdes.2016.07.136.

Yan-Shuai, Wang, Dai Jian-Guo, Ding Zhu, and Xu Wei-Ting. 2017. "Phosphate-Based Geopolymer: Formation Mechanism and Thermal Stability." *Materials Letters* 190: 209–12. https://doi.org/https://doi.org/10.1016/j.matlet.2017.01.022.

Yang, R. Y., R. P. Zou, and A. B. Yu. 2003. "Effect of Material Properties on the Packing of Fine Particles." *Journal of Applied Physics* 94 (5): 3025–34. https://doi.org/10.1063/1.1598638.

Yao, Yizhou, Chao Liu, Huawei Liu, Xianqin Chen, Xin Li, Tilin Wang, Yifei Wang, Wei Zhang, and Yiwen Wu. 2025. "Multiscale Study of Microstructural Evolution in Alkali-Activated Lunar Regolith Simulant under In-Situ Lunar Temperatures: Insight into the Reaction Mechanism." *Journal of Building Engineering* 103 (June): 112209. https://doi.org/10.1016/j.jobe.2025.112209.

Yun, Kyong-Ku, Pangil Choi, and Jung Heum Yeon. 2015. "Correlating Rheological Properties to the Pumpability and Shootability of Wet-Mix Shotcrete Mixtures." *Construction and Building Materials* 98 (November): 884–91. https://doi.org/10.1016/j.conbuildmat.2015.09.004.

Yun, Kyong-Ku, Pangil Choi, and Jung Heum Yeon. 2018. "Rheological Characteristics of Wet-Mix Shotcrete Mixtures with Crushed Aggregates and Mineral Admixtures." *KSCE Journal of Civil Engineering* 22 (7): 2469–79. https://doi.org/10.1007/s12205-017-0198-5.

Zain, Hariz, Mohd Mustafa Al Bakri Abdullah, Kamarudin Hussin, Nurliyana Ariffin, and Ridho Bayuaji. 2017. "Review on Various Types of Geopolymer Materials with the Environmental Impact Assessment." Edited by M. A. B. Abdullah, S. Z. Abd Rahim, M. E. Muhammad Suandi, M. N. Mat Saad, and M. F. Ghazali. *MATEC Web of Conferences* 97 (February): 01021. https://doi.org/10.1051/matecconf/20179701021.

Zareiyan, Babak, and Behrokh Khoshnevis. 2017. "Automation in Construction Effects of Interlocking on Interlayer Adhesion and Strength of Structures in 3D Printing of Concrete." *Automation in Construction* 83 (August): 212–21. https://doi.org/10.1016/j.autcon.2017.08.019.

Zhang, Chao, Venkatesh Naidu Nerella, Anurag Krishna, Shen Wang, Yamei Zhang, Viktor Mechtcherine, and Nemkumar Banthia. 2021. "Mix Design Concepts for 3D Printable Concrete: A Review." *Cement and Concrete Composites* 122 (June): 104155. https://doi.org/10.1016/j.cemconcomp.2021.104155.

Zhang, Dan, and Bin Wei, eds. 2017. *Mechatronics and Robotics Engineering for Advanced and Intelligent Manufacturing*. Lecture Notes in Mechanical Engineering. Cham: Springer International Publishing. https://doi.org/10.1007/978-3-319-33581-0.

Zhang, Zuhua, John L. Provis, Andrew Reid, and Hao Wang. 2014. "Geopolymer Foam Concrete: An Emerging Material for Sustainable Construction." *Construction and Building Materials* 56 (April): 113–27. https://doi.org/10.1016/j.conbuildmat.2014.01.081.

Zhao, Jihui, Liangyu Tong, Boen Li, Tianhao Chen, Chunpeng Wang, Gangqiang Yang, and Ye Zheng. 2021. "Eco-Friendly Geopolymer Materials: A Review of Performance Improvement, Potential Application and Sustainability Assessment." *Journal of Cleaner Production* 307 (July): 127085. https://doi.org/10.1016/j.jclepro.2021.127085.

Zhu, Jianying, Yougan Chen, Xiang Sun, Jianming Wu, Zhenwen Liu, and Xiaomei Xu. 2021. "ECRKQ: Machine Learning-Based Energy-Efficient Clustering and Cooperative Routing for Mobile Underwater Acoustic Sensor Networks." *IEEE Access* 9: 70843–55. https://doi.org/10.1109/ACCESS.2021.3078174.

Ziejewska, Celina, Joanna Marczyk, Kinga Korniejenko, Sebastian Bednarz, Piotr Sroczyk, Michał Łach, Janusz Mikuła, Beata Figiela, Magdalena Szechyńska-Hebda, and Marek Hebda. 2022. "3D Printing of Concrete-Geopolymer Hybrids." *Materials* 15 (8): 2819. https://doi.org/10.3390/ma15082819.

Zou, H. Y., W. L. Zhong, X. Zhao, and L. F. Fan. 2024. "Optimizing Geopolymer Mortar for Shotcrete Applications by Focusing on Flowability and Early-Age Mechanical Characteristics." *Construction and Building Materials* 418 (March): 135290. https://doi.org/10.1016/j.conbuildmat.2024.135290.

Zribi, M., and S. Baklouti. 2021. "Investigation of Phosphate Based Geopolymers Formation Mechanism." *Journal of Non-Crystalline Solids* 562 (June): 120777. https://doi.org/10.1016/j.jnoncrysol.2021.120777.

Zribi, M., and S. Baklouti. 2022. "Phosphate-Based Geopolymers: A Critical Review." *Polymer Bulletin* 79 (9): 6827–55. https://doi.org/10.1007/s00289-021-03829-0.

7 3D-Printable Limestone-Calcined Clay-Based Cementitious Materials

Adewumi John Babafemi and Jyotirmoy Mishra

7.1 INTRODUCTION

Cement has been the single most indispensable material in construction for nearly 2,000 years. Modern infrastructural needs resulting from rapid urbanization rely heavily on the cement industry to facilitate economic growth. However, this growth comes at a cost, with a significant environmental impact. Hence, the cement industry is under extreme scrutiny. The cement industry, which is responsible for nearly 7%–8% of global carbon dioxide (CO_2) emissions, is a primary contributor to environmental issues such as climate change (Massoumi Nejad, Enferadi, and Andrew, 2025). Furthermore, cement production is a highly energy-intensive process, primarily due to the burning of fuels and the production of clinker. Therefore, amid growing environmental concerns, a call for decarbonization in the cement industry has become urgent. However, decarbonizing the cement industry is not easy due to the generally conservative attitude among cement companies. Hence, a holistic approach is required, which can improve policy frameworks, global collaborative efforts, public awareness, and the pursuit of innovative green binders. Only then, in the 21st century, will attaining net-zero carbon emissions be a reality.

Over the last two decades, there has been a significant escalation in the research and development of innovative green binders. Undoubtedly, these green binders not only show tremendous potential as an alternative (partial or whole replacement) to ordinary portland cement (OPC) in making concrete but also have proven to be less aggressive in terms of environmental impact. To date, innovations in green binders, such as blended cements, one- and two-part geopolymers, and alkali-activated materials, have come into the picture as a means to partially or completely eradicate the use of OPC in manufacturing concrete (Gao et al. 2025; Jameel et al., 2025; Sona and Sangeetha, 2025). Blended cements are composed of a mixture of industrial pozzolanic wastes, such as ground granulated blast furnace slag (GGBFS) or fly ash (FA), and a portion of traditional cement. These blended cements are now popular and are widely used in concrete applications throughout the world. Out of these blended cements, the most promising is limestone calcined clay cement (LC^3), owing to the large volume availability and global spread of limestone and clay compared to other supplementary cementitious materials (SCMs), such as FA and GGBFS, which do not have these advantages. Also, most coal power plants in the Global North are already

 DOI: 10.1201/9781003434863-9

being shut down for greener energy sources (wind and solar), and slag from the steel industry is in limited supply.

Meanwhile, innovative construction technologies are essential to the future of sustainable construction (Tay et al., 2017). Over the past few decades, significant innovations in construction methodologies have emerged to address issues related to productivity, cost, and environmental sustainability. Waste valorization and cost reduction while simultaneously increasing productivity require the help of innovative digital construction technologies. The application of digital construction technology will transform traditional construction methodologies, thereby becoming a part of the digital economy (Xue et al., 2022). Meanwhile, there is a shortage of skilled human labor force in developed countries, and issues related to low labor productivity in the UK, Singapore, and Hong Kong have been reported extensively in a study (Wu, Wang, and Wang, 2016). In this regard, human-free construction is envisioned as the futuristic pathway by 2050. Digital construction technologies that employ drones, robots, and 3D printers could significantly transform the current construction industry into a sustainable and profitable industry (Woodhead, Stephenson, and Morrey, 2018). One of the most sought-after digital construction methods in the last decade is 3D concrete printing. This technology provides automation in construction, thereby increasing productivity and minimizing cost and waste.

This chapter focuses on 3D-printed limestone calcined clay cement ($3DPLC^3$) composite, which has been recognized as a novel and promising low-carbon product. An introduction to LC^3, an alternative to ordinary portland cement, is first provided, including its hydration process and production methods. Then, it discusses 3D-printed concrete (3DPC) technology, including its fundamentals, techniques, merits, and applications. Thereafter, an overview of the relevant literature on $3DPLC^3$ is presented. Finally, some research gaps and priorities are highlighted to provide future directions for this novel construction technique.

7.2 LIMESTONE CALCINED CLAY CEMENT

LC^3 is composed of three principal components: raw limestone powder, calcined clay, and cement clinker in pre-determined proportions (Antoni et al., 2012). This makes it a ternary blended cement. Typically, one part of limestone is added to two parts of calcined clay. A mixture of 15% limestone and 30% calcined clay is combined with 50% clinker cement to produce LC^3-50. However, other proportions are also developed, e.g., LC^3-65, where 65% clinker cement is used depending on performance requirements (Zunino and Scrivener, 2020). One study reported a ratio of 1:1:0.5 by weight of clinker, calcined clay, and limestone for making LC^3 (Bonavetti, Rahhal, and Irassar 2001). Besides, up to 5% gypsum is usually added to this ternary blended cement. It is crucial to understand the role of gypsum, as it influences the hydration, rheology, and mechanical performance of LC^3. Gypsum reacts with C_3S or calcium silicate hydrate (C-S-H) during hydration in addition to aluminates to promote early strength development of LC^3 (Yi et al., 2024). The typical constituents of LC^3 are presented in Figure 7.1. In this way, a synergistic effect is observed between the components, resulting in a reactive microstructure interface that leads to outstanding mechanical and durability properties in the resulting concrete. The aim

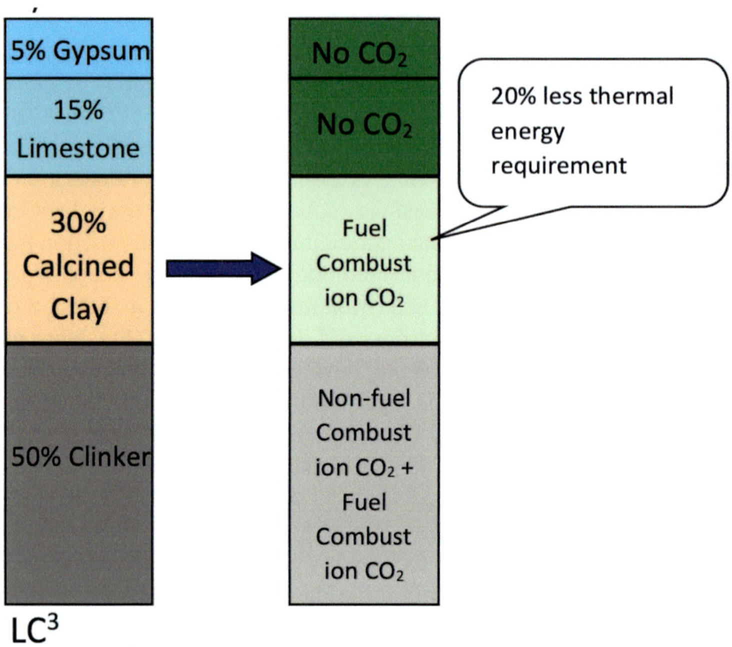

FIGURE 7.1 Constituents of LC^3. Source: Sharma et al. (2021) with permission.

TABLE 7.1
Merits and Demerits of Traditional Cement and LC^3

Cement Type	Merits	Demerits
Traditional cement	• Widely available • Reliable performance • Established market • Long-term durability • Relatively cost-effective	• High carbon emissions • Energy intensive • Waste generation • Noise pollution • Resource depletion
LC^3	• Low carbon emissions • Low clinker use • High durability • Use of natural clay • Low permeability • Low environmental impact	• Require quality control • Lack of long-term data • Absence of supply chain • Lacks wide adoption • May include initial cost regarding new setup

is to reduce the cement clinker content as much as possible; LC^3 has been reported to contain as little as 40% clinker (Antoni et al., 2012). Likewise, the proportions of calcined clay are limited to 25% replacement in the blended system. One important merit that LC^3 offers when compared with other green binders is that the raw materials, namely, clay and limestone, are abundantly available in most parts of the world.

Therefore, manufacturing LC^3 can reduce our dependency on traditional cement without requiring significant investment in deploying new manufacturing technologies. The already available technologies in the cement industry are sufficient for LC^3 production. When following the sustainability route, LC^3 may aid in attaining the goals of a low-carbon future. Table 7.1 presents the merits and demerits of both types of cements for a holistic understanding.

7.2.1 Hydration Process

The hydration of LC^3 is a complex chemical process that involves interaction among calcined clay, cement, and limestone. In LC^3, clay, which contains kaolinite, needs to be calcined first so that metakaolin (MK) will be formed. MK is an amorphous material containing alumina and silica phases. It reacts with calcium available in cement and limestone to generate C-S-H and calcium aluminosilicate hydrate (C-A-S-H) gels (Scrivener et al., 2018). Moreover, alumina from calcined clay reacts with the carbonate phases from limestone to produce carboaluminate compounds. Thus, the hydration products of LC^3 include C-S-H/C-A-S-H gel and carboaluminate compounds. These compounds are extremely crystalline and are considered to be the reason for enhanced microstructure, strength development, and durability in LC^3 when compared to traditional cement (Antoni et al., 2012).

The proportions of calcined clay, cement, and limestone are crucial for the effective hydration reaction to occur in the system. Besides, the physical and chemical properties of the clinker highly influence the hydration reaction. The quantity of calcined kaolinite available for the hydration reaction is proportional to the quantity of C_3S and C_2S in the clinker (Barbhuiya, Nepal, and Das, 2023). Furthermore, the extent of C-A-S-H gel formation is influenced by hydration temperature and the amount of calcined clay used. It is reported that C-A-S-H occupies nearly 50% of the total volume of hydration products formed in the system (Mishra, Emmanuel, and Bishnoi, 2019).

Additionally, the rate of hydration is highly dependent upon the alkali and sulfate content available in the system (Barbhuiya, Nepal, and Das, 2023). The amount of sulfate is crucial to the hydration phase, as it regulates the formation of aluminate phases essential for generating carboaluminate compounds. It is also important to note that during the hydration, not all the contents of the limestone participate in the reaction. Rather, it serves to act as filler as well. This reduces the overall matrix porosity (Bonavetti, Rahhal, and Irassar, 2001).

7.2.2 Production

The LC^3 production process requires the calcination of clays and the grinding of raw materials. The calcination temperature is usually around 700–850°C (Gowri and Rahim, 2024; Nasiri, Madandoust, and Ranjbar, 2023). The calcination process is performed using a rotary kiln, as is the case in clinker production. At this temperature, kaolinite converts into MK, a highly reactive component that is essential in the production process. Other types of clays, such as illite and montmorillonite, are not preferred because their reactivity is lower than that of kaolinite (Fernandez, Martirena, and Scrivener, 2011). Interestingly, the temperature requirement is significantly lower than that for clinker production, which is around 1,450°C. This production, therefore, saves energy and cost. The production cost can be further lowered by using low-grade fuels, such as biomass, in the rotary kilns. This will ensure that the cost of calcination in LC^3 is less than the cost of calcination in clinkering. Although several researchers have suggested different calcination methods, it is important to note that as long as the temperature remains the same, the method of calcination does not significantly impact the reactivity (Scrivener et al., 2018). Following calcination, a simultaneous co-grinding process is performed by mixing reactive clays, limestone, clinker, and gypsum into a consistent blend. Typically, the blend consists of 50% clinker only. The materials are allowed to cool, and then they are ground in a roller mill to achieve the required fineness. Since LC^3 has many components of different hardness, it is also recommended that separate grinding may be the best option.

Clays with high kaolinite content enhance the reactivity during co-grinding and influence the hydration of cement. Therefore, sometimes, other physical and chemical pretreatments are used to enhance the reactivity of the clays. It is also established that heat treatments, if done beyond 900°C, could lead to reduced reactivity. Therefore, an optimal range is preferred, which is 700–850°C (Barbhuiya, Nepal, and Das, 2023). During the LC^3production process, the clays that are used sometimes contain several impurities, such as iron content (Chandrasekhar and Ramaswamy, 2002; Lopes et al., 2024). These impurities highly impact the reactivity and color of the final product. Therefore, understanding the type, internal mineral structure, and composition of clay is very important before the production process starts. Clays with 40% or higher amounts of kaolinite are generally preferred by manufacturers and are widely available (Avet et al., 2016). In a nutshell, understanding the composition of clay, along with its impurities, is crucial to realizing the full potential of LC^3 and utilizing it as a green alternative to traditional cement.

7.3 3D-PRINTED CONCRETE

Driven by multiple objectives at each stage of a project, the traditional construction method suffers from numerous problems. The requirement of human assistance in decision-making and on-field physical work, combined with high costs and material wastage, necessitates the introduction of automotive construction systems. This ensures high productivity, reduced human involvement, lower waste generation, cost-effectiveness, and sustainability. In this context, the use of 3D printing technology in concrete construction is indeed revolutionary. Lately, the focus of the construction industry is rapidly shifting toward adopting 3D printing construction or 3DPC, and its applications are escalating worldwide. It is also gaining attention from the viewpoint of building habitats in space, not just on planet Earth (Siddika et al., 2020). Other important merits, such as formwork-less construction and a high degree of freedom to realize complex geometries, give 3DPC the edge over traditional construction methods (Paul et al., 2018).

Meanwhile, countries like China, India, and South Africa are rapidly developing their infrastructure to accommodate population growth and urbanization. Therefore, the construction market is expected to rise to USD 13.9 trillion by 2037 (Maretasari and Ahmadi, 2025). Given the growing demand for infrastructure, the construction industry is also in need of low-carbon binder materials, such as LC^3, to reduce carbon emissions. Therefore, in this new era of construction fueled by high demand for infrastructural developments, automation in the construction industry is highly essential, which will aid in waste management and promote sustainability. 3DPC offers all these advantages over traditional construction methods. Furthermore, it is reported that using 3DPC in construction projects results in overall cost and time reduction by 50%–80% and 50%–70%, respectively (Zhang et al., 2019).

7.3.1 3D Printing Techniques

3DPC technology ensures the complete automation of construction projects, eliminating the need for human involvement. The 3DPC construction process consists of two phases. First, a digital model is created using CAD software. Second, this model is transformed into a 3D object using a printer (Perrot, 2019). Currently, two types of printing techniques are being explored by the construction industry: (i) extrusion-based printing and (ii) powder-based printing, as shown in Figure 7.2. The extrusion-based printing technique involves placing the fresh concrete mix, extruded through a pipe–pump–nozzle system, in a layer-by-layer sequence until the entire 3D object is printed. On the other hand, the powder-based printing technique involves laying the dry concrete material mix on the bed in layers. Then, a liquid binder is added to allow the dry mix to harden. This process is continued until the entire design is printed as a 3D object. It is pertinent to mention here that of these two techniques, extrusion-based printing is widely used in the construction industry in the current scenario (Singh, Colangelo, and Farina, 2023). Some of the merits that the extrusion-based technique offers over powder-based printing are cost-effectiveness, time efficiency, and a better degree of freedom to print concrete elements.

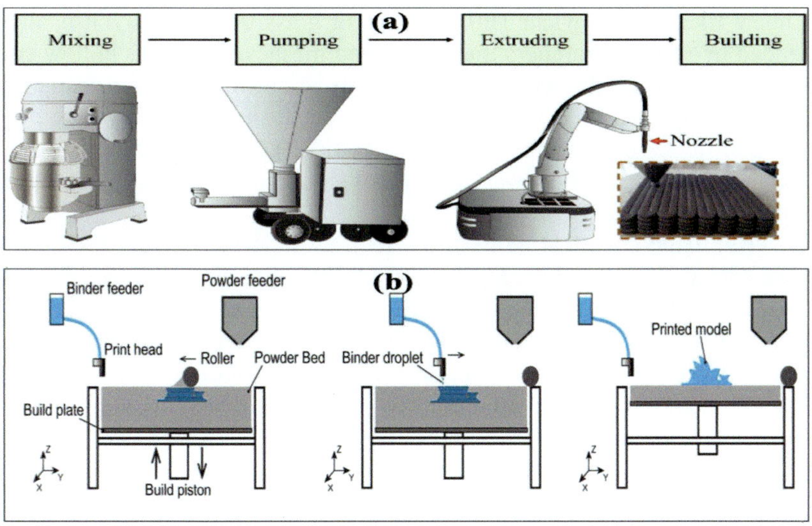

FIGURE 7.2 (a) Extrusion-based 3D printing; (b) powder-based 3D printing. Source: (a) Zhong and Zhang (2022), open access. (b) Nematollahi et al. (2019)

7.3.2 General Requirements for Successful Printing

Concrete can be viewed as the "*ink*" of a 3D printer, which is extruded from a nozzle (Zhang et al., 2025). This is why the rheology of concrete is crucial in the 3DPC construction method. Consequently, the rheology is highly dependent upon the mix design and raw material selection. Even if an initial mix has a higher slump value, it does not guarantee successful extrusion during printing. This is why the concrete mix needs to be judged from a detailed rheology point of view. Rheology holds high significance, and tests for plastic viscosity and yield stresses will provide information regarding the extrudability of the concrete mix (Gao et al., 2024). It is reported that for successful printing, a mix is required to have low dynamic but high static yield stress, along with high viscosity (Qaidi et al., 2022; Sakhare, Najar, and Deshpande, 2023). This will result in improved rheology, leading to smooth extrusion throughout the entire printing process. In addition to rheology, setting times and early-age strength are also crucial in concrete printing. Shorter setting times, coupled with high early-age strength, will ensure that layers can withstand the loads exerted on each other during the printing process. Since printing happens as a layer-by-layer sequence, there is a need for proper adhesion and interlayer bonding to ensure structural integrity and buildability without failure. Besides, the entire printer setup, like the shape and size of the nozzle, pipe length, pump flow rate, and degree of freedom, must be carefully customized as per application and design requirements (Panda et al., 2017).

The extrusion process and layer-by-layer deposition of mortar/concrete during printing are not favorable for introducing steel reinforcements in the printed elements, unlike in the case of traditional concrete construction. However, flexural strength is important for concrete members as concrete is weak in tension. In this regard, to increase the flexural strength, researchers have recommended using short fibers like polypropylene (PP), carbon, or glass. Additionally, it is also recommended by some researchers that reinforced meshes, as well as hollow components with external reinforcement, may be of great use for ensuring strong printed elements (Asprone et al., 2018; Hack and Lauer, 2014). One study revealed that the addition of 1% carbon fibers (by vol.) could lead to an increase in the flexural strength by 174.5%; however, it may not significantly affect the compressive strength (Zhang et al., 2019).

Likewise, using chemical admixtures such as superplasticizers (SPs) is indispensable for successful printing. The use of SPs not only enhances rheological properties but also ensures better mechanical properties and interlayer bonding (Babafemi et al., 2021), and has a significant impact on the flowability, viscosity, and buildability of the mix. The use of SPs allows the practitioner to lower the water–cement ratio while simultaneously ensuring better rheology for smooth extrusion during the printing process. These merits are highly crucial because they allow for optimizing yield stress and plastic viscosity according to the application type, thereby preventing any failure or collapse during printing. The chances of segregation in the layers are thus largely reduced, and the mechanical performance improves.

7.3.3 Current Applications

Publications on 3DPC have increased rapidly in the last decade. This indicates that 3DPC is currently a highly researched area worldwide. Notably, 3DPC is used not only for traditional cement-based concrete but also for low-carbon concrete applications. On the other hand, there is also rapid progress in translating lab-scale research into real-world implementation. Many projects, namely, buildings and pedestrian bridges, have been successfully constructed to date using 3DPC. In 2015, a five-story apartment was built in China by a company named WinSun. In Shanghai, a 3DPC bridge was constructed, measuring 26.3 m in length (Hou et al., 2021). Likewise, in Europe, the first 3D-printed building project was completed in Copenhagen, Denmark, in 2017, by the 3D construction printer company COBOD in collaboration with Saga Space Architects ("The BOD – Europe's First 3D Printed Building | COBOD International" 2017). Interesting 3DPC projects have also been accomplished in Dubai, where the first 3DPC office building was constructed in 2016. The project covered an area of 2,600 m² and utilized an 11-m × 3.7-m × 1.85-m 3D printer (Urbanist, 2016). Designed by the Institute of Advanced Architecture of Catalonia (IAAC), a 3DPC pedestrian bridge measuring 12 m in length and 1.75 m in width was built in Spain in 2016 ("Spain Builds World's First 3D Printed Concrete Pedestrian Bridge" 2016). The applications mentioned here highlight the rapid adoption of this technology and are poised to disrupt traditional construction practices in the future. Figure 7.3 presents some real-life 3DPC projects, as reported by Zhang et al. (2019).

Table 7.2 compares traditional construction and 3DPC construction based on some key aspects. As such, the use of formworks and vibrators, as in traditional

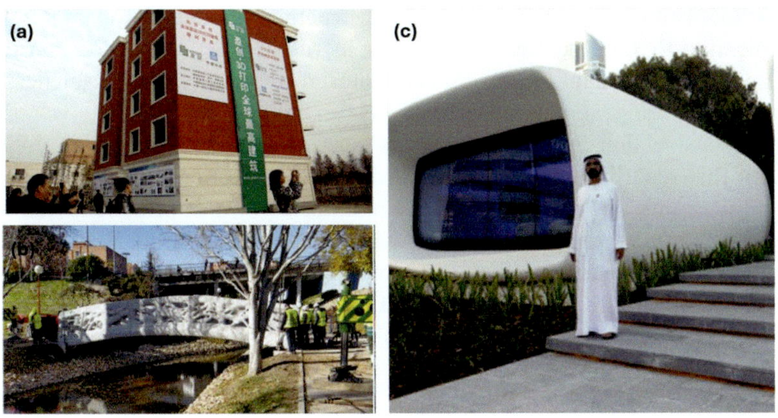

FIGURE 7.3 (a) World's first 3D-printed apartment; (b) 3D-printed bridges; (c) first 3DPC building. Source: Zhang et al. (2019), with permission.

TABLE 7.2
Comparison between 3DPC and Traditional Construction Methods

Aspects	3DPC	Traditional Construction
Speed	• Faster due to automation	• Lack of automation, time-consuming
Workforce requirement	• Less involvement of manual labor	• High requirement of manual labor at each stage
Sustainability	• Lower carbon emissions	• High carbon emissions due to material wastage
Cost	• Lower cost due to less material waste and reduced need for manual workforce	• High recurring cost over the timeline of the projects
Flexibility in design	• Enables printing complex geometries with ease, adding an aesthetic element as well	• Very low degree of flexibility to create complex structures, as it is limited by traditional formwork-based construction
Material wastage	• Minimum waste generation and optimized use of materials	• High degree of wastage due to manual handling of materials
On-field safety	• Allows automation, hence less involvement of laborers, thereby enhancing safety on site	• High risk of accidents, particularly in large-scale projects, due to the involvement of more human labor

construction, is not required in the 3DPC construction method. However, it is important to mention that, unlike the traditional construction method, 3DPC construction needs an interdisciplinary approach, i.e., material science, computer science, robotics, civil engineering, and architecture (Gosselin et al., 2016).

3DPC construction requires effective strategies at every stage of its implementation. In this regard, the scale of the project, the selection of raw materials, the buildability of printed structures, and the monitoring of post-print behavior become crucial. Furthermore, the post-print behavior is significantly influenced by the properties of materials used, along with the printing technique and environmental conditions. The prospective application of the 3DPC technology for large-scale structures under different loading conditions, therefore, needs to be assessed beforehand to ensure there is no failure. However, this technology reduces reliance on manual labor and minimizes waste generation compared to traditional construction methods, as shown in Figure 7.4. This is why it accelerates the construction process (Siddika et al., 2020).

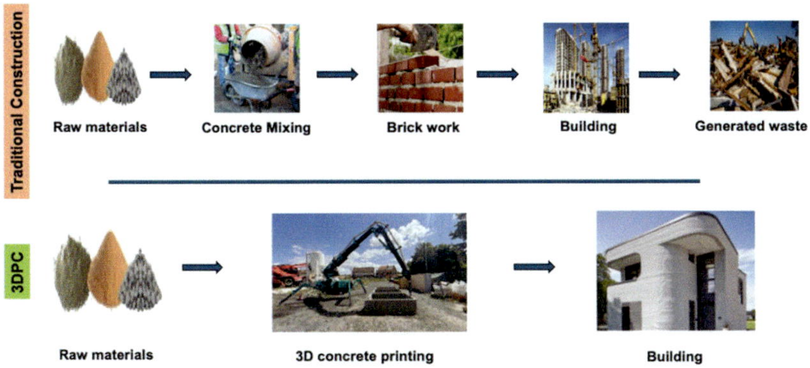

FIGURE 7.4 Stages of construction in traditional construction and 3D printing.

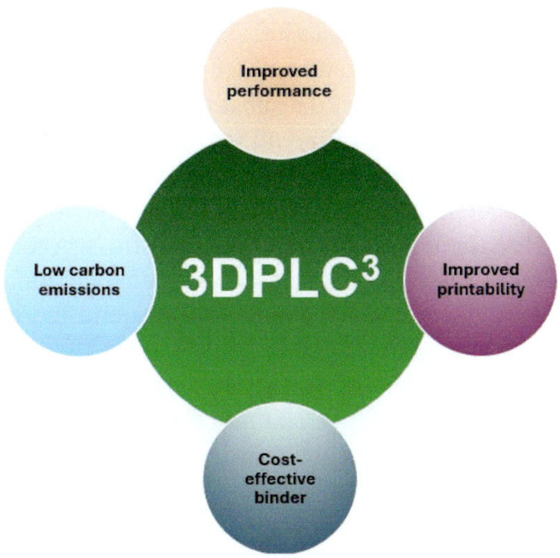

FIGURE 7.5 Merits of 3DPLC³.

7.4 3DPLC³ – A PATHWAY TOWARD A LOW-CARBON FUTURE

3DPLC³ combines low-clinker-based cement (LC³) and digital construction (3D printing). The individual merits of LC³ and 3DPC have already been highlighted in the preceding sections of this chapter. LC³ and 3DPC have been studied separately before; however, only in the past 5 years have researchers attempted to study 3DPLC³ as an innovative construction strategy, which offers several merits, as presented in Figure 7.5. Interesting results have been reported by researchers so far in 3DPLC³ projects. Therefore, this section provides an overview of the relevant literature on the rheology, mechanical properties, durability, and creep behavior of 3DPLC^3 composites.

7.4.1 Rheology

The rheological parameters of a fresh concrete mix are very crucial while printing. An accurate understanding of rheological properties, such as yield stresses (static and dynamic) and thixotropy behavior, largely influences parameters such as extrudability, printability, buildability, and structural integrity of the printed samples. It is worthwhile to mention that in 3DPC, quantifying thixotropy behavior is essential. In this regard, indices such as the structuration rate (A_{thix}) and re-flocculation rate (R_{thix}) provide an idea about how the concrete mix restores after application of shear. A_{thix} is the measurement of structural build-up with respect to time at rest conditions, whereas R_{thix} is the measurement of recovery after extrusion (under shear), which directly affects interlayer bonding.

The rheological behavior of OPC is greatly influenced by the C-S-H gel links and colloidal particle flocculation (Roussel et al., 2012). However, rheological behavior is different in the case of LC^3 mixes, wherein the interaction between limestone and calcined dominates. It is reported that the substitution of OPC by limestone-calcined clay greatly enhances the rheology of the mixes, influencing the cohesion, viscosity, yield stresses, and thixotropy behavior (Muzenda et al., 2020). This is attributed to the fact that particle flocculation is governed by the morphological characteristics of limestone-calcined clay, thereby increasing the yield stress. Another consideration is that the calcined clay, as colloidal particles, induces flocculation and demonstrates shear-thinning behavior, thereby decreasing viscosity and increasing extrudability during printing (Al-Noaimat et al., 2023).

In a study by Ibrahim et al. (2023), the rheology performance of the 3DPLC^3 concrete mix was evaluated. The mix contained micro-synthetic PP fiber as reinforcement. The LC^2 contained limestone powder and kaolinitic clay at a ratio of 1:2 (by mass). An International Center for Aggregates Research (ICAR) rheometer was used to investigate the yield shear stress and thixotropic behavior of freshly prepared concrete. The enhanced initial static yield stress, influenced by LC^3, significantly improved the buildability of fresh mixes. This is attributed to the inclusion of PP fibers, which resulted in high water demand and consequently led to flocculation in the mixes. Significant improvement was thus noted in the yield shear stresses (initial static and dynamic yield stresses) of the mixes, which necessitated successful 3D printing. When compared with OPC-based mixes, the LC^3 mixes exhibited better cohesive behavior, establishing LC^3 as a sustainable binder for effective 3DPC applications.

De Matos et al. (2023) investigated the effect of SP addition on the rheological properties of 3DPLC^3 concrete mixes. In particular, the timing of adding the SP was investigated. A Viscotester iQ Air rheometer was used to evaluate rheological parameters. The binder in the study consisted of LC^3 cement, which had proportions of (OPC + calcium sulfate), calcined clay, and limestone powder at the ratio of 55:30:15 (by weight). The results revealed that the direct addition of SP enhanced the yield stress and viscosity of LC^3 mixes. On the other hand, when the addition of SP was delayed (after 10 minutes), the yield stress and viscosity values showed a slower rate of increase. These values were also compared with those of control OPC-based mixes, which revealed that the LC^3 mix showed lower A_{thix}. Thus, there was a consequent slow increase in yield stress over time. This behavior, regarding a slower rate of increase, was also attributed to the high degree of fineness of the calcined clays. The fineness of the particles led to high water requirements, thereby influencing the SP–matrix interactions. It was also noted that delayed SP addition significantly improved the SP dispersion in the matrix by lessening the encapsulation effect of hydration products within the matrix.

The use of recycled plastic aggregates (RPAs) in $3DPLC^3$ concrete and their impact on rheology were investigated by Babafemi et al. (2024). In this study, the binder consisted of 50% LC^2 by weight along with OPC, while RESIN8 (RPA) replaced 10% of the fine aggregate (by vol.). RESIN8 is a recycled plastic eco-aggregate produced from resins 1–7 with the addition of some mineral admixtures. The rheology test was done using an ICAR rheometer. Based on the yield stress (initial and dynamic), the R_{thix} and A_{thix} were attained. These parameters provided the data for the evolution of yield stress build-up over time. It was observed that the LC^3 mix exhibited higher plasticity. The reason was attributed to the use of SP content. Additionally, the LC^3 mix, both with and without RESIN8, showed higher slump values compared to the control (FA + OPC mix). There was an increment in the yield stress build-up over time in the case of LC^3 mixed with RESIN8. This behavior was ascribed to the unique interaction of RESIN8 in the binder mix. Higher R_{thix} and A_{thix} were recorded for LC^3 mixes with and without RESIN8. Therefore, there was an improvement in thixotropic behavior, while simultaneously the setting time was reduced. The authors concluded that the LC^3 mix provided a dual benefit; i.e., it improved pumpability and buildability during printing, and, at the same time, with RESIN8, it provided better shape stability. Additionally, Table 7.3 reports the rheological parameters for $3DPLC^3$. This table provides an overview of the evolution of yield stress, viscosity, and thixotropy parameters under the influence of binder composition, serving as guidance for researchers when formulating $3DPLC^3$.

7.4.2 Mechanical Performance

In a study by Wang et al. (2024), the compressive strength of $3DPLC^3$ composites was evaluated, and their performance was compared with that of normal mold-cast specimens. The binder consisted of OPC, limestone, and calcined clay, along with FA in the ratio of 0.55:0.15:0.30:1.2 by weight. The water-to-binder ratio was kept at 0.25. Furthermore, polyvinyl alcohol (PVA) fibers were added as reinforcement at 2% (by vol.) in the binder, while silica sand was utilized as fine aggregate in this study. For assessing the compressive strength, the loading directions adopted for this study were X (printing direction), Y (laying direction), and Z (building direction). It was found that at 28 days, the printed samples exhibited the highest strength of 45.3 MPa in the Z direction. This value was found to be 26.2% and 14.7% more than the values obtained in the X and Y directions, respectively. The reason for this behavior was attributed to the denser microstructures in the Z direction. On the other hand, the mold-cast specimens showed lower strength but more uniform strength

TABLE 7.3
Reported Rheological Parameters for 3DPLC[3]

	Rheology Parameter			
Mix Composition	**Yield Stress (Pa)**	**Visocity (Pa.s)**	**Thixotrophy (Pa/min)**	**Reference**
34.8% OPC + 18.97% metakolin + 9.49% limestone + 34.8% GGBFS + 1.93% gypsum	$\tau_{y,static}$: 116.56 $\tau_{y,dynamic}$: 423.03	11.49	-	Yan et al. (2024)
57.31% OPC + 30% calcined clay + 11.69% limestone + 1% $CaSO_4$	$\tau_{y,static}$: 347 $\tau_{y,dynamic}$: -	4.0 ± 1.3	A_{thix}: 12.6	de Matos et al. (2023)
45% OPC + 33.33% calcined clay + 16.37% limestone + 5% silica fume	$\tau_{y,static}$: 800 $\tau_{y,dynamic}$: 1400 (resting time: 30 minutes)	14.5 (resting time: 30 minutes)	-	Long et al. (2021)
40% OPC, 40% calcined clay + 20% limestone	τ_y: 13000	-	-	Chen et al. (2019a)
15% limestone + 30% calcined clay + 5% gypsum + 50% OPC	$\tau_{y,static}$:1368.7 $\tau_{y,dynamic}$: 900.7	-	A_{thix}: 0.51 R_{thix}: 3.73 (Pa/s)	Ibrahim, van Zijl, and Babafemi (2024)
15% limestone + 30% calcined clay + 5% gypsum + 50% OPC	$\tau_{y,static}$: 4499 $\tau_{y,dynamic}$: 2281 (Pa/s)		A_{thix}: 1.43 R_{thix}: 16.24 (Pa/s)	Babafemi et al. (2024)

trends. The authors concluded that at 28 days, the effect of LC[3] hydration in 3DPLC[3] was beneficial for high-strength performance, thereby establishing the superiority of 3DPLC[3] over mold-cast specimens.

One study by Chen et al. (2019b) investigated the influence of different grades of calcined clay with low (40%–50%), medium (62.5%), and high (75%) levels of MK dosage on the compressive strength of mortar samples to assess their suitability in 3DPLC[3]. The binder composition was 40% OPC, 40% calcined clay, and 20% limestone. All mixes were mold-cast. Additionally, a control mix was also cast without calcined clay to understand its role in strength development. It was noticed that as the level of MK increased, the strength increased as well. The mix containing a high dosage of MK attained the highest compressive strength of above 45 MPa in comparison to all other mixes, signifying the importance of using calcined clays in the binder system. The authors concluded that higher MK content aids in faster hydration and rapid microstructural development.

Likewise, the nominal flexural strength of 3DPLC[3] composite was studied by Yan et al. (2024), where the binder was composed of OPC, MK, limestone, GGBFS, and semi-hydrated gypsum. Furthermore, FA cenosphere was utilized as a lightweight filler material in place of sand to enhance sustainability and cost-effectiveness. Water-reducing agents such as hydroxypropyl methylcellulose (HPMC) and SP were used. The optimal mix in the study incorporated 1.25% high-strength polyethene (PE) fiber. The flexural strength was assessed through a fracture toughness test, and the results were compared with mold-cast specimens. The printed specimens were loaded in the *Y* and *Z* directions. The results revealed that the nominal flexural strength was enhanced by 13.94% and 23.49% in the *Y* and *Z* directions, respectively, compared with mold-cast specimens, which had a nominal flexural strength of 27.54 kN·m. The authors concluded that the enhanced flexural strength was due to the inclusion of high-strength PE fibers and their orientation during the loading. Consequently, this resulted in a more uniform distribution of cracks within the matrix. In the same study, it was also noted that the elastic modulus of the composites was significantly improved when compared with that of mold-cast specimens. The modulus of 22.09 GPa and 21.44 GPa was attained by the specimens loaded in the *X* and *Z* directions, respectively, thus showing increments of about 18.95% and 15.45%, respectively, over mold-cast values. It was due to the denser and compact microstructure of the composites, which in turn reduced air voids. It is equally important to mention here that the strength characteristics of 3DPLC³ are directly influenced by the anisotropy behavior of the specimen. In this study, the compressive strength and flexural strength properties demonstrated minimal anisotropy of about 0.067 and less than 1, respectively. Interestingly, in the study reported by Wang et al. (2024), the strength properties exhibited significant anisotropic behavior when loaded in different directions, with no fiber reinforcement incorporated.

The tensile strength behavior of 3DPLC³ composite was investigated by Wang et al. (2024), where the binder was composed of OPC, limestone powder, calcined clay, and FA. Furthermore, 2% PVA fibers (by vol.), silica sand, and SP (0.55% of the binder) were added during the mixing procedure. The tensile test was performed on dog-bone-shaped samples as per Japan Society of Civil Engineers (JSCE) recommendations. Only the *X* direction was considered for loading, and the results were compared with mold-cast specimens. The results revealed that at 28 days, the printed specimens exhibited tensile strength in the range of 2.1–2.4 MPa. When compared with mold-cast specimens, it was found that there was a reduction in ultimate tensile strength by 16.6%–22.7%. The strength loss, as reported in the study, was due to the poor orientation of fibers within the matrix of 3DPLC³ specimens, which led to a weak fiber bridging effect.

7.4.3 Durability

While mechanical performance, such as compressive or flexural strength, matters from the viewpoint of loading conditions, durability relates to how well the 3DPLC³ can resist deterioration against aggressive environmental conditions over its service life. As discussed in the previous section, 3DPLC³ exhibits an impressive mechanical performance when compared with respective mold-cast specimens; however, these properties do not guarantee the durability of these specimens. Therefore, evaluation of durability properties becomes necessary in the case of 3DPLC³.

The literature regarding the durability properties of 3DPLC^3 composites is scarce. However, in a recent study by de la Flor Juncal et al. (2024), the durability properties of 3DPLC^3 mortar were reported. The impact of printing parameters such as filament overlaps and nozzle offset on porosity and oxygen permeability was studied. An electrical resistivity test was conducted to evaluate chloride ion penetration. The binder consisted of OPC, MK, limestone, and gypsum. Furthermore, reinforcement, such as PVA fibers and polycarboxylate SP, was incorporated into the mix. To compare the performance, the control mix had a nozzle offset of 13 mm and an overlapping of 1 mm. The results revealed that the mix showed improved durability when the filament overlap was increased to 4 mm. After 90 days of curing, this mix also exhibited porosity below 10%, while the electrical resistivity remained greater than 134 kΩ·cm. This indicated that the matrix had a dense microstructure with interconnected pores, and chloride ion penetration was considered negligible. The oxygen permeability index (OPI) in the *X* and *Y* directions for the optimal mix exhibited lower permeability (by 24%) compared to the control mix. This interesting study highlighted the significance of optimizing filament overlap and nozzle offset values to ensure better durability of 3DPLC^3 specimens.

7.4.4 Creep

Creep behavior in 3DPLC^3 specimens is also of great concern. 3D printing involves a layer-by-layer construction method, due to which there is weak interlayer bond strength in general. Consequently, it might lead to microcracking of the structures during the service period of the structure. Excessive creep lowers the load-carrying ability with respect to time. Therefore, evaluating creep in 3D-printed structures becomes highly relevant, as it provides insights into deformation and stability, thereby ensuring structural longevity. Ibrahim et al. (2024) investigated the creep behavior of 3DPLC^3 concrete under sustained loadings for up to 225 days. The loading conditions involved both tensile and flexural aspects. This study was the first to report the creep behavior of 3DPLC^3. The binder consisted of OPC (50%), limestone (15%), calcined clay (30%), and gypsum (5%). Further PP microfibers were added as reinforcements (1% by vol). To assess tensile and flexural creep, the specimens were exposed to stress levels of 40%, 60%, and 80% of their tensile and flexural strengths. Creep tests (quasi-static) were conducted in parallel (D1) and perpendicular (D3) directions to the printing. The highest tensile creep strain at the age of 225 days was found to be 556.08 $\times 10^{-5}$ µm/µm at 80% loading in D1. In D3, at the same loading level, it was found to be 99.67 $\times 10^{-5}$ µm/µm.

Regarding flexural creep, it was found that the values in D1 and D3 were 60.8 and 16.4 µm/MPa, respectively, at the age of 215 days. This indicated that flexural creep was higher in D1 compared to D3. Interestingly, no creep fracture was noticed even after exposure to 80% of sustained loading. The authors attributed this observation to the self-healing behavior of LC^3, which was initiated by the effect of pozzolanic reactions.

7.4.5 Sustainability

The adoption of low-carbon binder materials, such as LC^3, provides the prospect of reducing environmental impact, thereby ensuring high sustainability. LC^3, coupled with 3D printing, offers an innovative approach to further reduce environmental impacts in the modern construction scenario. However, only a few studies have reported the sustainability aspects of $3DPLC^3$ mortar. A study by Long et al. (2021) reported the results of a life-cycle assessment (LCA) performed on $3DPLC^3$. The binder consisted of limestone powder, OPC, calcined clay, and silica fume. In this study, the optimal mix was found to be the one that consisted of 50% OPC, 33.33% calcined clay, 16.7% limestone powder, and 5% silica fume (by weight). The results of LCA studies in terms of embodied energy (EE) and embodied carbon (EC) were compared with control 3D-printable OPC-based mixes. It was found that the LC^3 mix exhibited reductions of 45.6% and 40.2% in EE and EC, respectively, compared to the control mix. These reductions were attributed to the inclusion of clay and silica fume. The authors concluded that using silica fume led to a denser microstructure that improved printability while simultaneously reducing the environmental footprint of $3DPLC^3$.

Another study by Jin et al. (2024) conducted in Canada and France reported on the LCA of $3DPLC^3$ concrete based on several impact categories. The binder was composed of limestone, calcined clay, OPC, and silica fume. The results were compared with 3D-printable OPC-based mixes. The results revealed that LC^3 mixes exhibited a reduction in global warming potential (GWP) by 36% and 46% in the context of Quebec and France, respectively. For the LC^3 mix, it was found that indicators such as ozone depletion and ionizing radiation increased slightly. However, in terms of human health and ecosystem impacts, reductions of 31%–39% and 30%–43%, respectively, were observed when compared with the control mixes. The authors also conducted a sensitivity analysis, which revealed that if calcined clay could be obtained locally, then GWP could be further reduced by 15%.

7.5 RESEARCH GAPS AND PRIORITIES

The preceding section presented an overview of some recently published research in $3DPLC^3$, indicating that the fusion of LC^3 with 3D printing construction is still in its early stages. Following the overview, this section highlights some of the key research gaps in $3DPLC^3$, which require further investigations to make $3DPLC^3$ a truly commercially viable construction material in the near future.

(i) Durability aspects of $3DPLC^3$ mortar/concrete remain highly underexplored, particularly related to acid and sulfate attacks. To ensure the wide adoption of $3DPLC^3$, it is important to investigate the behavior of $3DPLC^3$ when exposed to these aggressive conditions. Considering the interlayers in $3DPLC^3$, there can be ingress of acids, salts, and sulfates into the structures, potentially leading to degradation and thus compromising durability performance. Apart from these, an understanding of the nature and extent of carbonation depth and rate of $3DPLC^3$ is also crucial. Therefore, more research is required to investigate the durability aspects of 3DPLC^3, and simultaneously, it is essential to develop suitable strategies to improve durability.

(ii) To ensure better extrudability, the use of chemical admixtures is inevitable in the 3D printing process. The chemical admixtures and additives highly influence the rheology and mechanical properties of 3D-printable mixes. In this regard, studies on the effects of different additives and admixtures on 3DPLC^3 mixes are limited. Therefore, more efforts should be made in this direction, utilizing different types of chemical admixtures to figure out their compatibility with 3DPLC^3. Strength-enhancing additives in the binder, such as micro- and nano-silica fume and rice husk ash, should be investigated to densify the microstructure of printed specimens and mitigate the risk of microcracking during service life.

(iii) LC^3 mix provides a unique opportunity to customize the mix design as per application requirements. However, the studies related to the effect of the mix design on rheology and mechanical properties have not been explored in detail. Particular emphasis on this aspect is crucial to the development of standards and guidelines to enhance the implementation rate of 3DPLC^3 at the local levels.

(iv) More studies are needed to correlate the printing parameters and systems with the end properties of 3DPLC^3 specimens. Establishing these parameters and comparing the properties of 3DPLC^3 and 3D-printed conventional concrete will reflect the degree to which 3DPLC^3 can be customized as per the application requirements. Comparison studies on this aspect are currently lacking. Printing parameters, such as open time, significantly impact the printability and buildability of specimens. Therefore, future research should focus on comparison studies and exploring the effect of printing parameters more exhaustively.

(v) The main factor that hinders the adoption of 3DPLC^3 is the absence of standards and codes. Understandably, this technology is still new, and it will take time to develop a uniform code of practice. Currently, 3DPLC^3 studies are done from the viewpoint of conventional cement mixes. LC^3 largely differs from conventional cement; hence, the properties of 3DPLC^3 mix will vary from those of 3D-printed conventional cement mixes, as noted in this chapter. Therefore, the need of the hour is developing a uniform code of practice with global collaboration.

(vi) Only two studies have been conducted on the environmental footprints of 3DPLC^3 so far. Future research should focus more on quantifying environmental footprints, and comparisons should be made with 3D printing of traditional cement-based composites and other low-carbon alternatives such as geopolymers. By doing so, a holistic understanding would be attained, which may encourage the rapid adoption of LC^3 into 3D printing applications at national and international levels.

(vi) As discussed in Section 4.3, assessing long-term creep behavior is crucial to ensure the structural longevity of 3D-printed structures throughout their service life. Only one study to date has reported the tensile and flexural creep behavior of 3DPLC^3. Hence, there is limited understanding of this aspect of $3DPLC^3$. Since $3DPLC^3$ involves printing in a layer-by-layer sequence, anisotropy may impact creep behavior very differently than in LC^3-based mold-cast specimens. More investigations are thus needed to understand how the printed structures deform under sustained load-bearing conditions.

(vii) The inclusion of reinforcement in 3D printing concrete has been a challenge so far, especially due to the lack of formwork and the layer-by-layer method of printing. Although some researchers have reported the use of fibers such as PP, PVA, and PE as reinforcements to enhance the strength behavior of 3DPLC3, more studies should focus on using different types of natural fibers to lower the cost and simultaneously assess the compatibility between these fibers and $3DPLC^3$.

7.6 SUMMARY

The fusion of 3D printing and LC^3 provides an innovative and sustainable route to modern construction needs. Using LC^3 in 3D printing offers several advantages over traditional cement-based printing methods. In this context, this comprehensive chapter provides an overview of the $3DPLC^3$ system and highlights its key aspects. LC^3 is currently emerging as an alternative to traditional cement. When combined with 3D printing construction, it can be a way forward to addressing issues relating to carbon emissions and wastage of materials while also increasing productivity in construction.

The chapter laid a strong foundation for the application of LC^3 and 3D printing in today's construction scenario. Then, the hydration process and production of LC^3 were elaborated. Following the discussion on LC^3, the techniques, requirements, and applications of 3DPC technology were detailed. Thereafter, the overview of $3DPLC^3$ properties, such as rheology, mechanical, durability, and creep properties, was presented. To shed light on the sustainability of $3DPLC^3$, the EE and EC, including some environmental impact indicators, were presented as well. Finally, based on the overview, some key research gaps were highlighted. At the same time, to address these gaps, specific priorities were identified that would encourage further research on $3DPLC^3$ to ensure its widespread adoption. In conclusion, $3DPLC^3$ offers remarkable benefits in terms of its performance and sustainability. It would not be an exaggeration to say that $3DPLC^3$ could transform the construction sector while limiting the use of traditional cement and traditional construction methods. However, research and development in $3DPLC^3$ is still in its early stages. Therefore, continued research on key aspects, as highlighted in this chapter, will unlock its full potential.

REFERENCES

Al-Noaimat, Yazeed A., Mehdi Chougan, Mazen J. Al-Kheetan, Othman Al-Mandhari, Waleed Al-Saidi, Marwan Al-Maqbali, Haitham Al-Hosni, and Seyed Hamidreza Ghaffar. 2023. "3D Printing of Limestone-Calcined Clay Cement: A Review of Its Potential Implementation in the Construction Industry." *Results in Engineering* 18 (June): 101115. https://doi.org/10.1016/j.rineng.2023.101115.

Antoni, M., J. Rossen, F. Martirena, and K. Scrivener. 2012. "Cement Substitution by a Combination of Metakaolin and Limestone." *Cement and Concrete Research* 42 (12): 1579–89. https://doi.org/10.1016/j.cemconres.2012.09.006.

Asprone, Domenico, Ferdinando Auricchio, Costantino Menna, and Valentina Mercuri. 2018. "3D Printing of Reinforced Concrete Elements: Technology and Design Approach." *Construction and Building Materials* 165 (March): 218–31. https://doi.org/10.1016/j.conbuildmat.2018.01.018.

Avet, François, Ruben Snellings, Adrian Alujas Diaz, Mohsen Ben Haha, and Karen Scrivener. 2016. "Development of a New Rapid, Relevant and Reliable (R3) Test Method to Evaluate the Pozzolanic Reactivity of Calcined Kaolinitic Clays." *Cement and Concrete Research* 85 (July): 1–11. https://doi.org/10.1016/j.cemconres.2016.02.015.

Babafemi, Adewumi John, John Temitope Kolawole, Md Jihad Miah, Suvash Chandra Paul, and Biranchi Panda. 2021. "A Concise Review on Interlayer Bond Strength in 3D Concrete Printing." *Sustainability* 13 (13). Multidisciplinary Digital Publishing Institute: 7137. https://doi.org/10.3390/su13137137.

Babafemi, Adewumi John, Chris Norval, John Temitope Kolawole, Suvash Chandra Paul, and Kamoru Ademola Ibrahim. 2024. "3D-Printed Limestone Calcined Clay Cement Concrete Incorporating Recycled Plastic Waste (RESIN8)." *Results in Engineering* 22 (June): 102112. https://doi.org/10.1016/j.rineng.2024.102112.

Barbhuiya, Salim, Jaya Nepal, and Bibhuti Bhusan Das. 2023. "Properties, Compatibility, Environmental Benefits and Future Directions of Limestone Calcined Clay Cement (LC3) Concrete: A Review." *Journal of Building Engineering* 79 (November): 107794. https://doi.org/10.1016/j.jobe.2023.107794.

Bonavetti, V. L, V. F Rahhal, and E. F Irassar. 2001. "Studies on the Carboaluminate Formation in Limestone Filler-Blended Cements." *Cement and Concrete Research* 31 (6): 853–59. https://doi.org/10.1016/S0008-8846(01)00491-4.

Chandrasekhar, Sathy, and S Ramaswamy. 2002. "Influence of Mineral Impurities on the Properties of Kaolin and Its Thermally Treated Products." *Applied Clay Science* 21 (3): 133–42. https://doi.org/10.1016/S0169-1317(01)00083-7.

Chen, Yu, Stefan Chaves Figueiredo, Çağlar Yalçinkaya, Oğuzhan Çopuroğlu, Fred Veer, and Erik Schlangen. 2019a. "The Effect of Viscosity-Modifying Admixture on the Extrudability of Limestone and Calcined Clay-Based Cementitious Material for Extrusion-Based 3D Concrete Printing." *Materials* 12 (9). Multidisciplinary Digital Publishing Institute: 1374. https://doi.org/10.3390/ma12091374.

Chen, Yu, Zhenming Li, Stefan Chaves Figueiredo, Oğuzhan Çopuroğlu, Fred Veer, and Erik Schlangen. 2019b. "Limestone and Calcined Clay-Based Sustainable Cementitious Materials for 3D Concrete Printing: A Fundamental Study of Extrudability and Early-Age Strength Development." *Applied Sciences* 9 (9). Multidisciplinary Digital Publishing Institute: 1809. https://doi.org/10.3390/app9091809.

Fernandez, Rodrigo, Fernando Martirena, and Karen L. Scrivener. 2011. "The Origin of the Pozzolanic Activity of Calcined Clay Minerals: A Comparison between Kaolinite, Illite and Montmorillonite." *Cement and Concrete Research* 41 (1): 113–22. https://doi.org/10.1016/j.cemconres.2010.09.013.

Flor Juncal, L. de la, G. Loporcaro, A. Scott, and D. Clucas. 2024. "Influence of Printing Parameters on the Durability of 3D-Printed Limestone Calcined Clay Cement Mortar: Overlap between Filaments and Nozzle Offset." *Materials and Structures* 57 (8): 191. https://doi.org/10.1617/s11527-024-02451-6.

Gao, Huaxing, Lang Jin, Yuxuan Chen, Qian Chen, Xiaopeng Liu, and Qingliang Yu. 2024. "Rheological Behavior of 3D Printed Concrete: Influential Factors and Printability Prediction Scheme." *Journal of Building Engineering* 91 (August): 109626. https://doi.org/10.1016/j.jobe.2024.109626.

Gao, Peng, Wei Zha, Yuting Chu, Yang Yu, Binggen Zhan, Qijun Yu, Faquan Xie, and Haijun Lu. 2025. "Calculation Model for CO2 Emissions of Blended Cement Production."

Journal of Cleaner Production 489 (January): 144646. https://doi.org/10.1016/j.jclepro.2024.144646.

Gosselin, C., R. Duballet, Ph. Roux, N. Gaudillière, J. Dirrenberger, and Ph. Morel. 2016. "Large-Scale 3D Printing of Ultra-High Performance Concrete – a New Processing Route for Architects and Builders." *Materials & Design* 100 (June): 102–9. https://doi.org/10.1016/j.matdes.2016.03.097.

Gowri, K., and Abdul Rahim. A. 2024. "Studies on Hybrid Quaternary Blended Limestone Calcined Clay Cement (LC3) for Sustainable Concrete." *Innovative Infrastructure Solutions* 9 (11): 430. https://doi.org/10.1007/s41062-024-01742-9.

Hack, Norman, and Willi Viktor Lauer. 2014. "Mesh-Mould: Robotically Fabricated Spatial Meshes as Reinforced Concrete Formwork." *Architectural Design* 84 (3): 44–53. https://doi.org/10.1002/ad.1753.

Hou, Shaodan, Zhenhua Duan, Jianzhuang Xiao, and Jun Ye. 2021. "A Review of 3D Printed Concrete: Performance Requirements, Testing Measurements and Mix Design." *Construction and Building Materials* 273 (March): 121745. https://doi.org/10.1016/j.conbuildmat.2020.121745.

Ibrahim, Kamoru A., Gideon P. A. G. van Zijl, and Adewumi J. Babafemi. 2023. "Influence of Limestone Calcined Clay Cement on Properties of 3D Printed Concrete for Sustainable Construction." *Journal of Building Engineering* 69 (June): 106186. https://doi.org/10.1016/j.jobe.2023.106186.

Ibrahim, Kamoru, Gideon P. A. G. van Zijl, and Adewumi John Babafemi. 2024. "Time-Dependent Behaviour of 3D Printed Fibre-Reinforced Limestone Calcined Clay Cement Concrete under Sustained Loadings." *Construction and Building Materials* 446 (October): 138001. https://doi.org/10.1016/j.conbuildmat.2024.138001.

Jameel, Minha, Thongchai Suteerasak, Poopatai Chumpol, Avirut Puttiwongrak, Worathep Sae-Long, and Piti Sukontasukkul. 2025. "Properties of One-Part Geopolymer Pedestrian Blocks Made Using 100 % Waste Materials." *Case Studies in Construction Materials* 22 (July): e04538. https://doi.org/10.1016/j.cscm.2025.e04538.

Jin, Willy, Charlotte Roux, Claudiane Ouellet-Plamondon, and Jean-François Caron. 2024. "Life Cycle Assessment of Limestone Calcined Clay Concrete: Potential for Low-Carbon 3D Printing." *Sustainable Materials and Technologies* 41 (September): e01119. https://doi.org/10.1016/j.susmat.2024.e01119.

Long, Wu-Jian, Can Lin, Jie-Lin Tao, Tao-Hua Ye, and Yuan Fang. 2021. "Printability and Particle Packing of 3D-Printable Limestone Calcined Clay Cement Composites." *Construction and Building Materials* 282 (May): 122647. https://doi.org/10.1016/j.conbuildmat.2021.122647.

Lopes, Joaldo S., Marcelo L. Morais, Alan I. S. Morais, Roosevelt D. S. Bezerra, Josy A. O. Furtini, and Edson C. Silva Filho. 2024. "A New Method for Reducing Impurities in Clay Minerals." *Applied Clay Science* 260 (November): 107539. https://doi.org/10.1016/j.clay.2024.107539.

Maretasari, Lina, and Slamet Ahmadi. 2025. "The Influence of Salesmanship Competency and Service Quality on Competitiveness and Business Performance in Mitra Karya Sukses." *GoodWill Journal of Economics, Management, and Accounting* 5 (1): 31–42.

Massoumi Nejad, Bahman, Sara Enferadi, and Robbie Andrew. 2025. "A Comprehensive Analysis of Process-Related CO2 Emissions from Iran's Cement Industry." *Cleaner Environmental Systems* 16 (March): 100251. https://doi.org/10.1016/j.cesys.2024.100251.

Matos, Paulo Ricardo de, Tuani Zat, Marcelo Moraes Lima, José da Silva Andrade Neto, Marcelo Tramontin Souza, and Erich David Rodríguez. 2023. "Effect of the Superplasticizer Addition Time on the Fresh Properties of 3D Printed Limestone Calcined

Clay Cement (LC3) Concrete." *Case Studies in Construction Materials* 19 (December): e02419. https://doi.org/10.1016/j.cscm.2023.e02419.

Mishra, Geetika, Arun C. Emmanuel, and Shashank Bishnoi. 2019. "Influence of Temperature on Hydration and Microstructure Properties of Limestone-Calcined Clay Blended Cement." *Materials and Structures* 52 (5): 91. https://doi.org/10.1617/s11527-019-1390-5.

Muzenda, Tafadzwa Ronald, Pengkun Hou, Shiho Kawashima, Tongbo Sui, and Xin Cheng. 2020. "The Role of Limestone and Calcined Clay on the Rheological Properties of LC^3." *Cement and Concrete Composites* 107 (March): 103516. https://doi.org/10.1016/j.cemconcomp.2020.103516.

Nasiri, Sina, Rahmat Madandoust, and Malek Mohammad Ranjbar. 2023. "Investigating the Calcination Temperature and Grinding Time of Calcined Clay on the Mechanical Properties and Durability of LC^3 Concrete." *Infrastructures* 8 (10). Multidisciplinary Digital Publishing Institute: 139. https://doi.org/10.3390/infrastructures8100139.

Nematollahi, Behzad, Ming Xia, and Jay Sanjayan. 2019. "Post-Processing Methods to Improve Strength of Particle-Bed 3D Printed Geopolymer for Digital Construction Applications." *Frontiers in Materials* 6 (July). Frontiers. https://doi.org/10.3389/fmats.2019.00160.

Panda, Biranchi, Suvash Chandra Paul, Lim Jian Hui, Yi Wei Daniel Tay, and Ming Jen Tan. 2017. "Additive Manufacturing of Geopolymer for Sustainable Built Environment." *Journal of Cleaner Production* 167 (November): 281–88. https://doi.org/10.1016/j.jclepro.2017.08.165.

Paul, Suvash Chandra, Gideon P. A. G. van Zijl, Ming Jen Tan, and Ian Gibson. 2018. "A Review of 3D Concrete Printing Systems and Materials Properties: Current Status and Future Research Prospects." *Rapid Prototyping Journal* 24 (4). Emerald Publishing Limited: 784–98. https://doi.org/10.1108/RPJ-09-2016-0154.

Perrot, Arnaud. 2019. *3D Printing of Concrete: State of the Art and Challenges of the Digital Construction Revolution*. John Wiley & Sons.

Qaidi, S., A. Yahia, B. A. Tayeh, H. Unis, R. Faraj, and A. Mohammed. 2022. "3D Printed Geopolymer Composites: A Review." *Materials Today Sustainability* 20 (December): 100240. https://doi.org/10.1016/j.mtsust.2022.100240.

Roussel, N., G. Ovarlez, S. Garrault, and C. Brumaud. 2012. "The Origins of Thixotropy of Fresh Cement Pastes." *Cement and Concrete Research* 42 (1): 148–57. https://doi.org/10.1016/j.cemconres.2011.09.004.

Sakhare, Vishakha, Najar, Mohamed, and Sachin and Deshpande. 2023. "Printing Performance of 3D-Printed Geopolymer through Pumpability – Extrudability – Buildability Properties – A Review." *Architectural Science Review* 0 (0). Taylor & Francis: 1–25. https://doi.org/10.1080/00038628.2024.2445050.

Scrivener, Karen, Fernando Martirena, Shashank Bishnoi, and Soumen Maity. 2018. "Calcined Clay Limestone Cements (LC^3)." *Cement and Concrete Research*, Report of UNEP SBCI Working Group on Low-CO2 Eco-Efficient Cement-Based Materials, 114 (December): 49–56. https://doi.org/10.1016/j.cemconres.2017.08.017.

Sharma, Meenakshi, Shashank Bishnoi, Fernando Martirena, and Karen Scrivener. 2021. "Limestone Calcined Clay Cement and Concrete: A State-of-the-Art Review." *Cement and Concrete Research* 149 (November): 106564. https://doi.org/10.1016/j.cemconres.2021.106564.

Siddika, Ayesha, Mamun, Md. Abdullah Al, Ferdous, Wahid, Saha, Ashish Kumer, and Rayed and Alyousef. 2020. "3D-Printed Concrete: Applications, Performance, and Challenges." *Journal of Sustainable Cement-Based Materials* 9 (3). Taylor & Francis: 127–64. https://doi.org/10.1080/21650373.2019.1705199.

Singh, Narinder, Francesco Colangelo, and Ilenia Farina. 2023. "Sustainable Non-Conventional Concrete 3D Printing – A Review." *Sustainability* 15 (13). Multidisciplinary Digital Publishing Institute: 10121. https://doi.org/10.3390/su151310121.

Sona, S., and S. P. Sangeetha. 2025. "Eco-Friendly Alternative Activators Derived from Industrial Wastes for the Sustainable Production of Two-Part Geopolymer Concrete at Low Cost." *Construction and Building Materials* 467 (March): 140374. https://doi.org/10.1016/j.conbuildmat.2025.140374.

"Spain Builds World's First 3D Printed Concrete Pedestrian Bridge." 2016. https://www.sohu.com/a/www.sohu.com/a/121659264_181700.

Tay, Yi Wei Daniel, Biranchi Panda, Suvash Chandra Paul, Nisar Ahamed Noor Mohamed, Ming Jen Tan, and Kah Fai Leong. 2017. "3D Printing Trends in Building and Construction Industry: A Review." *Virtual and Physical Prototyping* 12 (3). Taylor & Francis: 261–76. https://doi.org/10.1080/17452759.2017.1326724.

"The BOD – Europe's First 3D Printed Building | COBOD International." 2017. *COBOD*. https://cobod.com/projects-customers/the-bod-building/.

Urbanist. 2016. "World's First 3D-Printed Office Building Unveiled in Dubai – WebUrbanist." May 27. https://weburbanist.com/2016/05/26/worlds-first-3d-printed-office-building-unveiled-in-dubai/.

Wang, Yuting, Meng Chen, Tong Zhang, and Mingzhong Zhang. 2024. "Hardening Properties and Microstructure of 3D Printed Engineered Cementitious Composites Based on Limestone Calcined Clay Cement." *Cement and Concrete Composites* 152 (September): 105641. https://doi.org/10.1016/j.cemconcomp.2024.105641.

Woodhead, Roy, Paul Stephenson, and Denise Morrey. 2018. "Digital Construction: From Point Solutions to IoT Ecosystem." *Automation in Construction* 93 (September): 35–46. https://doi.org/10.1016/j.autcon.2018.05.004.

Wu, Peng, Jun Wang, and Xiangyu Wang. 2016. "A Critical Review of the Use of 3-D Printing in the Construction Industry." *Automation in Construction* 68 (August): 21–31. https://doi.org/10.1016/j.autcon.2016.04.005.

Xue, Xiaolong, Xianyu Tan, Qiongyu Huang, Hui Zhu, and Jianshuo Chen. 2022. "Exploring the Innovation Path of the Digital Construction Industry Using Mixed Methods." *Buildings* 12 (11). Multidisciplinary Digital Publishing Institute: 1840. https://doi.org/10.3390/buildings12111840.

Yan, Kang-Tai, Xian-Peng Wang, Yao Ding, Ling-Zhi Li, Dilshod Bazarov, Bo-Yu Deng, Kovshar-Sergey Nikolayevich, and Ke-Quan Yu. 2024. "3D-Printed LC³-Based Lightweight Engineered Cementitious Composites: Fresh State, Harden Material Properties and Beam Performance." *Journal of Building Engineering* 93 (September): 109838. https://doi.org/10.1016/j.jobe.2024.109838.

Yi, Peng, Hangyu Du, Chengzhang Chai, Yongqiang Li, Yikai Hu, Xiaohui Sun, Wei Liu, and Feng Xing. 2024. "Potential of Desulfurized Gypsum as a Sulfate Alternative in Preparing Solid Waste-Based Limestone Calcined Clay Cement (LC³) and Its Effects on Early Properties." *Journal of Building Engineering* 98 (December): 111335. https://doi.org/10.1016/j.jobe.2024.111335.

Zhang, Jingchuan, Jialiang Wang, Sufen Dong, Xun Yu, and Baoguo Han. 2019. "A Review of the Current Progress and Application of 3D Printed Concrete." *Composites Part A: Applied Science and Manufacturing* 125 (October): 105533. https://doi.org/10.1016/j.compositesa.2019.105533.

Zhang, Yuying, Xiaohong Zhu, Muduo Li, Chao Zhang, Yamei Zhang, Xiuli Du, Nemkumar Banthia, et al. 2025. "3D Printing Technology in Concrete Construction." *Nature Reviews Clean Technology* 1 (4). Nature Publishing Group: 288–303. https://doi.org/10.1038/s44359-025-00047-z.

Zhong, Hui, and Mingzhong Zhang. 2022. "3D Printing Geopolymers: A Review." *Cement and Concrete Composites* 128 (April): 104455. https://doi.org/10.1016/j.cemconcomp.2022.104455.

Zunino, Franco, and Karen L. Scrivener. 2020. "Influence of Kaolinite Content, Limestone Particle Size and Mixture Design on Early-Age Properties of Limestone Calcined Clay Cements (LC3)." In *Calcined Clays for Sustainable Concrete*, edited by Shashank Bishnoi, 331–37. Singapore: Springer. https://doi.org/10.1007/978-981-15-2806-4_39.

8 3D-Printable Concrete with Recycled Materials

Dhrutiman Dey, Li Wang, and Biranchi Panda

8.1 INTRODUCTION

With the rapid advancement of computer control systems, 3D printing technology has undergone significant development. As a new industrial manufacturing process, 3D printing is considered an important production tool for the Third Industrial Revolution (Dey, Srinivas, Panda, et al. 2022; Tay et al. 2017). 3D concrete printing (3DCP) is based on digital models, using cementitious materials, admixtures, additives, fibers, and aggregates as "ink"; employing robotic or gantry printers; and achieving rapid prototyping through layered extrusion and stacking in either in situ or assembly printing modes (Warsi, Panda, et al. 2023; Dey et al. 2023). Due to its ability to reduce construction waste by 60%, labor costs by 70%, and production time by 70%, as well as waste and greenhouse gas emissions, 3DCP has garnered widespread attention in the construction industry (Singh et al. 2022; Panda et al. 2017). Buildings and infrastructure continuously consume resources and generate waste throughout their life cycle. Additionally, the high labor demand and material costs lead to high construction expenses. The construction industry is the third-largest contributor to greenhouse gas emissions due to the massive demand for concrete. Within the framework of a low-carbon economy, choosing renewable and recycled building materials is a key step toward more sustainable construction practices (Bhattacherjee et al. 2021; Dey and Panda 2023). Using recycled building materials helps reduce construction waste and reintroduces resources into the production cycle. Recycling construction waste, plastics, rubber, and other materials reduces the demand for raw materials and lowers the environmental impact of the production processes. Despite the continuous development of digital construction technology, the performance of 3D printing materials remains crucial for the quality of printed products (Dey et al. 2025). To achieve good printing results, high material performance (such as flowability, extrudability, workability, and open time) is required (Buswell et al. 2018). Concrete for 3D printing must have good rheological properties. An ideal 3D printing mix should have a high static yield stress and a low dynamic yield stress (Kruger et al. 2019; Ding et al. 2021; Ding, Xiao, Zou, et al. 2020). The static yield stress is the point at which the material begins to flow, while the dynamic yield stress is the minimum stress needed to maintain this flow (Ding, Xiao, Zou, et al. 2020). Due to the formation of molecular forces, the viscosity decreases when shear force is applied and gradually recovers after the shear force disappears, known as flocculation. By applying external forces to the material to break the binding forces and

DOI: 10.1201/9781003434863-10

achieve de-flocculation, the viscosity decreases and the material starts to flow. Once the external force/stirring ceases, flocculation reoccurs (Boddepalli et al. 2023). The yield stress of the mix is crucial to ensure that the bottom concrete layer does not collapse during the printing process and changes over time with the hardening of the concrete and the application of shear forces during pumping/transportation (Dey, Srinivas, Boddepalli, et al. 2022; Srinivas et al. 2022). The deposited layer needs to be strong enough to support subsequent layers without deformation, so the material should have sufficient early stiffness. Therefore, exploring the rheology of concrete is important for printing work.

From the perspective of environmental protection and sustainable development, replacing cement and natural sand with recycled materials in 3D printing not only reduces the dependency on traditional materials but also lowers the cost of 3D printing materials, energy consumption, and CO_2 emissions. This recycling approach aligns with the concept of a low-carbon economy and promotes the construction industry toward a more sustainable circular economy model.

8.2 RECYCLED MATERIALS AND THEIR CHARACTERISTICS

Recycled materials are those that have been recovered and reprocessed from waste products, diverting them from landfills and reintroducing them into the production cycle as valuable raw materials. In the context of concrete, these materials typically come from the construction and demolition process, and various industrial waste are reused to replace natural resources in new concrete mixes. Among these, 3DCP has embraced a range of recycled materials to enhance sustainability and reduce reliance on natural resources. Each of these recycled materials offers unique properties that can enhance the performance and environmental profile of 3D-printed concrete. This practice not only helps in waste management but also contributes to sustainability by reducing the environmental footprint of digital construction activities (Han et al. 2021).

8.2.1 Classification of Recycled Material

Upon classification according to the source of waste, recycled materials are largely divided into two categories: (i) construction and demolition waste and (ii) industrial waste, which are further classified as follows.

8.2.2 Recycled Construction Powder

Recycled construction powders (RCPs) are ultra-fine (finer than cement) particles derived from demolition waste, which mainly contain waste concrete and waste brick (Xiao and Ma 2022). These materials are increasingly being used in concrete production, including 3D-printed concrete (3DPC), due to their ability to fill voids and enhance the properties of concrete mixtures. The production involves crushing larger waste materials into fine particles, followed by screening to achieve the desired particle size distribution (Zhang et al. 2023). The particle size of recycled powder typically measures less than 150 μ, although its distribution can be controlled to achieve

FIGURE 8.1 Classification of recycled material used in 3DCP. Reproduced with permission from Hou et al. (2021), Wu et al. (2023), H. Liu et al. (2022), Al-Noaimat et al. (2023), Ting et al. (2019), Skibicki et al. (2022a).

uniformity, which is essential for optimizing concrete properties. The angular nature of the particles, resulting from the crushing process, contributes to improved packing density and better interlocking within the concrete matrix. The high surface area of these fine particles enhances the pozzolanic reaction, thereby boosting the concrete's overall strength and durability (Hou et al. 2021). Details on the chemical compositions and microstructures of RCPs are provided in Table 8.1 and Figure 8.1.

8.2.3 Recycled Fine Aggregate

As the availability of natural sand diminishes due to extensive extraction and environmental regulations, the construction industry is compelled to seek alternative materials. Recycled fine aggregates (RFAs) derived from the processing of construction and demolition waste, such as crushed concrete, bricks, and masonry, offer a practical and cost-effective solution to this. The production process of RFAs involves crushing larger waste materials into finer particles, followed by screening to ensure uniform particle size distribution (Rifa et al. 2023). RFAs typically range from 75 μ to 4.75 mm, similar to natural sand. The particle size distribution can be adjusted during processing to meet specific grading requirements. It mainly consists of silica, similar to natural sand, but the chemical composition can vary depending on the source material. RFA particles tend to be more angular with rougher surfaces and are reported to be less durable compared to natural fine aggregate owing to the existence of adhered mortar, cement paste, clay, and other impurities like metal, salts, and organic matter (Chandru et al. 2023). Thus, the chemical composition can vary depending on the source and processing of the recycled material. The fineness modulus of recycled sand typically ranges from 2.0 to 3.5, which is similar to that of natural sand; however, RFAs have a lower apparent density compared to natural sand. Aside from this, RFAs often exhibit a higher water absorption rate (~12%) and moisture content (~0.6%) compared to natural sand due to their porous nature and the presence of residual cement paste, which absorbs more water (M. Bai et al. 2023).

8.2.4 Recycled Coarse Aggregate

While materials utilized in 3DPC are often referred to as concrete, they are primarily mortar lacking coarse aggregates. This composition means that conventional cement constitutes a significant portion of the mix, necessitating an increase in fine aggregates (such as natural sand) to compensate for the high cement content (Warsi, Srinivas, et al. 2023). Therefore, incorporating coarse aggregates, particularly recycled ones from construction and demolition waste, is crucial for advancing the sustainability of 3DPC. Recycled coarse aggregate (RCA) comprises three main aggregate types: old mortar aggregate, old aggregate with residual mortar, and old aggregate without attached mortar, of which old mortar aggregate constitutes nearly 50% of the total (G. Bai et al. 2020). RCA typically exhibits a more angular shape and a rougher surface compared to natural coarse aggregate, as a result of physical damage by crushing, collision, extrusion, and abrasion during its production. For optimized bulk density and performance in 3D printing, RCA is broadly categorized into particle sizes of 5–7 mm, 7–10 mm, and 10–12 mm (H. Liu et al. 2022). Such precise gradation is crucial for ensuring mix uniformity, which directly influences layer adhesion and the structural integrity of 3D-printed components. In line with T/CECS 786–2020 ("GB/T 14685–2022 English PDF," n.d.), the maximum particle size for coarse aggregates used in 3DPC is stipulated to be no greater than one-third of the nozzle diameter. To comply with this standard, researchers are employing larger print heads with nozzle diameters reaching up to 40 mm. This adaptation facilitates the extrusion of concrete mixes containing aggregate particles up to 12 mm in size (Tong et al. 2023). The chemical composition of RCA includes silica, calcium oxide, and various oxides from the cement paste. It generally contains contaminants like gypsum, bricks, glass, wood, and metals, which must be minimized through effective processing (Medina et al. 2015).

8.2.5 Recycled Brick Aggregate

Recycled brick aggregate (RBA) is derived from the systematic crushing and processing of waste bricks, originating from demolished structures and excess or defective bricks from manufacturing plants. This aggregate is characterized by its distinct angular and porous texture, which is seen to influence the fresh state and hardened state properties of the concrete matrix (Wong et al. 2018). The production process involves crushing the bricks into desired aggregate sizes, screening to remove fines, and washing to eliminate dust and contaminants. Physically, RBA particles typically range from fine to coarse aggregates, with sizes tailored for specific concrete applications. The bulk density of RBA is generally lower than that of natural coarse aggregates, averaging between 1,100 and 1,300 kg/m^3, due to its porous nature (Zheng et al. 2018). Additionally, RBA exhibits higher water absorption rates, ranging from 10% to 20%, which necessitates careful adjustments in mix design to ensure appropriate workability, consistency, and strength of the concrete mix. Chemically, RBA is primarily composed of silica and alumina, reflecting the original composition of brick materials. It may also contain minor impurities, such as calcium oxide and iron oxide, depending on the source of the bricks (Ji and Zhang 2022).

8.2.6 Recycled Glass

Recycled glass (RG) is created by processing waste glass from various sources such as discarded bottles, windows, and industrial glass waste. The production involves crushing the glass into specific sizes, resulting in an aggregate with an angular and smooth texture within the concrete matrix (Shayan and Xu 2004). RG particles can range from fine to coarse, depending on the application requirements. With a bulk density typically between 2,400 and 2,500 kg/m³, RG is denser than many traditional recycled aggregates due to the intrinsic density of glass. Additionally, RG's low water absorption rates, usually below 1%, contribute to a more consistent and workable concrete mix by reducing water content variability (Mardani-Aghabaglou et al. 2015). Chemically, RG is predominantly composed of silica (SiO_2), with minor quantities of other oxides such as calcium oxide (CaO), magnesium oxide (MgO), and sodium oxide (Na_2O), varying based on the glass source. This chemical makeup makes RG inert and enhances its durability and resistance to chemical attacks in concrete applications (Zhou et al. 2024). The smooth surface of RG particles can pose challenges for bond strength with cement paste, but these can be addressed by incorporating supplementary cementitious materials (SCMs).

8.2.7 Recycled Plastic

Recycled plastic (RP) is produced by processing materials such as crumb rubber, waste tires, face masks, containers, and packaging waste. The production process involves shredding these plastics into small, uniform fragments, which are then mixed at temperatures ranging from 190to 200°C. After mixing, the material is extruded and cooled in water, and then further milled to achieve a sand-like texture (Almeshal et al. 2020). RP is characterized by its lightweight nature, low density, and versatility in particle size, which can range from fine to coarse aggregates depending on the specific requirements of the concrete mix. The bulk density of RP is significantly lower than that of traditional aggregates, typically ranging from 500 to 700 kg/m³, which can substantially reduce the weight of the concrete (Iucolano et al. 2013). Additionally, RP exhibits negligible water absorption, which helps maintain the water–cement ratio and consistency of the 3D-printable concrete mix. Chemically, RPs primarily include polyethylene (PE), polypropylene (PP), and polyethylene terephthalate (PET), each contributing different mechanical properties to the concrete mix. These plastics are inert and resistant to chemical reactions, enhancing the durability and longevity of the concrete (Skibicki et al. 2022b). The incorporation of RP in concrete can pose challenges in terms of bonding with the cement matrix due to the smooth and nonporous surface of plastic particles. However, these challenges can be mitigated by using surface treatments or coupling agents to improve the interfacial adhesion between the plastic aggregates and the cement paste.

TABLE 8.1
Chemical Composition (wt.%) of Different Recycled Material Discussed

Recycled Material	CaO	SiO_2	Al_2O_3	Fe_2O_3	SO_3	MgO	K_2O	Na_2O	Ref
RCP	21.99	45.71	15.83	6.77	2.29	2.71	2.56	1.14	Hou et al. (2021)
RFA	18.2	51.3	6.45	2.20	-	2.57	1.88	1.03	Ulsen et al. (2021)
RCA	13.27	58.29	7.69	6.12	0.92	2.28	0.80	1.45	Medina et al. (2015)
RBA	12.70	50.91	15.29	8.97	0.20	4.06	0.76	0.83	Ji and Zhang (2022)
RG	9.80	73.3	1.5	0.06	0.20	0.35	0.60	14.20	Zhou et al. (2024)

8.3 ROLE OF RECYCLED MATERIALS (AS BINDER AND AGGREGATE) IN FRESH PROPERTIES OF 3D-PRINTABLE CONCRETE

8.3.1 Recycled Construction Powder

Duan et al. (2020) investigated the physical and chemical properties of RCP derived from various construction and demolition sources to assess their impact on the rheological properties of mortar. Their findings indicate that the inclusion of RCP significantly increases the yield stress, plastic viscosity, and thixotropy of the mortar. This suggests that RCP is suitable for use in concrete applications that demand higher viscosity and thixotropy, such as 3D-printed mortar (3DPM). Supporting this, Hou et al. (2021) revealed that the incorporation of RCP into 3DPM significantly accelerates the rate of flowability loss, resulting in a substantial reduction in open time. Specifically, the open time decreased from 60 minutes to 15 minutes as the RCP replacement ratio increased from 0% to 30%. Furthermore, the 3DPM with RCP showed a rapid increase in penetration resistance stress over time, resulting in shorter initial and final setting times. Specifically, replacing 30% of the cement with RCP reduced the initial setting time from 6 hours to 3.2 hours and the final setting time from 8 hours to 4.6 hours. The inclusion of RCP also positively impacted the stress–strain behavior of 3DPM. When the RCP replacement ratio was below 10%, the maximum stress and Young modulus increased linearly over time. However, at higher RCP ratios (above 20%), these properties exhibited rapid growth (see Figure 8.2a). Compared to the reference 3DPM (3DPM-REF), the maximum stress and Young modulus of 3DPM with 30% RP (3DPM-RP30) increased by 226.5% and 252.2% at 180 minutes, respectively. Additionally, 3DPM containing 10%–30% RCP demonstrated excellent extrudability and buildability, allowing it to be extruded continuously and smoothly without blockages or crack, as shown in Figure 8.2b.

8.3.2 Recycled Fine Aggregate

Ding et al. (Ding, Xiao, Qin, et al. 2020) investigated the early-age performance of 3DPM by partially replacing natural sand with recycled sand at 25% and 50%

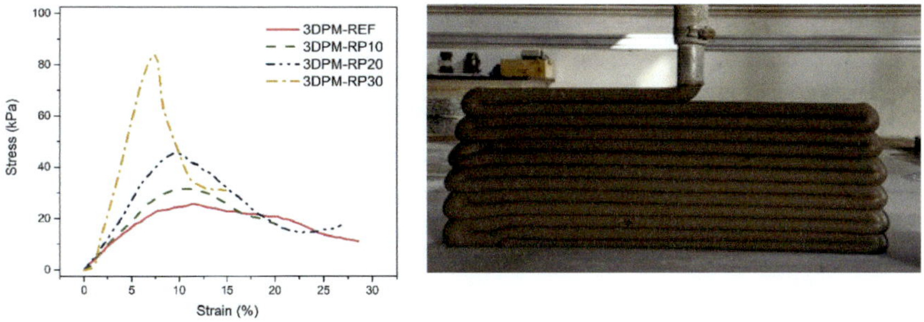

(a) (b)

FIGURE 8.2 (a) Stress–strain curve; (b) printing of 3D-printable mortar with RCP (Hou et al. 2021).

ratios. The study revealed that incorporating recycled sand significantly affects the fresh mechanical properties of the mortar. As the recycled sand content increased, the peak load capacity of the mortar also increased, while the corresponding lateral deformation decreased. This indicates that higher amounts of recycled sand enhance the material's green strength and Young modulus, thereby accelerating early-age strength development and significantly improving the buildability of 3D-printed structures. In terms of failure modes, the research identified two distinct patterns in mortar containing recycled sand under compression at different maturity stages (see Figure 8.3). For immature specimens, tested between 30 minutes and 60 minutes after extrusion, the material exhibited plastic and deformable behavior. Conversely, for mature specimens tested between 90 minutes and 150 minutes after extrusion, the mortar displayed characteristics of a rigid and solid material. This transition from a deformable to a stiffer state can also be seen in the vertical load-displacement relationship and lateral deformation behavior.

In another study, De Vlieger et al. (2023) investigated the effects of 100% replacement of natural sand with recycled sand in mortar mixtures and found that such a substitution leads to insufficient pumpability and extrudability. This issue is attributed to the lower sphericity and roundness of recycled sand particles, which cause increased internal friction and higher dynamic yield stress, adversely affecting the rheological properties of the mortar. Furthermore, the incorporation of recycled sand decreases the initial setting time of the mortar for 3D printing. This is likely due to the presence of non-hydrated cement particles in the recycled sand, which lowers the water-to-cement ratio and accelerates the hydration process. Zou et al. (Zou, Xiao, Ding, et al. 2021) also reported similar results where 100% replacement of natural sand with recycled sand not only decreases the initial fluidity of printing mortar but also increases the rate at which fluidity is lost, thereby shortening the printability window. To address this issue, sodium gluconate was added to the printing mortar. This adjustment allowed the printability requirements to be met and the printability window to be extended. Additionally, the modified mortar exhibited significantly higher green strength at early ages compared to the mortar with 100% natural sand. This increased early-age strength is beneficial for the bottom printed layers, enabling

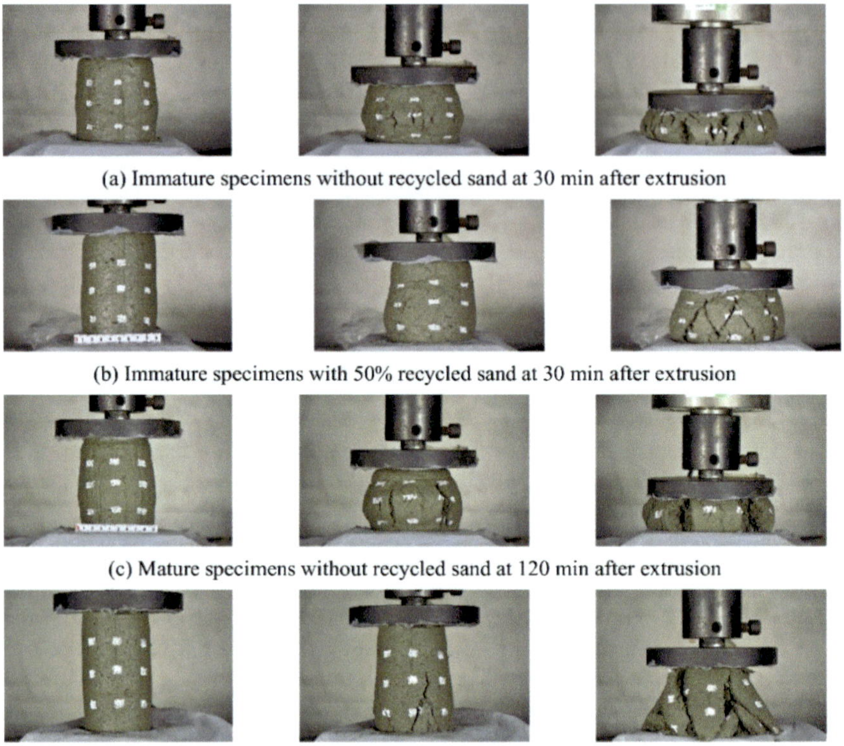
(a) Immature specimens without recycled sand at 30 min after extrusion

(b) Immature specimens with 50% recycled sand at 30 min after extrusion

(c) Mature specimens without recycled sand at 120 min after extrusion

(d) Mature specimens with 50% recycled sand at 120 min after extrusion

FIGURE 8.3 Fresh stage compression behavior of 3D-printable mixes (with and without RFA) at (a, b) 30 minutes and (c, d) 120 minutes after extrusion (Ding, Xiao, Qin, et al. 2020).

them to bear more load in a shorter time, thus facilitating continuous 3D printing (Zou, Xiao, Duan, et al. 2021).

8.3.3 Recycled Coarse Aggregate

Rahul et al. (2022) studied the rheological properties of mortar and found that incorporating coarse aggregates notably decreases yield stress while only slightly reducing plastic viscosity. Research supports that larger aggregates have smaller friction angles, higher fineness modulus, and a lower uncompact void ratio, all of which contribute to the observed reduction in yield stress. Additionally, the lower specific surface area of coarse aggregates compared to fine aggregates resulted in increased paste film thickness and water film thickness, further reducing the yield stress. This reduction in yield stress due to recycled aggregates leads to a significant decrease in buildability. However, by adjusting the superplasticizer dosage (to achieve a water film thickness similar to that of the control mixture), comparable buildable heights (170.7 mm) could be achieved to that of the control mixture (174.7 mm).

To further investigate, Wu et al. (2021) formulated mixtures with varying replacement rates of RCA at 0%, 50%, and 100% for 3D printing applications. They observed a linear decrease in flowability over time, with higher RCA replacement rates leading to a more rapid loss of flowability. Additionally, the yield stress of fresh recycled aggregate concrete (RAC) was found to be higher than that of the mortar (reference mix without coarse aggregate), which further increased with an increase in RCA replacement rate. This is likely due to the rougher surface texture and higher water absorption of RCA compared to natural sand and coarse aggregates. The rheological behavior (shear stress–shear strain curve) of RAC also differed significantly (see Figure 8.4e,f) from that of mortar, exhibiting multiple "plastic flow segments" caused by mechanical interlocking between coarse aggregates. This indicated that the residual strength of RAC was greater than that of mortar. In addition, the shear modulus of fresh concrete increased with the RCA replacement rate, with a gradual decrement in the growth rate, which impacted the buildability of recycled concrete. Notably, within the first 15 minutes of printing, RAC exhibited better buildability with higher replacement rates, despite the shortened open time, as shown in Figure 8.4a–d.

8.3.4 Recycled Brick Aggregate

Christen et al. (Christen, Cho, et al. 2022) developed a sustainable 3D-printable concrete by replacing 64% of natural aggregate with RBA. The inclusion of RBA in the 3D-printable concrete mix increases the water demand due to the porous nature and high water absorption characteristics of the brick aggregates. Despite this, the mix exhibited good printability and reasonable surface finish, achieving up to 32 layers. Furthermore, the study demonstrated that the addition of phase change material to the pores of the RBA via a vacuum impregnation method significantly improved the printability, allowing the structure to support up to 72 layers without failure. This

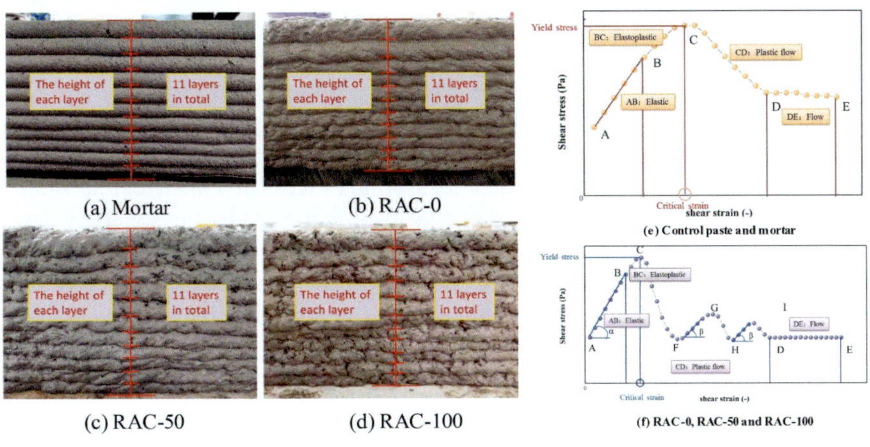

FIGURE 8.4 Printability of 3D-printable (a) mortar, (b) concrete with natural aggregate, (c) concrete with 50% recycled aggregate, (d) concrete with 100% recycled aggregate, (e, f) stress–strain behavior of (e) paste and mortar, (f) recycled aggregate concrete (Wu et al. 2021).

FIGURE 8.5 Buildability of alkali-activated material with (a) 0% RBA, (b) 30% RBA, (c) 40% RBA, and (d) 50% RBA (Al-Noaimat et al. 2023).

indicates that RBA can be effectively used in 3DPC with modifications to enhance its performance.

Al-Noaimat et al. (2023) explored the use of end-of-life bricks as both a binder in powder form and as 1- to 2-mm particles replacing natural aggregate to create alkali-activated materials for 3D printing. The mixtures consisted of 60% brick dust and 40% ground granulated blast-furnace slag as a binder. The study found that incorporating RBA decreased slump measurements, flowability, and setting time, while increasing green strength and Young's modulus. The reduction in flowability and setting time is attributed to the high water absorption of RBA and the irregular shape of brick particles, which decreases the ball-bearing effect compared to spherical natural aggregates. Despite these changes in fresh properties, the buildability of the mixtures improved with the inclusion of RBA. The mixtures with RBA exhibited a higher number of layers without failure compared to the control mixture, which showed significant deformation in the bottom layer. However, it was noted that increasing the RBA content to 50% resulted in discontinuity after depositing 12 layers, indicating a threshold beyond which printability is compromised (see Figure 8.5).

8.3.5 Recycled Glass

Zhou et al. (2024) investigated the effects of using recycled ultrafine glass powder (UFGP) as a partial cement replacement in concrete mixtures, with replacement rates

FIGURE 8.6 Buildability of 3DPM with RGA replacement from 0% to 100% (Ting et al. 2021).

between 0% and 25%. Their findings revealed that a 25% incorporation of UFGP led to a 27.45% increase in slump flow. This enhancement in workability was attributed to the hydrophobic and smooth surface of the glass powder, which increased the availability of free water, thereby improving lubrication between particles and reducing friction. The findings suggest that UFGP can effectively improve the fresh performance of 3D-printable concrete by enhancing its flowability. Another study by Andrew Ting et al. (2022) focused on the effects of varying the glass-to-binder ratio and the fineness modulus of RGA on filament deformation in 3DCP. The results showed that increasing the glass-to-binder ratio and decreasing the fineness modulus of RGA led to reduced filament deformation. This suggests that optimizing the proportion and particle size distribution of RG can enhance the dimensional stability of 3D-printed structures.

Ting et al. (2019) compared mixtures containing fine river sand and those with RG aggregates. The mixture with RG exhibited lower dynamic yield stress and plastic viscosity than the river sand mixture, indicating better flow properties. However, the static yield stress was much lower for mixtures with RG, potentially leading to weaker buildability. In a separate study, Ting et al. (2021) studied the application of RG cullets as fine aggregates in concrete for 3DCP. It is observed that increasing the RG cullet content from 25 wt.% to 100 wt.% led to a slight reduction in static yield stress. Despite this reduction, there were no significant changes in buildability across the mixtures, with all displaying similar buildability levels. However, the overall buildability was relatively low, supporting only 8–9 layers, which is suboptimal for building applications. The study further observed that as the RG content increased, both the dynamic yield stress and plastic viscosity also increased. This effect was attributed to the angular shape and sharp edges of the glass particles, which hindered pumpability.

8.3.6 Recycled Plastic

Skibicki et al. (2022b) explored the feasibility of modifying a mortar for 3D printing by incorporating PET granules as a replacement (10%, 30%, and 50% by volume) for fine aggregate. The study found that the addition of PET significantly decreased the Young modulus of the fresh mixture up to 30.1% for PET50 mixtures. Rajeev et al. (2023) investigated the use of shredded face masks (FMs) as short fibers in 3D-printable concrete. The inclusion of FMs (1% and 2% by volume) significantly altered the rheological behavior, increasing the mix's apparent viscosity and dynamic yield strength. Despite these increases, the mixes with FMs maintained good extrudability and fell within the pumpability range. Interestingly, when comparing FMs to commercial PP fibers, it was found that FM mixes exhibited similar rheological behavior at 1% fiber content. Moreover, unlike PP fibers, increasing the FM content did not lead to a sudden rise in viscosity and yield strength, indicating improved ease of extrusion. This study suggests that FMs could be a viable substitute for PP fibers in 3DCP applications.

Liu et al. (2023) conducted research on the inclusion of milled recycled carbon fibers (rCFs) in 3D-printed mortar and observed that as the fiber content increased from 0 vol% to 1.5 vol%, the filament width decreased by 5%, due to a slight decrease in flowability. In terms of extrudability and buildability, all mixtures with up to 1.5 vol% rCFs could be printed continuously without material blockages or structural collapses. The study also noted that filament deformation primarily occurred within the first ten layers, with the bottom layer thickness ranging from 6.5 mm (CF0) to 7.5 mm (CF1.5). Daniel Oosthuizen et al. (2023) investigated the use of RP waste (Resin8), as a replacement for fine aggregate in 3D-printed concrete, with replacement levels varying from 5% to 15%. The inclusion of Resin8 was observed to mitigate thixotropic behavior, demonstrated by reduced static and dynamic yield stresses relative to the control mix. This reduction in thixotropic behavior adversely affected the buildability performance of the concrete. While Resin8 contributed to sustainability, its impact on the mechanical and rheological properties suggests that further optimization is required to balance environmental benefits with performance requirements.

8.4 HARDENED PROPERTIES OF 3D-PRINTED CONCRETE WITH RECYCLED MATERIALS

8.4.1 Recycled Construction Powder

Qian et al. (2022) studied the impact of RCP on the hardening performance of materials by optimizing the gradation of printable materials, conducting rheological tests, and initial failure height tests. The results indicated that although the addition RCP reduced the total heat release of 3DPM, it also decreased the percentage of large pores (>1000 nm) in 3DPM from 10.7% to 7.1%. Through gradation optimization, it was found that when the RCP replacement rate was 20%, the compressive strength decreased from 49.5 MPa to 44.1 MPa, remaining within an acceptable range. Hou et al. (2023) discovered that at low RCP replacement rates, the filling effect and secondary hydration characteristics of RCP reduced the proportion of harmful pores in the cement matrix, resulting in a slight improvement in the mechanical properties. However, when the replacement rate exceeded 20%, the fluidity loss rate increased significantly, leading to weaker interlayer bonding and reduced concrete mechanical strength. Additionally, the inclusion of RCP significantly increased the anisotropy coefficient of the 3DPM. Although the chemical composition of RCP is similar to that of cement, its CaO content is much lower, resulting in lower hydration reaction activity and fewer hydration products. Consequently, this can adversely affect the hardening performance of concrete.

8.4.2 Recycled Fine Aggregate

Bai et al. (2023) found that the high water absorption rate of RFA improves the consistency of 3D printing mortar, making it more suitable for 3D printing applications. The porous nature of recycled sand and its inherent micro-cracks facilitate the development of multiple cracks in 3DPC, which contributes to enhanced ductility. When the replacement rate of RFA reached 100%, the energy absorption capacity increased by 30.3%. Moreover, the substitution of quartz sand with RFA with a lower fineness modulus increases the bulk density, and RFA, composed of adhered old mortar and partially unhydrated cement, undergoes secondary hydration within the RFA particles, which is beneficial for improving the compressive strength of concrete (Delsaute and Staquet 2020; "Effects of Using Recycled Concrete Aggregates on the Shrinkage of High Performance Concrete | FUTUR," n.d.). Ding et al. (Ding, Xiao, Zou, et al. 2020) found that a high amount of unhydrated cement paste attached to recycled sand, coupled with the abundance of pores in the old cement paste and its loose microstructure, significantly reduced the density of 3D-printed recycled mortar. In their study, the average compressive strength of 3DPC with 25% RFA decreased by 38.4%, 54.8%, and 13.2% in the filament printing direction (*X*), filament translation direction (*Y*), and filament deposition direction (*Z*), respectively. The addition of RFA to 3D printing materials increases the amount of additional cement, ensuring that the mechanical strength does not decrease excessively, thus meeting engineering requirements.

RFA particles are found to be angular and have rougher surfaces than natural fine aggregates. Owing to the presence of adhered mortar, cement paste, clay, and other impurities (such as metals, salts, and organic matter), their durability and mechanical properties are poor.

8.4.3 Recycled Coarse Aggregate

RCA, which constitutes the largest portion of the current construction waste, has been widely applied in traditional construction fields. As a major component of construction waste, RCA provides an effective solution to the dual challenges of natural resource depletion and increased volume of construction waste. Xiao et al. (2022) found that the addition of RCA significantly altered the anisotropic compressive strength of 3DPC, but in any direction, the compressive strength was lower than that of the cast samples. Rahul et al. (2022) attributed this strength reduction to the interfacial transition zone (ITZ) caused by the old mortar attached to the RCA. Additionally, the inclusion of coarse aggregates significantly increases the uncertainty of interlayer bonding in 3D-printed RAC, leading to a substantial increase in interlayer porosity, which in turn causes a decrease in strength. Sun et al. (2022) studied the effects of carbonated RCA on the mechanical properties and pore structure of 3DPC. The results showed that carbonated RCA effectively optimized the pore defects in 3D-printed recycled aggregate concrete, reducing the proportion of small pores and overall porosity, while improving the compressive strength, flexural strength, and splitting tensile strength. Moreover, the anisotropy of the material was effectively improved. Owing to the continuous physical damage during production, such as crushing, impact, compression, and abrasion, the surface morphology of RCA is rougher than that of natural coarse aggregates. Consequently, the physical degradation of RCA leads to a noticeable decline in the mechanical properties of the concrete.

8.4.4 Recycled Brick Aggregate

Christen et al. (Christen, van Zijl, et al. 2022) replaced 64% of the natural aggregates with RBA in 3DCP mixtures and found that the overall porosity of the concrete increased with higher RBA content, leading to a 25% reduction in compressive strength and a 20% reduction in the 3DPC interlayer tensile strength. This reason for such a decrement in strength is ascribed to the porous nature of RBA. Also, RBAs are made by crushing waste concrete, and because of the production process, aggregate particles often have initial defects such as microcracks and residual old mortar. The old mortar adhering to the RBA weakens the bonding strength of the cementitious materials, adversely affecting the mechanical properties of the hardened concrete. Additionally, RBAs are porous and tend to retain residual cement paste, resulting in lower density and higher water absorption rates (Cao et al. 2022; Ma et al. 2022). The mechanical crushing process used before reuse introduces microcracks on the surface, leading to defects in the printed material. As the replacement rate of natural aggregates with RBA increases, the porosity introduced also increases, which can negatively impact the hardening performance of the mortar.

8.4.5 Recycled Glass

Ting et al. (2021) investigated the use of RG fragments to replace natural river sand at mass fractions of 25%, 50%, 75%, and 100% in 3D printing. The study found that owing to the smooth surface and sharp edges of the RG particles, the ITZ between the RG particles and the binder matrix lacked adhesion (Tan and Du 2013). Moreover, because of the hydrophobic nature of RG fragments, excess water in the binder matrix can lead to segregation of the glass aggregate, further reducing the strength of concrete (de Castro and de Brito 2013). The compressive strength of the printed samples decreased by 20.2% to 25.8% at 100% replacement compared to the control samples.

Using RG fragments to replace fine aggregates in 3D printing can potentially reduce the annual demand for construction sand, making it more advantageous from environmental and sustainability perspectives than traditional 3D printing materials. However, the hydrophobicity of the glass particles leads to a higher water content in the binder matrix, which may cause potential defects and stress concentrations during the layering process, creating weak points in the structure. In future studies, the surface treatment of glass particles could increase their hydrophilicity and improve their bonding performance with cementitious materials. Additionally, the incorporation of appropriate modifiers or additives into the mixture can enhance the bonding strength and uniformity of the material, thereby satisfying the requirements for 3D printing.

8.4.6 Recycled Plastic

Novel modified concrete materials can be created by incorporating RP pellets of varying sizes and types into concrete aggregates at different proportions. The mechanical and physical properties of RP concrete may differ significantly from those of traditional concretes. However, owing to the ability of plastic pellets to fill voids within concrete and hinder crack formation, coupled with their inherent impact resistance and wear resistance, this type of concrete can exhibit enhanced ductility and other potentially beneficial yet unexplored properties.

Daniel Oosthuizen et al. (2023) studied the applicability of incorporating RP waste, specifically an eco-aggregate called Resin8, into 3D-printable concrete. Their research found that the hydrophobic nature of the waste plastic led to increased surface moisture in the interlayer regions, thereby improving interlayer bonding and cement hydration. Consequently, the porosity in the interlayer of the Resin8 mixtures was reduced, resulting in an increased interlayer bond strength. Although the overall porosity of the 3D-printed samples increased from 5.8% in the control group to 10.1%, 14.5%, and 16.8%, owing to the air entrapped by the porous particles, the flexural and compressive strengths of all the Resin8 mixtures decreased compared to the control group. However, the performance still met the standard structural requirements, making the substitution of fine aggregates with Resin8 in 3D-printable concrete a feasible solution to effectively reduce plastic waste in the environment.

Sambucci et al. (2020) converted waste tires into coarse rubber granules (RGs, 2–4 mm) and fine rubber powder (less than 1 mm), incorporated these rubber particles into concrete to replace natural aggregates, and studied the mechanical properties of 3D-printed rubber concrete. They found that when river sand was replaced by rubber powder at 50 and 100 vol%, the compressive strength decreased by 65% and 77%, respectively, owing to weaker filler–matrix adhesion and reduced material density. Through four-point bending tests, they observed that at the same replacement levels, the flexural strength decreased by 30% and 67%, respectively. In another study, Sambucci et al. ("Rubber-Cement Composites for Additive Manufacturing: Physical, Mechanical and Thermo-Acoustic Characterization | SpringerLink," n.d.) further demonstrated that despite the reduction in mechanical strength, adding rubber particles to 3DPC positively affected the ductility, sound insulation, and thermal insulation properties. The partial replacement of natural aggregates with rubber aggregates led to a reduction in the mechanical strength of the hardened material because of the decreased overall density and cement filler adhesion. However, the viscoelasticity of the polymer aggregates significantly improves the deformability of the cement matrix (with a decrease in Young's modulus and an increase in elongation at break), and the sound and thermal insulation properties. Additionally, fine rubber aggregates maintained good compaction and thermal insulation requirements in the mixture, whereas coarse fillers positively impacted the ductility, microstructural characteristics, and sound damping behavior. This synergy is beneficial from an engineering perspective, making rubber aggregates suitable for lightweight components (blocks, bricks, insulation barriers, and pavements), potentially reducing structural dead loads, lowering construction costs, and meeting thermal and sound insulation requirements.

Rajeev et al. (2023) evaluated the potential of using recycled fibers from discarded masks as fiber reinforcement in 3DCP applications to improve printability and reduce landfill waste. The study found that owing to the bridging effect of the fibers inhibiting crack formation, the compressive strength at 1% mask content increased by 41% compared with the control group. Additionally, the flexural strength along the weaker interfaces increased by 74% and 82% with 1% and 2% mask content, respectively, and the interlayer bonding strength with 1% mask content increased by 21%. In addition to the bridging effect, Liu et al. (2023) found that grooves or channels (with an apparent channel width of ≤0.5 μm) on the surface of rCFs provided additional nucleation sites, promoting the growth of early hydration products. When the proportion of ground rCFs was 1.0 vol% and 1.5 vol%, the mechanical strength of the mortar mix slightly improved compared to the control group.

8.5 CHALLENGES AND OPPORTUNITIES

Ensuring the consistency of raw material quality is crucial when using recyclable materials for construction 3D printing. However, there are many challenges, including the diversity of material sources, cleanliness, mixing processes, aging phenomena, and changes during processing. To address these issues, strict material testing and classification systems should be established, cleaning and mixing techniques optimized, aging tests conducted, real-time monitoring of the processing implemented, and relevant standards and regulations promoted to ensure the safety and stability of printed structures.

When using large quantities of raw materials, performance can be improved through optimized selection and pretreatment of raw materials, precise proportioning and uniform mixing, the use of performance enhancers, control of processing parameters, the establishment of a comprehensive quality monitoring system, effective recycling and reuse of materials, and continuous research and innovation. These methods ensure the reliability and stability of the final product in practical applications.

Recycled construction aggregates typically have higher porosity and water absorption characteristics. Thus, incorporating RCA in practical applications forms a noticeable ITZ, adversely affecting the mechanical performance and long-term durability of fresh concrete. Carbonated recycled construction aggregates have emerged as a promising approach to address these issues. By reacting carbon dioxide (CO_2) with alkaline materials (such as calcium-based or magnesium-based materials), the resulting carbonates fill the interior and interface of the aggregates, enhancing the interface bonding strength between the aggregate and the matrix. This process improves the compressive strength of the material and promotes CO_2 sequestration, offering unique advantages in terms of environmental sustainability and economic feasibility.

8.6 SUMMARY

The use of recycled materials in 3DCP has great potential, especially where industrial wastes are increasing with population and economic growth, and landfill space is becoming limited. Based on published literature on 3DPC, this study discusses the fresh and hardened properties of 3DPC materials using recycled materials from construction and other industrial wastes. For 3DPC, recycled wastes can significantly contribute to improving the rheology, either in terms of extrudability or buildability depending on material composition and dosages. Due to particle shape and size, some of these materials are found to increase the ease of extrudability, while negatively affecting hardened properties. Compared with natural sand, recycled materials usually have low density and high water absorption due to their porous characteristics. In this regard, both physical and chemical interactions must be studied to consider recycled materials as a sustainable replacement for natural aggregates in concrete printing applications. In the future, new technologies to incorporate a high volume of recycled materials should be explored as pilot projects, while including environmental and economic assessment. This will lead to further development of an industry-scale setup for digital fabrication of sustainable concrete.

REFERENCES

Al-Noaimat, Yazeed A., Mehdi Chougan, Abdulrahman Albar, et al. 2023. "Recycled Brick Aggregates in One-Part Alkali-Activated Materials: Impact on 3D Printing Performance and Material Properties." *Developments in the Built Environment* 16 (December): 100248. https://doi.org/10.1016/j.dibe.2023.100248.

Almeshal, Ibrahim, Bassam A. Tayeh, Rayed Alyousef, Hisham Alabduljabbar, Abdeliazim Mustafa Mohamed, and Abdulaziz Alaskar. 2020. "Use of Recycled Plastic as Fine Aggregate in Cementitious Composites: A Review." *Construction and Building Materials* 253 (August): 119146. https://doi.org/10.1016/j.conbuildmat.2020.119146.

Andrew Ting, Guan Heng, Tan Kai Noel Quah, Jian Hui Lim, Yi Wei Daniel Tay, and Ming Jen Tan. 2022. "Extrudable Region Parametrical Study of 3D Printable Concrete Using Recycled Glass Concrete." *Journal of Building Engineering* 50 (June): 104091. https://doi.org/10.1016/j.jobe.2022.104091.

Bai, Guoliang, Chao Zhu, Chao Liu, and Biao Liu. 2020. "An Evaluation of the Recycled Aggregate Characteristics and the Recycled Aggregate Concrete Mechanical Properties." *Construction and Building Materials* 240 (April): 117978. https://doi.org/10.1016/j.conbuildmat.2019.117978.

Bai, Meiyan, Yuching Wu, Jianzhuang Xiao, Tao Ding, and Kequan Yu. 2023. "Workability and Hardened Properties of 3D Printed Engineered Cementitious Composites Incorporating Recycled Sand and PE Fibers." *Journal of Building Engineering* 71 (July): 106477. https://doi.org/10.1016/j.jobe.2023.106477.

Bhattacherjee, Shantanu, Anusha S. Basavaraj, A. V. Rahul, et al. 2021. "Sustainable Materials for 3D Concrete Printing." *Cement and Concrete Composites* 122 (September): 104156. https://doi.org/10.1016/j.cemconcomp.2021.104156.

Boddepalli, Uday, Biranchi Panda, and Indu Siva Ranjani Gandhi. 2023. "Rheology and Printability of Portland Cement Based Materials: A Review." *Journal of Sustainable Cement-Based Materials* 12 (7): 789–807. https://doi.org/10.1080/21650373.2022.2119620.

Buswell, R. A., W. R. Leal de Silva, S. Z. Jones, and J. Dirrenberger. 2018. "3D Printing Using Concrete Extrusion: A Roadmap for Research." *Cement and Concrete Research*, SI: Digital Concrete 2018, vol. 112 (October): 37–49. https://doi.org/10.1016/j.cemconres.2018.05.006.

Cao, Yubin, Yanru Wang, Zuhua Zhang, and Hao Wang. 2022. "Recycled Sand from Sandstone Waste: A New Source of High-Quality Fine Aggregate." *Resources, Conservation and Recycling* 179 (April): 106116. https://doi.org/10.1016/j.resconrec.2021.106116.

Chandru, Uma, A. Bahurudeen, and R. Senthilkumar. 2023. "Systematic Comparison of Different Recycled Fine Aggregates from Construction and Demolition Wastes in OPC Concrete and PPC Concrete." *Journal of Building Engineering* 75 (September): 106768. https://doi.org/10.1016/j.jobe.2023.106768.

Christen, Heidi, Seung Cho, Gideon van Zijl, and Wibke de Villiers. 2022. "Phase Change Material Infused Recycled Brick Aggregate in 3D Printed Concrete." *Heliyon* 8 (11): e11598. https://doi.org/10.1016/j.heliyon.2022.e11598.

Christen, Heidi, Gideon van Zijl, and Wibke de Villiers. 2022. "The Incorporation of Recycled Brick Aggregate in 3D Printed Concrete." *Cleaner Materials* 4 (June): 100090. https://doi.org/10.1016/j.clema.2022.100090.

Daniel Oosthuizen, Jandré, Adewumi John Babafemi, and Richard Shaun Walls. 2023. "3D-Printed Recycled Plastic Eco-Aggregate (Resin8) Concrete." *Construction and Building Materials* 408 (December): 133712. https://doi.org/10.1016/j.conbuildmat.2023.133712.

de Castro, Sara, and Jorge de Brito. 2013. "Evaluation of the Durability of Concrete Made with Crushed Glass Aggregates." *Journal of Cleaner Production* 41 (February): 7–14. https://doi.org/10.1016/j.jclepro.2012.09.021.

De Vlieger, Jentel, Luc Boehme, Jan Blaakmeer, and Jiabin Li. 2023. "Buildability Assessment of Mortar with Fine Recycled Aggregates for 3D Printing." *Construction and Building Materials* 367 (February): 130313. https://doi.org/10.1016/j.conbuildmat.2023.130313.

Delsaute, Brice, and Stéphanie Staquet. 2020. "Impact of Recycled Sand and Gravels in Concrete on Volume Change." *Construction and Building Materials* 232 (January): 117279. https://doi.org/10.1016/j.conbuildmat.2019.117279.

Dey, Dhrutiman, and Biranchi Panda. 2023. "An Experimental Study of Thermal Performance of 3D Printed Concrete Slabs." *Materials Letters* 330 (January): 133273. https://doi.org/10.1016/j.matlet.2022.133273.

Dey, Dhrutiman, Biranchi Panda, Yash Shukla, and Rajan Rawal. 2025. "A Comprehensive Assessment of Thermal Performance of 3D Printed Concrete Lattice Walls." *Progress in Additive Manufacturing*, ahead of print, March 16. https://doi.org/10.1007/s40964-025-01042-3.

Dey, Dhrutiman, Dodda Srinivas, Uday Boddepalli, Biranchi Panda, Indu Siva Ranjani Gandhi, and T. G. Sitharam. 2022. "3D Printability of Ternary Portland Cement Mixes Containing Fly Ash and Limestone." *Materials Today: Proceedings*, The International Conference on Additive Manufacturing for a Better World (AMBW 2022), vol. 70 (January): 195–200. https://doi.org/10.1016/j.matpr.2022.09.020.

Dey, Dhrutiman, Dodda Srinivas, Biranchi Panda, Prannoy Suraneni, and T. G. Sitharam. 2022. "Use of Industrial Waste Materials for 3D Printing of Sustainable Concrete: A Review." *Journal of Cleaner Production*, 130749.

Dey, Dhrutiman, Vuong Nguyen Van, H. Nguyen Xuan, Dodda Srinivas, Biranchi Panda, and Phuong Tran. 2023. "Flexural Performance of 3D Printed Concrete Structure with Lattice Infills." *Developments in the Built Environment* 16: 100297.

Ding, Tao, Jianzhuang Xiao, Fei Qin, and Zhenhua Duan. 2020. "Mechanical Behavior of 3D Printed Mortar with Recycled Sand at Early Ages." *Construction and Building Materials* 248 (July): 118654. https://doi.org/10.1016/j.conbuildmat.2020.118654.

Ding, Tao, Jianzhuang Xiao, Shuai Zou, and Yu Wang. 2020. "Hardened Properties of Layered 3D Printed Concrete with Recycled Sand." *Cement and Concrete Composites* 113 (October): 103724. https://doi.org/10.1016/j.cemconcomp.2020.103724.

Ding, Tao, Jianzhuang Xiao, Shuai Zou, and Jiangtao Yu. 2021. "Flexural Properties of 3D Printed Fibre-Reinforced Concrete with Recycled Sand." *Construction and Building Materials* 288 (June): 123077. https://doi.org/10.1016/j.conbuildmat.2021.123077.

Duan, Zhenhua, Shaodan Hou, Jianzhuang Xiao, and Amardeep Singh. 2020. "Rheological Properties of Mortar Containing Recycled Powders from Construction and Demolition

Wastes." *Construction and Building Materials* 237 (March): 117622. https://doi.org/10.1016/j.conbuildmat.2019.117622.

"Effects of Using Recycled Concrete Aggregates on the Shrinkage of High Performance Concrete | FUTUR." n.d. Accessed August 11, 2024. https://futur.upc.edu/18723252.

"GB/T 14685–2022 English PDF." n.d. Accessed August 11, 2024. https://www.chinesestandard.net/PDF/English.aspx/GBT14685-2022.

Han, Yilong, Zhihan Yang, Tao Ding, and Jianzhuang Xiao. 2021. "Environmental and Economic Assessment on 3D Printed Buildings with Recycled Concrete." *Journal of Cleaner Production* 278 (January): 123884. https://doi.org/10.1016/j.jclepro.2020.123884.

Hou, Shaodan, Zhenhua Duan, Taohua Ye, Shuai Zou, and Jianzhuang Xiao. 2023. "Mechanical Properties and Pore Structure of 3D Printed Mortar with Recycled Powder." *Construction and Building Materials* 394 (August): 132068. https://doi.org/10.1016/j.conbuildmat.2023.132068.

Hou, Shaodan, Jianzhuang Xiao, Zhenhua Duan, and Guowei Ma. 2021. "Fresh Properties of 3D Printed Mortar with Recycled Powder." *Construction and Building Materials* 309 (November): 125186. https://doi.org/10.1016/j.conbuildmat.2021.125186.

Iucolano, F., B. Liguori, D. Caputo, F. Colangelo, and R. Cioffi. 2013. "Recycled Plastic Aggregate in Mortars Composition: Effect on Physical and Mechanical Properties." *Materials & Design (1980–2015)* 52 (December): 916–22. https://doi.org/10.1016/j.matdes.2013.06.025.

Ji, Yongcheng, and Hongrui Zhang. 2022. "Frost Resistance Investigation of Fiber Reinforced Recycled Brick Aggregate Cementitious Materials." *Scientific Reports* 12 (1): 15311. https://doi.org/10.1038/s41598-022-19006-w.

Kruger, Jacques, Stephan Zeranka, and Gideon van Zijl. 2019. "An Ab Initio Approach for Thixotropy Characterisation of (Nanoparticle-Infused) 3D Printable Concrete." *Construction and Building Materials* 224: 372–86.

Liu, Huawei, Chao Liu, Yiwen Wu, et al. 2022. "3D Printing Concrete with Recycled Coarse Aggregates: The Influence of Pore Structure on Interlayer Adhesion." *Cement and Concrete Composites* 134 (November): 104742. https://doi.org/10.1016/j.cemconcomp.2022.104742.

Liu, Junli, Phuong Tran, Vuong Nguyen Van, Chamila Gunasekara, and Sujeeva Setunge. 2023. "3D Printing of Cementitious Mortar with Milled Recycled Carbon Fibres: Influences of Filament Offset on Mechanical Properties." *Cement and Concrete Composites* 142 (September): 105169. https://doi.org/10.1016/j.cemconcomp.2023.105169.

Ma, Zhiming, Jiaxin Shen, Changqing Wang, and Huixia Wu. 2022. "Characterization of Sustainable Mortar Containing High-Quality Recycled Manufactured Sand Crushed from Recycled Coarse Aggregate." *Cement and Concrete Composites* 132 (September): 104629. https://doi.org/10.1016/j.cemconcomp.2022.104629.

Mardani-Aghabaglou, Ali, Murat Tuyan, and Kambiz Ramyar. 2015. "Mechanical and Durability Performance of Concrete Incorporating Fine Recycled Concrete and Glass Aggregates." *Materials and Structures* 48 (8): 2629–40. https://doi.org/10.1617/s11527-014-0342-3.

Medina, C., W. Zhu, T. Howind, M. Frías, and M. I. Sánchez de Rojas. 2015. "Effect of the Constituents (Asphalt, Clay Materials, Floating Particles and Fines) of Construction and Demolition Waste on the Properties of Recycled Concretes." *Construction and Building Materials* 79 (March): 22–33. https://doi.org/10.1016/j.conbuildmat.2014.12.070.

Panda, Biranchi, Suvash Chandra Paul, Lim Jian Hui, Yi Wei Daniel Tay, and Ming Jen Tan. 2017. "Additive Manufacturing of Geopolymer for Sustainable Built Environment." *Journal of Cleaner Production* 167 (November): 281–88. https://doi.org/10.1016/j.jclepro.2017.08.165.

Qian, Hao, Sudong Hua, Hongfei Yue, et al. 2022. "Utilization of Recycled Construction Powder in 3D Concrete Printable Materials through Particle Packing Optimization."

Journal of Building Engineering 61 (December): 105236. https://doi.org/10.1016/j.jobe.2022.105236.

Rahul, A. V., Manu K. Mohan, Geert De Schutter, and Kim Van Tittelboom. 2022. "3D Printable Concrete with Natural and Recycled Coarse Aggregates: Rheological, Mechanical and Shrinkage Behaviour." *Cement and Concrete Composites* 125 (January): 104311. https://doi.org/10.1016/j.cemconcomp.2021.104311.

Rajeev, Pathmanathan, Akilesh Ramesh, Satheeskumar Navaratnam, and Jay Sanjayan. 2023. "Using Fibre Recovered from Face Mask Waste to Improve Printability in 3D Concrete Printing." *Cement and Concrete Composites* 139 (May): 105047. https://doi.org/10.1016/j.cemconcomp.2023.105047.

Rifa, Asim, Sk M. Subhani, A. Bahurudeen, and Kumar Gedela Santhosh. 2023. "A Systematic Comparison of Performance of Recycled Concrete Fine Aggregates with Other Alternative Fine Aggregates: An Approach to Find a Sustainable Alternative to River Sand." *Journal of Building Engineering* 78 (November): 107695. https://doi.org/10.1016/j.jobe.2023.107695.

"Rubber-Cement Composites for Additive Manufacturing: Physical, Mechanical and Thermo-Acoustic Characterization | SpringerLink." n.d. Accessed August 11, 2024. https://link.springer.com/chapter/10.1007/978-3-030-49916-7_12.

Sambucci, Matteo, Danilo Marini, Abbas Sibai, and Marco Valente. 2020. "Preliminary Mechanical Analysis of Rubber-Cement Composites Suitable for Additive Process Construction." *Journal of Composites Science* 4 (3): 3. https://doi.org/10.3390/jcs4030120.

Shayan, Ahmad, and Aimin Xu. 2004. "Value-Added Utilisation of Waste Glass in Concrete." *Cement and Concrete Research* 34 (1): 81–89. https://doi.org/10.1016/S0008-8846(03)00251-5.

Singh, Amardeep, Qiong Liu, Jianzhuang Xiao, and Qifeng Lyu. 2022. "Mechanical and Macrostructural Properties of 3D Printed Concrete Dosed with Steel Fibers under Different Loading Direction." *Construction and Building Materials* 323 (March): 126616. https://doi.org/10.1016/j.conbuildmat.2022.126616.

Skibicki, Szymon, Monika Pułtorak, Maria Kaszyńska, Marcin Hoffmann, Ewa Ekiert, and Daniel Sibera. 2022a. "The Effect of Using Recycled PET Aggregates on Mechanical and Durability Properties of 3D Printed Mortar." *Construction and Building Materials* 335 (June): 127443. https://doi.org/10.1016/j.conbuildmat.2022.127443.

Skibicki, Szymon, Monika Pułtorak, Maria Kaszyńska, Marcin Hoffmann, Ewa Ekiert, and Daniel Sibera. 2022b. "The Effect of Using Recycled PET Aggregates on Mechanical and Durability Properties of 3D Printed Mortar." *Construction and Building Materials* 335 (June): 127443. https://doi.org/10.1016/j.conbuildmat.2022.127443.

Srinivas, Dodda, Dhrutiman Dey, Biranchi Panda, and Thallak G. Sitharam. 2022. "Printability, Thermal and Compressive Strength Properties of Cementitious Materials: A Comparative Study with Silica Fume and Limestone." *Materials* 15 (23): 8607.

Sun, Bochao, Qiang Zeng, Dianchao Wang, and Weijian Zhao. 2022. "Sustainable 3D Printed Mortar with CO2 Pretreated Recycled Fine Aggregates." *Cement and Concrete Composites* 134 (November): 104800. https://doi.org/10.1016/j.cemconcomp.2022.104800.

Tan, Kiang Hwee, and Hongjian Du. 2013. "Use of Waste Glass as Sand in Mortar: Part I – Fresh, Mechanical and Durability Properties." *Cement and Concrete Composites* 35 (1): 109–17. https://doi.org/10.1016/j.cemconcomp.2012.08.028.

Tay, Yi Wei Daniel, Biranchi Panda, Suvash Chandra Paul, Nisar Ahamed Noor Mohamed, Ming Jen Tan, and Kah Fai Leong. 2017. "3D Printing Trends in Building and Construction Industry: A Review." *Virtual and Physical Prototyping* 12 (3): 261–76.

Ting, Guan Heng Andrew, Yi Wei Daniel Tay, Ye Qian, and Ming Jen Tan. 2019. "Utilization of Recycled Glass for 3D Concrete Printing: Rheological and Mechanical Properties." *Journal of Material Cycles and Waste Management* 21 (4): 994–1003. https://doi.org/10.1007/s10163-019-00857-x.

Ting, Guan Heng Andrew, Yi Wei Daniel Tay, and Ming Jen Tan. 2021. "Experimental Measurement on the Effects of Recycled Glass Cullets as Aggregates for Construction 3D Printing." *Journal of Cleaner Production* 300 (June): 126919. https://doi.org/10.1016/j.jclepro.2021.126919.

Tong, Jiaqi, Yahong Ding, Xiuwen Lv, and Wei Ning. 2023. "Effect of Carbonated Recycled Coarse Aggregates on the Mechanical Properties of 3D Printed Recycled Concrete." *Journal of Building Engineering* 80 (December): 107959. https://doi.org/10.1016/j.jobe.2023.107959.

Ulsen, Carina, Juliana Lívi Antoniassi, Isabel M. Martins, and Henrique Kahn. 2021. "High Quality Recycled Sand from Mixed CDW – Is That Possible?" *Journal of Materials Research and Technology* 12 (May): 29–42. https://doi.org/10.1016/j.jmrt.2021.02.057.

Warsi, Syed Bustan Fatima, Biranchi Panda, and Pankaj Biswas. 2023. "Exploring Fibre Addition Methods and Mechanical Properties of Fibre-Reinforced 3D Printed Concrete: A Review." *Developments in the Built Environment* 16 (December): 100295. https://doi.org/10.1016/j.dibe.2023.100295.

Warsi, Syed Bustan Fatima, Dodda Srinivas, Biranchi Panda, and Pankaj Biswas. 2023. "Investigating the Impact of Coarse Aggregate Dosage on the Mechanical Performance of 3D Printable Concrete." *Innovative Infrastructure Solutions* 9 (1): 5. https://doi.org/10.1007/s41062-023-01317-0.

Wong, Chee Lum, Kim Hung Mo, Soon Poh Yap, U. Johnson Alengaram, and Tung-Chai Ling. 2018. "Potential Use of Brick Waste as Alternate Concrete-Making Materials: A Review." *Journal of Cleaner Production* 195 (September): 226–39. https://doi.org/10.1016/j.jclepro.2018.05.193.

Wu, Yiwen, Chao Liu, Guoliang Bai, Huawei Liu, Yisheng Meng, and Zhihui Wang. 2023. "3D Printed Concrete with Recycled Sand: Pore Structures and Triaxial Compression Properties." *Cement and Concrete Composites* 139 (May): 105048. https://doi.org/10.1016/j.cemconcomp.2023.105048.

Wu, Yiwen, Chao Liu, Huawei Liu, et al. 2021. "Study on the Rheology and Buildability of 3D Printed Concrete with Recycled Coarse Aggregates." *Journal of Building Engineering* 42 (October): 103030. https://doi.org/10.1016/j.jobe.2021.103030.

Xiao, Jianzhuang, Zhenyuan Lv, Zhenhua Duan, and Shaodan Hou. 2022. "Study on Preparation and Mechanical Properties of 3D Printed Concrete with Different Aggregate Combinations." *Journal of Building Engineering* 51 (July): 104282. https://doi.org/10.1016/j.jobe.2022.104282.

Xiao, Jianzhuang, and Zhiming Ma. 2022. "Chapter 18 – Utilization of Recycled Powder from Construction and Demolition Waste." In *Low Carbon Stabilization and Solidification of Hazardous Wastes*, edited by Daniel C. W. Tsang and Lei Wang. Elsevier. https://doi.org/10.1016/B978-0-12-824004-5.00022-0.

Zhang, Hongru, Cuirong Zhang, Bingjian He, Shifan Yi, and Lingyan Tang. 2023. "Recycling Fine Powder Collected from Construction and Demolition Wastes as Partial Alternatives to Cement: A Comprehensive Analysis on Effects, Mechanism, Cost and CO2 Emission." *Journal of Building Engineering* 71 (July): 106507. https://doi.org/10.1016/j.jobe.2023.106507.

Zheng, Chaocan, Cong Lou, Geng Du, Xiaozhen Li, Zhiwu Liu, and Liqin Li. 2018. "Mechanical Properties of Recycled Concrete with Demolished Waste Concrete Aggregate and Clay Brick Aggregate." *Results in Physics* 9 (June): 1317–22. https://doi.org/10.1016/j.rinp.2018.04.061.

Zhou, Yiyi, Haoran Luo, Kamal Anand, Amardeep Singh, and Yi Min Xie. 2024. "Sustainable Use of Ultrafine Recycled Glass in Additive Manufactured (3D Printed) Reactive Powder Concrete." *Construction and Building Materials* 419 (March): 135556. https://doi.org/10.1016/j.conbuildmat.2024.135556.

Zou, Shuai, Jianzhuang Xiao, Tao Ding, Zhenhua Duan, and Qingtian Zhang. 2021. "Printability and Advantages of 3D Printing Mortar with 100% Recycled Sand." *Construction and Building Materials* 273 (March): 121699. https://doi.org/10.1016/j.conbuildmat.2020.121699.

Zou, Shuai, Jianzhuang Xiao, Zhenhua Duan, Tao Ding, and Shaodan Hou. 2021. "On Rheology of Mortar with Recycled Fine Aggregate for 3D Printing." *Construction and Building Materials* 311 (December): 125312. https://doi.org/10.1016/j.conbuildmat.2021.125312.

9 3D-Printable Engineered Cementitious Composites

Binrong Zhu, Jinlong Pan, and Mingzhong Zhang

9.1 INTRODUCTION

Driven by the demand for improved production efficiency, sustainable construction materials, and enhanced durability of engineering structures, the construction industry is undergoing significant transformation. Traditional construction methods, limited by high labor intensity, material waste, and low design flexibility, are increasingly inadequate to meet modern engineering needs (Tay et al. 2017). 3D printing technology, also known as additive manufacturing, has emerged as a promising solution, offering high efficiency, precision, and resource-saving advantages for constructing complex structures (Bos et al. 2022; Li et al. 2023b). Currently, extrusion-based 3D concrete printing (3DCP) is the dominant approach, where material is continuously extruded through a nozzle in a filament form and systematically layered to construct 3D structures (De et al. 2018; Buswell et al. 2018). However, due to the layer-by-layer deposition process, extrusion-based 3D printing faces challenges such as weak interlayer bonding, poor toughness, and difficulties in reinforcing implementation (Bos et al. 2016; Mechtcherine et al. 2021). Addressing these issues requires the development of novel high-performance materials to enhance the overall performance of printed structures.

Engineered cementitious composites (ECCs), also called strain-hardening cementitious composites (SHCCs), are a class of high-performance cement-based materials known for high tensile strength, superior ductility, and multiple cracking characteristics (Li 2019a, 2019b; Zhu et al. 2024). This material has been widely applied in engineering fields such as buildings, bridges, water and energy infrastructures, and tunnels. Combining 3D printing technology with ECCs enables structural toughening through material self-reinforcement, eliminating the complexity of traditional reinforcement processes and demonstrating the potential for constructing safe and durable structures with minimal reinforcement. In 2018, Soltan and Li (2018) pioneered the concept of printable ECCs, developing 3D-printable ECCs based on thixotropic design principles, which laid the foundation for subsequent research. Since then, extensive studies have been conducted on material mix design, workability control, mechanical performance characterization, and potential structural applications of 3D-printed ECCs (3DP-ECCs), achieving significant progress. A bibliometric analysis through VOSviewer based on the Web of Science database (Figure 9.1) reveals that from 2018 to 2024, research hotspots in 3DP-ECCs focused on keywords such as 3D concrete printing, ECC, SHCC, mechanical properties, extrusion, fiber, rheology,

DOI: 10.1201/9781003434863-11

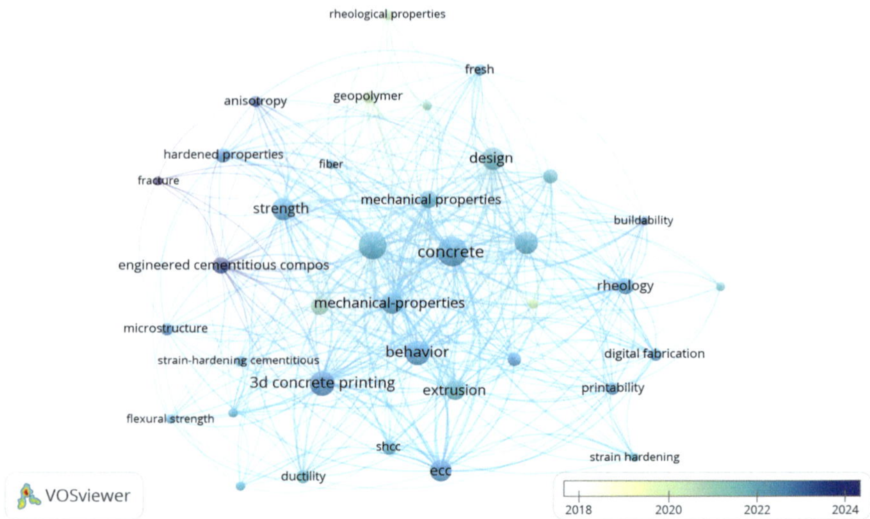

FIGURE 9.1 Keyword co-occurrence network for 3DP-ECCs.

ductility, steel reinforcement, anisotropy, microstructure, and fresh properties. The temporal evolution of research themes indicates that emerging directions such as 3DP-ECCs, printable mortar, and tensile property have gradually gained attention.

Despite the remarkable progress in 3DP-ECC research, comprehensive review articles remain scarce. Only two publications have summarized this field, while they fail to fully capture its rapid development in recent years (Li et al. 2020; Asghari et al. 2024). From material development to structural applications (Zhu et al. 2021a; Hass et al. 2023; Ivaniuk et al. 2024; Ye et al. 2024c), from macro-mechanical properties to micro-mechanical mechanisms (Ye et al. 2021b; Pang et al. 2023), and from the incorporation of recycled aggregates to the use of low-carbon LC^3 cement (Zhu et al. 2022a; Wang et al. 2024), the scope of 3DP-ECC research continues to expand. Therefore, a systematic review of the latest advancements in this field is of great significance. This chapter provides a comprehensive overview of 3DP-ECC research progress, focusing on raw materials, production processes, printability, mix design, fresh properties, hardened mechanical performance, and sustainability evaluation. It highlights the influences of different factors such as matrix composition, fiber type, and printing parameters on material performance. Additionally, the chapter discusses future opportunities and challenges for 3DP-ECC, aiming to provide valuable insights for further research.

9.2 3D PRINTING OF ECCs

9.2.1 ECC Design Theory

Based on the theoretical design principles of micromechanics and fracture mechanics, ECC is a class of high-performance cementitious composites reinforced with

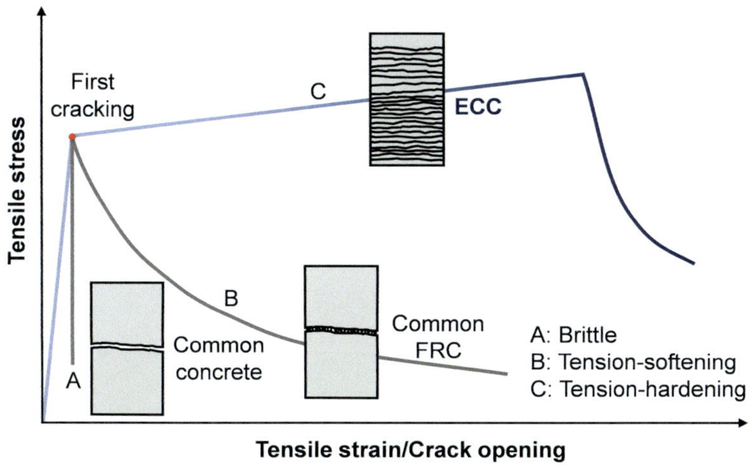

FIGURE 9.2 Typical tensile behavior of common concrete, common FRC, and ECCs. Adapted from Ref. (Fischer and Li 2003).

randomly distributed short-cut fibers. Compared to conventional concrete or fiber-reinforced concrete (FRC), ECCs with 2% fiber volume fraction exhibit remarkable tensile ductility, typically exceeding 2%, which is more than 200 times that of conventional concrete or FRC. While conventional FRC demonstrates tensile softening behavior after crack initiation, characterized by localized deformation and continuous stress reduction, ECCs exhibit steady-state crack propagation across multiple regions under tensile loading. Through fiber debonding and slip mechanisms, ECCs bridge microcracks and continue to sustain higher loads while increasing strain. This repeated process of microcrack formation and fiber-bridging results in distinct strain-hardening behavior (Li et al. 2002, 2023a; Dai et al. 2021; Zhu et al. 2022b). Figure 9.2 illustrates the typical tensile behavior of conventional concrete, FRC, and ECCs.

According to the micromechanical design theory of ECCs, achieving tensile strain-hardening and multiple cracking behavior requires satisfying both the strength criterion (Eq. (9.1)) and the energy criterion (Eq. (9.2)). The strength criterion stipulates that the composite's cracking strength (σ_{fc}) must not exceed the fiber-bridging stress (σ_0), which governs crack initiation in ECCs. The energy criterion, based on the fiber-bridging stress–crack opening (σ-δ) curve (Figure 9.3), requires that the crack tip toughness (J_{tip}) is lower than the complementary energy ($J_b^{'}$). The expressions are given below:

$$\sigma_{fc} \leq \sigma_0, \tag{9.1}$$

$$J_{tip} = \sigma_{ss}\delta_{ss} - \int_0^{\delta_{ss}} \sigma(\delta)\,d\delta = \frac{K_m^2}{E_m} \leq J_b^{'} = \sigma_0\delta_0 - \int_0^{\delta_0} \sigma(\delta)\,d\delta, \tag{9.2}$$

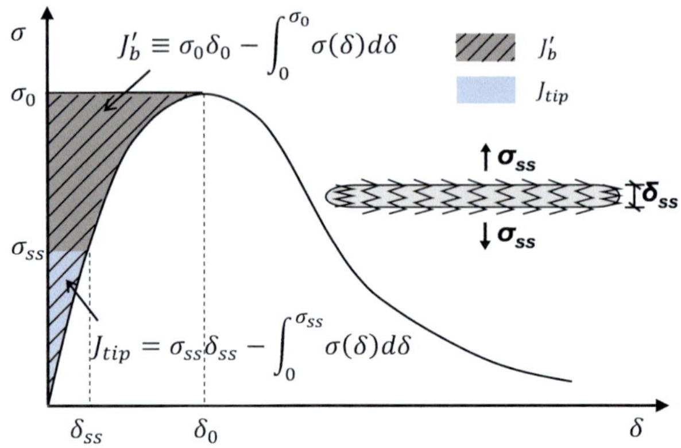

FIGURE 9.3 Typical fiber-bridging σ-δ relation for ECCs. Adapted from Wang and Li (2007), Li (2019b).

where δ_0 is the crack opening at maximum bridging stress (σ_0), δ_{ss} is the crack opening corresponding to steady-state bridging stress (σ_{ss}), K_m is the fracture toughness of the fiber-free ECC matrix, and E_m is the elastic modulus of the matrix obtained from uniaxial compressive stress–strain curves.

The cracking strength (σ_{fc}) and fracture toughness (K_m) can be calculated as follows:

$$\sigma_{fc} = \frac{K_m}{2}\sqrt{\frac{\pi}{c}}, \tag{9.3}$$

$$K_m = \frac{P_s L}{HW^{3/2}} \cdot 3\sqrt{\frac{a}{W}} \cdot \frac{1.99 - \left(\frac{a}{W}\right)\left(1-\frac{a}{W}\right)\left[2.15 - 3.93\frac{a}{W} + 2.7\left(\frac{a}{W}\right)^2\right]}{2\left(1+2\frac{a}{W}\right)\left(1-\frac{a}{W}\right)^{3/2}}, \tag{9.4}$$

where P_s is the peak load, L is the span of the notched specimen, H is the specimen thickness, W is the specimen width, and a is the crack length. K_m can be determined through three-point bending tests on fiber-free ECC matrix specimens.

Additionally, the pseudo strain-hardening (PSH) index, comprising $PSH_{strength}$ (σ_0 / σ_{fc}) and PSH_{energy} ($J_b^{'} / J_{tip}$), quantifies the reserve capacity of tensile ductility. In practice, for PVA fibers, $PSH_{strength} \geq 1.45$ and $PSH_{energy} \geq 3.0$ are required; for polyethylene (PE) fibers, $PSH_{strength} \geq 1.2$ and $PSH_{energy} \geq 3.0$ are necessary to ensure high reserve factors and superior mechanical performance (Nematollahi et al. 2017; Kanda and Li 2006).

9.2.2 Raw Materials of ECCs

Unlike conventional ECCs, 3DP-ECC must not only satisfy the strength and energy criteria required for strain-hardening performance but also meet printability requirements, including pumpability, extrudability, buildability, and a certain open time. Simultaneously fulfilling these performance requirements poses significant challenges, and the selection of raw materials plays a crucial role in determining printability. Table 9.1 summarizes the raw materials of the 3DP-ECC matrix in existing studies. The raw materials for 3DP-ECCs primarily include cementitious materials, supplementary cementitious materials, aggregates, admixtures, water, and fibers. To reduce matrix brittleness, ECCs typically omit conventional coarse aggregates and use only fine aggregates, with the cement content approximately three times that of conventional concrete. Mineral admixtures such as fly ash (FA), ground granulated blast-furnace slag (GGBFS), calcined clay, silica fume (SF), and limestone are widely used in 3DP-ECCs because of their lower cost compared to cement and their ability to regulate hardened strength and durability. These industrial by-products not only reduce material cost but also mitigate environmental impact, promoting the large-scale application of green 3DP-ECCs. Additionally, mineral admixtures play a significant role in improving the workability of fresh concrete and regulating the fracture toughness and durability of hardened matrix.

Fibers, as essential reinforcement materials in ECCs, effectively transfer tensile stress at crack interfaces, meeting the complementary energy requirement for fiber bridging and enabling crack width control and tensile ductility. Currently, fibers such as polyvinyl alcohol (PVA), PE, high-density polyethylene microfibers (HDPEs), and high-tenacity polypropylene (PP) have been used in ECC development (Zhang et al. 2020; Yoo et al. 2022). Table 9.2 details the parameters of fibers used in 3DP-ECCs. As shown, the fiber content in 3DP-ECCs is typically equal to or below 2%, with PVA and PE fibers being the most widely used. PVA/PE fibers have diameters of 12–40 μm, tensile strength of 600–3,000 MPa, elastic modulus of 8–116 GPa, and tensile strain of 1%–7%. Fiber selection is optimized based on matrix material properties. For ECCs reinforced with PVA, PP, or PET fibers, fine aggregates are usually quartz sand with specific gradations, typically with an average and maximum size of 110 μm and 200 μm, respectively (Li 2019b). For ECCs reinforced with PE or HDPE fibers, given the balance between high-strength matrices and high-tensile-strength fibers, recent studies have also used river sand or a combination of river sand and artificial aggregates with particle sizes up to 4.75 mm (Li et al. 2022; Xu et al. 2022a; Hou et al. 2023; Zhu et al. 2024). For 3DP-ECCs, larger sand particles can also regulate the rheological properties of the material, enhancing the yield stress of the fresh mixture and improving shape retention.

In addition to the basic raw materials of ECCs (binder materials, fine aggregates, fibers, and water), admixtures such as nano-clay (NC), hydroxypropyl methylcellulose (HPMC), and superplasticizers (SPs) were incorporated to adjust the rheology of the printable material, which ensures uniform dispersion of short-cut fibers and enables continuous pumping and smooth extrusion.

TABLE 9.1
Raw Materials of the 3DP-ECC Matrix

Authors	OPC	SC/SAC/ CAC	FA	Sand	GGBFS	SF	LP	MK	SP/ HRWR	VMA (NC/ HPMC)	Retarder	Others
Ogura et al. (2018); Weng et al. (2018, 2019); Zhou et al. (2022b)	√		√	√		√			√			
Soltan and Li (2018); Zhu et al. (2021a, 2022a, 2023a, 2023b)	√	√	√	√		√			√	√		
Chaves Figueiredo et al. (2019, 2020a, 2020b)	√		√	√	√		√		√	√		
Zhu et al. (2019); Pang et al. (2023); Pi et al. (2024)	√	√	√	√					√	√		
Ding et al. (2020); Tawfek et al. (2023)	√			√					√	√		
Ding et al. (2021)	√			RS+NS					√	√		
Sun et al. (2020); Sun et al. (2022)		√		√	√	√					SG	MWCNTs
Zhang et al. (2021b)	√					√			√			
Zhu et al. (2021b)	√						√	√	√			CC
Ye et al. (2021a, 2021b, 2024b, 2024c); Aslani et al. (2022)	√		√	√		√			√			CR
Yu et al. (2021); Ge et al. (2021)	√		√	√					√	√		
Zhang and Aslani (2021)	√		√	√		√			√	√	√	ACP
(Sun et al. (2021)	√		√	√		√			√	√		HGM
Chu et al. (2021)	√			√		√		√	√			$NCaCO_3$
Van Overmeir et al. (2022, 2023a, 2023b); Nefs et al. (2024a, 2024b)	√			√	√	√	√		√	√		
Ivaniuk et al. (2022b)	√		√	√		√			√	SE+CE		
Ivaniuk et al. (2022a, 2024)	√		√	√		√			√			MP
Liu et al. (2022)	√	√		√		√			√	CMC-Na		EA+AFs+AAs
Zhou et al. (2022a, 2023a)	√		√	√					√			
Xu et al. (2022b, 2023); Liu et al. (2023); Qiu et al. (2024a, 2024b); Du et al. (2024); Du and Qian (2024); Teng et al. (2024)	√		√	√		√			√	√		

(Continued)

TABLE 9.1 *(Continued)*
Raw Materials of the 3DP-ECC Matrix

Authors	OPC	SC/SAC/ CAC	FA	Sand	GGBFS	SF	LP	MK	SP/ HRWR	VMA (NC/ HPMC)	Retarder	Others
Zafar et al. (2022, 2023)	√		√	√	√	√		√	√	√		
Hass et al. (2023)	√			√	√	√	√		√	√		
Pi et al. (2023)	√	√	√	√					√	√		CR
Luo et al. (2023)	√		√	√	√				√	√		
Wang et al. (2023)	√	√		√		√			√			
Bai et al. (2023, 2024; Xiao et al. (2024)	√		√	√		√			√			RS
Ye et al. (2023)	√		√	√	√	√			√	√		IBA
Yu et al. (2024b)	√		√	√	√	√			√	√		
Chen et al. (2024b)	√		√	√		√			√	√		CR
Anandaraj et al. (2024)	√		√	√						√		NS
Wang et al. (2024)	√		√	√			√		√	√		CC
Luo et al. (2024)	√		√	√					√	CE	SG	
Zhou et al. (2024)	PLC	√		√		√	√	√	√			CR
Yan et al. (2024a, 2024b)	√		√		√		√	√	√	√	SG	
Gou et al. (2024)	√		√	√		√			√			LSF
Ding et al. (2024)	√			√		√	√		√			
Şahin et al. (2025)	√		√	CLA					√			

Note: OPC = ordinary portland cement; CAC = calcium aluminate cement; SAC = sulfoaluminate cement; PLC = portland limestone cement; MK = metakaolin; LP = limestone powder; SP = superplasticizer; HRWR = high-range water reducing admixture; CC = calcined clay; viscosity modifying agent (VMA) = nano-clay (NC) and/or hydroxypropyl methylcellulose (HPMC); MP = white marble powder; $NCaCO_3$ = nano calcium carbonate; retarder = sodium gluconate (SG); HGM = hollow glass microsphere; ACP = activated carbon powder; CR = crumb rubber; MWCNTs = multi-walled carbon nanotubes; IBA = incineration bottom ash; SE + CE = starch ether and cellulose ether; NS = nano-silica; EA = expansion agent; AFs = antifoams; AAs = air-entraining agents; CMC-Na = carboxymethylcellulose sodium; RS = recycled concrete sand; RS + NS = recycled sand and natural sand; LSF = lightweight sand fines; CLA = crushed limestone aggregate.

TABLE 9.2
Summary of Discontinuous Micro-Fibers Used for 3DP-ECCs in the Literature

Year	Ref.	Fiber Type	Content (by Volume)	Length (mm)	Diameter (μm)	Density (g/cm³)	Tensile Strength (MPa)	Elastic Modulus (GPa)	Elongation at Break
2018	Ogura et al. (2018)	HDPE	0.3%, 1.0%, 1.5%	6	12	0.97	3000		
2018	Weng et al. (2018)	PVA	1%	8	40		1560	41	6.5%
2019	Zhu et al. (2019)	HDPE	1%, 1.5%, 2%	12	24	0.97	3000	116	2.5%
2020	Ding et al. (2020)	PE	0.25%, 0.5%, 1%, 1.4%	6/12	20	0.97	2400	100	
2020	Sun et al. (2020)	PVA	0.8%, 1.2%, 1.5%	9	31	1.3	1600	43	
2021	Zhu et al. (2021a)	HDPE	1.5%/0.5%	6/12	24	0.97	3000	116	
2021	Zhang et al. (2021b)	GF	0~2%	12	9	2.54	2800	71	
2021	Zhu et al. (2021b)	PVA	2%	8	39	1.3	1600	42.8	6%
2021	Ding et al. (2021)	PE	0.25%, 0.5%, 1%, 1.4%	6/12		0.97	2400	100	
2021	Ye et al. (2021a)	PE	1.0%, 1.5%, 2.0%	12	25	0.97	2900		
2021	Ye et al. (2021b)	PE	2%	12	25	0.97	2900		
2021	Yu et al. (2021)	PVA	2%	8	39		1600	30	
2021	Zhang and Aslani (2021)	PVA	1%,1.5%	12	39		1600	41	
2021	Sun et al. (2021)	PVA	1.75%	18	39	1.3		41	
2021	Chu et al. (2021)	CF,SF,GF	0.50%	6, 13, 12	7, 200,7	1.8, 7.85, 2.68	3000, 2000, 1700	250, 210, 75	
2022	Aslani et al. (2022)	PVA	1.75%	12	39	1.3	1600	42.8	
2022	Van Overmeir et al. (2022)	PVA	2%	8	40		1600	41	6%
2022	Zhu et al. (2022a)	PE	1.5%/0.5%	6/12	24	0.97	3000	110	2.7%
2022	Zhou et al. (2022b)	PE	2%	12	24	0.97	3000	116	1~3%
2022	Zhou et al. (2022a)	PVA	1.50%	8	39	1.3	1600	42.8	6%

(Continued)

TABLE 9.2 *(Continued)*
Summary of Discontinuous Micro-Fibers Used for 3DP-ECCs in the Literature

Year	Ref.	Fiber Type	Content (by Volume)	Length (mm)	Diameter (μm)	Density (g/cm³)	Tensile Strength (MPa)	Elastic Modulus (GPa)	Elongation at Break
2022	Bong et al. (2022)	PVA	2%	8	40	1.3	1600	41	6%
2022	Xu et al. (2022b)	PE	0.5%, 1.0%, 1.5%	12	24	0.97	3000	110	2~3%
2022	Zafar et al. (2022)	PVA, PE	1.0%, 1.5%, 2.0%	8	38,15	1.3,0.97	1600, 3000	40, 100	
2022	Sun et al. (2022)	PVA	0, 0.8%, 1.0%, 1.2%, 1.4%, 1.6%	9	31	1.3	1600	43	7%
2023	Van Overmeir et al. (2023a, 2023b)	PVA	2%	8	40	1.3	1600	41	6%
2023	Pang et al. (2023)	PVA	2%	12	39		600	18	7%
2023	Pi et al. (2023)	PVA	2%	12	40		1200	39	6.5%
2023	Luo et al. (2023)	PVA	2%	12	40	1.3	1700	39	6.5%
2023	Wang et al. (2023)	PP/PVA	0.3%, 0.2%	9, 9	22, 14	0.91, 1.3	568, 1700	4.06, 33.4	19%, 6%
2023	Zhou et al. (2023b)	PVA	2%	8	39		1600	42.8	6%
2023	Bai et al. (2023)	PE	1.50%	12	24	0.97	2900	116	2.4%
2023	Liu et al. (2023); Ye et al. (2023)	PE	1.50%	12	20	0.97	2900	116	
2023	Xu et al. (2023)	PE	1.0%, 1.5%, 2.0%	9, 12, 18	24	0.97	3000	110	2~3%
2023	Zafar et al. (2023)	PVA	1.50%	8	38	1.3	1600	40	
2024	Ye et al. (2024b)	PE	1.50%	12	24	0.97	2900	116	
2024	Wang et al. (2024)	PVA	2.00%	12	39	1.28	1600	40	
2024	Teng et al. (2024)	PP+PE	0.5%,1.0%	6/12	12, 24		910, 2900	9, 116	
2024	Nefs et al. (2024a)	PVA		8	40		1600	41	

9.2.3 Extrusion-Based Printing of ECCs

Researchers have successfully applied ECCs to 3D printing technology using various types of printing devices. As shown in Figure 9.4, the commonly used extrusion-based 3DP-ECC systems include the following four types: caulking guns (manual style based on piston extrusion or electric-driven style based on screw

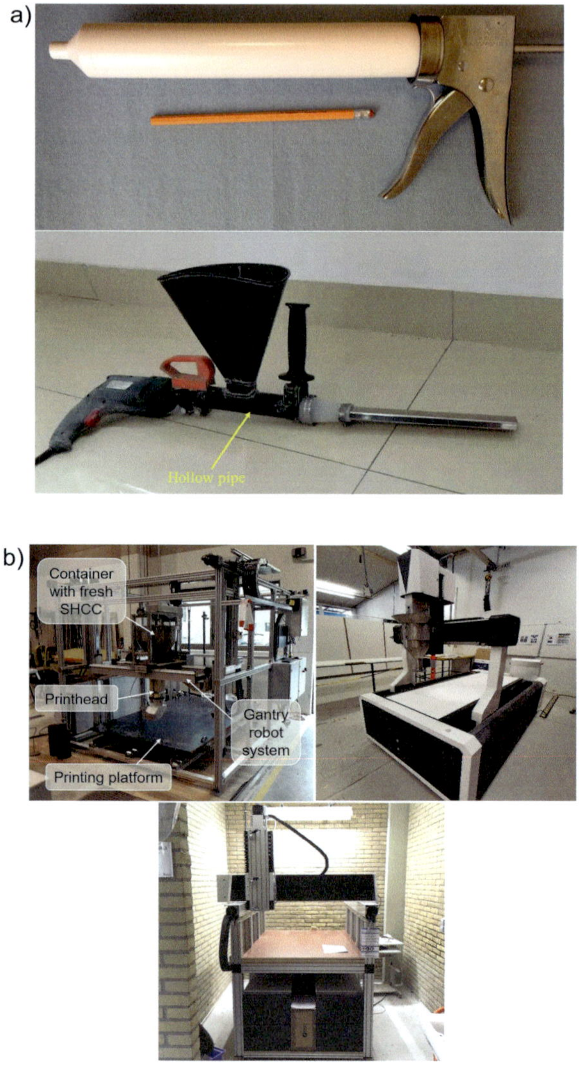

FIGURE 9.4 Typical printing setups for 3DP-ECCs: (a) caulk gun or piston tube extrusion (Soltan and Li 2018; Ge et al. 2021); (b) desktop 3D concrete printer (Chaves Figueiredo et al. 2020a; Ivaniuk et al. 2022b; Sun et al. 2021); (c) robotic-arm 3D concrete printer (Zhou et al. 2022a); (d) large-scale 3D concrete printer (gantry-type and frame-type) (Ding et al. 2020; Yan et al. 2024b).

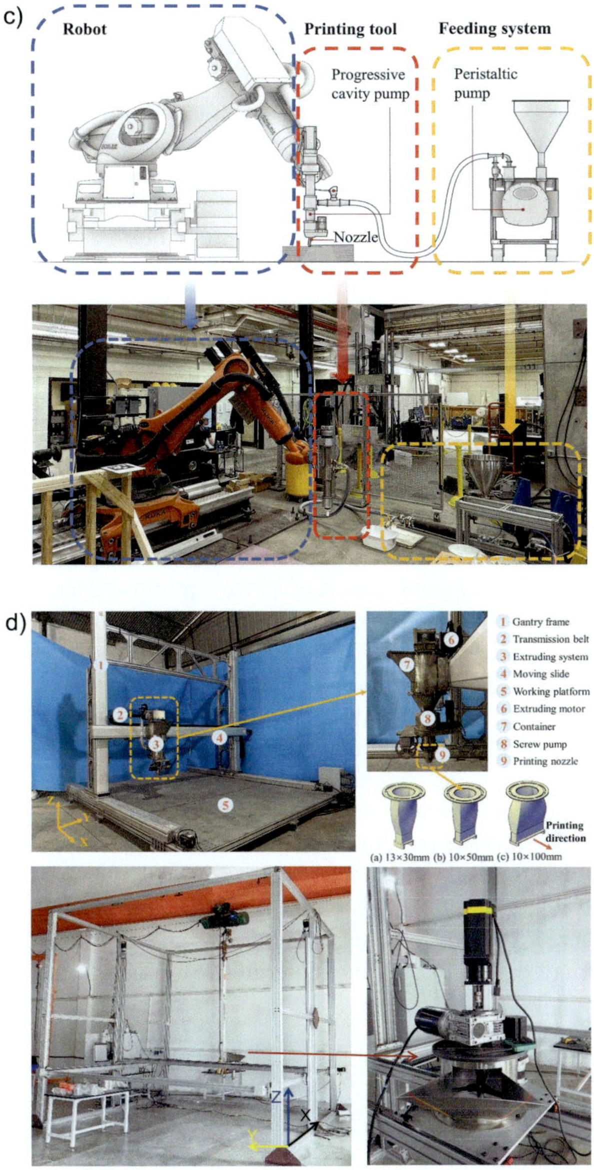

FIGURE 9.4 (Continued)

extrusion), desktop small-scale printers, gantry frame printers, and robotic-arm 3D concrete printers. Among these, manual caulking guns are simple and low-cost but offer poor printing precision. Simulating the 3D printing process using a manual caulking gun may result in unstable material flow during transportation and extrusion, leading to uneven material distribution and ultimately affecting the stability of the printed material's properties. Desktop small-scale 3D concrete printers (gantry or frame type) are cost-effective and suitable for laboratory-scale experiments but have limited printing sizes. Robotic-arm 3D concrete printers are expensive, highly flexible, and complex to maintain but offer strong scalability. Some industrial robotic arms are serial rotary and can exhibit up to 6 rotational degrees of freedom (DOFs) to print more complex geometries. In addition, by mounting the arm on a rail or converting it into a tracked, mobile robotic-arm printer and working with multiple arms, the workspace of a single stationary robotic-arm printer can be greatly extended. Large-scale 3D concrete printers (gantry or frame type) are easy to assemble and maintain but have limited scalability. Factors such as the size of the print object, the size of the site, the climate, and the cost of construction need to be considered to determine which printer is preferable.

Like 3DCP, 3DP-ECC printing systems typically contain a material transport process, a material extrusion process, and a control unit. In current 3DP-ECC practices, the material transport process often lacks a pumping step. Instead, the mixed material is manually placed into the feed hopper in batches and then extruded through a screw mechanism. In robotic-arm 3D concrete printing, pumping processes are rarely integrated. Only a limited number of cases utilize a screw pump (Xu et al. 2022b; Van Overmeir et al. 2023b) or a combination of a screw pump and a peristaltic pump (Yu et al. 2021; Zhou et al. 2024) for material delivery during printing. Pumps suitable for pumping mortar materials typically include models such as the Pictor pump, m-tec P20 connect, and PFT Swing L. The following uses a 3D concrete printer equipped with a screw pump and a screw extrusion type printing head to further explain the working principles of the screw pump and screw extrusion, as shown in Figure 9.5. As the screw rotor rotates inside the stator, one cavity opens as the other closes. The cavities progress along the stator while the rotor turns, allowing a constant volume of material to pass through the pump. The screw pump enables continuous delivery of material to the printing head's hopper. The screw then extrudes the material through nozzles with various shapes onto the work platform. The axial thrust generated by the screw rotation provides shear force and downward pressure on the material, which not only improves the material's extrudability but also enhances its

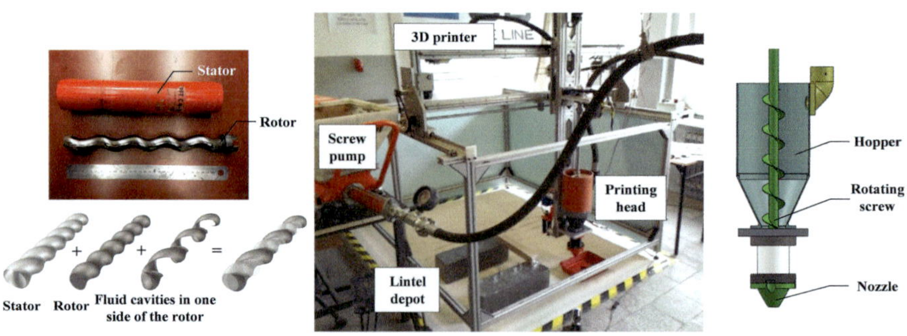

FIGURE 9.5 Principles of the screw pump and screw extrusion type printing head. Taken from Refs. (Hoffmann et al. 2020; Van Overmeir et al. 2022).

compactness. The extrusion capacity of the screw is mainly related to parameters such as screw diameter, pitch, channel depth, screw speed, screw length, flight shape, and compression ratio.

Currently, the shapes of print nozzle outlets used for 3DP-ECCs are mainly rectangular and circular, with the circular nozzle being the most popular. This is because the circular nozzle does not require additional steering devices when path deflection occurs, as the nozzle direction is always tangent to the printing path. Additionally, the angle between the nozzle outlet's normal vector and the horizontal plane is typically 0° or 90°, with angles between these two values not yet applied in 3DP-ECCs. Figure 9.6 illustrates the typical shapes and angles of printer nozzles used in 3DP-ECCs. The flow and extrusion process of materials in the transporting pipeline and print nozzle may form a lubricating layer due to the shear-induced particle migration (SIPM) mechanism, potentially leading to uneven radial distribution of the printing material particles (Breedveld 2000; Mechtcherine et al. 2020). Extrusion deposition process parameters, such as nozzle shape size and the angle between the nozzle outlet's normal vector and the horizontal plane (i.e., flow direction), influence the flow direction of the printing material, thereby affecting the fiber orientation distribution. The fiber orientation distribution in the extruded strip is crucial for its strain-hardening performance. Therefore, it is necessary to further study the effects of nozzle shape size and material flow direction on the extrusion flow characteristics of heterogeneous ECC mixtures. Given the relatively high fiber dosage in 3DP-ECCs (usually 1–2% by volume) and the necessary fiber length (usually 6–12 mm), it is recommended that the printer nozzle size (e.g., circular nozzle) should not be less than 20 mm in diameter to ensure smooth extrusion of ECCs and no clogging of fibers.

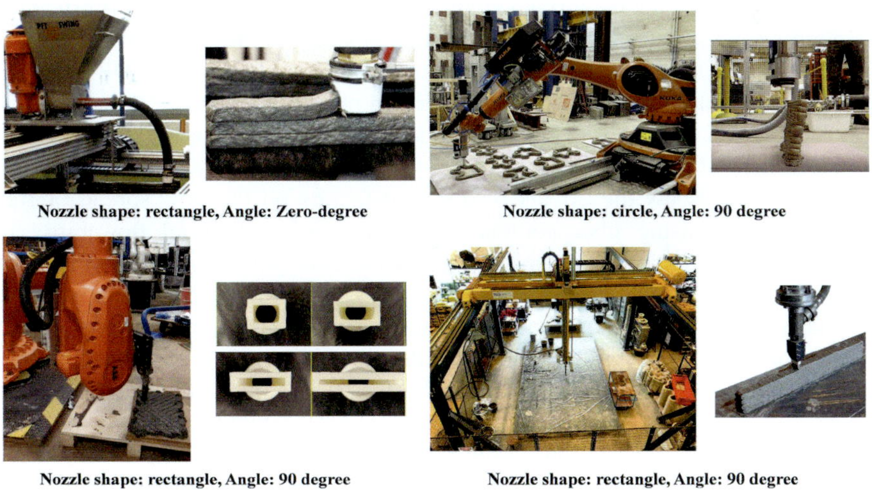

FIGURE 9.6 Typical nozzle shape and the angle between the normal vector of the nozzle outlet surface and the horizontal plane. Images from Refs. (Ogura et al. 2018; Chaves Figueiredo et al. 2020a; Zhou et al. 2022a; Xu et al. 2022b; Zhou 2023a; Nefs et al. 2023).

9.3 FRESH PROPERTIES OF 3DP-ECCs

The fresh properties of 3DP-ECCs primarily include workability, rheological properties, and printability, which are influenced by various factors such as material composition, fiber characteristics, extrusion processes, and aggregate types. The behavior of the material in its fresh state directly affects its flow, stacking continuity, and shape retention during the 3D printing process, thereby impacting the interlayer bond strength, structural integrity, and hardened mechanical properties of the printed object. A thorough understanding of these fresh properties is essential for optimizing the mix design of 3DP-ECCs. This section systematically analyzes and summarizes these influencing factors from the perspectives of workability, rheological properties, and their correlation with printability. Table 9.3 summarizes the evaluation metrics related to the fresh properties of 3DP-ECCs.

9.3.1 WORKABILITY

Workability is one of the key characteristics of 3DP-ECCs, ensuring smooth pumping, extrusion, and shape retention during the printing process. It is typically evaluated through indicators, e.g., flowability. Insufficient flowability may lead to material clogging or delamination, while excessive flowability can cause structural collapse. As seen in Table 9.3, the most used method to characterize the flowability of 3DP-ECCs is the flow table test, which records the time-dependent changes in flow spread, followed by the mini-slump test. Additionally, the printable window is mainly characterized by setting time and penetration depth.

Many studies have highlighted the significant influence of factors such as fiber content, fiber type, water-to-binder (w/b) ratio, and admixtures on workability. Ogura et al. (2018) evaluated the workability of 3DP-ECCs using flow tests and found that lower flow values typically correspond to better shape retention, which is crucial for maintaining the structural integrity of printed objects. Optimization of particle size distribution (PSD) can also significantly improve the workability of 3DP-ECCs. By applying the modified Andreasen and Andersen model to optimize PSD, better particle packing can be achieved, enhancing the flowability and stability of the material (Van Overmeir et al. 2022).

Figure 9.7 presents the effects of different factors on the flow spread of fresh 3DP-ECCs. Increasing fiber content generally reduces material flowability but enhances shape retention. Excessively low flowability may hinder pumping and extrusion during 3D printing. For instance, when the PE fiber volume fraction is increased from 1.0% to 2.0%, the flow spread diameter measured by the flow table test drops by 14.8%, from 169 mm to 144 mm (Ye et al. 2021a). Adding 1.75% PVA fibers reduces the slump from 229 mm to 124 mm and shortens the setting time from 70 minutes to 51 minutes (Sun et al. 2021). When the PVA fiber content increases from 0 wt% to 1.1 wt%, the slump goes down from 88 mm to 25 mm. However, the addition of fibers forms a network structure in the matrix, increasing the interparticle friction and enhancing the material's cohesion and shape retention through bridging effects. Hydrophobic PE fibers have a lower impact on ECC flowability compared to hydrophilic PVA fibers. Fiber length is also a critical factor affecting workability. It was

TABLE 9.3
Summary of Existing Parameters for Charactering the Fresh Behavior of 3DP-ECCs

Fresh Properties	Evaluation Indicators	Test Methods
Workability	Slump and spread flow/diameter Relative slump	Mini-slump test (Ogura et al. 2018; Zhu et al. 2019; Zhang and Aslani 2021; Ivaniuk et al. 2022b; Zhou et al. 2022b; Xu et al. 2022b; Van Overmeir et al. 2023a; Chen et al. 2024b; Wang et al. 2024; Yan et al. 2024b) and flow table test (Soltan and Li 2018; Zhu et al. 2019, 2021b, 2023b; Sun et al. 2020, 2021, 2022; Ye et al. 2021a, 2021b, 2023; Yu et al. 2021; Zhang and Aslani 2021; Aslani et al. 2022; Zhou et al. 2022b, 2024; Xu et al. 2022b, 2023; Zafar et al. 2022, 2023; Bai et al. 2023; Chen et al. 2024b; Wang et al. 2024; Yan et al. 2024b; Gou et al. 2024)
Workability (open time/printability window)	Setting time, penetration depth	Automatic Vicat apparatus (Weng et al. 2019; Van Overmeir et al. 2022), needle gauge (Sun et al. 2020, 2021; Xu et al. 2022b), consistency test (Ye et al. 2021a, 2021b; Chen et al. 2024b; Yan et al. 2024b; Gou et al. 2024)
Printability: extrudability	Whether filaments can be printed without ruptures or significant defects Cracking number and cracking distribution Extrudability coefficient	3D concrete printing test (Ogura et al. 2018; Sun et al. 2021; Ivaniuk et al. 2022b; Zhu et al. 2023b; Xu et al. 2023; Chen et al. 2024b; Zafar et al. 2023; Gou et al. 2024)
	Ram extrusion force Static extrusion pressure: extrusion pressure needed to initiate the extrusion Nonlinear Benbow–Bridgwater model: total pressure drop	Ram extrusion test (Ogura et al. 2018; Chu et al. 2021; Van Overmeir et al. 2022)

(Continued)

TABLE 9.3 *(Continued)*
Summary of Existing Parameters for Charactering the Fresh Behavior of 3DP-ECCs

Fresh Properties	Evaluation Indicators	Test Methods
Printability: buildability	Down-scaled wall retained its geometry and shape without observable deformations	Direct printing of ECC walls (Ogura et al. 2018)
	Dimensionless parameter equations	Direct printing of a hollow cylindrical model (Zhu et al. 2019), direct printing of a 10-layer structure using the rectangular printing nozzle (Ye et al. 2021a), a printed 24-layer structure at a fixed printing speed (Sun et al. 2021), printing a straight wall and a hollow cylinder (Zhou et al. 2022a), printing a linear wall model measuring 300 mm × 34 mm × 90 mm (Zhu et al. 2023a), printing a 10-layer sample stacking to 100 mm height (Chen et al. 2024b), direct printing of a hollow prism specimen (Gou et al. 2024)
	Height or number of stacking layers of fresh mortar filament before visual deformation or collapse occurs	Direct printing of mortar with several layers (Sun et al. 2020, 2022; Zhu et al. 2023b; Bai et al. 2023; Zafar et al. 2023; Yan et al. 2024b)
	Heights of the whole element and each layer were measured at four edges separately and an average value was obtained	Box-shaped element stacking to approximately a height of 300 mm (Zhou et al. 2022b)
	Compressive yield stress at failure	Printing a slender wall construction in subsequent layers of 800 mm length until it fails or a maximum element height of the printer gantry is reached (Van Overmeir et al. 2022, 2023a; Gou et al. 2024; Suiker 2018)
	Weight resistance	Shape-retention-ability test: applying load on top of the specimen (Zhu et al. 2021b, 2023b; Ye et al. 2021b; Yu et al. 2021; Zhou et al. 2024)
	Green strength–deformation relationship Initial yield strength and its development over time Initial apparent Young's modulus and its time development	Uniaxial unconfined compression tests (Yu et al. 2021; Van Overmeir et al. 2022, 2023a; Zhu et al. 2023a, 2023b)
	Shape retention factor: ratio of the base area of the mixture while still inside the slump cone to that of the mixture after the slump-flow test	Buildability test (Chu et al. 2021)
	Penetration resistance	Penetration resistance testing (Bai et al. 2023)

(Continued)

TABLE 9.3 *(Continued)*
Summary of Existing Parameters for Charactering the Fresh Behavior of 3DP-ECCs

Fresh Properties	Evaluation Indicators	Test Methods
Rheological properties	Fitting models between rheological parameters (plastic viscosity and dynamic yield stress) and material compositions	Response surface methodology (Zhu et al. 2023a)
	Flow resistance model, torque viscosity model, and thixotropy model	Rheological tests using a Viskomat XL with a fish-bone probe (Weng et al. 2018)
	Benbow–Bridgwater model	Rheology tests using a ram extruder (Chaves Figueiredo et al. 2019)
	Plastic viscosity, static yield stress, and dynamic yield stress	Rheological tests using a Viskomat XL with a six-blade vane probe and a cage (Weng et al. 2019), Anton Paar MCR 302 rheometer with cruciform blades measuring 20 mm in diameter and 30 mm in height (Zhu et al. 2023a, 2023b), Brookfield RST-SST rheometer with a four-blade vane (VT-40–20MB) (Ye et al. 2023; Zafar et al. 2023), and commercial rheometer with a custom-designed coaxial cylinder cup (cylinder of height 100 mm and diameter 74 mm) (Xu et al. 2023)

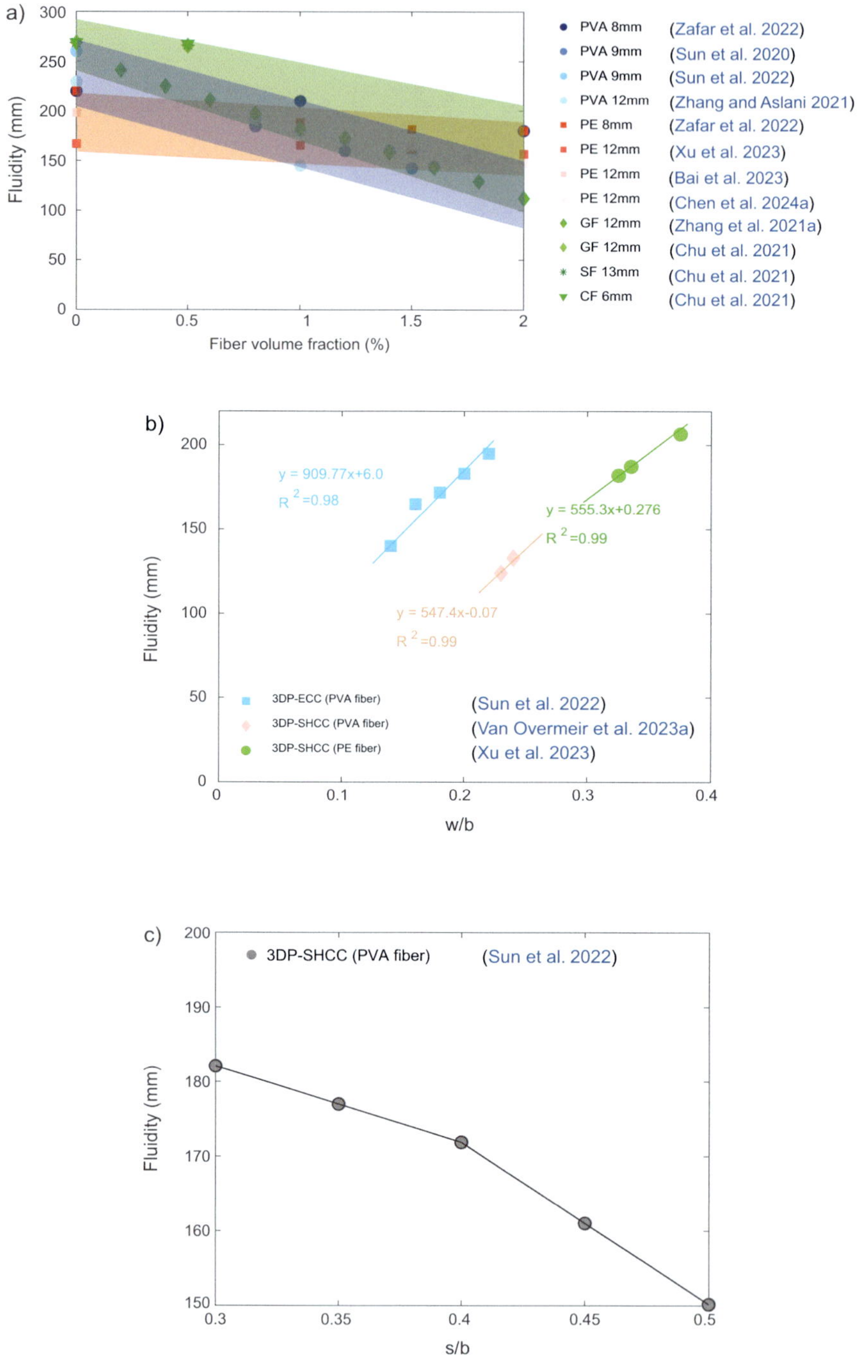

FIGURE 9.7 Effects of different typical factors on the fluidity of fresh 3DP-ECCs: (a) fiber content; (b) w/b ratio; (c) s/b ratio; (d) ratio in fine aggregate; (e) admixture ratio.

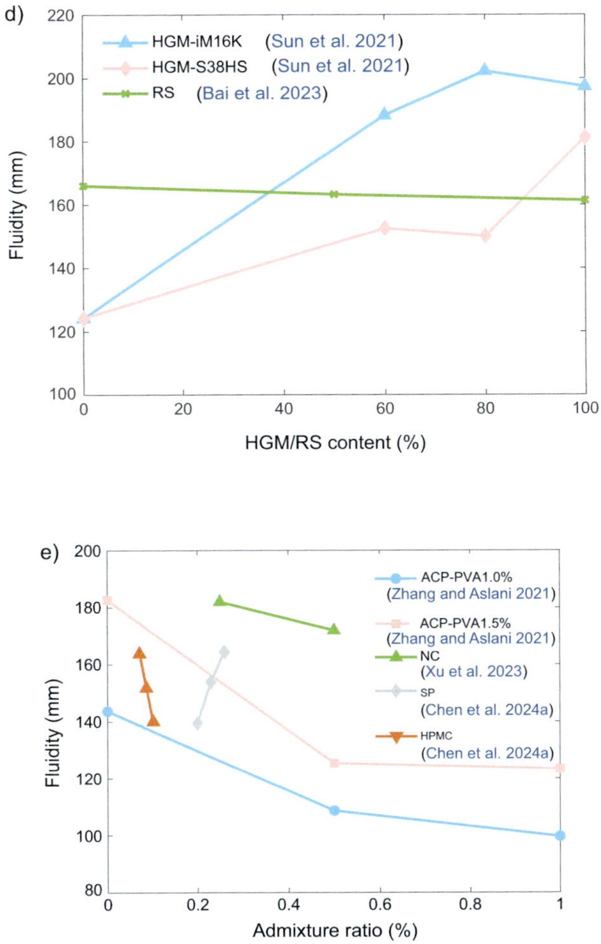

FIGURE 9.7 (Continued)

found that 6-mm-long fibers provide the best workability, while 12-mm-long fibers significantly reduce workability (Xu et al. 2023; Şahin et al. 2025). Considering the strain-hardening requirements of ECCs, fiber length must be carefully balanced, as using only 6-mm fibers may lead to insufficient bridging capacity, affecting multiple cracking behavior. Thus, a combination of long and short fibers can balance printability and strain-hardening performance (Zhu et al. 2021a, 2022a, 2022b).

Increasing the w/b ratio reduces flow resistance, thereby improving workability. It was observed that flowability increases approximately linearly with the w/b ratio, but this may also reduce shape retention and increase the risk of collapse (Ogura et al. 2018; Sun et al. 2022; Van Overmeir et al. 2023a; Xu et al. 2023). Adjusting the sand-to-binder (s/b) ratio also plays a significant role in shape retention. The flow spread of 3DP-ECCs decreases with an increasing s/b ratio (Sun et al. 2022). An appropriate s/b ratio can reduce internal friction and improve flowability, but higher sand content increases the material's yield stress, enhancing stability during layer stacking (Chu et al. 2021). Therefore, designing the mix proportion of 3DP-ECCs requires balancing the w/b and s/b ratios to achieve optimal flowability and shape retention. Changing the aggregate type is another effective method to adjust the flowability of 3DP-ECCs. It was found that the flowability of ECCs goes up linearly with the substitution rate of hollow glass microspheres (HGMs) (Sun et al. 2021). The incorporation of recycled sand with high water absorption and rough surfaces reduces flowability but to a lesser extent (Bai et al. 2023).

Additives such as cellulose ether, micro-silica (MS), and NC also significantly affect workability. For instance, HPMC and attapulgite nano-clay (ANC) can prevent segregation and increase the viscosity and thixotropy of fresh mixtures, effectively regulating initial flowability (Soltan and Li 2018). The content of calcium aluminate cement (CAC) significantly affects both hardening rate and initial flowability. The introduction of NC can enhance the cohesion of ECCs, improving their shape retention. Meanwhile, the addition of MS significantly accelerates the initial hardening of ECCs, shortening the initial setting time and making them suitable for printing tasks that require high structural stability. Zhu et al. (2019) studied the effects of HPMC and ANC on the workability of 3DP-ECCs. The results showed that adding HPMC and ANC reduced the slump and flow spread from 64 mm and 176 mm to 55 mm and 155 mm, respectively, and this lower workability contributed to better shape retention. Ivaniuk et al. (2022b) used chemical admixtures such as cellulose ether and starch ether to modify printed ECCs, improving their stability, cohesion, and robustness, thereby extending the open time, reducing shrinkage, and enhancing crack resistance. Adding 0.5% NC (by weight of binder) provided satisfactory 3D printing performance for ECCs (Xu et al. 2022b). The addition of nano-calcium carbonate ($NCaCO_3$) and fibers significantly improved the shape retention of the mixture, enhancing its buildability (Chu et al. 2021). Specifically, adding 0.5% fibers and 4% nano-clay (NC) slightly increased the demand for SPs to 2.5%, which remains within an acceptable range. Both HPMC and fibers reduced the width of extruded filaments due to decreased workability, with HPMC having a more pronounced effect (Chen et al. 2024b). Furthermore, adding more HPMC led to a decrease in filament width over time, indicating that HPMC can extend the open time (i.e., the maximum time interval during which the mixture can be smoothly extruded) (Chen et al. 2024b). Increasing SP dosage improved the workability of 3DP-ECCs, making the extrusion process smoother. However, excessive SPs may lead to overly wide filaments that cannot maintain the specified shape, affecting the buildability of printed components (Chen et al. 2024b). Additionally, excessive SPs can make the material more adhesive (Aïtcin and Flatt 2015). The addition of 0.15% NS significantly improved the workability of ECCs, primarily due to their high surface area and pozzolanic reactivity. It is important to precisely control the dosage of nano-scale additives to avoid excessive viscosity, which may cause clogging during printing (Anandaraj et al. 2024). In summary, appropriate adjustments to these components are crucial for optimizing the workability of 3DP-ECCs.

9.3.2 Rheological Properties

The rheological properties of 3DP-ECCs, including static yield stress (SYS), dynamic yield stress (DYS), and plastic viscosity, are critical for its pumping, extrusion, and strain-hardening performance (Yang et al. 2009; Roussel 2018; Zhang et al. 2021a; Pi et al. 2024). These parameters typically measured via commercial rheometers under varied shear protocols reflect the material's internal structural evolution and flow-deformation behavior under external forces.

Some commonly used linear and nonlinear models can be used to describe the relationship between shear stress and shear rate for the non-Newtonian fluid of 3DP-ECCs, e.g., Bingham model, Herschel–Bulkley model, and modified Bingham model, as given in Eqs. (9.3–9.5) (Zhaidarbek et al. 2023). Among these, the Bingham model is the most widely applied. For instance, Weng et al. (2019) used the Bingham model to describe the rheological behavior of materials, emphasizing that the rational design of SYS and plastic viscosity is crucial for ensuring the pumpability and buildability of materials during the printing process. Notably, excessive SYS may suppress structural collapse but risks triggering pumping resistance (e.g., arching effects), highlighting the need to balance flowability and stability (Van Overmeir et al. 2023a).

$$\begin{cases} \tau = \tau_0 + \mu_\mathrm{p}\dot{\gamma} & \text{for } \tau > \tau_0 \\ \dot{\gamma} = 0 & \text{for } \tau \le \tau_0 \end{cases}, \tag{9.3}$$

$$\begin{cases} \tau = \tau_0 + K\dot{\gamma}^n & \text{for } \tau > \tau_0 \\ \dot{\gamma} = 0 & \text{for } \tau \le \tau_0 \end{cases}, \tag{9.4}$$

$$\begin{cases} \tau = \tau_0 + \mu_\mathrm{p}\dot{\gamma} + A_2\dot{\gamma}^2 & \text{for } \tau > \tau_0 \\ \dot{\gamma} = 0 & \text{for } \tau \le \tau_0 \end{cases}, \tag{9.5}$$

where τ (Pa) is the shear stress; τ_0 (Pa) is the DYS; μ_p (Pa·s) is the plastic viscosity; $\dot{\gamma}$ (1/s) is the shear rate; K is the consistency; n is the power index that represents the deviation from a Bingham behavior; the shear-thickened fluid exceeds one, and the shear-thinning fluid is less than one; and A_2 is a second-order coefficient. The Herschel–Bulkley model leads to the Bingham model when n is equal to 1, where the consistency can represent plastic viscosity. Shear thickening occurs when the second-order coefficient A_2 is greater than zero, while shear thinning occurs when A_2 is less than zero.

Figure 9.8 illustrates the effects of different factors on the rheological parameters of 3DP-ECCs. Numerous studies have shown that ECCs exhibit higher initial yield stress compared to fiber-free printable cement-based materials. Increasing fiber content simultaneously increases torque viscosity and flow resistance, leading to significant improvements in SYS and DYS, thereby enhancing shape retention and deformation resistance (Weng et al. 2018; Chaves Figueiredo et al. 2019; Xu et al.

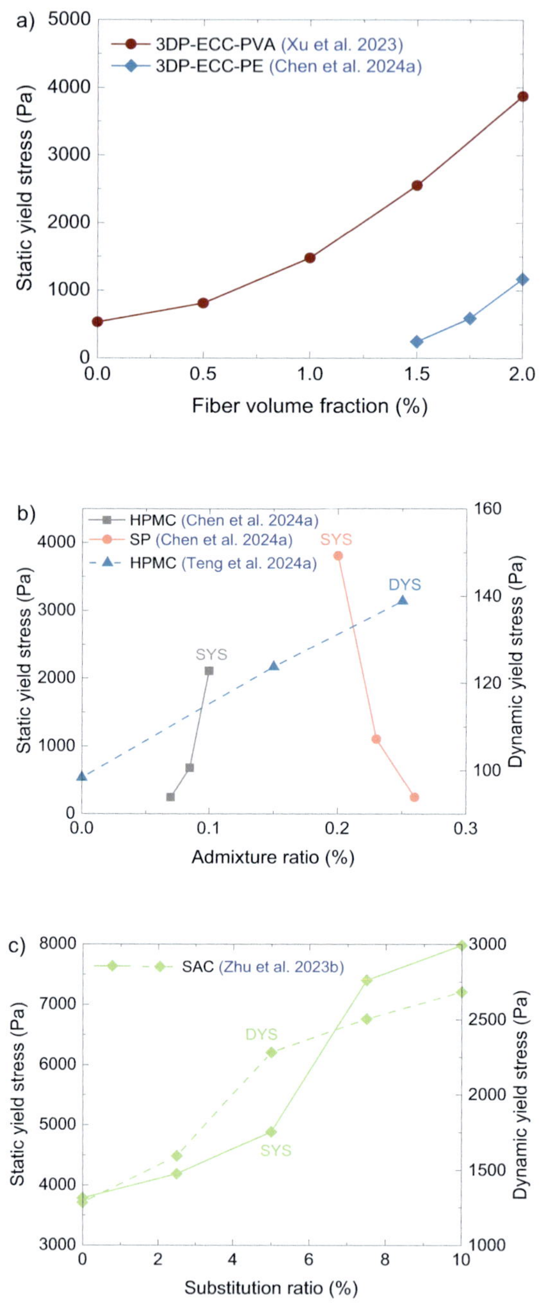

FIGURE 9.8 Effects of different typical factors on the rheological parameter of fresh 3DP-ECCs: (a) fiber content vs. SYS; (b) admixture ratio vs. SYS; (c) SAC substitution ratio vs. SYS and DYS; (d) SAC substitution ratio vs. plastic viscosity; (e) IBA and FA substitution ratio vs. SYS and DYS; (f) IBA and FA substitution ratio vs. plastic viscosity.

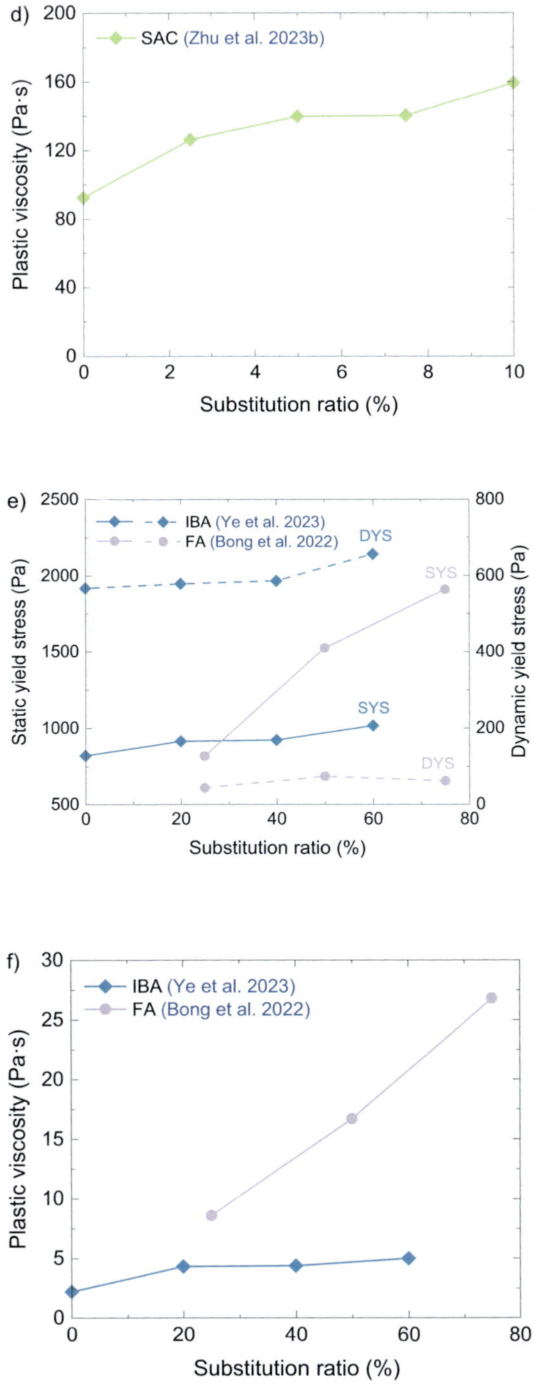

FIGURE 9.8 (Continued)

2023; Chen et al. 2024b). Moreover, longer fibers result in higher yield stress and plastic viscosity, likely due to their tendency to entangle and form network structures (Xu et al. 2023).

Viscosity modifying agent (e.g., HPMC) (Soltan and Li 2018; Chaves Figueiredo et al. 2019; Zhu et al. 2019, 2023a) and nanomaterials (e.g., ANC) (Soltan and Li 2018; Zhu et al. 2019; Xu et al. 2022b, 2023) synergistically improve rheology. For instance, it was reported that the incorporation of HPMC and ANC induced shear-thinning behavior in 3DP-ECCs, achieving 83.1% viscosity recovery (Zhu et al. 2019). This can be ascribed to the surface charge of ANC particles and the thickening effect of HPMC, which enhances the thixotropy and cohesion of the material. HPMC adsorption on cement particles increases internal friction, thereby enhancing DYS and viscosity (Zhu et al. 2023a).

Even a small dosage of polycarboxylate SP can markedly change (usually lower) the SYS by de-flocculating cementitious particles via electrosteric dispersion (Chen et al. 2024b), breaking weak floc networks, and reducing interparticle attractive forces; thus, the structural build-up at rest is sensitively tuned at low admixture levels. Owing to its rapid hydration, small-dose sulfoaluminate cement (SAC) is a powerful rheology modifier: early ettringite quickly forms a rigid particle–gel network, markedly increasing yield stress and thixotropy (Zhu et al. 2023b).

When sand is partially replaced by incineration bottom ash (IBA), the impact on rheological parameters is typically smaller than that caused by replacing slag with FA in an alkali-activated ECC matrix (Ye et al. 2023; Bong et al. 2022) because IBA mainly perturbs the granular skeleton (shape/texture effects) with limited surface reactivity, whereas FA-for-slag substitution alters binder chemistry and early gel formation, thus more strongly modifying yield stress and plastic viscosity. FA and SF exhibit opposite effects on rheological properties: increasing FA dosage increases torque viscosity but reduces flow resistance, while increasing SF reduces torque viscosity but increases flow resistance (Weng et al. 2018). These differences can be attributed to their particle shapes and surface characteristics. Generally, as the sand content increases, the SYS and DYS of 3DP-ECCs gradually increase (Zhu et al. 2023a). Optimized particle size distribution can enhance the thixotropy of the material, thereby improving SYS and DYS (Van Overmeir et al. 2022).

Time-dependent rheological behavior significantly impacts printing processes. It was found that nozzle speed and standoff distance alter the material's microstructure and rheological response, requiring real-time parameter adjustments to mitigate time-dependent effects and ensure print quality (Zhou et al. 2022b).

9.3.3 Printability

Printability that comprises pumpability, extrudability, and buildability is a critical determinant of 3DP-ECCs' compatibility with additive manufacturing processes. Current evaluation frameworks, as summarized in Table 9.3, primarily focus on extrudability (e.g., filament morphology or extrusion pressure) and buildability (e.g., structural stability or maximum printing height). However, methodological limitations persist: laboratory-scale extruders may deviate from industrial screw-based systems, while self-weight resistance tests oversimplify interlayer bonding dynamics in complex geometries.

There is a close correlation among workability, rheological properties, and printability of ECCs. Each parameter must fall within a specific range to ensure a smooth printing process. Flowability directly affects the pumping and extrusion performance of the material. DYS and plastic viscosity are closely related to extrudability, while SYS and thixotropy are highly correlated with buildability. These rheological parameters determine whether the material can rapidly support its own weight without collapsing after extrusion.

Based on a comprehensive analysis of existing research, to achieve optimal printability, the flowability of 3DP-ECCs should be controlled within an appropriate range, with a recommended flow spread of 110 mm–190 mm (Sun et al. 2020, 2022; Van Overmeir et al. 2023a; Zhu et al. 2023b; Ye et al. 2023; Chen et al. 2024b; Zhou et al. 2024; Yan et al. 2024b; Pi et al. 2024). The setting time should be maintained between 20–120 minutes (Sun et al. 2020, 2022), and the open time is suggested to be 30–60 minutes (Zhu et al. 2023b; Wang et al. 2024; Zhou et al. 2024; Yan et al. 2024b). Variations in flowability-related parameters and open time can be ascribed to differences in fiber types, nozzle sizes, and extruder power. SYS and DYS should be within reasonable ranges to ensure smooth extrusion and shape retention during printing. For SYS, a range of 1,450–3,170 Pa is recommended to provide sufficient deformation resistance, while DYS should be between 237 and 1,080 Pa to ensure smooth extrusion and good interlayer bonding (Zafar et al. 2023). Plastic viscosity significantly affects pumpability and extrudability. Materials with plastic viscosities between 8.9 and 11.5 Pa·s exhibit better pumping performance and shape retention (Zafar et al. 2023). Optimizing these rheological parameters can effectively improve the printing quality and structural stability of 3DP-ECCs. Due to large variations in rheological property indicators across different studies, the above parameter ranges are based on specific reports from individual studies. The variations may result from (i) parameter variability: disparate testing protocols (e.g., shear rate ranges and fiber alignment quantification) limit cross-study comparability; (2) scale effects: lab-optimized parameters (e.g., nozzle size) often fail in large-scale printing due to altered fiber dynamics; and (3) dynamic monitoring: offline rheological measurements inadequately capture transient printing behavior. Future studies on the effects of different fiber types, cementitious material systems, and printer parameters are required.

To optimize workability, rheological properties, and printability, an appropriate balance must be achieved. For example, adjusting the fiber content and w/b ratio can ensure the material meets pumping and extrusion requirements while maintaining structural stability. Enhancing printability from a process perspective is also an efficient approach, such as setting on-demand control (Reiter et al. 2020; Muthukrishnan et al. 2022), vibration-induced control (Sanjayan et al. 2021), magnetorheological control (Jiao et al. 2022), and coupling of temperature and viscosity modifiers (Zhang et al. 2024). With the advancement of numerical simulation technologies, finite element method or computational fluid dynamics can simulate and predict potential failure modes of printed structures before actual printing, thereby providing more precise performance indicators for specific materials and significantly reducing the number of trials (An et al. 2024; Chen et al. 2024a; Yu et al. 2024a; Wolfs et al. 2021).

Printability optimization requires synergistic material–process integration anchored in rheological control. Future efforts should transcend empirical trial-and-error to enable model-driven design, accelerating the scalable application of 3DP-ECCs in complex engineering scenarios.

9.4 HARDENED MECHANICAL PROPERTIES OF 3DP-ECCS

The hardened mechanical properties of 3DP-ECCs are critical indicators for evaluating the safety and reliability of printed structures. These properties are influenced by various factors, including fiber content, fiber length, fiber hybridization, w/b ratio, s/b ratio, aggregate type and content, curing age, preparation method (mold casting or 3D printing), printing setting parameters, and loading direction (anisotropy). Among these, the material composition, the direction of the force (anisotropy), and the preparation method are three factors worthy of special attention.

9.4.1 Compressive and Flexural Strengths

The mechanical properties of ECCs are critically influenced by material composition, with key variables including fiber content, s/b ratio, crumb rubber (CR) replacement, multi-walled carbon nanotube (MWCNT) dosage, and FA content. Figure 9.9 presents

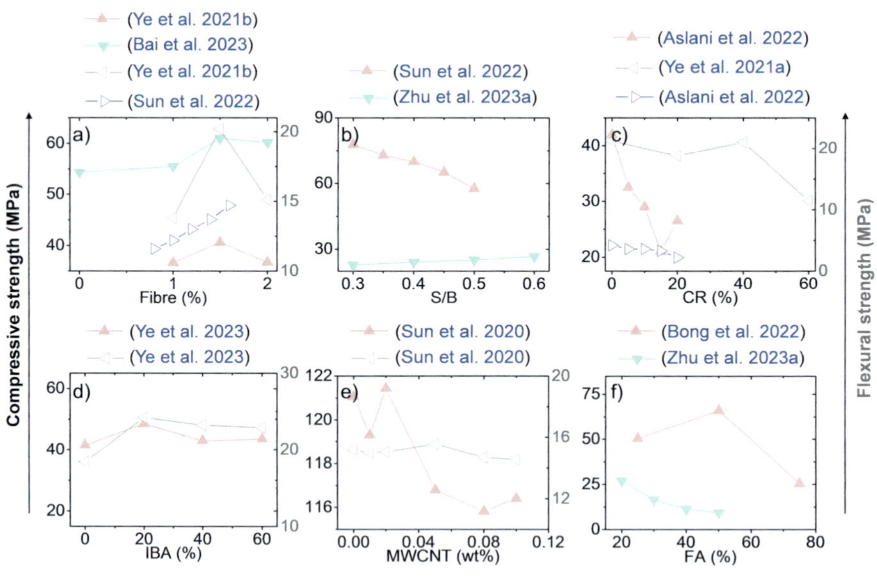

FIGURE 9.9 Effects of different typical factors on hardened strengths of ECCs: (a) fiber content; (b) s/b; (c) CR content; (d) IBA content; (e) MWCNT content; (f) FA content. Note: the solid symbol lines correspond to compressive strength, while the hollow symbol lines correspond to flexural strength.

the effects of these factors on compressive and flexural strengths. Variations in fiber type and content yield distinct strength evolution patterns. For PE fiber-reinforced ECCs, compressive and flexural strengths initially increase and then decline as the fiber content increases from 1% to 2%, with flexural strength exhibiting greater sensitivity to fiber agglomeration at higher dosages (Ye et al. 2021a; Bai et al. 2023). This can be ascribed to the higher risk of fiber agglomeration within the matrix at higher fiber contents. In contrast, PVA fiber-reinforced ECCs demonstrate a continuous increase in flexural strength across a narrower fiber content range (0.8%–1.6%), owing to improved fiber dispersion and reduced agglomeration risks (Sun et al. 2022).

The impact of s/b ratio reveals complexity across studies. Sun et al. (2022) observed a decrease in 28-day compressive strength of PVA-ECCs from 77.84 MPa to 57.56 MPa as the s/b ratio increased from 0.3 to 0.5, while Zhu et al. (2023a) reported a reverse trend at 7-day curing, with strength increasing from 22.9 MPa to 26.71 MPa for s/b ratios of 0.3–0.6. Such divergence may stem from various testing ages and the negligible effect of sand content on workability in the latter study.

Environmental substitution strategies, e.g., CR incorporation, generally degrade mechanical performance. Volume-based CR replacement of FA in PVA-ECCs reduces the 28-day compressive strength due to weak interfacial transition zones (ITZs) and increased matrix porosity (Aslani et al. 2022). Flexural strength follows a threshold-dependent pattern: CR substitution of quartz sand in printable PE-ECCs causes no significant strength loss below 40% replacement, but a drastic 52% reduction occurs at 60% CR (Ye et al. 2021b). Similarly, IBA as an aggregate replacement in PE-ECCs achieves peak strength at 20% IBA content through enhanced cement hydration, while higher IBA dosages induce matrix porosity and strength decline (Ye et al. 2023).

Nanomaterial integration shows nuanced effects. MWCNTs at 0.1% mass fraction minimally affect the 28-day strength in 3D-printed PVA-ECCs but enhance the 3-day early strength through accelerated hydration when optimally dispersed (0.01–0.05 wt%) (Sun et al. 2020). In alkali-activated systems, FA substitution for slag exhibits a strength maximum at 50% FA content, with higher FA levels suppressing geopolymerization (Bong et al. 2022). Another study (Zhu et al. 2023a) indicated that increasing FA content (20%–50%) in PVA-ECCs progressively lowers the 7-day strength. HPMC demonstrates a dosage threshold of 0.30%, where moderate additions enhance fiber dispersion and mechanical performance, while excessive HPMC disrupts matrix integrity.

These findings collectively highlight that optimizing ECC strength requires balancing component interactions, dispersion thresholds, and age-dependent effects, rather than isolated parameter adjustments.

9.4.2 Mechanical Anisotropy

The anisotropy of hardened mechanical properties in 3D-printed concrete is one of the critical factors affecting its engineering applications. Due to the layer-by-layer deposition process, distinct interlayer and inter-filament interfaces are formed, leading to significant differences in mechanical properties along different directions. Understanding this anisotropy is crucial for optimizing printing processes, enhancing material performance, and ensuring structural safety. Particularly for 3DP-ECCs, their high ductility and strain-hardening characteristics may result in more complex effects of anisotropy on overall performance. To facilitate further discussion, three loading directions for compressive and flexural tests and two typical interfaces are defined, as displayed in Figure 9.10. The horizontal printing plane is defined as the *X*–*Y* plane, where the *X*-axis is parallel to the printing filament direction (nozzle movement direction), and the *Y*-axis is perpendicular to the *X*-axis within the printing plane. The vertical direction across layers is defined as the *Z*-axis. This coordinate system helps clarify the differences in mechanical properties along different directions. The 3D-printed specimens exhibited two primary interface types: interface-D (deposition interface), which forms between filaments during the deposition process, and interface-T (translation interface), which arises between printed filaments during the translation process.

According to the existing literature data, the compressive properties of 3DP-ECCs exhibit significant anisotropy, as seen in Figure 9.11. The general trend indicates that the compressive strengths in the *X* and *Y* directions are similar but lower than that in the *Z* direction. However, among the 15 datasets compiled in Figure 9.11, five still

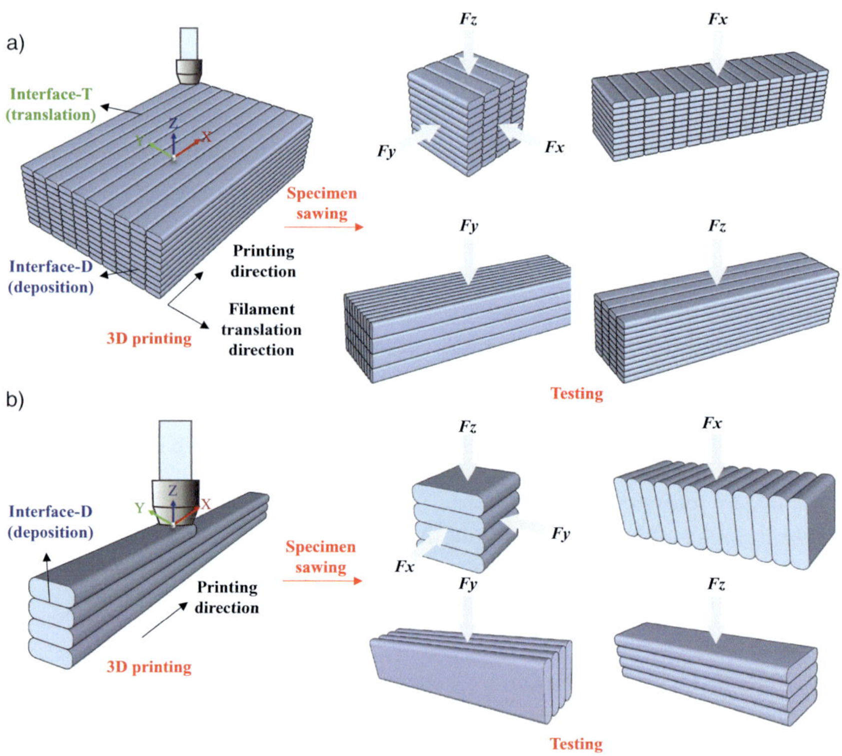

FIGURE 9.10 Schematic illustration of sample preparation and loading directions under compression and flexure: (a) two interfaces and (b) single interface.

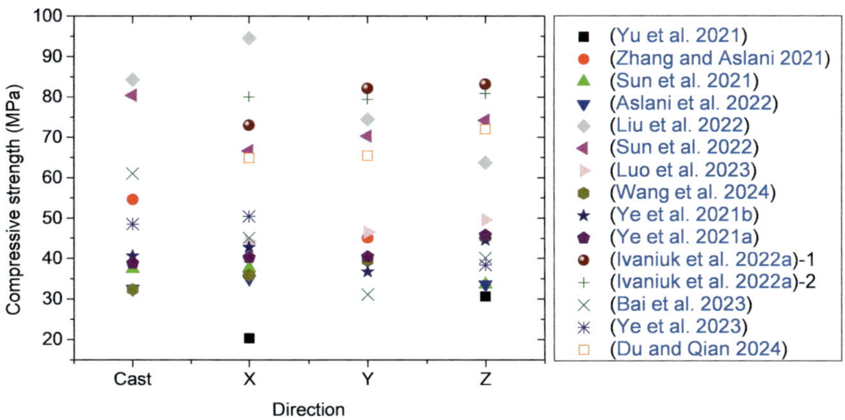

FIGURE 9.11 Anisotropic behavior of hardened 3DP-ECCs under compression.

show the *X* direction compressive strength as the highest of the three directions (Sun et al. 2021; Aslani et al. 2022; Liu et al. 2022; Bai et al. 2023; Ye et al. 2023); therefore, no clear universal pattern can be established for the anisotropic compressive strength of 3DP-ECCs. The compressive strength of cast specimens may be either the highest or the lowest among the four scenarios. This anisotropy can be primarily ascribed to the interlayer (i.e., interface-D) and inter-filament (i.e., interface-T) interfaces formed during printing. The higher compressive strength in the *Z* direction is due to the compressive forces acting mainly on the interlayer interfaces, which are compacted during printing and exhibit higher density. When compression is applied perpendicular to the fiber orientation (*Z* direction), fibers can more effectively bridge cracks, enhancing compressive strength. In contrast, the compressive strengths in the *X* and *Y* directions are lower as stresses in these directions are more likely to initiate cracking along the interlayer and inter-filament interfaces. Moreover, the presence of more interfaces parallel to the loading direction in the *X* and *Y* directions makes it difficult for fibers to bridge cracks across interfaces. In cases where the *X* direction strength is the highest, the specimen is formed by being surrounded by multiple printed filaments, providing a continuous and well-integrated load path under vertical compression; both interface-D and interface-T are parallel to the loading direction; moreover, stress concentration around pores is the lowest in *X*-oriented specimens (Davis et al. 2017). These combined factors lead to the highest compressive strength in the *X* direction. Furthermore, the compressive strength of cast specimens may be affected by surface and internal quality defects. If cast specimens have significant surface or internal defects, their compressive strength may be lower than that of printed specimens; conversely, if cast specimens are well-formed, their compressive strength may exceed that of printed specimens. Cast specimens, due to their monolithic casting process, have better interface continuity and integrity compared to layer-by-layer printed specimens. However, issues such as layering, honeycombing, and surface treatment during casting can lead to unstable compressive strength.

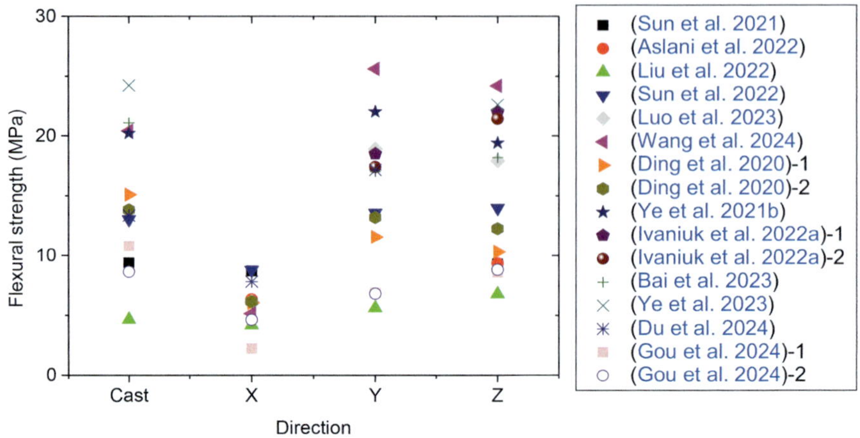

FIGURE 9.12 Anisotropic behavior of hardened 3DP-ECCs under bending.

The variations in cast specimen performance across different studies may be due to differences in these processes and quality control factors.

As seen in Figure 9.12, 3DP-ECCs also exhibit anisotropy in flexural properties. The general trend indicates that the flexural strength in the *X* direction is the lowest, while the flexural strengths in the *Y* and *Z* directions are similar, with most data showing slightly lower strength in the *Y* direction compared to the *Z* direction. Different nozzle shapes and sizes may affect the flexural strengths in the *Y* and *Z* directions. Some printed specimens even exhibit higher flexural strength than cast specimens, depending on the quality of the cast specimens. The lowest flexural strength in the *X* direction is primarily due to the weak bonding between printing filaments, with *X* direction specimens having weak interfaces perpendicular to the principal tensile stress direction, making cracks more likely to grow along filament interfaces and reducing overall flexural capacity. The higher flexural strengths in the *Y* and *Z* directions can be ascribed to the more uniform interface distribution in these directions, with no weak interfaces in the tensile direction. The layered stacking, resembling the natural nacre structure, helps distribute stress and enhances material strength and toughness. Additionally, the favorable fiber orientation and optimized microstructure during 3D printing extrusion result in higher crack-bridging capacity in the tensile direction of continuous filaments. The differences in flexural performance between the *Y* and *Z* directions mainly stem from variations in interlayer and intralayer bonding strengths.

The anisotropy in compressive and flexural properties of 3DP-ECCs results from the combined effects of multiple factors, with complex patterns and mechanisms influenced by various factors. Table 9.4 summarizes the specimen sizes and printer nozzle sizes involved in the testing of hardened mechanical properties of 3DP-ECCs in the existing literature. Significant variations in specimen sizes (e.g., cubes of 50 × 50 × 50 mm, 70.7 × 70.7 × 70.7 mm, 100 × 100 × 100 mm, prisms, and cylinders) and nozzle sizes (rectangular, circular, etc.) across different studies may lead

TABLE 9.4
Summary of Dimensional Information for Testing the Mechanical Properties of Hardened 3DP-ECCs

Mechanical Test	Specimen	Specimen Size (mm)	Printer Nozzle Size (mm)
Compressive test	PVA-ECC	50 × 50 × 50 (Weng et al. 2019; Yu et al. 2021; Zhang and Aslani 2021; Sun et al. 2021; Aslani et al. 2022; Bong et al.2022; Wang et al. 2024), 50 × 50 × 100 (Yu et al. 2021), 35 × 35 × 35 (Chaves Figueiredo et al. 2020a, 2020b; Van Overmeir et al. 2023a), 40 × 40 × 40 (Sun et al. 2020), 100 × 100 × 100 (Liu et al. 2022), 70.7 × 70.7 × 70.7 (Sun et al. 2022; Luo et al. 2023)	30 × 15 (Weng et al. 2019), 50 × 10 (Chaves Figueiredo et al. 2020a, 2020b; Yu et al. 2021), / (Zhang and Aslani 2021), Φ12 (Sun et al. 2021), Φ 20 (Aslani et al. 2022; Liu et al.2022; Bong et al. 2022; Luo et al. 2023), Φ 15 (Wang et al. 2024), 40 × 14 (Van Overmeir et al. 2023a), 40 × 10 (Sun et al. 2020; Sun et al. 2022)
	PE-ECC	40 × 40 × 40 (Zhu et al. 2019; Ivaniuk et al. 2022b), 50 × 50 × 100 (Ye et al. 2021a, 2021b), 70.7 ×70.7 × 70.7 (Bai et al. 2023), 50 × 50 × 50 (Ye et al. 2023; Gou et al. 2024), 60 × 60 × 60 (Zhu et al. 2022a; Du and Qian 2024), 50 × 50 × 150 (Yan et al. 2024a), 70 × 70 × 210 (Yan et al. 2024a), 100 × 100 × 300 (Yan et al. 2024a)	Φ 20 (Zhu et al. 2019), 50 × 25 (Ivaniuk et al. 2022b), *X* and *Z* directions: 50 × 10, *Y* direction: 100 × 10 (Ye et al. 2021a, 2021b), Φ 30 (Bai et al. 2023), 50 × 10 (Ye et al. 2023), Φ 30 (Gou et al. 2024), 30 × 16 (Zhu et al. (2022a), 30 × 5 (Du and Qian 2024), 100 × 10 (Yan et al. 2024a)
Uniaxial tensile test	PVA-ECC	Print rectangular flatbed by caulk-gun apparatus (Soltan and Li 2018), dog-bone sample based on rectangle long filament (Yu et al. 2021; Bong et al. 2022; Van Overmeir et al. 2023a), dog-bone sample based on rectangle long filament with 12.7 × 30 (Pang et al. 2023), dog-bone sample sawed from printed flatbed (Wang et al. 2024)	Φ 8–13 (Soltan and Li 2018), 30 × 13 (Yu et al. 2021), Φ 20, Bong et al. (2022), 40 × 14 (Van Overmeir et al. 2023a), Φ 20 and Φ 8 (Pang et al. 2023), Φ15 (Wang et al. 2024)
	PE-ECC	Saw-cut prisms (250 × 24 × 40) with eventually strengthened ends (Ogura et al. 2018), dog-bone sample sawed from printed flatbed (Zhu et al. 2019; Zhou et al. 2022b), sawed and grinded rectangular specimens with 240 × 30 × 25 (Xu et al. 2022b), dog-bone sample based on rectangle long filament (Ye et al. 2021a, 2021b, 2023; Ivaniuk et al. 2022b; Zhu et al. 2022a; Yan et al. 2024a, 2024b), dog-bone sample sawed from printed element (Gou et al. 2024)	18.72 × 30 (Ogura et al. (2018), Φ 15 (Zhu et al. 2019; Zhou et al. 2022b), 3 × 50 (Xu et al. 2022b), 5 × 30 (Xu et al. 2022b), 7.5 × 20 (Xu et al. 2022b), 10 × 15 (Xu et al. (2022b), 30 × 15 (Ye et al. 2021a), 50 × 25 (Ivaniuk et al. 2022b), 30 × 16 (Zhu et al. 2022a), 30 × 13 (Ye et al. 2021b, 2023; Yan et al. 2024a, 2024b), Φ 30 (Gou et al. 2024)

(Continued)

TABLE 9.4 *(Continued)*
Summary of Dimensional Information for Testing the Mechanical Properties of Hardened 3DP-ECCs

Mechanical Test	Specimen	Specimen Size (mm)	Printer Nozzle Size (mm)
Flexural test	PVA-ECC	30 × 30 × 350 (Weng et al. 2019), 40 × 40 × 160 (Sun et al. 2020, 2021, 2022; Zhang and Aslani 2021; Luo et al. 2023; Wang et al. 2024), 50 × 50 × 150 (Aslani et al. 2022), 100 × 100 × 400 (Liu et al. 2022), 350 × 60 × 20 (Bong et al. 2022), 150 × 30 × 8 (Van Overmeir et al. 2023a), 240 × 60 × 60 (Pi et al. 2023), 400 × 100 × 15 (Luo et al. 2023), 350 × 50 × 15 (Ge et al. 2021)	30 × 15 (Weng et al. 2019), 40 × 10 (Sun et al. 2020), / (Zhang and Aslani 2021), Φ 12 (Sun et al. 2021), Φ 20 (Aslani et al. 2022; Liu et al. 2022; Bong et al. 2022; Pi et al. 2023; Luo et al. 2023), / (Sun et al. 2022), Φ 15 (Wang et al. 2024), 40 × 14 (Van Overmeir et al. 2023a), extrusion machine with a steel tube of 50 × 10 (Ge et al. 2021)
	PE-ECC	90 × 90 × 360 (Ding et al. 2020), 100 × 100 × 400 (Bai et al. 2023), 50 × 50 × 200 (Ye et al. 2021a, 2023), 50 × 50 × 300 (Ye et al. 2021b), 60 × 60 × 240 (Du et al. 2024; Du and Qian 2024), 40 × 40 × 160 (Zhu et al. 2019; Ivaniuk et al. 2022b; Yan et al. 2024a; Gou et al. 2024), 60 × 60 × 240 (Yan et al. 2024a), 80 × 80 × 320 (Yan et al. 2024a), 40 × 40 × 200 (Yan et al. 2024b)	Φ 30 (Ding et al. 2020; Bai et al. 2023; Gou et al. 2024), 50 × 10 (Ye et al. 2021a, 2021b, 2023), 30 × 5 (Du et al. 2024; Du and Qian 2024), 100 × 10 (Yan et al. 2024a, 2024b), Φ 20 (Zhu et al. 2019), 50 × 25 (Ivaniuk et al. 2022b)

to inconsistencies in experimental results. Factors such as printed filament size, fiber type and content, interfacial modifying materials, the number of inter-filament and interlayer interfaces (the number of interfacial interfaces contained in a compressive specimen is different for the same size, as shown in Figure 9.10), differences in interface bonding properties, and fabrication quality defects in cast specimens further complicate the anisotropy patterns. For instance, larger specimen sizes or wider printing filaments may introduce more interface defects, reducing compressive strength, while smaller specimen sizes or narrower filaments may reduce interface defects and improve compressive strength. Moreover, quality defects in cast specimens (e.g., air bubbles and cracks) can significantly affect their compressive strengths. Certain nozzle shapes (e.g., rectangular nozzles that are more effective than circular nozzles in eliminating gaps between adjacent layers and filaments) may affect filament density and interface bonding strength, thereby altering flexural performance. Furthermore, an increase in the number of inter-filament and interlayer interfaces may lead to more stress concentration points, reducing flexural strength, while optimizing interface mechanical properties (e.g., through improved material mix designs or printing processes) may enhance flexural strength. To mitigate the pronounced anisotropy in 3DP-ECCs, some recent studies have reported bio-inspired cross-printing techniques inspired by natural structures to ensure safety under multidirectional loading (Du et al. 2024; Du and Qian 2024).

In summary, the contradictory patterns of anisotropy in the mechanical properties of 3DP-ECCs are closely related to factors such as specimen size, printing filament size, nozzle shape and size, fiber orientation, number and performance of interfaces, and forming quality. Future research is required to systematically examine the effects of these variables and clarify their interaction mechanisms to provide more reliable theoretical foundations and practical guidance for the material design and engineering applications of 3DP-ECCs. Standardizing specimen sizes and printing parameters, optimizing interface performance, and improving the fabrication quality of cast specimens may be key directions for resolving the contradictions in anisotropy patterns.

9.5 UNIAXIAL TENSILE PERFORMANCE

9.5.1 Differences in Uniaxial Tensile Performance between Cast and 3DP-ECCs

Strain-hardening behavior under tensile loading is one of the most distinctive features of ECCs, and the differences in tensile performance between 3D-printed and cast ECCs have attracted significant attention. Figure 9.13 illustrates the uniaxial tensile performance of cast and printed ECCs. It should be noted that the fiber content of 3D-printed (3DP) PVA-ECCs is 2%, with fiber lengths mostly being 8 mm and a few at 12 mm, considering extrudability constraints. For 3D-printed PE-ECCs, the fiber content is ≥1.25%, with fiber lengths including 6 mm, 12 mm, and 18 mm. As seen in Figure 9.13, the literature data points for 3DP-PVA-ECCs are relatively concentrated, primarily because PVA fiber characteristics make PVA-ECCs suitable for materials with medium compressive strength (around 40 MPa). However, due to the ultra-high

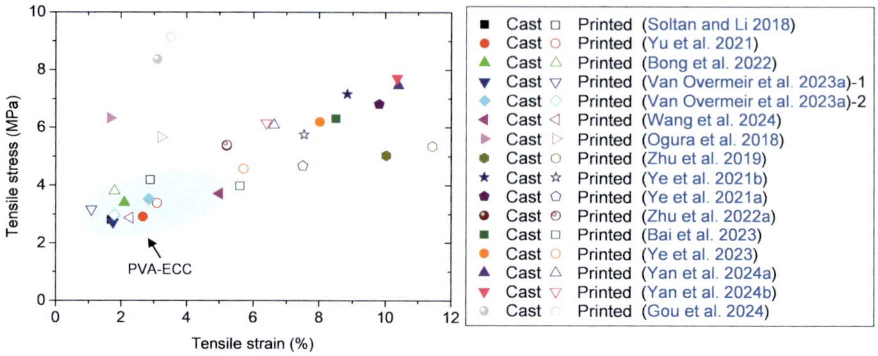

FIGURE 9.13 Uniaxial tensile performance of cast and 3DP-ECCs.

tensile strength of PE fibers, which enables significant plastic deformation, 3DP-PE-ECCs exhibit higher tensile strength and ductility (with maximum tensile strength and ductility reaching 9.13 MPa and 11.43%, respectively), and the data distribution is more scattered. Therefore, PE fibers are suitable for preparing high-strength (≥60 MPa) printable materials.

Overall, affected by factors such as fiber type and content, sampling methods, specimen forming quality, and printing processes, the comparative data on tensile performance between cast and printed ECCs indicate inconsistent trends. For 3DP-PVA-ECCs, the printed specimens generally exhibit higher tensile ductility and strength than cast specimens. However, for 3DP-PE-ECCs, the trend is the opposite. Thus, multiple factors need to be considered. The weakening of tensile performance may be attributed to the use of longer fibers (e.g., 18 mm) in some studies, where flexible PE fibers at around 2% content could hardly be smoothly extruded from a 20-mm-diameter nozzle, leading to fiber agglomeration within the filaments and significantly reducing the number of fibers effectively bridging cracks (i.e., stress transfer efficiency is reduced). Poor fiber dispersion in the printed ECCs can be observed in Figure 9.14. Additionally, from a sampling perspective, some studies cut dog-bone specimens from printed elements, and the cutting process may shorten the fiber lengths on the specimen surface, affecting tensile performance. Although some studies reported weakened tensile performance in 3DP-ECC specimens, 3DP-ECCs still maintains good ductility and strength, with multiple microcracks and strain-hardening characteristics, offering potential for lightly reinforced 3D-printed structures. The optimization of tensile performance in printed specimens may be due to the favorable orientation of fibers along the printing direction and more uniform dispersion within the filaments. Moreover, the extrusion pressure during printing results in denser internal matrices and fewer large pores (large harmful pores can lead to incomplete cracking and localized failure).

Therefore, to achieve satisfactory tensile performance in 3DP-ECCs, a balance between strength and ductility must be achieved based on engineering requirements while ensuring printability. On the one hand, with constant fiber content, it is recommended to use a mix of short and long fibers (e.g., 6 mm and 12 mm) or fibers of

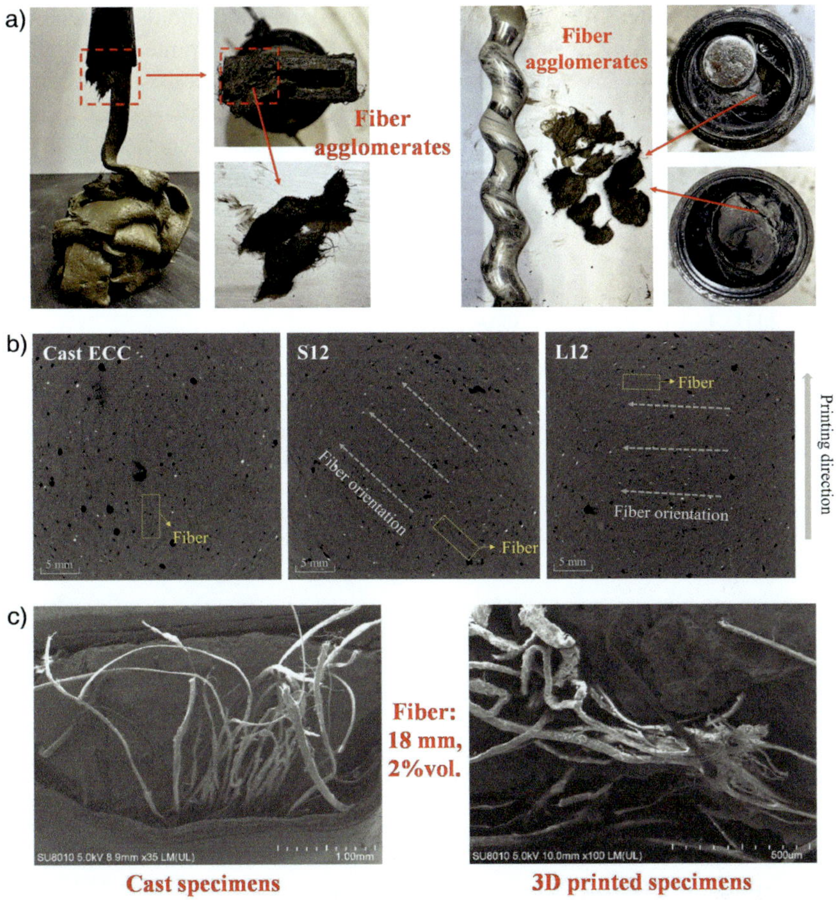

FIGURE 9.14 Poor fiber dispersion in 3DP-ECCs: (a) fiber agglomerates clogging in the nozzle and the pump screw (Xu et al. 2023); (b) fiber orientation in X-ray micro-CT scanning images for cast vs. 3D-printed specimens (Zhou et al. 2022b); (c) fibers twisted together in printed ECCs observed from scanning electron microscope (SEM) (Ye et al. 2021b).

moderate length (e.g., 8 mm or 9 mm). Fibers should not be too short, as sufficient embedment length is needed to provide bridging stress. On the other hand, if longer fibers (e.g., 12 mm) are used, the fiber content can be appropriately reduced to achieve a balance in micromechanical performance.

Differences in tensile performance among various printed ECCs may also be related to inconsistencies in printer process parameters and sampling methods. Existing studies generally follow the testing methods for fiber-reinforced concrete and ECC uniaxial tensile tests, but the sampling methods and specimen dimensions for printed ECC tensile tests are inconsistent (some studies cut dog-bone specimens directly from hardened printed elements, some directly test printed filaments, and others cast high-strength material around printed filaments in molds to prepare dog-bone specimens), leading to significant variations in tensile parameter results. Since tensile parameters are crucial in ECC structural design, inconsistencies in these parameters will affect the accuracy of the constitutive model for 3DP-ECCs, resulting in suboptimal structural design. Therefore, there is an urgent need to establish standardized testing methods specifically defining specimen geometry, sampling methods, testing procedures, and analysis methods for 3DP-ECCs with different types of fiber.

9.5.2 Effects of Printing Parameters

As seen in Figure 9.15, the tensile strength and strain of 3DP ECCs are highly sensitive to printing parameter configurations, including nozzle speed, standoff distance, nozzle geometry, and printing rotation angle (i.e., rotation angle of odd and even layer

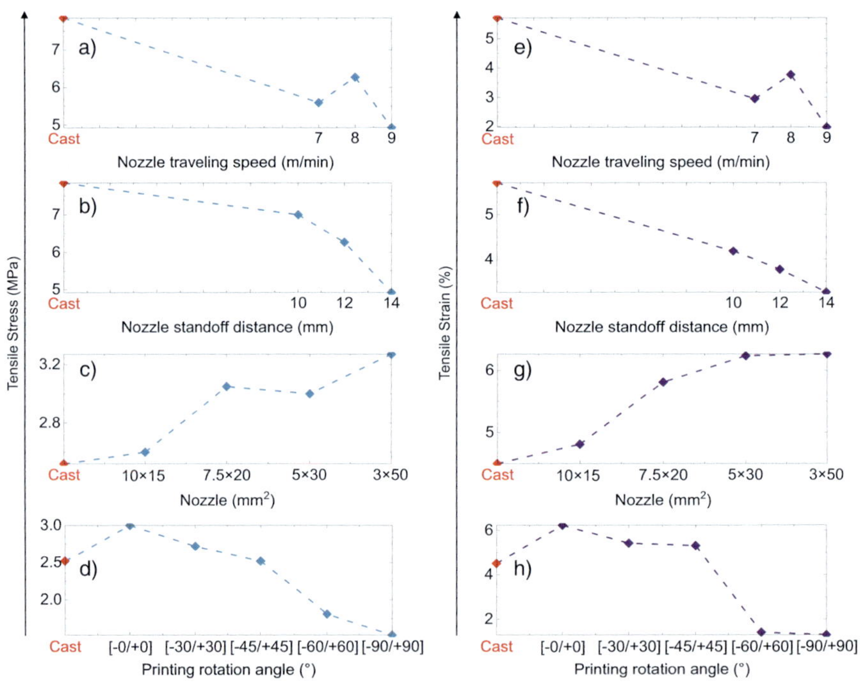

FIGURE 9.15 Effects of printing parameters on the tensile properties of 3DP-PE-ECCs: (a) nozzle traveling speed vs. tensile stress (Zhou et al. 2022b); (b) nozzle standoff distance vs. tensile stress (Zhou et al. 2022b); (c) nozzle size vs. tensile stress (Xu et al. 2022b); (d) printing rotation angle vs. tensile stress (Xu et al. 2022b); (e) nozzle traveling speed vs. tensile strain (Zhou et al. 2022b); (f) nozzle standoff distance vs. tensile strain (Zhou et al. 2022b); (g) nozzle size vs. tensile strain (Xu et al. 2022b); (h) printing rotation angle vs. tensile strain (Xu et al. 2022b).

filaments with respect to the loading direction) (Zhou et al. 2022b; Xu et al. 2022b). These parameters collectively govern fiber alignment, matrix density, and interfacial bonding, ultimately determining mechanical anisotropy and defect distribution.

Nozzle speed critically balances fiber–matrix interaction and extrusion quality. At 8 m/minute, the tensile strength and strain peak due to optimal fiber–matrix bonding and controlled fiber orientation. Deviations from this speed degrade performance: lower speeds induce material accumulation, forcing fibers to align perpendicularly to the printing direction, while higher speeds increase out-of-plane fiber angles, disrupting load transfer (Figure 9.15a). Similarly, nozzle standoff distance modulates compaction effects. A reduced distance enhances in-plane fiber alignment and matrix densification, improving the tensile strength and strain by up to 39% and 30%, respectively (Figure 9.15b,f). However, excessive proximity compromises uniformity by creating localized stress concentrations.

Wider rectangular nozzles (e.g., 3 × 50 mm^2) outperform narrower counterparts by promoting material homogeneity and reducing voids (Figure 9.15c). For instance, a [5 × 30] nozzle yields 1.3× higher ultimate tensile strain than a [10 × 15] nozzle, attributed to improved fiber distribution and minimized interfacial defects. This suggests that nozzle width-to-height ratio, rather than cross-sectional areas alone, should be prioritized for defect-sensitive applications. Printing rotation angle significantly alters load-path alignment. Angles ≤45° (e.g., [−0/+0], [−30/+30]) enhance tensile properties by aligning fibers closer to the loading direction, surpassing cast ECC performance. In contrast, angles ≥60° introduce unfavorable fiber orientations, weakening the load-bearing capacity (Figure 9.15d). Rotational strategies are particularly advantageous for multidirectional loading scenarios, effectively mitigating anisotropic behavior.

Tensile strain trends (Figure 9.15e–h) mirror strength responses, with nozzle speed and standoff distance dominating ductility. Wider nozzles further enhance strain capacity by optimizing fiber orientation distribution, especially when nozzle height remains below fiber length (e.g., 12-mm fibers). This interdependence highlights the need for holistic parameter optimization rather than isolated adjustments.

9.6 SUSTAINABILITY ASSESSMENT

Existing studies predominantly adopt a cradle-to-gate boundary system to assess the environmental impact of proposed materials. Based on the available literature (Joshi et al. 2004; Choi et al. 2012; Yu et al. 2019; Tapper et al. 2020; Long et al. 2022; Lao et al. 2023; Chen et al. 2023; Zhou et al. 2024; Wang et al. 2024; Ye et al. 2024a), Table 9.5 summarizes the embodied carbon, embodied energy, and material costs of key raw materials in 3DP-ECCs over their life cycle. These materials can be categorized into five groups: cement, supplementary cementitious materials, fine aggregates, fiber types, and admixtures. From a sustainability perspective, cement and fibers have the highest embodied carbon and energy, with fibers also being the most expensive in terms of economic cost. Therefore, optimizing material selection to minimize carbon footprint and costs while meeting performance requirements is crucial.

TABLE 9.5
Embodied Carbon, Embodied Energy, and Material Costs of 3DP-ECCs

Raw Material		Embodied Carbon ($t\ CO_2 \cdot eq/t$)	Embodied Energy (MJ/t)	Material Costs (CNY/t)
Cement	Ordinary portland cement (OPC) (Choi et al. 2012)	0.87	5500	800
	Portland limestone cement (PLC) (Zhou et al. 2024)	0.811	5000	210
	Calcium sulfoaluminate cement (CSA) (Zhou et al. 2024)	0.36	2200	1200
Supplementary cementitious materials	FA (Lao et al. 2023)	0.004	100	120
	Semi-hydrated gypsum (SG) (Long et al. 2022)	0.12	1800	200
	GGBFS (Lao et al. 2023)	0.042	200	190
	SF (Lao et al. 2023)	0.024	100	1200
	Metakaolin (MK) (Zhou et al. 2024)	0.33	3480	350
	Limestone powder (LP/LS) (Zhou et al. 2024)	0.012	620	80
	Calcined clay (CA) (Wang et al. 2024)	0.196	2734	282
Fine aggregates	Silica sand (SS) (Wang et al. 2024)	0.023	67	400
	River sand (RS) (Ye et al. 2024a)	0.0024	80	78
Fiber type	Glass fiber (GF) (Joshi et al. 2004)	2.04	48330	4300
	PP fiber (Joshi et al. 2004)	1.85	77190	7900
	Carbon fibers (CFs) (Tapper et al. 2020)	24.4	286000	7700
	Steel fibers (SFs) (Yu et al. 2019)	2.83	36000	14000
	PVA fiber (Choi et al. 2012)	3.4	101000	26000
	PE fiber (Lao et al. 2023)	2	94500	200000
Admixtures	SP (Choi et al. 2012)	1.667	35000	8000
	Viscosity modifying agent (VMA) (Wang et al. 2024)	0.24	2560	22000
	HPMC (Long et al. 2022)	0.24	2560	2360
	Anhydrous sodium metasilicate and GD grade sodium silicate powders (Na_2SiO_3) (Chen et al. 2023)	1.514	9400	4500
	Water (Yu et al. 2019)	0.001	0.01	2.6

Figure 9.16 further compares the embodied carbon, embodied energy, and material costs of different mixtures of 3DP-ECCs reported in the existing studies, categorized by PVA fibers, PE fibers, and other fibers. There exist significant variations in raw material compositions. In PVA-ECCs, cement is the primary contributor to embodied carbon and energy, followed by fibers. For example, ECCs in reference (Van Overmeir et al. 2023b) exhibit the highest embodied carbon, reaching 868 kg CO_2/m^3, with cement accounting for over 91% of this value. In contrast, a cement-free 3DP alkali-activated composite in reference (Bong et al. 2022) has an embodied carbon of only 267 kg CO_2/m^3. Among PE-ECC mixtures, ECCs in reference (Ivaniuk et al. 2022b) have the highest embodied carbon at 911 kg CO_2/m^3, despite the use of recycled sand, which did not significantly reduce embodied carbon. The use of eco-friendly LC^3 cement reduced embodied carbon to 407 kg CO_2/m^3, further highlighting the significant environmental impact of cement (Yan et al. 2024a). Therefore, to enhance environmental benefits, priority should be given to using supplementary cementitious materials or cement-free binders (e.g., alkali-activated materials) to optimize mix design. Additionally, material cost analysis reveals that PE fibers are relatively expensive. If mechanical performance can be maintained, PVA fibers or other alternatives should be considered. In summary, optimizing mix proportions plays a vital role in reducing the carbon footprint and economic cost of 3DP-ECCs.

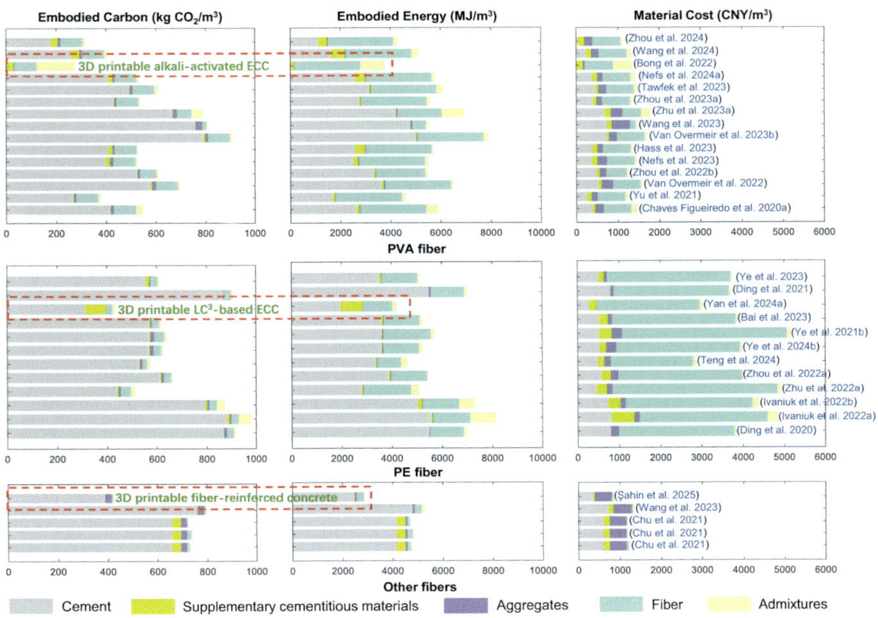

FIGURE 9.16 Comparison of embodied carbon, embodied energy, and material costs of different mixtures of 3DP-ECCs reported in the literature.

9.7 CONCLUDING REMARKS

In recent years, the rapid development of extrusion-based 3DCP technology has significantly advanced the research and application of 3DP-ECCs. This review systematically summarizes recent advances in manufacturing processes, printability requirements, mix design, fresh and hardened properties, and sustainability. The main conclusions can be drawn as follows:

(i) Fresh properties are critical for printability. Fresh ECCs must exhibit adequate flowability, extrudability, and buildability to ensure continuous printing and strong interlayer bonding. Adjusting the w/b ratio, thickening agents, and fiber content can effectively optimize the fresh properties. However, fiber incorporation may negatively impact flowability, necessitating a balance between flowability and fiber dispersion. Additionally, setting time and open time are key factors affecting print quality and should be tailored to the specific printing process.

(ii) Hardened mechanical properties are influenced by multiple factors. The interlayer bond strength, compressive strength, and tensile strength of 3DP-ECCs are significantly affected by printing paths, interlayer interval time, and fiber distribution. Although the tensile strain-hardening capacity of 3DP-ECCs may be slightly lower than that of conventionally cast ECCs, they still exhibit excellent crack control and toughness. Micromechanical analysis indicates that fiber–matrix interface performance and fiber dispersion uniformity are crucial to mechanical performance. Moreover, anisotropy introduced during printing remains a challenge that requires further research and improvement.

(iii) 3DP-ECCs can reduce cement consumption and carbon footprint by incorporating industrial by-products or waste such as FA and slag, by making use of abundant natural resources such as limestone powder and calcined clay, or by utilizing alkali-activated materials instead of cement. Process optimizations (e.g., printing speed, layer thickness, and path design) and structural optimizations (e.g., topology optimization and biomimetic design) can improve printing efficiency and minimize material waste. Moreover, 3D printing inherently reduces material waste and construction energy consumption, further enhancing its environmental benefits. However, comprehensive sustainability assessments must consider the environmental impact of fiber production and energy consumption during printing.

(iv) Future efforts can be focused on material optimization and mix design, including the development of novel fibers and fiber surface treatments to improve interfacial performance. Research into multifunctional ECCs (e.g., self-healing ECCs and conductive ECCs) is also required to expand its application potential. Furthermore, printing process parameters (e.g., printing speed, layer thickness, and extrusion pressure) should be optimized, and real-time monitoring and control technologies can be explored to enhance print quality and stability. To address interlayer bonding issues, it is vital to explore the interface treatment methods and the impact of environmental conditions on bond performance. From a sustainability perspective, life-cycle assessment, carbonation curing techniques, and the use of renewable resources (e.g., bio-based materials and recycled fibers) are required to quantify and minimize environmental impacts. Efforts should be made to standardize testing and acceptance criteria for 3DP-ECCs and to promote engineering-scale demonstration projects to facilitate broader practical applications.

REFERENCES

Aïtcin, Pierre-Claude, and Robert J. Flatt, eds. *Science and Technology of Concrete Admixtures*. Woodhead Publishing, 2015.

An, Dong, Y. X. Zhang, and Richard Chunhui Yang. "Numerical modelling of 3D concrete printing: Material models, boundary conditions and failure identification." *Engineering Structures* 299 (2024): 117104.

Anandaraj, S., S. S. Sakthi, A. R. Krishnaraja, and T. Mithun. "Development of green engineered cementitious composite using fly ash and nanosilica for 3D printing." *Journal of Environmental Nanotechnology* 13, no. 1 (2024): 117–124.

Asghari, Y., S. E. Mohammadyan-Yasouj, M. Petrů, H. Ghandvar, and S. S. R. Koloor. "3D printing and implementation of engineered cementitious composites – A review." *Case Studies in Construction Materials* 21 (2024): e03462.

Aslani, Farhad, Ryan Dale, Fatemeh Hamidi, and Afsaneh Valizadeh. "Mechanical and shrinkage performance of 3D-printed rubberised engineered cementitious composites." *Construction and Building Materials* 339 (2022): 127665.

Bai, Meiyan, Yuching Wu, Jianzhuang Xiao, Tao Ding, and Kequan Yu. "Workability and hardened properties of 3D printed engineered cementitious composites incorporating recycled sand and PE fibers." *Journal of Building Engineering* 71 (2023): 106477.

Bai, Meiyan, Jianzhuang Xiao, Yuching Wu, and Tao Ding. "Experimental and numerical study on the flexural behavior of 3D-printed composite beams with U-shaped ECC formwork." *Engineering Structures* 312 (2024): 118225.

Bong, Shin Hau, Behzad Nematollahi, Venkatesh Naidu Nerella, and Viktor Mechtcherine. "Method of formulating 3D-printable strain-hardening alkali-activated composites for additive construction." *Cement and Concrete Composites* 134 (2022): 104780.

Bos, Freek P., Costantino Menna, Mauricio Pradena, Eric Kreiger, W. R. Leal da Silva, Atta Ur Rehman, Daniel Weger et al. "The realities of additively manufactured concrete structures in practice." *Cement and Concrete Research* 156 (2022): 106746.

Bos, Freek P., Rob Wolfs, Zeeshan Ahmed, and Theo Salet. "Additive manufacturing of concrete in construction: Potentials and challenges of 3D concrete printing." *Virtual and Physical Prototyping* 11, no. 3 (2016): 209–225.

Breedveld, Laurens Victor Adriaan. "Shear-induced self-diffusion in concentrated suspensions." PhD diss., 2000.

Buswell, Richard A., W. R. Leal De Silva, Scott Z. Jones, and Justin Dirrenberger. "3D printing using concrete extrusion: A roadmap for research." *Cement and Concrete Research* 112 (2018): 37–49.

Chaves Figueiredo, Stefan, Anne L. van Overmeir, Karsten Nefs, Erik Schlangen, Theo A. M. Salet, Branko Šavija, Akke S. J. Suiker, and Freek P. Bos. "Quality assessment of printable strain hardening cementitious composites manufactured in two different printing facilities." In *RILEM International Conference on Concrete and Digital Fabrication*, pp. 824–838. Cham: Springer International Publishing, 2020a.

Chaves Figueiredo, Stefan, Claudia Romero Rodríguez, Zeeshan Y. Ahmed, Derk H. Bos, Yading Xu, Theo M. Salet, Oğuzhan Çopuroğlu, Erik Schlangen, and Freek P. Bos. "An approach to develop printable strain hardening cementitious composites." *Materials & Design* 169 (2019): 107651.

Chaves Figueiredo, Stefan, Claudia Romero Rodríguez, Zeeshan Y. Ahmed, Derk H. Bos, Yading Xu, Theo M. Salet, Oğuzhan Çopuroğlu, Erik Schlangen, and Freek P. Bos. "Mechanical behavior of printed strain hardening cementitious composites." *Materials* 13, no. 10 (2020b): 2253.

Chen, Baixi, Xueqi Zhao, and Xiaoping Qian. "Voxel-based path-driven 3D concrete printing process simulation framework embedding interlayer behavior." *Automation in Construction* 168 (2024a): 105776.

Chen, Gai, Dong-ping Zheng, Yi-wu Chen, Jia-Xiang Lin, Wei-jian Lao, Yong-chang Guo, Zhan-biao Chen, and Xue-wei Lan. "Development of high performance geopolymer concrete with waste rubber and recycle steel fiber: A study on compressive behavior, carbon emissions and economical performance." *Construction and Building Materials* 393 (2023): 131988.

Chen, Wenguang, Junhong Ye, Fangming Jiang, Roman Fediuk, Jiangtao Yu, Victor C. Li, and Kequan Yu. "Printability region for 3D-printable engineered cementitious composites (3DP-ECC)." *Cement and Concrete Composites* 151 (2024b): 105599.

Choi, Won-Chang, Hyun-Do Yun, Joo-Won Kang, and Sun-Woo Kim. "Development of recycled strain-hardening cement-based composite (SHCC) for sustainable infrastructures." *Composites Part B: Engineering* 43, no. 2 (2012): 627–635.

Chu, S. H., L. G. Li, and A. K. H. Kwan. "Development of extrudable high strength fiber reinforced concrete incorporating nano calcium carbonate." *Additive Manufacturing* 37 (2021): 101617.

Dai, Jian-Guo, Bo-Tao Huang, and Surendra P. Shah. "Recent advances in strain-hardening UHPC with synthetic fibers." *Journal of Composites Science* 5, no. 10 (2021): 283.

Davis, Timothy, D. Healy, A. Bubeck, and Richard Walker. "Stress concentrations around voids in three dimensions: The roots of failure." *Journal of Structural Geology* 102 (2017): 193–207.

De Schutter, Geert, Karel Lesage, Viktor Mechtcherine, Venkatesh Naidu Nerella, Guillaume Habert, and Isolda Agusti-Juan. "Vision of 3D printing with concrete – Technical, economic and environmental potentials." *Cement and Concrete Research* 112 (2018): 25–36.

Ding, Tao, Jianzhuang Xiao, Shuai Zou, and Jiangtao Yu. "Flexural properties of 3D printed fibre-reinforced concrete with recycled sand." *Construction and Building Materials* 288 (2021): 123077.

Ding, Tao, Jianzhuang Xiao, Shuai Zou, and Xinji Zhou. "Anisotropic behavior in bending of 3D printed concrete reinforced with fibers." *Composite Structures* 254 (2020): 112808.

Ding, Yao, Jiepeng Liu, Xingjian Ou, Tomoya Nishiwaki, Gang Xiong, Xin Chen, Kequan Yu, and Hongtuo Qi. "3D printing hybrid-fiber reinforced engineered cementitious composites (3DP-HECC): Feasibility in long-open-time applications." *Construction and Building Materials* 445 (2024): 137960.

Du, Guoqiang, and Ye Qian. "Effects of printing patterns and loading directions on fracture behavior of 3D printed strain-hardening cementitious composites." *Engineering Fracture Mechanics* 304 (2024): 110155.

Du, Guoqiang, Yan Sun, and Ye Qian. "Flexural performance of nature-inspired 3D-printed strain-hardening cementitious composites (3DP-SHCC) with Bouligand structures." *Cement and Concrete Composites* 149 (2024): 105494.

Fischer, Gregor, and Victor C. Li. "Deformation behavior of fiber-reinforced polymer reinforced engineered cementitious composite (ECC) flexural members under reversed cyclic loading conditions." *Structural Journal* 100, no. 1 (2003): 25–35.

Ge, Zhi, Abdullah M. Tawfek, Hongzhi Zhang, Yongwei Yang, Huaqiang Yuan, Renjuan Sun, and Zheng Wang. "Influence of an extrusion approach on the fiber orientation and mechanical properties of engineering cementitious composite." *Construction and Building Materials* 306 (2021): 124876.

Gou, Hongxiang, Massoud Sofi, Zipeng Zhang, Mintao Zhu, Hongbo Zhu, and Priyan Mendis. "Combined printable and mechanical analysis of 3D-printed green high-strength, lightweight engineered cementitious composites." *Cement and Concrete Composites* 149 (2024): 105506.

Hass, Lauri, Karsten Nefs, F. P. Bos, and T. A. M. Salet. "Application potential of combining strain hardening cementitious composites and helical reinforcement for 3D concrete printed structures: Case study of a spiral staircase." *Journal of Building Engineering* 80 (2023): 107926.

Hoffmann, Marcin, Szymon Skibicki, Paweł Pankratow, Adam Zieliński, Mirosław Pajor, and Mateusz Techman. "Automation in the construction of a 3D-printed concrete wall with the use of a lintel gripper." *Materials* 13, no. 8 (2020): 1800.

Hou, Mengjun, Ke Xu, Paulo JM Monteiro, and Victor C. Li. "Rubber particle bridging effect on crack width control of low carbon engineered cementitious composites (ECC)." *Cement and Concrete Composites* 140 (2023): 105106.

Ivaniuk, Egor, Martin Friedrich Eichenauer, Zlata Tošić, Steffen Müller, Daniel Lordick, and Viktor Mechtcherine. "3D printing and assembling of frame modules using printable strain-hardening cement-based composites (SHCC)." *Materials & Design* 219 (2022a): 110757.

Ivaniuk, Egor, Irina Ivanova, Dmitrii Sokolov, Zlata Tošić, Martin Friedrich Eichenauer, Daniel Lordick, and Viktor Mechtcherine. "Application-driven material design of printable strain hardening cementitious composites (SHCC)." *Materials* 15, no. 5 (2022b): 1631.

Ivaniuk, Egor, Zlata Tošić, Steffen Mueller, Daniel Lordick, and Viktor Mechtcherine. "Automated manufacturing of reinforced modules of segmented shells based on 3D printing with strain-hardening cementitious composites." *Automation in Construction* 166 (2024): 105591.

Jiao, Dengwu, Caijun Shi, and Geert De Schutter. "Magneto-rheology control in 3D concrete printing: A rheological attempt." *Materials Letters* 309 (2022): 131374.

Joshi, Satish V., Lawrence T. Drzal, Amar K. Mohanty, and Satish Arora. "Are natural fiber composites environmentally superior to glass fiber reinforced composites?" *Composites Part A: Applied Science and Manufacturing* 35, no. 3 (2004): 371–376.

Kanda, Tetsushi, and Victor C. Li. "Practical design criteria for saturated pseudo strain hardening behavior in ECC." *Journal of Advanced Concrete Technology* 4, no. 1 (2006): 59–72.

Lao, Jian-Cong, Bo-Tao Huang, Ling-Yu Xu, Mehran Khan, Yi Fang, and Jian-Guo Dai. "Seawater sea-sand engineered geopolymer composites (EGC) with high strength and high ductility." *Cement and Concrete Composites* 138 (2023): 104998.

Li, Junxia, Jishen Qiu, Jian Weng, and En-Hua Yang. "Micromechanics of engineered cementitious composites (ECC): A critical review and new insights." *Construction and Building Materials* 362 (2023a): 129765.

Li, Shuai, Hung Nguyen-Xuan, and Phuong Tran. "Digital design and parametric study of 3D concrete printing on non-planar surfaces." *Automation in Construction* 145 (2023b): 104624.

Li, Victor C. "High-performance and multifunctional cement-based composite material." *Engineering* 5, no. 2 (2019a): 250–260.

Li, Victor C. *Engineered cementitious composites (ECC): Bendable Concrete for Sustainable and Resilient Infrastructure*. Springer, 2019b.

Li, Victor C., Freek P. Bos, Kequan Yu, Wes McGee, Tsz Yan Ng, Stefan Chaves Figueiredo, Karsten Nefs et al. "On the emergence of 3D printable engineered, strain hardening cementitious composites (ECC/SHCC)." *Cement and Concrete Research* 132 (2020): 106038.

Li, Victor C., Cynthia Wu, Shuxin Wang, Atsuhisa Ogawa, and Tadashi Saito. "Interface tailoring for strain-hardening polyvinyl alcohol-engineered cementitious composite (PVA-ECC)." *Materials Journal* 99, no. 5 (2002): 463–472.

Li, Yazhao, Junxia Li, En-Hua Yang, and Xinchun Guan. "Mechanism study of crack propagation in river sand engineered cementitious composites (ECC)." *Cement and Concrete Composites* 128 (2022): 104434.

Liu, Bing, Xiaoyan Liu, Guangtao Li, Songyuan Geng, Zhi Li, Yunhao Weng, and Kai Qian. "Study on anisotropy of 3D printing PVA fiber reinforced concrete using destructive and non-destructive testing methods." *Case Studies in Construction Materials* 17 (2022): e01519.

Liu, Haoran, Jianzhuang Xiao, and Tao Ding. "Flexural performance of 3D-printed composite beams with ECC and recycled fine aggregate concrete: Experimental and numerical analysis." *Engineering Structures* 283 (2023): 115865.

Long, Wu-Jian, Zhuorui Wu, Kamal H. Khayat, Jingjie Wei, Biqin Dong, Feng Xing, and Jinrui Zhang. "Design, dynamic performance and ecological efficiency of fiber-reinforced mortars with different binder systems: Ordinary portland cement, limestone calcined clay cement and alkali-activated slag." *Journal of Cleaner Production* 337 (2022): 130478.

Luo, Surong, Wenqiang Li, and Dehui Wang. "Study on bending performance of 3D printed PVA fiber reinforced cement-based material." *Construction and Building Materials* 433 (2024): 136637.

Luo, Surong, Qian Lin, Wei Xu, and Dehui Wang. "Effects of interval time and interfacial agents on the mechanical characteristics of ultra-high toughness cementitious composites under 3D-printed technology." *Construction and Building Materials* 374 (2023): 130936.

Mechtcherine, Viktor, Richard Buswell, Harald Kloft, Freek P. Bos, Norman Hack, Rob Wolfs, Jay Sanjayan, Behzad Nematollahi, Egor Ivaniuk, and Tobias Neef. "Integrating reinforcement in digital fabrication with concrete: A review and classification framework." *Cement and Concrete Composites* 119 (2021): 103964.

Mechtcherine, Viktor, Freek P. Bos, Arnaud Perrot, W. R. Leal Da Silva, V. N. Nerella, Shirin Fataei, Rob J. M. Wolfs, Mohammed Sonebi, and Nicolas Roussel. "Extrusion-based additive manufacturing with cement-based materials – Production steps, processes, and their underlying physics: A review." *Cement and Concrete Research* 132 (2020): 106037.

Muthukrishnan, Shravan, Sayanthan Ramakrishnan, and Jay Sanjayan. "Set on demand geopolymer using print head mixing for 3D concrete printing." *Cement and Concrete Composites* 128 (2022): 104451.

Nefs, K., K. C. J. de Kroon, J. Sloots, F. P. Bos, and T. A. M. Salet. "Orientation dependency of 3D printed SHCC at increasing length scale." *Cement and Concrete Composites* 149 (2024a): 105515.

Nefs, Karsten, V. Menkovski, Freek P. Bos, Akke S. J. Suiker, and Theo A. M. Salet. "Automated image segmentation of 3D printed fibrous composite micro-structures using a neural network." *Construction and Building Materials* 365 (2023): 130099.

Nefs, K., J. Sloots, K. C. J. de Kroon, F. P. Bos, and T. A. M. Salet. "Analytical modelling of the orientation dependency of 3D printed SHCC at increasing levels of scale." *Materials & Design* 242 (2024b): 113018.

Nematollahi, Behzad, Jishen Qiu, En-Hua Yang, and Jay Sanjayan. "Microscale investigation of fiber-matrix interface properties of strain-hardening geopolymer composite." *Ceramics International* 43, no. 17 (2017): 15616–15625.

Ogura, Hiroki, Venkatesh Naidu Nerella, and Viktor Mechtcherine. "Developing and testing of strain-hardening cement-based composites (SHCC) in the context of 3D-printing." *Materials* 11, no. 8 (2018): 1375.

Pang, Zhiming, Cong Lu, Baoshan Li, and Jiajie Wang. "A multiscale model for quantifying fiber orientation effects on the tensile properties of 3D printed engineered cementitious composites (3DP-ECC)." *Journal of Building Engineering* 68 (2023): 106090.

Pi, Yilin, Cong Lu, Baoshan Li, and Junhui Zhou. "Crack propagation and failure mechanism of 3D printing engineered cementitious composites (3DP-ECC) under bending loads." *Construction and Building Materials* 408 (2023): 133809.

Pi, Yilin, Cong Lu, Yiming Yao, and Baoshan Li. "A rheological-based printability assessment method for 3D printing engineered cementitious composites considering fiber dispersion." *Case Studies in Construction Materials* 20 (2024).

Qiu, Minghong, Ye Qian, and Jian-Guo Dai. "Enhancing the flexural performance of concrete beams with 3D-printed UHP-SHCC permanent formwork via graded fiber volume fraction." *Composite Structures* 341 (2024a): 118211.

Qiu, Minghong, Yan Sun, and Ye Qian. "Interfacial bonding performance of 3D-printed ultra-high performance strain-hardening cementitious composites (UHP-SHCC) and cast normal concrete." *Journal of Building Engineering* 82 (2024b): 108268.

Reiter, Lex, Timothy Wangler, Ana Anton, and Robert J. Flatt. "Setting on demand for digital concrete – Principles, measurements, chemistry, validation." *Cement and Concrete Research* 132 (2020): 106047.

Roussel, Nicolas. "Rheological requirements for printable concretes." *Cement and Concrete Research* 112 (2018): 76–85.

Şahin, Hatice Gizem, Fatih Eren Akgümüş, and Ali Mardani. "Mechanical and rheological properties of fiber-reinforced 3D printable concrete; in terms of fiber content and aspect ratio." *Structural Concrete* 26, no. 2 (2025): 1597–1612.

Sanjayan, Jay Gopalakrishnan, Rasanga Jayathilakage, and Pathmanathan Rajeev. "Vibration induced active rheology control for 3D concrete printing." *Cement and Concrete Research* 140 (2021): 106293.

Soltan, Daniel G., and Victor C. Li. "A self-reinforced cementitious composite for building-scale 3D printing." *Cement and Concrete Composites* 90 (2018): 1–13.

Suiker, A. S. J. "Mechanical performance of wall structures in 3D printing processes: Theory, design tools and experiments." *International Journal of Mechanical Sciences* 137 (2018): 145–170.

Sun, Junbo, Farhad Aslani, Jenny Lu, Lining Wang, Yimiao Huang, and Guowei Ma. "Fibre-reinforced lightweight engineered cementitious composites for 3D concrete printing." *Ceramics International* 47, no. 19 (2021): 27107–27121.

Sun, Xiaoyan, Qun Wang, Hailong Wang, and Long Chen. "Influence of multi-walled nanotubes on the fresh and hardened properties of a 3D printing PVA mortar ink." *Construction and Building Materials* 247 (2020): 118590.

Sun, Xiaoyan, Jiawei Zhou, Qun Wang, Jiangpeng Shi, and Hailong Wang. "PVA fibre reinforced high-strength cementitious composite for 3D printing: Mechanical properties and durability." *Additive Manufacturing* 49 (2022): 102500.

Tapper, Rhys J., Marco L. Longana, Andrew Norton, Kevin D. Potter, and Ian Hamerton. "An evaluation of life cycle assessment and its application to the closed-loop recycling of carbon fibre reinforced polymers." *Composites Part B: Engineering* 184 (2020): 107665.

Tawfek, Abdullah M., Zhi Ge, Jian Li, Kangkang Zhang, Nengdong Jiang, Yingxuan Shao, Yifeng Ling, and Branko Šavija. "Influence of fiber orientation on the water and ions transportation of engineered cementitious composite (ECC)." *Materials* 16, no. 13 (2023): 4877.

Tay, Yi Wei Daniel, Biranchi Panda, Suvash Chandra Paul, Nisar Ahamed Noor Mohamed, Ming Jen Tan, and Kah Fai Leong. "3D printing trends in building and construction industry: a review." *Virtual and Physical Prototyping* 12, no. 3 (2017): 261–276.

Teng, Fei, Junhong Ye, Jie Yu, Heng Li, Yiwei Weng, and Viktor Mechtcherine. "Development of strain-hardening cementitious composites (SHCC) as bonding materials to enhance interlayer and flexural performance of 3D printed concrete." *Cement and Concrete Composites* 152 (2024): 105657.

Van Overmeir, Anne Linde, Stefan C. Figueiredo, Branko Šavija, Freek P. Bos, and Erik Schlangen. "Design and analyses of printable strain hardening cementitious composites with optimized particle size distribution." *Construction and Building Materials* 324 (2022): 126411.

Van Overmeir, Anne Linde, Branko Šavija, Freek P. Bos, and Erik Schlangen. "3D printable strain hardening cementitious composites (3DP-SHCC), tailoring fresh and hardened state properties." *Construction and Building Materials* 403 (2023a): 132924.

Van Overmeir, Anne Linde, Branko Šavija, Freek P. Bos, and Erik Schlangen. "Effects of 3D concrete printing phases on the mechanical performance of printable strain-hardening cementitious composites." *Buildings* 13, no. 10 (2023b): 2483.

Wang, Li, Yuanyuan Hu, Qiao Wang, and Tianlong Cui. "Shrinkage and cracking performance of PP/PVA fiber-reinforced 3D-printed mortar." *Journal of Materials in Civil Engineering* 35, no. 6 (2023): 04023132.

Wang, Shuxin, and Victor C. Li. "Engineered cementitious composites with high-volume fly ash." *ACI Materials Journal-American Concrete Institute* 104, no. 3 (2007): 233–241.

Wang, Yuting, Meng Chen, Tong Zhang, and Mingzhong Zhang. "Hardening properties and microstructure of 3D printed engineered cementitious composites based on limestone calcined clay cement." *Cement and Concrete Composites* 152 (2024): 105641.

Weng, Yiwei, Mingyang Li, Zhixin Liu, Wenxin Lao, Bing Lu, Dong Zhang, and Ming Jen Tan. "Printability and fire performance of a developed 3D printable fibre reinforced cementitious composites under elevated temperatures." *Virtual and Physical Prototyping* 14, no. 3 (2019): 284–292.

Weng, Yiwei, Bing Lu, Mingyang Li, Zhixin Liu, Ming Jen Tan, and Shunzhi Qian. "Empirical models to predict rheological properties of fiber reinforced cementitious composites for 3D printing." *Construction and Building Materials* 189 (2018): 676–685.

Wolfs, Rob J. M., Theo A. M. Salet, and Nicolas Roussel. "Filament geometry control in extrusion-based additive manufacturing of concrete: The good, the bad and the ugly." *Cement and Concrete Research* 150 (2021): 106615.

Xiao, Jianzhuang, Meiyan Bai, Yuching Wu, Zhenhua Duan, and Jifeng Qin. "Interlayer bonding strength and pore characteristics of 3D printed engineered cementitious composites (ECC)." *Journal of Building Engineering* 84 (2024): 108559.

Xu, Ling-Yu, Bo-Tao Huang, Victor C. Li, and Jian-Guo Dai. "High-strength high-ductility engineered/strain-hardening cementitious composites (ECC/SHCC) incorporating geopolymer fine aggregates." *Cement and Concrete Composites* 125 (2022a): 104296.

Xu, Nuoyan, and Ye Qian. "Effects of fiber volume fraction, fiber length, water-binder ratio, and nanoclay addition on the 3D printability of strain-hardening cementitious composites (SHCC)." *Cement and Concrete Composites* 139 (2023): 105066.

Xu, Nuoyan, Ye Qian, Jing Yu, and Christopher KY Leung. "Tensile performance of 3D-printed strain-hardening cementitious composites (SHCC) considering material parameters, nozzle size and printing pattern." *Cement and Concrete Composites* 132 (2022b): 104601.

Yan, Kang-Tai, Ling-Zhi Li, Jun-Hong Ye, Dilshod Bazarov, Bo-Yu Deng, and Ke-Quan Yu. "Anisotropic size effect of 3D printed LC3-based engineered cementitious composites (LC3-ECC)." *Journal of Building Engineering* 92 (2024a): 109668.

Yan, Kang-Tai, Xian-Peng Wang, Yao Ding, Ling-Zhi Li, Dilshod Bazarov, Bo-Yu Deng, Kovshar-Sergey Nikolayevich, and Ke-Quan Yu. "3D-printed LC3-based lightweight engineered cementitious composites: Fresh state, harden material properties and beam performance." *Journal of Building Engineering* 93 (2024b): 109838.

Yang, En-Hua, Mustafa Sahmaran, Yingzi Yang, and Victor C. Li. "Rheological control in production of engineered cementitious composites." *ACI Materials Journal* 106, no. 4 (2009): 357.

Ye, Huzi, Binrong Zhu, Pengxin Ping, Yuanzheng Lin, Jingming Cai, and Jinlong Pan. "Mechanical performance, impact behavior and environmental assessment of coal furnace slag based low-carbon ultra-high performance engineered cementitious composites (UHP-ECC)." *Journal of Cleaner Production* 450 (2024a): 141921.

Ye, Junhong, Can Cui, Jiangtao Yu, Kequan Yu, and Fangyuan Dong. "Effect of polyethylene fiber content on workability and mechanical-anisotropic properties of 3D printed ultra-high ductile concrete." *Construction and Building Materials* 281 (2021a): 122586.

Ye, Junhong, Can Cui, Jiangtao Yu, Kequan Yu, and Jianzhuang Xiao. "Fresh and anisotropic-mechanical properties of 3D printable ultra-high ductile concrete with crumb rubber." *Composites Part B: Engineering* 211 (2021b): 108639.

Ye, Junhong, Fei Teng, Jie Yu, Shiwei Yu, Hongjian Du, Dong Zhang, Shaoqin Ruan, and Yiwei Weng. "Development of 3D printable engineered cementitious composites with incineration bottom ash (IBA) for sustainable and digital construction." *Journal of Cleaner Production* 422 (2023): 138639.

Ye, Junhong, Minxin Yang, Jiangtao Yu, Yecheng Dai, B. B. Yin, and Yiwei Weng. "Size effect on flexural and fracture behaviors of 3D printed engineered cementitious composites: Experimental and numerical studies." *Engineering Structures* 298 (2024b): 117062.

Ye, Junhong, Jiangdi Zhang, Jie Yu, Jiangtao Yu, and Kequan Yu. "Flexural behaviors of 3D printed lightweight engineered cementitious composites (ECC) slab with hollow sections." *Engineering Structures* 299 (2024c): 117113.

Yoo, Doo-Yeol, and Nemkumar Banthia. "High-performance strain-hardening cementitious composites with tensile strain capacity exceeding 4%: A review." *Cement and Concrete Composites* 125 (2022): 104325.

Yu, Hao, Weiwei Zhang, Binbin Yin, Weikang Sun, Arslan Akbar, Yang Zhang, and K. M. Liew. "Modeling extrusion process and layer deformation in 3D concrete printing via smoothed particle hydrodynamics." *Computer Methods in Applied Mechanics and Engineering* 420 (2024a): 116761.

Yu, Jing, Yixin Chen, and Christopher KY Leung. "Mechanical performance of strain-hardening cementitious composites (SHCC) with hybrid polyvinyl alcohol and steel fibers." *Composite Structures* 226 (2019): 111198.

Yu, Jie, Fei Teng, Junhong Ye, Dong Zhang, Kequan Yu, Jiangtao Yu, Jian-Guo Dai, and Yiwei Weng. "Size-dependent model to predict the flexural strength of 3D printed ECC beams." Available at *SSRN* (2024b) *4720889*.

Yu, Kequan, Wes McGee, Tsz Yan Ng, He Zhu, and Victor C. Li. "3D-printable engineered cementitious composites (3DP-ECC): Fresh and hardened properties." *Cement and Concrete Research* 143 (2021): 106388.

Zafar, Muhammad Saeed, Amir Bakhshi, and Maryam Hojati. "Toward 3D-printable engineered cementitious composites: Mix design proportioning, flowability, and mechanical performance." In *Tran-SET 2022*, pp. 102–114. Reston, VA: American Society of Civil Engineers, 2022.

Zafar, Muhammad Saeed, Amir Bakhshi, and Maryam Hojati. "Printability and shape fidelity evaluation of self-reinforced engineered cementitious composites." *Construction and Building Materials* 408 (2023): 133676.

Zhaidarbek, Balnur, Aruzhan Tleubek, Galymbek Berdibek, and Yanwei Wang. "Analytical predictions of concrete pumping: Extending the Khatib–Khayat model to Herschel–Bulkley and modified Bingham fluids." *Cement and Concrete Research* 163 (2023): 107035.

Zhang, Chao, Venkatesh Naidu Nerella, Anurag Krishna, Shen Wang, Yamei Zhang, Viktor Mechtcherine, and Nemkumar Banthia. "Mix design concepts for 3D printable concrete: A review." *Cement and Concrete Composites* 122 (2021a): 104155.

Zhang, Duo, Jing Yu, Haoliang Wu, Beata Jaworska, Brian R. Ellis, and Victor C. Li. "Discontinuous micro-fibers as intrinsic reinforcement for ductile engineered cementitious composites (ECC)." *Composites Part B: Engineering* 184 (2020): 107741.

Zhang, Hao, Liming Zhu, Fan Zhang, and Mijia Yang. "Effect of fiber content and alignment on the mechanical properties of 3D printing cementitious composites." *Materials* 14, no. 9 (2021b): 2223.

Zhang, Yi, Qiang Ren, Xiaodi Dai, Yaxin Tao, Yiyuan Zhang, Zhengwu Jiang, Kim Van Tittelboom, and Geert De Schutter. "A potential active rheology control approach for 3D printable cement-based materials: Coupling of temperature and viscosity modifiers." *Cement and Concrete Composites* 149 (2024): 105496.

Zhang, Yifan, and Farhad Aslani. "Development of fibre reinforced engineered cementitious composite using polyvinyl alcohol fibre and activated carbon powder for 3D concrete printing." *Construction and Building Materials* 303 (2021): 124453.

Zhou, Wen. "3D printing of engineered cementitious composites (3DP-ECC)." PhD diss., 2023a.

Zhou, Wen, Wes McGee, H. Süleyman Gökçe, and Victor C. Li. "A bio-inspired solution to alleviate anisotropy of 3D printed engineered cementitious composites (3DP-ECC): Knitting/tilting filaments." *Automation in Construction* 155 (2023b): 105051.

Zhou, Wen, Wes McGee, He Zhu, H. Süleyman Gökçe, and Victor C. Li. "Time-dependent fresh properties characterization of 3D printing engineered cementitious composites (3DP-ECC): On the evaluation of buildability." *Cement and Concrete Composites* 133 (2022a): 104704.

Zhou, Wen, Yamei Zhang, Lei Ma, and Victor C. Li. "Influence of printing parameters on 3D printing engineered cementitious composites (3DP-ECC)." *Cement and Concrete Composites* 130 (2022b): 104562.

Zhou, Wen, He Zhu, Wei-Hsiu Hu, Ryan Wollaston, and Victor C. Li. "Low-carbon, expansive engineered cementitious composites (ECC) in the context of 3D printing." *Cement and Concrete Composites* 148 (2024): 105473.

Zhu, Binrong, Jinlong Pan, Junrui Li, Penghui Wang, and Mingzhong Zhang. "Relationship between microstructure and strain-hardening behaviour of 3D printed engineered cementitious composites." *Cement and Concrete Composites* 133 (2022a): 104677.

Zhu, Binrong, Jinlong Pan, Behzad Nematollahi, Zhenxin Zhou, Yang Zhang, and Jay Sanjayan. "Development of 3D printable engineered cementitious composites with ultra-high tensile ductility for digital construction." *Materials & Design* 181 (2019): 108088.

Zhu, Binrong, Jinlong Pan, Mingzhong Zhang, and Christopher K. Y. Leung. "Predicting the strain-hardening behaviour of polyethylene fibre reinforced engineered cementitious composites accounting for fibre-matrix interaction." *Cement and Concrete Composites* 134 (2022b): 104770.

Zhu, Binrong, Jinlong Pan, Zhenxin Zhou, and Jingming Cai. "Mechanical properties of engineered cementitious composites beams fabricated by extrusion-based 3D printing." *Engineering Structures* 238 (2021a): 112201.

Zhu, Binrong, Yang Wei, Hongyan Chu, Huzi Ye, Jingming Cai, and Jinlong Pan. "Dynamic response of ultra-high performance engineered cementitious composites (UHP-ECC) under low-velocity impact: Effect of waste rubber incorporation and low temperatures." *Cement and Concrete Composites* 151 (2024): 105576.

Zhu, He, Kequan Yu, Wesley McGee, Tsz Yan Ng, and Victor C. Li. "Limestone calcined clay cement for three-dimensional-printed engineered cementitious composites." *ACI Materials Journal* 118, no. 6 (2021b): 111–122.

Zhu, LingLi, Jie Yao, Yu Zhao, Wenqiang Ruan, Yali Ge, and XueMao Guan. "Printability and early mechanical properties of material composition modified 3D printing engineered cementitious composites based on the response surface methodology." *Journal of Building Engineering* 79 (2023a): 107837.

Zhu, LingLi, Jie Yao, Yu Zhao, Wenqiang Ruan, Guang Yang, and XueMao Guan. "Effects of composite cementation system on rheological and working performances of fresh 3D-printable engineered cementitious composites." *Journal of Building Engineering* 65 (2023b): 105801.

10 3D-Printable Earth-Based Materials

Maryam Hojati and Muhammad Saeed Zafar

10.1 INTRODUCTION: EMERGENCE OF EARTHEN MATERIALS IN 3D PRINTING

The construction industry is undergoing a technological and philosophical transformation, spurred by the emergence of 3D printing, also known as additive manufacturing (AM). This digital fabrication method enables the layer-by-layer construction of structures directly from computer models, promising to revolutionize traditional building practices. The potential advantages are numerous: improved design flexibility, reduced material waste, faster construction times, and the possibility of fully automated building processes (Dilberoglu et al. 2017; Wangler et al. 2016; Hojati et al. 2022). These capabilities are particularly attractive in the context of global challenges such as labor shortages, housing crises, and the need for infrastructure in remote or rapidly developing regions (Wangler et al. 2019; Nazarian et al. 2021).

Despite the promise of 3D printing, its current material practices raise significant concerns regarding sustainability. Most printable mixtures used today are based on cementitious systems, and due to limitations in extrusion and pumpability, they exclude coarse aggregates. This constraint results in a disproportionate reliance on cement or other cement-rich compositions, which not only drives up costs (commercial 3D-printed (3DP) concrete is typically 5–8 times more expensive than conventional concrete) but also aggravates environmental impacts (Tinoco et al. 2022; Batikha et al. 2022). Embodied carbon of these materials is one to two times higher than that of standard concrete, further undermining the sustainability claims of construction-scale 3D printing (Bhattacherjee et al. 2021). Given that ordinary portland cement (OPC) production alone accounts for nearly 8% of global CO_2 emissions, the reliance on high-cement-content formulations risks replicating the same resource-intensive patterns of the past, even as the technology itself advances (Miller et al. 2018).

This growing awareness has led researchers to explore alternative binder systems that are more environmentally responsible. Among these, geopolymers, cement-free, aluminosilicate-based binders activated by alkaline solutions, have gained traction for their lower carbon footprint and comparable or superior mechanical performance (Ricciotti et al. 2023; Shilar et al. 2023; Hojati et al. 2018). However, even geopolymers often rely on industrial by-products like fly ash

DOI: 10.1201/9781003434863-12

and slag, which are regionally limited and subject to supply chain volatility. This has prompted a renewed interest in more abundant, naturally occurring materials such as earth.

Earthen materials, including clay, silt, sand, and combinations, offer a compelling alternative that aligns with environmental and cultural goals (Zhai and Previtali 2010). Used in vernacular architecture for millennia, earth is one of the oldest and most widely available construction materials on the planet (Bajpayee et al. 2020; McNally 1998; Balsam, Deaton, and Adler 2007). Its thermal mass, breathability, and low embodied energy make it attractive for passive and climate-adapted building design. When paired with digital construction technologies like 3D printing, earth-based materials can be reimagined for the future by offering low-carbon, site-responsive, and scalable solutions for sustainable development (Zozaya et al. 2024, 2026; Alqenaee, Memari, and Hojati 2022; Bhusal 2023; Zozaya 2024).

The re-emergence of earthen construction through new construction technology such as 3D printing also opens exciting new possibilities for circularity, localization, and resilience. Earthen materials can often be harvested directly from construction sites or surrounding areas, reducing the need for long-distance transport and minimizing environmental disruption (Faleschini et al. 2023; Curth et al. 2024; Bhusal, Uviña, and Hojati 2022; Bhusal, Sedghi, and Hojati 2023). Their compatibility with natural cycles and potential for reuse or return to the earth at the end of a structure's life make them ideal candidates for regenerative construction strategies. Furthermore, using locally sourced earthen materials fosters community engagement and promotes knowledge exchange between modern engineers and traditional builders.

As global construction demands increase, especially in regions facing climate vulnerability, housing shortages, or infrastructure gaps, the need for sustainable, scalable, and adaptable building materials is more urgent than ever. Integrating earth-based materials with 3D printing presents a rare opportunity to bridge the past and the future: to draw from traditional wisdom while leveraging cutting-edge technology. This chapter explores how earthen materials can be designed, processed, and deployed through 3D printing to redefine sustainable construction in the 21st century and beyond.

This chapter explores the growing potential of new construction technologies (such as 3D printing) in using earth-based materials for construction. It begins by examining the historical context of traditional earthen construction methods and their relevance to modern sustainable practices. The discussion then transitions to the fundamental principles of 3D printing in the built environment, highlighting how these technologies are being adapted for nonconventional materials like soil. Key considerations in material selection and characterization are addressed, with a focus on understanding the properties that influence printability and structural performance. The chapter also covers strategies for mix design, including the incorporation of natural or engineered additives to enhance workability and strength. Process parameters such as layer deposition rate, moisture content, and shape stability are analyzed for their impact on the overall success of printed structures.

10.2 HISTORICAL CONTEXT AND TRADITIONAL EARTH CONSTRUCTION

Archaeological evidence demonstrates that humanity's transition from nomadic hunter-gatherer societies to settled agricultural communities during the Neolithic Revolution (10,000–3,000 BC) (Zeiher 1996; P. Jaquin 2012) catalyzed the first widespread use of earthen construction. As early civilizations emerged across fertile regions, they independently developed sophisticated soil-based building techniques adapted to local materials and climates (P. A. Jaquin and Augarde 2012).

Some of the earliest known examples originate from the Fertile Crescent, where the proto-city of Jericho (8300 BC) in modern Palestine featured circular dwellings of compacted earth and primitive adobe bricks (Sir Banister Fletcher and Cruickshank 1996). Concurrently, ancient Mesopotamian civilizations in present-day Iraq and Iran constructed monumental ziggurats using sun-dried mudbricks, while the Elamite civilization in southwestern Iran (3400–550 BC) developed advanced rammed earth techniques evident in the ruins of Chogha Zanbil (Beek and Beek 2013; Crawford 2017). Parallel developments occurred across ancient civilizations:

- Chinese cultures along the Yellow River perfected rammed earth (hangtu) construction as early as 5000 BC (Xie et al. 2021).
- The Indus Valley civilization (3300–1300 BC) in modern Pakistan/India created standardized fired bricks for urban centers like Mohenjo-Daro (Jarrige and Meadow 1980).
- Nubian civilizations in northeast Africa developed vaulted mudbrick architecture requiring no formwork (Bats and Licitra 2023).
- Ancient Egyptians (circa 3000 BC–500 AD) used sun-dried and fired mudbricks extensively in residential, administrative, and funerary architecture, including large-scale complexes and storage facilities (Spencer 1979).
- The Ancestral Puebloans (700–1600 AD) of the American Southwest built multistory adobe complexes like Pueblo Bonito using local soils and stone masonry (Paul Dana and Brent Anderson 2007).

In the Near East, Persian architecture notably advanced earthen construction through the following:

- The Achaemenid Empire's (550–330 BC) use of mudbrick for monumental structures at Persepolis (Oudbashi, Shekofteh, and Fadaei 2023);
- Sassanian innovations in clay mortar and vaulting techniques (Oates 1990);
- Vernacular earth architecture that persists in Iranian villages today (Azizi-Bondarabadi and Haji Sadeghi 2022).

The Spanish colonial period (1598–1848 AD) in New Mexico later synthesized European and Indigenous Pueblo adobe techniques (Smith and Austin 1989), while across the Americas, civilizations from the Mississippi Mound Builders to the Inca demonstrated remarkable adaptability in earth construction. These diverse traditions collectively represent humanity's longest-running architectural experiment – one that continues to inform sustainable building practices today. This section presents a classification of the historical use of earth as a construction material, based on archaeological findings and architectural research. The classification is organized according to material composition and construction methodology, distinguishing between monolithic and unit-based techniques. Various forms within each category are discussed.

10.2.1 Monolithic Earth Construction Techniques

Monolithic earth construction encompasses building techniques in which structures are formed as continuous, unified masses of earthen material, rather than being assembled from discrete, modular units such as bricks or blocks. These methods utilize the cohesive and plastic properties of natural soils, often stabilized through moisture content or additives, to create seamless, load-bearing elements. This section provides a detailed examination of the principal monolithic construction techniques, highlighting their historical significance, material formulations, and structural performance characteristics. A summary of these methods is listed in Table 10.1.

10.2.1.1 Cob (Stacked Earth)

Cob is a monolithic earthen construction technique that uses a hand-applied mixture of clay-rich subsoil, sand, straw, and water to form solid, load-bearing walls without the need for formwork. This sculptural method allows builders to shape organic forms and integrate architectural features directly into the wall mass during construction. The material composition is carefully balanced: clay (15%–30%) provides cohesion, sand (50%–70%) minimizes shrinkage, and straw (5%–10%) adds tensile strength, with water adjusted for workability. Mixed manually or mechanically to a dough-like consistency, the cob is applied in lifts of 30–60 cm, each left to stiffen before the next is added. Once cured over several weeks, the walls gain structural integrity through slow drying (Haddad, Lannon, and Latif 2024; Gounni and Louahlia 2020; Hamard et al. 2016; Dente, Smith, and Burke 2024).

Historically, cob has been used across continents, from medieval cottages in England to the tower homes of Yemen and adobe-like structures in the American Southwest, demonstrating its adaptability to local materials and climates. Technically, cob offers moderate structural performance: compressive strength ranges from 0.1 to 2 MPa, tensile strength from 0.01 to 0.75 MPa, and shear strength between 0.4 and 0.6 MPa. Its thermal conductivity varies from 0.12 to 1.06 W/m·K, while its high moisture buffering capacity (1.06–1.74 $g/m^2 \cdot \%RH$) supports comfortable indoor environments. Cob construction is labor-intensive and requires moisture protection, but it remains a low-cost, low-carbon method well-suited for sustainable, small-scale architecture (Haddad, Lannon, and Latif 2024).

10.2.1.2 Rammed Earth (Pisé de Terre)

Rammed earth (Pisé de Terre) is a monolithic construction method where damp soil is compacted in temporary formwork to create thick, load-bearing walls. The mixture typically consists of 15%–20% clay as a binder and 60%–75% sand and gravel

TABLE 10.1
Comparison of Cob, Puddled Mud, and Rammed Earth Construction Techniques

Aspect	Cob	Rammed Earth	Puddled Mud
Material composition	- Clay (15%–30%) - Sand (50%–70%) - Straw (5%–10%) - Water (15%–25%)	- Clay (15%–20%) - Sand/gravel (60%–75%) - Stabilizers (optional: cement, lime, and fibers)	- Clay (30i–50%) - Minimal sand - Straw/manure (optional) - High water content
Construction process	- Hand-mixed to stiff consistency - Built in 12–24" lifts - No formwork - Slow drying (days between layers)	- Soil compacted in formwork (6–8" layers) - Pneumatic/hand tamping - Fast curing (days)	- Wet, slurry-like mix - Poured/puddled in 12–18" layers - No formwork - Very slow drying (months)
Structural performance	- Moderate compressive strength (1–2 MPa) - Good thermal mass - Vulnerable to water erosion	- High compressive strength (2–5 MPa) - Excellent durability (if stabilized) - Fireproof	- Low strength (erodes easily) - High shrinkage cracks - Excellent thermal mass
Durability	- Requires protective plaster/roof - Susceptible to rain erosion - Needs maintenance	- Resistant to weathering (if stabilized) - Long lifespan (centuries in dry climates)	- Highly erosion-prone - Requires thick protective coatings - Short lifespan in wet climates
Thermal properties	- High thermal mass (slow heat transfer) - Passive temperature regulation	- Excellent thermal mass - Reduces energy needs for heating/cooling	- High thermal mass - Effective in arid climates
Historical use	- UK (Devon cottages) - Yemen (Shibam) - American Southwest (Taos Pueblo)	- China (Great Wall) - North Africa (Carthage) - Modern sustainable builds	- Iran (Yazd) - American Southwest (Pueblo pit houses)
Modern applications	- Eco-homes (Oregon Cob) - Artistic/sculptural walls	- Contemporary sustainable housing - 3DP earth projects	- Rare (limited to restoration/archaeology)
Advantages	- Sculptural flexibility - Low-tech (no tools needed) - Low cost	- High strength - Fast construction - Code-compliant (if stabilized)	- Ultra-simple (no tools) - Good for temporary structures
Disadvantages	- Labor-intensive - Slow drying - Not code-approved in many regions	- Requires formwork/skill - Stabilizers increase carbon footprint	- Extremely slow drying - Weak against rain/wind

for structural integrity. Stabilizers like cement or lime (5%–10%) are sometimes added to improve strength and moisture resistance. Construction involves placing soil in 15–20 cm layers and compacting each layer to about half its height using manual or mechanical rammers. As layers accumulate, the formwork is moved upward, and walls are left to cure gradually over several weeks (Ciancio and Beckett 2015; Canivell et al. 2020; Laborel-Préneron et al. 2016).

This ancient method has been used in structures such as parts of the Great Wall of China, North African fortifications, and Spain's Alhambra. Today, it is experiencing a revival in sustainable architecture for its low embodied energy, fire resistance, and high thermal mass. Unstabilized rammed earth can reach compressive strengths of 2–5 MPa, while stabilized variants may exceed 8 MPa (Canivell et al. 2020; Thompson, Augarde, and Osorio 2022; Ávila, Puertas, and Gallego, 2021). However, its low tensile strength and ductility limit its use in seismic zones unless reinforced. The technique also demands skilled labor, precise moisture control, and formwork setup, which can increase construction time and complexity. Despite these challenges, rammed earth remains a durable, eco-friendly solution for low- to mid-rise structures (Canivell et al. 2020; Thompson, Augarde, and Osorio 2022; Ávila, Puertas, and Gallego, 2021).

10.2.1.3 Puddled Mud Construction (Chineh or Chunam)

Puddled mud is one of the oldest forms of earthen construction, involving the hand-shaping of wet, clay-rich soil into thick, load-bearing walls without formwork. Common in arid regions like Yazd in Iran and ancestral Puebloan sites in the American Southwest, this technique relies on high clay content (30%–50%) for plasticity, with sand and organic fibers like straw or manure added to enhance workability and reduce cracking. The mud is applied in successive lifts (30–45 cm), shaped while wet, and left to air-dry over weeks or months. Although labor-intensive and slow to cure, especially in humid climates, puddled mud offers excellent thermal mass and remains a sustainable, low-tech building method when properly protected from moisture (Kita et al. 2024).

10.2.2 Unit-Based Techniques (Modular Components)

Unlike monolithic earth construction methods such as cob and rammed earth, which involve shaping continuous masses of material directly on site, unit-based construction relies on assembling prefabricated components like bricks, blocks, or sod units to form structural walls. This modular approach offers several benefits, including faster construction, since units can be produced and cured in advance, and greater design flexibility, allowing for easier incorporation of openings, corners, and architectural features. Additionally, damaged sections can be repaired by replacing individual units, and standardized dimensions can improve alignment and overall structural stability. However, unlike monolithic techniques that depend on the natural cohesion of the material, unit-based methods typically require binding agents such as mortar or stabilizers to ensure durability and structural integrity. The following sections explore three key techniques that exemplify this approach. A summary of these methods is presented in Table 10.2.

TABLE 10.2
Comparison of Cob, Puddled Mud, and Rammed Earth Construction Techniques

Aspect	Adobe (Sun-Dried Bricks)	Compressed Earth Blocks	Cut Sod/Turf Blocks (Terrón)
Material composition	- Clay (15%–30%) - Sand (50%–70%) - Straw (5%–10%) - Water (15%–25%)	- Clay (10%–30%) - Sand/gravel (60%–80%) - Stabilizers (5%–10% cement/lime)	- Clay-rich soil (30%–50%) - Grass roots (natural reinforcement) - Organic matter
Production method	- Hand-mixed, sun-dried in molds - 3–4 weeks of curing	- Mechanically pressed (manual/hydraulic) - 7–14 days of curing	- Cut from grasslands/riverbanks - 1–2 weeks of air-drying
Compressive strength	1–2 MPa	3–10 MPa (with stabilizers)	0.5–1.5 Mpa
Thermal properties	- High thermal mass (0.3 BTU/hr·ft·°F) - Slow heat transfer	- Similar to adobe but denser - Improved with fiber additives	- Superior insulation (R–2.5 per inch) - Grass enhances thermal resistance
Water resistance	- Poor; requires plaster/roof protection	- Moderate (with stabilizers) - Lime/cement reduces erosion	- Highly vulnerable; needs thick coatings
Historical use	- Mesopotamia; Puebloan, SW USA; Yemen	- Post-1950s (CINVA-Ram) - Modern eco-housing (Auroville)	- Icelandic turf houses, Rio Grande Valley missions
Modern applications	- Heritage restoration - Eco-homes in arid climates	- Affordable housing - Schools (UN-Habitat projects)	- Green roofs - Niche eco-builds
Advantages	- Low-tech, sculptural - Excellent thermal mass	- Code-compliant, faster than adobe - Scalable production	- Ultra-local materials - High insulation
Limitations	- Slow drying, labor-intensive - Weak in earthquakes	- Equipment/stabilizers increase cost - Skill-dependent	- Short lifespan in rain - Irregular shapes complicate building

10.2.2.1 Adobe (Sun-Dried Mudbricks)

Adobe is one of the oldest and most enduring unit-based earthen construction techniques, with a history of over 10,000 years (Patrice Doat et al. 1991). These sun-dried mudbricks are composed primarily of clay (15%–30%) to provide plasticity and cohesion, sand (50%–70%) to minimize shrinkage and enhance stability (Hohn 2011), and straw or other organic fibers (5%–10%) to add tensile reinforcement and promote even drying (Brinson 2023). Water is incorporated to achieve a stiff, moldable consistency, typically around 15%–25% by volume. In some regions, traditional additives like manure or asphalt are used for waterproofing (Patrice Doat et al. 1991), while optional stabilizers such as lime or cement (5%–10%) can enhance water resistance at the cost of increased embodied energy (Downton and Clarke 2013). Structurally, adobe demonstrates compressive strength values ranging from 1 to 2 MPa, sufficient for code compliance when properly mixed (Hohn 2011), but it has limited tensile capacity, making it prone to cracking without reinforcement. Due to its low ductility, adobe is also vulnerable in seismic zones unless augmented with bamboo, geomesh, or timber frameworks (Wang and Ng 2023). Thermally, adobe walls exhibit high thermal mass, which helps buffer temperature swings and maintain interior comfort in arid climates (Downton and Clarke 2013). Moreover, adobe has a low embodied energy profile, as it requires no kiln-firing and can be produced using local materials with minimal processing (Ramos Huarachi et al. 2020). Its breathable clay composition supports indoor humidity regulation and improved air quality. Despite these strengths, adobe is sensitive to moisture: unprotected bricks degrade in heavy rain, and freeze–thaw cycles in colder climates can cause structural failure without stabilization. The construction process involves mixing the raw ingredients by foot or machine, molding the bricks in wooden forms (typically 25 × 36 × 10 cm), sun-drying them for 3–4 weeks (Brinson 2023), and stacking them with mud mortar in staggered patterns for added stability. Modern innovations include stabilizers like cement or lime (Communities_of_Architects 2023) and tools such as the CINVA-Ram press for higher-density bricks and faster drying (Bourgeois 2021).

10.2.2.2 Compressed Earth Blocks

Compressed earth blocks (CEBs) are a mechanized evolution of traditional adobe, designed for improved structural integrity and efficiency. Unlike sun-dried bricks, CEBs are compacted under high pressure using mechanical presses, resulting in dense, uniform blocks (Riza and Rahman 2015). The typical material mix includes 10%–30% clay for cohesion (Riza and Rahman 2015), 60%–80% sand or gravel for strength and shrinkage control (Valenzuela et al. 2024), and 10%–15% water for optimal compaction (Egenti and Khatib 2016). Optional stabilizers, such as 5%–10% cement or lime, enhance water resistance and durability in wet or freeze-prone environments (Egenti and Khatib 2016), while polymers or bitumen may be used for added waterproofing (Valenzuela et al. 2024). Ideal soils contain 15%–30% clay, 50%–70% sand, and 10%–20% silt, with basic field tests (e.g., ribbon or drop test) guiding mix suitability (Valenzuela et al. 2024; Riza and Rahman 2015).

CEBs offer high compressive strength (3–10 MPa) (Riza and Rahman 2015) and moderate tensile capacity (0.5–1.5 MPa), making them suitable for structural and even multistory applications with reinforcement. Thermally, their high mass helps regulate indoor temperatures, while their breathable nature aids humidity control (Giuffrida et al. 2024; Shoukat et al. 2025). CEBs also have a 50%–70% lower carbon footprint compared to fired bricks (Giuffrida et al. 2024). Stabilized CEBs can last over 50 years with minimal maintenance (Bailly et al. 2024), although unstabilized ones require protective coatings in wet climates. In cold regions, cement stabilization is essential to resist freeze–thaw cycles (Bailly et al. 2024), making CEBs a durable and sustainable option when used appropriately.

10.2.2.3 Cut Sod/Turf Blocks (Terrón)

Terrón, or cut sod blocks, is a traditional unit-based earthen construction method using naturally cohesive, grass-root-reinforced soil. Unlike manufactured adobe or CEBs, terrón blocks are directly harvested from riverbanks or grasslands, requiring minimal processing (Caporuscio 2024). These blocks typically consist of 30%–50% clay-rich soil for cohesion, grass roots for tensile reinforcement, and decomposed organic matter for insulation and durability (Caporuscio 2024).

Structurally, terrón has a compressive strength of 0.5–1.5 MPa, lower than adobe or CEBs due to irregularity, and offers limited seismic resistance without reinforcement (Caporuscio 2024). However, it provides superior insulation and extremely low embodied energy since it requires no firing or mechanical pressing. Biodegradability is another advantage, making it an eco-friendly material. The blocks are air-cured for 1–2 weeks and laid using mud mortar or dry stacking (Caporuscio 2024). Due to susceptibility to water erosion, they need protective finishes like lime plaster. In wet climates, their lifespan is limited to 10–30 years unless stabilized with cement or lime (Caporuscio 2024).

Historically, terrón was used in Puebloan architecture in New Mexico and turf houses in Iceland, with examples like Taos Pueblo and UNESCO-listed Icelandic turf homes demonstrating its cultural and functional value (Caporuscio 2024). Today, it remains relevant in green building, particularly for green roofs and stabilized sod applications (Caporuscio 2024; Artisan-Outdoor 2025).

10.3 PRINCIPLES OF 3D PRINTING AND AUTOMATION FOR EARTH-BASED CONSTRUCTION

3D printing is at the forefront of a technological revolution in the construction industry and is an integral part of Construction 4.0. This method departs fundamentally from conventional subtractive or form-based building practices by enabling the layer-by-layer deposition of material directly from digital models. In the context of earth-based construction, this approach introduces unprecedented opportunities for design customization, material efficiency, and on-site/off-site automation.

This section presents three potential approaches to introducing new technologies into traditional earthen construction, aimed at enhancing sustainability, promoting circular economy principles, and increasing automation in residential building practices.

10.3.1 In Situ 3D Printing with Local Earth

In situ 3D printing with locally sourced earthen materials or excavated earth represents a paradigm shift in sustainable construction, offering a decentralized, resource-efficient alternative to conventional building practices. By directly utilizing native soils at or near the construction site, this approach minimizes the environmental impact of conventional building materials' extraction, transportation, and processing. The whole process, starting from material processing to the 3D printing of actual geometry displayed in Figure 10.1, will be reviewed in the next sections

FIGURE 10.1 3D printing process of earthen material: from material processing to the final 3DP geometry.

of this chapter. Notably, projects like WASP's *Gaia House* (Alberto Chiusol 2018) have demonstrated the technical feasibility and ecological benefits of this method by incorporating a mix of 75% on-site soil, combined with 15% rice husk and 10% lime as stabilizers, thus forming a fully biocompatible and low-carbon composite (Chiusol 2018). This form of material localization directly supports circular economy principles and significantly reduces the embodied carbon footprint of building systems (Curth et al. 2024).

Material extrusion, as classified under the ISO/ASTM 52900 standard (ISO/ASTM 52900 2015), is currently the most viable 3D printing process for earth-based construction due to its compatibility with wet clay, adobe, and other fine-grained composites. In this method, a prepared earthen mixture is extruded in a continuous bead through a nozzle mounted on a gantry or robotic arms, enabling precise layer-by-layer deposition of the structure. Unlike traditional masonry systems, such as CEBs or rammed earth, that depend on manual labor, modular units, or fixed formwork, 3D printing enables seamless fabrication of monolithic walls with both rectilinear and curvilinear geometries, unconstrained by conventional formwork limitations (Curth et al. 2024; Gomaa, Jabi, et al. 2021).

One of the most transformative advantages of 3D printing in the fabrication of in situ earth-based construction is its capacity to produce structurally optimized curvilinear/rectilinear geometries, forging a link between time-tested architectural wisdom and modern digital fabrication. Historical examples demonstrate how ancient builders intuitively employed form to enhance durability and performance. The rectilinear, trapezoidal tower houses of Shibam, Yemen (16th century) – often called the "Manhattan of the Desert" – achieve remarkable resilience through their straight-edged, tapered forms that efficiently manage structural loads in harsh desert climates (Lemmin-Woolfrey 2021). For fully curvilinear examples, the Nubian vaults of Egypt and Sudan demonstrate perfect parabolic arches that channel compressive forces efficiently (Lemmin-Woolfrey 2021; Ranjan 2020; Borrero 2024), while the cob cottages of Devon, England (15th century), showcase completely organic, curved walls that eliminate stress concentrations (Jones 2012). The Taos Pueblo in the American Southwest (c. 1200 CE) presents an intermediate case – primarily rectilinear in form but with intentionally rounded corners and occasional curved walls that mitigate environmental stresses (Reynolds 1981). Similarly, the Trulli of Puglia, Italy (pre-16th century), employ conical dry-stone roofs with finely tuned curvature to enhance rainwater runoff and maintain thermal stability in variable Mediterranean climates

(Beguin, Pioline, and Rienzi 2023). These precedents collectively validate how both pure curvilinear and hybrid rectilinear-curved geometries, when properly executed, can optimize earthen material performance.

Additionally, digital fabrication offers an unprecedented improvement in material efficiency. Traditional masonry techniques often result in up to 20%–30% material losses due to breakage, trimming, and handling. In contrast, digitally controlled extrusion achieves approximately 98% material utilization, as the deposition path is precisely defined through G-code programming, minimizing overuse and rework (Curth et al. 2024). This efficiency becomes particularly advantageous when working with earth mixtures that are sourced and processed on-site, as it ensures maximum output from limited input volumes.

A compelling demonstration of these capabilities is the *TECLA House* in Italy (Samarasinghe 2024), developed by Mario Cucinella Architects in partnership with WASP. The structure employed a site-optimized mix composed of 30% clay, 40% silt, and 30% sand, stabilized with lime and rice husk (Samarasinghe 2024). Fabricated using a crane-mounted dual-arm 3D printer, the house was completed in approximately 200 printing hours. It showcases both the scalability and architectural potential of in situ 3D printing, producing doubly curved, load-bearing walls with integrated thermal mass and minimal environmental impact.

10.3.2 Integration with Automated CEB Production

The integration of CEBs with 3D printing technologies represents a promising hybrid approach that combines the efficiency of mechanized unit production with the geometric flexibility of 3D printing. While traditional CEB production relies on hydraulic presses (e.g., CINVA-Ram or Appropriate Earth Construction Technologies (AECT) machines) to create standardized blocks at rates of 240–2,000 units/day (Runge 2024), recent advancements explore synergies with digital fabrication. For instance, projects like Egypt's Ecofordable House have demonstrated the potential of combining machine-pressed interlocking CEBs with digitally controlled mortar deposition for joints, enhancing structural continuity while reducing manual labor (Curth et al. 2024; Ahmad, Jabi, and Sosa 2024). However, fully digitally fabricated CEBs, where robotic systems handle both compaction and placement, remain experimental, with research focusing on automated soil grading, robotic stacking, and 3DP mortar infill (Ahmad, Jabi, and Sosa 2024; Puzatova et al. 2022).

Material compatibility is a key advantage: CEBs (with 3–10 MPa compressive strength when stabilized) complement the lower strength of in situ 3DP earth (1–3 MPa), enabling load-bearing hybrid walls (Curth et al. 2024; Cruz et al. 2024). Innovations like lime-stabilized earth mortars or geopolymer binders further improve water resistance and durability, addressing a major limitation of unstabilized earthen materials (Cruz et al. 2024). Despite progress, challenges persist, including the need for standardized reinforcement methods (e.g., integrating natural fibers or steel rebars) and scalable robotic systems for on-site CEB assembly (Puzatova et al. 2022).

This hybrid approach aligns with sustainable construction goals by minimizing material waste and transportation emissions, as seen in projects like Nubian Vaults, where CEBs are paired with lime mortar for rapid, low-carbon construction

(Curth et al. 2024; Ahmad, Jabi, and Sosa 2024). Future directions include fully automated CEB production lines and artificial intelligence (AI)-driven quality control to optimize soil mixtures and block performance (Runge 2024; Cruz et al. 2024), and CEB blocks will then be assembled using an automated robotic construction system.

10.3.3 Robotic Automation in Rammed Earth Construction

Recent advancements in robotic rammed earth fabrication have demonstrated significant potential to transform traditional construction methods through automation. Researchers at Technische Universität Braunschweig have developed a robotic system combining a six-axis robotic arm with a mobile slipform and pneumatic rammers, enabling precise layer-by-layer compaction of earth without the need for fixed formwork. This method reduces formwork-related labor by 60%–70% while maintaining consistent material density and structural integrity (Schmitz et al. 2024; Gosslar 2023; Ing. Harald Kloft 2019). The system optimizes compaction energy by adjusting layer height (typically 5–10 cm, vs. traditional 15 cm) and ramming frequency, achieving compressive strengths of 2–4 MPa, comparable to conventional rammed earth but with higher reproducibility (Schmitz et al. 2024; Eksi Akbulut and Naz Alibeyoglu 2025). Projects like the GOLEHM initiative further explore mobile robotic units for on-site construction using locally sourced soil, eliminating transportation emissions (Gosslar 2023). However, challenges remain, including material scattering during robotic compaction and the need for real-time moisture content monitoring to ensure optimal compaction, and a detailed exploration of this topic will be presented in the future work section.

10.4 MATERIAL SELECTION AND CHARACTERIZATION FOR AUTOMATED EARTH-BASED CONSTRUCTION

As mentioned earlier, various traditional techniques are used to exploit soil as a building material, such as cob, CEB, and rammed earth (McNally 1998; Smith and Austin 1989; Bajpayee et al. 2020; Bui et al. 2016; Liuzzi and Stefanizzi 2014; Côté and Konrad 2005; Beckett et al. 2018; Hall, Najim, and Keikhaei Dehdezi 2012; Mai et al. 2024; Bhusal, Sedghi, and Hojati 2023). Recently, there has been a growing interest and attention in the field of 3D printing of soil. This innovative approach combines the principles of 3D printing technology with the use of soil as a construction material (Bhusal, Uviña, and Hojati 2022; Shiva Bhusal 2023; Arrieta-Escobar et al. 2020; Wi et al. 2020). The soil mixture for each application should be carefully prepared to achieve the right consistency and workability. The clay fraction in the soil is essential as it provides cohesion, binding properties, and formability to the soil. When it comes to new buildings, supplementary hydraulic binders, like lime or cement, or both, may be added, particularly when the soil does not contain enough clay or when the type of clay is inappropriate for the selected earth construction technique. This section examines the material specifications necessary for 3D printing applications, with particular emphasis on optimizing locally available resources for this fabrication.

10.4.1 Material Selection and Sourcing

One of the critical stages in earthen construction is selecting suitable material, processing, and characterization, mainly because the properties of local soils and resources tend to vary significantly. It is crucial to identify the suitable layers of the soil profile for construction (Figure 10.2) (Sapriza-Azuri et al. 2018; Natural Resources Conservation Service 2025). The soil profile, a vertical cross section, includes the O layer (organic) and A layer (topsoil), both unsuitable for construction due to high organic content. The B layer (subsoil) is suitable for its mineral content and stability. The C layer (parent material) and R layer (bedrock) are generally not recommended. In summary, after removing organic matter, the B and sometimes the A layer can be used for earth-based construction mixes. Local soil varies based on location and environmental conditions, generally consisting of mineral particles, organic matter, water, and air. The mineral particles, which include sand, silt, and clay, differ in size and texture (Meyer and Read 2015; Murthy 2002). Sand particles, the largest, range from 0.05 to 2 mm in diameter and are visible to the naked eye. Silt particles are smaller, ranging from 0.002 to 0.075 mm, and have a smooth, flour-like texture. Clay particles, the smallest, measure less than 0.002 mm and have a sticky, plastic-like texture when wet. By examining the size and texture of these particles, we can differentiate among sand, silt, and clay in the soil (Murthy 2002).

For construction, both the fraction and type of clay are vital. Clay fraction refers to the proportion of clay particles within a soil sample, typically expressed as a percentage, which impacts its strength, stability, and workability. An optimal clay

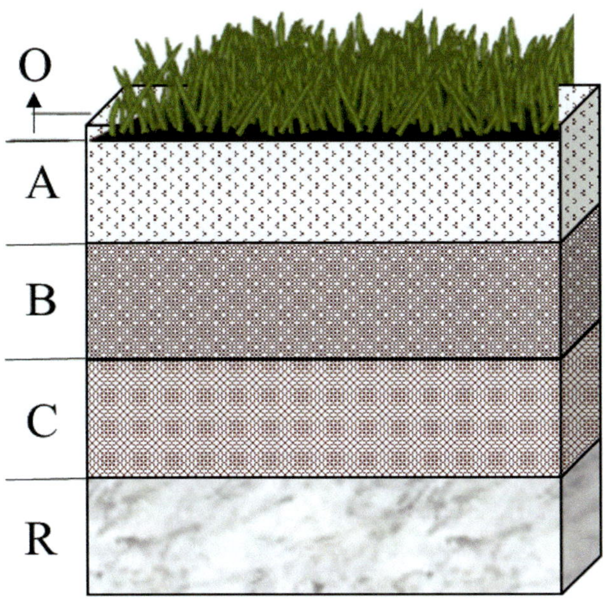

FIGURE 10.2 Soil profile: O (organic), A (topsoil), B (subsoil), C (parent material), and R (bedrock).

fraction for earthen construction is presented in Tables 10.1 and 10.2. While a higher clay fraction can enhance cohesion and compaction, it may also cause problems such as low workability, excessive shrinkage, and cracking after drying. The type of clay, determined by its mineral composition, also affects soil behavior. Clay minerals are phyllosilicates with a layered structure of tetrahedral and octahedral sheets, resulting in flat, platy particles with a high aspect ratio (Bergaya and Lagaly 2006). Approximately 15 minerals are generally classified as clay minerals and categorized into three main groups: kaolin, illite, and smectite. These groups represent the most frequently occurring soil clay minerals (Grim 1968). In addition to clay minerals, local soil typically contains non-clay minerals resulting from weathering conditions, such as quartz (SiO_2), goethite (FeO(OH)), siderite ($FeCO_3$), and calcite ($CaCO_3$). These minerals generally do not react with water.

With regard to the distance between the layers of a clay mineral, the variation in basal spacing is influenced by the quantity and nature (valence and hydration potential) of the exchangeable cations present in the interlayer spaces. Clay minerals, based on their cation exchange capacity (CEC), can be broadly categorized into two groups (Chiappone et al. 2004):

(i) Inert Clays: This group includes minerals, e.g., chlorite, illite, and kaolinite. These clays are characterized by a CEC ranging from 3 to 25 mequiv./100 g, a specific surface area of 10–100 m²/100 g, a liquid limit (LL) of 50–120, and a plasticity index (PI) of 20–60.

(ii) Active Clays: This category encompasses minerals like smectite and vermiculite, which have a significantly higher CEC of 80–150 mequiv./100 g, a specific surface area ranging from 100 to 700 m²/100 g, an LL between 100 and 700, and a PI from 80 to 600.

These classifications are based on their distinct physicochemical properties, which influence their behavior and suitability for various applications. Clay has been widely utilized in various industrial and construction applications. The three main groups of naturally occurring clays widely used in various commercial applications include bentonite, kaolin (china clay), and other clays (Willett 2022).

(i) Bentonite: Bentonite is primarily composed of montmorillonite, a smectite group clay mineral formed from the devitrification of volcanic ash (Brindley and Brown 1980). It may also include minerals like feldspar, cristobalite, and crystalline quartz. Bentonite's unique properties, such as forming thixotropic gels with water, high water absorption, and significant cation exchange capacity, stem from its crystal structure of octahedral alumina sheets between tetrahedral silica sheets (Gualtieri et al. 2008; Ding et al. 2008; Gaudette 1964; Al-Atroush et al. 2021). These properties make bentonite suitable for various commercial applications. The molecular formula for montmorillonite is $(M_x . nH_2O)\ (Al_{2y}Mg_x)\ Si_4O_{10}(OH)_2$, with M being Na^+, K^+, Mg^{2+}, or Ca^{2+}. Bentonite is classified into three types: high swelling (sodium), low swelling (calcium), and moderate swelling (intermediate sodium calcium) (Grim 1968; Brindley and Brown 1980).

(ii) Kaolin: Kaolin, also known as china clay, is primarily composed of kaolinite along with quartz, mica, feldspar, illite, and sometimes montmorillonite. It forms through rock weathering, featuring a pseudohexagonal structure of small triclinic crystal sheets. Kaolinite has limited cation exchange capacity. Its structure involves alternating tetrahedral silica sheets and octahedral alumina sheets, sharing two-thirds of oxygen atoms between silicon and aluminum in a balanced O configuration (Alrubaye, Hasan, and Fattah 2017; Clarke 1951). Analytical studies indicate minimal lattice substitution in kaolinite samples. The common molecular formula for the kaolinite group (including kaolinite, nacrite, and dickite) is dickite, which is $Al_2Si_2O_5(OH)_4$ (Grim 1968; Brindley and Brown 1980).

(iii) Other Commercial Clays: Other significant types of commercial clays, alongside bentonite and kaolin, include ball clay, common clay and shale, and fire clay. Ball clay is primarily composed of kaolinite with minor amounts of illite, chlorite, smectite minerals, quartz, and organic materials (Grim 1968; Brindley and Brown 1980; Dixon 1989). Common clay and shale are characterized by their main constituents, illite and chlorite. Fire clay consists mainly of kaolinite, halloysite, and/or diaspore. Illite, a group of clay minerals resembling mica (Dixon 1989), shows minimal lattice expansion with distinct 1.0-nm (001) and 0.33-nm (003) peaks that remain unchanged even under solvation, saturation, or heating conditions (Dixon 1989).

Local soil compositions often incorporate a mix of these clay types, influenced by geological and environmental factors. Understanding their properties and applications is essential for effectively utilizing local soils in industrial and construction contexts. The United States Department of Agriculture (USDA) provides valuable resources for identifying clay minerals in specific US regions through its comprehensive soil survey reports and maps. These documents detail soil types, properties, and characteristics across different areas. Similar geological and soil data are typically available from government agencies or research institutions in other countries as well.

When evaluating 3D printability, the plasticity of the soil is a crucial factor. Plasticity refers to how easily soil can be molded or shaped before breaking apart, impacting pumpability, flowability, extrudability, and the overall quality of printed soil mixtures (Bhusal, Sedghi, and Hojati 2023; Alqenaee, Memari, and Hojati 2022; Mai et al. 2024). The plasticity of soil is primarily influenced by its clay fraction, the size and shape of clay particles, and the type of clay present (Andrade, Al-Qureshi, and Hotza 2011). Clay minerals like smectite exhibit high plasticity due to their ability to retain and absorb water, enhancing cohesion and moldability (Grim 1968). In contrast, kaolinite and illite have lower plasticity because of their plate-like structure, which results in reduced water absorption and weaker cohesion. Consequently, soils with higher proportions of kaolinite or illite tend to have lower plasticity and may be more prone to cracking or crumbling when dry (Grim 1968).

While printability is key for 3DP construction, a building material's strength and durability are equally important. Clay mineral composition and fraction play a major role in this regard. Notably, expansive clays with high smectite content swell when wet but shrink and crack when dry. Even small quantities of smectite clays can cause significant swelling in soils, unlike illite or kaolinite, which exhibit minimal swelling behavior (Seed, Woodward Jr., and Lundgren 1962). This swelling could affect both the mechanical strength and dimensional stability of hardened soil mixes (Ammari et al. 2017; Narloch et al. 2020). Previous research has demonstrated that the swelling capacity of soil is substantially affected by both the clay mineralogy and content, highlighting their critical role in the design of earth-based construction materials (Seed, Woodward Jr., and Lundgren 1962; Narloch et al. 2020; Minke 2009). However, factors like compaction and moisture content also influence strength. Thus, soils rich in smectite boast higher binding capacity and provide high plasticity to the soil mix due to their water-binding abilities, which could affect strength (positively or negatively) and boast volume change, while kaolinite or illite clays provide greater stability but lower plasticity (Nagaraj 2016; Minke 2009; Narloch et al. 2020). Achieving a balance between plasticity for optimal printability and low swelling for durable properties poses a central challenge in using earth-based materials for construction.

Local soils typically consist of various clay minerals and fractions rather than a pure variety. Analyzing a soil sample's mineral composition and clay fraction is crucial for accurately characterizing it for 3D printing applications. This assessment helps determine the soil's suitability for 3D printing, assessing its strength, durability, and printability. Our study analyzes the types and fractions of clay for their application in 3D printing. It aims to find the optimal balance of clay types and fractions within the soil mix to ensure optimal printability, strength, and minimal volume change, which are essential for successful 3D printing applications.

Clay minerals are formed through the process of weathering, and local soils typically consist of combinations of different clay types rather than pure clay. It is rare to find a single, pure clay type. To effectively characterize local soil, it is essential first to determine the minerals present in each soil sample. Subsequently, a comprehensive evaluation can be conducted to assess the soil's suitability for 3D printing and its mechanical and durability performance. Existing evidence suggests that maintaining clay proportions within optimal ranges is critical for achieving desirable printability characteristics and structural performance in 3D printing applications.

10.4.2 Material Characterization and Processing

Before 3D printing, soils must be carefully characterized and processed to ensure they meet specific project requirements and are compatible with the chosen extrusion system. Initial screening may involve rapid field tests to assess basic material suitability. For a more detailed evaluation, particularly of key physical and chemical properties such as particle size distribution, plasticity, and clay content, more advanced laboratory testing is recommended. These tests are especially beneficial for regional material assessments and large-scale applications where efficiency and accuracy are critical. This section provides a comprehensive list of relevant tests; however, it is the responsibility of the user to select and implement the appropriate testing methods based on the intended application and project context.

Before conducting any geotechnical or material characterization tests, proper soil preparation is critical to ensure reliable results for various earth construction techniques, including rammed earth, CEBs, and in situ 3D printing applications. The preparation process begins with mechanical disaggregation to break down soil aggregates into uniform particles, which is particularly important for rammed earth, where compaction quality affects structural integrity, and for CEB production, where particle size distribution influences block strength. Following disaggregation, organic matter removal may be necessary as it can affect mechanical properties and durability, especially for stabilized CEBs requiring precise cement–soil ratios. The final homogenization creates a uniformly mixed sample suitable for comprehensive characterization, including particle size analysis and Atterberg limits testing. For in situ earth printing applications, this standardized preparation ensures the soil mixture achieves optimal rheological properties for consistent extrusion and layer adhesion during the 3D printing process. The prepared samples can then proceed to specific tests based on their intended construction method, whether evaluating compaction characteristics for rammed earth, compressive strength for CEBs, or printability parameters for 3D earth printing.

For more detailed analysis, advanced laboratory testing can provide comprehensive data on clay mineralogy, particle gradation, swelling potential, and morphological properties. While simplified field tests are commonly employed in current large-scale 3D printing applications, advanced material characterization, including physicochemical, mineralogical, and microstructural analyses, enables a deeper understanding of soil behavior and ensures optimal performance in construction. Depending on project needs and soil application, a combination of the following tests may be conducted.

10.4.2.1 Grain Size Analysis

Grain size analysis determines the particle distribution within soil samples, classifying them into constituent parts such as clay, silt, sand, and gravel, and achieves particle size distribution for the local soil. Gradation analysis of soils can be conducted using a combination of field and laboratory techniques to determine particle size distribution accurately.

As a proposed methodology, a modified approach was recommended for assessing the grain size distribution of local soils, particularly for applications in earth-based 3D printing. Building on the findings of Zozaya et al. (2024), this method addresses the limitations of the standard ASTM C136 dry sieving procedure, which has proven insufficient for accurately quantifying the clay fraction in soil samples. Instead, a combined strategy employing both a wet sieve analysis and a traditional shake jar test is advocated to improve the resolution of fine particle fractions. The wet sieve method allows better separation of silt and clay particles through dispersion and controlled washing, while the shake jar test offers a practical, low-cost field solution for estimating relative proportions of sand, silt, and clay. The shake jar technique involves mixing soil with water and a dispersant in a transparent jar, allowing particle layers to settle over time for visual measurement. Although less precise than hydrometer or laser diffraction methods, especially for very fine clays, this approach remains widely used in permaculture and preliminary site investigations due to its accessibility and minimal equipment requirements. For classification purposes, layer thicknesses observed in the jar test will be used to calculate particle percentages, which are then plotted on the USDA soil texture triangle to determine the soil type. This dual-method approach enables reliable in situ soil assessment and supports selecting and preparing site-specific materials for additive construction.

10.4.2.2 Plastic Properties

To further characterize the suitability of local soils for 3D printing applications, the plastic behavior of the soil will be evaluated using the Casagrande method, which determines the Atterberg limits, specifically the LL, plastic limit (PL), and PI, in accordance with ASTM D4318 (ASTM_D4318 2016). These parameters provide critical insights into the workability, cohesion, and deformation characteristics of fine-grained soils. The LL represents the moisture content at which the soil transitions from a plastic to a liquid state, while the PL defines the threshold between the plastic and semi-solid states. The PI, calculated as the difference between LL and PL (PI = LL − PL), quantifies the range of moisture contents over which the soil remains plastic.

Soils with higher PI and LL values typically exhibit greater clay content and a corresponding increase in plasticity and water retention, which are important attributes for achieving cohesive, extrudable mixtures in earth-based 3D printing. However, excessively high plasticity may also indicate a higher risk of volumetric instability and shrinkage upon drying, necessitating careful optimization. In this study, LL and PL values will not only support the classification of the soil using the Unified Soil Classification System (USCS) and USDA texture triangle but will also inform the design of soil–water ratios for printable mixes, ensuring sufficient flowability without compromising buildability or dimensional stability.

10.4.2.3 Reactivity and Swell Potential Assessment

Several standardized test methods exist to evaluate the swelling characteristics of soils, with the methylene blue (MB) test emerging as one of the most widely adopted techniques. This versatile method provides quantitative measurements of key soil properties, including (i) specific surface area (SSA), which indicates the reactive clay content; (ii) CEC, reflecting the soil's chemical activity; (iii) swell index, a direct measure of volumetric expansion potential; and (iv) overall swelling behavior under moisture variation (Verhoef 1992; AFNOR, 1993; Murat Türköz and Tosun 2011). The MB test is particularly valuable for expansive clay-rich soils where swelling potential significantly impacts construction suitability, offering advantages of rapid execution and reliable correlation with more complex laboratory swelling tests.

When mixed with water and soil, MB, a cationic dye (C16H18N3SCI), allows chloride ions to be replaced by cations on clay minerals, which are then adsorbed onto their surfaces. The amount of MB adsorbed varies based on the clay minerals present, their cation exchange capacity, and SSA. This test is popular due to its simplicity and because it does not require expensive or specialized equipment, making it a reliable method for assessing clay minerals in soils, particularly during the initial research stages (Verhoef 1992; AFNOR, 1993; Türköz and Tosun 2011). This measurement offers a quick assessment of cohesion and swelling characteristics within the soil's fine fraction, potentially replacing complex mineralogy tests with simpler field or lab procedures.

10.4.2.4 Mineralogical and Chemical Composition

Mineralogical and chemical analysis of clay (particles smaller than 0.002 mm) can be performed using techniques such as X-ray diffraction (XRD) and X-ray fluorescence (XRF). XRD is typically used to determine the mineral composition of natural clays and identify them based on their unique diffraction patterns. However, quantifying clay mineral composition and interpreting XRD data for multi-mineral soils are complex and lack standardized procedures. This chapter recommends using an established technique that combines mineral intensity factors (MIFs) (Moore and Reynolds 1997) from previous studies to approximate the quantities of the main clay minerals (kaolinite, illite, and smectite) present in the soil samples (Środoń 2006; Zhou et al. 2018; Zwell and Danko 1975). Despite this, quantitative analysis remains challenging due to various factors such as chemical composition, preferred orientation, structural disorder, and structural diversity of clay minerals. In addition to XRD, XRF will gather more data on morphology and elemental composition, providing further insights into the potential clay minerals. While XRF alone may not directly identify clay minerals, it can offer valuable information about elemental composition that, when combined with other techniques like XRD, can help determine clay mineralogy. Furthermore, scanning electron microscope (SEM) testing will provide insights into the microstructure, surface morphology, particle shape, and impurities. SEM may also be used to interpret data for some inputs for special applications.

10.4.2.5 Soil Classification

According to the characterization results, collected soil samples could be classified based on their clay fraction, clay minerals, and their elemental composition (especially SiO_2, Al_2O, and Fe_2O_3, CaO, and Na_2O), and their suitability for specific applications will be evaluated.

10.4.3 Additives and Stabilizers for Earth-Based Mixtures

In addition to clay content and minerals, soil stabilizers or additives have been used since ancient times to enhance soil strength and volumetric stability (Al-Atroush et al. 2021). These substances are incorporated into the soil matrix to improve its engineering properties. Common stabilizers include cement, lime, fly ash, natural or synthetic fibers, and chemical additives. The choice of stabilizer depends on the desired outcome and the soil characteristics. These materials can enhance the soil's compressive strength, printability, and resistance to volumetric changes such as swelling or shrinkage. Various stabilizers like lime, cement, fly ash, and polymer-based additives can be employed to improve soil performance. When considering 3D printing with earth-based materials, the use of additives must be carefully evaluated based on the specific capabilities of the printing system, particularly the nozzle dimensions, pump capacity, and extrusion mechanics. These technical constraints can limit the type, quantity, and particle size of stabilizers or fibers that can be effectively integrated into printable mixes. Currently, research on the compatibility of various additives with different earth-printing technologies remains limited. The following sections of this chapter will examine key printability requirements and the implications of additive selection within this context. The additives listed below are categorized into different classes and should be selected judiciously based on the specific characteristics of the local soil and their applications.

10.4.3.1 Incorporation of CEC Modifiers

CEC in clay soils (e.g., smectite group) leads to excessive cation retention, causing swelling, shrinkage, and instability. The CEC modifiers can effectively address the high CEC of certain clay types, such as those in the smectite group. Examples of soil stabilizers include lime, cement, fly ash, cement kiln dust, and gypsum, which can mitigate the excessive absorption and retention of cations by clay particles. By altering the soil's chemical and physical properties, these stabilizers help reduce swelling, shrinkage, and instability commonly associated with high CEC clay soils (Chiappone et al. 2004).

A Nigerian study demonstrated that incorporating 6% cement with 12% clinoptilolite zeolite into expansive clay significantly reduced the free swell index by 60% and swell pressure by 54% in high-plasticity soils (Umar et al. 2025). The stabilization effect arises from a dual mechanism: zeolite's high surface area and CEC facilitate the adsorption of excess exchangeable cations, particularly Ca^{2+}, while cement induces pozzolanic reactions that densify and chemically bind the clay matrix (Barman and Dash 2022; Umar et al. 2025).

For kaolinitic clays, which naturally possess a lower CEC (typically 3–15 meq/100 g), another study reported that combining lime with calcined clamshell ash (a source of reactive CaO derived from thermally treated $CaCO_3$) resulted in a 50% reduction in shrinkage and a marked improvement in unconfined compressive strength over a 30-day curing period (Zaini et al. 2024). This performance enhancement is attributed to lime's cation exchange and flocculation effects, which are particularly effective in low-CEC systems (Zaini et al. 2024; Barman and Dash 2022).

While not directly explored in the aforementioned studies, analogous stabilizing agents such as cement kiln dust (CKD) and Class C fly ash, both rich in calcium oxide, are hypothesized to reduce the swelling potential of smectitic clays via ion exchange and Ca^{2+} precipitation mechanisms (Barman and Dash 2022). Similarly, gypsum ($CaSO_4 \cdot 2H_2O$), widely used for ameliorating sodic soils, has shown promise in reducing swell behavior in high-CEC clays by facilitating the replacement of Na^{2+} ions with calcium, thereby stabilizing interlayer cations and improving soil structure (Barman and Dash 2022; Yang et al. 2024).

10.4.3.2 Incorporation of Fibers

Fibers have emerged as effective reinforcements in soil stabilization, significantly enhancing tensile strength and minimizing deformation and cracking by forming a 3D matrix within the soil. Both natural and synthetic fibers contribute to mechanical improvement and durability, with recent studies exploring their performance in various soil types and conditions. Among natural fibers, wheat straw and bamboo have shown promising results. A study demonstrated that incorporating 0.3% wheat straw improved slope stability by 30%, while 0.6% bamboo fibers increased it by

120% in silty clay, with bamboo proving more effective due to its higher density and cellulose content, especially at smaller slope angles (Zhang et al. 2025). Jute fiber, when combined with 15% fly ash, was found to reduce the PI of clayey soils by 25% and increase unconfined compressive strength (UCS) by 40%, while also mitigating shrinkage cracks and improving resistance to wet–dry cycles (Gupta and Kumar 2022). Similarly, sisal fibers at 0.9% content demonstrated a 50% reduction in freeze–thaw deformation in frost-prone soils by inhibiting crack propagation, with SEM analysis confirming effective fiber–soil interlocking and microstructural preservation (Deng et al. 2025). Hemp fiber has also been studied in combination with lime kiln dust (6%) and nano-silica (10%), where a 0.4% hemp addition yielded a UCS of 1,979 kPa and improved freeze–thaw resistance by 35% (Moayed, Arabani, and Ahmadi 2025). Studies by Gomaa et al. (Gomaa, Vaculik, et al. 2021) and Giacomobono et al. (2024) further confirmed that incorporating 2% wheat straw fibers by weight significantly improved the tensile strength of earth-based mixtures, reinforcing their suitability for 3D printing. Expanding on this concept, Faleschini et al. (2023) incorporated a complex blend of lime, rice husk, marble dust, cement, and municipal waste ash to enhance both the workability and mechanical strength of soil-based mixtures.

Synthetic fibers also show strong potential. Recron-3s polyester fibers, used with recycled construction materials such as fly ash and brick dust, increased the California bearing ratio (CBR) by 63.7% and provided long-term stability under humid conditions due to their hydrophobic nature (Nanda, Priya, and Kumar 2025). A hybrid system incorporating nylon (synthetic) and jute (natural) fibers along with cement resulted in a remarkable 1,380% increase in UCS and a 90% reduction in swell pressure, leveraging nylon's tensile capacity and jute's bonding ability (Paul, Tasnim, and Majumder 2025).

The improvements brought by fiber reinforcement are attributed to multiple mechanisms. Tensile reinforcement is achieved through friction and mechanical interlocking between fibers and soil, which redistributes stress and limits crack development. On a microstructural level, natural fibers like sisal and hemp catalyze the formation of calcium silicate hydrate (C-S-H) gels when used with lime or cement, effectively binding soil particles (Moayed, Arabani, and Ahmadi 2025; Deng et al. 2025). Moreover, the use of natural fibers such as wheat straw and bamboo not only contributes to mechanical strength but also offers environmental benefits by sequestering carbon and reducing the demand for high-embodied-energy materials like cement, ultimately lowering the carbon footprint of stabilization practices (Bhusal, Uviña, and Hojati. 2022; Bhusal, Sedghi, and Hojati 2023).

10.4.3.3 Incorporation of Bio-Based Modifiers

Several bio-based additives have been explored for enhancing the performance of earth mixes in 3D printing applications. Perrot et al. (Perrot, Rangeard, and Courteille 2018) investigated the use of alginate, a seaweed-derived biopolymer, to improve the rheological behavior of raw earth mixtures, facilitating smoother extrusion and providing early structural integrity during the printing process. Carcassi et al. (2024) evaluated a variety of natural gums, including sodium alginate, methylcellulose, xanthan gum, guar gum, and locust bean gum, alongside natural fibers such as wheat straw, hemp, banana, and kenaf, aiming to optimize both printability and post-print mechanical properties.

Additionally, biochar is a carbon-rich, fine-grained, and porous solid produced through the thermochemical decomposition of biomass in an oxygen-limited environment (Garg et al. 2020; Wani et al. 2022, 2023). Depending on the feedstock and production conditions, biochar can be categorized into various types, which influence its physical and chemical properties, such as porosity, surface area, and nutrient content. In this research, we focus specifically on biochar derived from wood. This type of biochar is created through the pyrolysis of wood biomass in an oxygen-limited environment, resulting in a carbon-rich material with unique properties. Wood-derived biochar is particularly valued for its stability and high surface area, making it beneficial for soil enhancement and environmental applications. Previous research (Garg et al. 2020; Wani et al. 2022, 2023) has demonstrated the positive impacts of biochar on lean clay and expansive soils.

10.5 MATERIAL REQUIREMENTS AND PERFORMANCE FOR 3D PRINTING WITH EARTH-BASED MIXTURES

This section presents current advancements in the material requirements for 3D printing with earth-based mixtures, with a focus on recent developments in both printability and mechanical performance. It is structured in two interrelated parts: the first outlines essential printability parameters, such as flowability, buildability, and setting behavior, while the second delves into the physical and structural properties required for long-term performance, including cohesion, internal friction, particle gradation, rheology, water absorption, shrinkage, green strength, and compressive strength. The analysis centers on in situ and prefabricated earth-based materials specifically developed for 3D printing. Automated fabrication systems for CEBs and rammed earth, while promising, fall outside the scope of this section due to the limited research available; those technologies have been discussed separately and warrant further investigation. Importantly, the performance of 3DP earthen structures relies on the complex interplay of multiple material attributes, each influencing both the print process and the final structural integrity. These interdependencies are illustrated in Figure 10.3, which maps the critical relationships that govern successful earth-based 3D printing. Table 10.3 presents a summary of some of the current research on 3D printing of earth.

10.5.1 PRINTABILITY OF EARTH-BASED MIXTURES

Printability is governed by both the rheological properties of the material and the operational parameters of the printing system. Successful 3D printing with earth-based materials requires a careful balance between material flow behavior and precise process control. As the field progresses, there is a growing need for standardized testing protocols and optimization frameworks that can accommodate diverse local soils and bio-based additives, enabling scalable and sustainable construction. To assess printability, previous studies have recommended evaluating extrudability and buildability through controlled testing procedures.

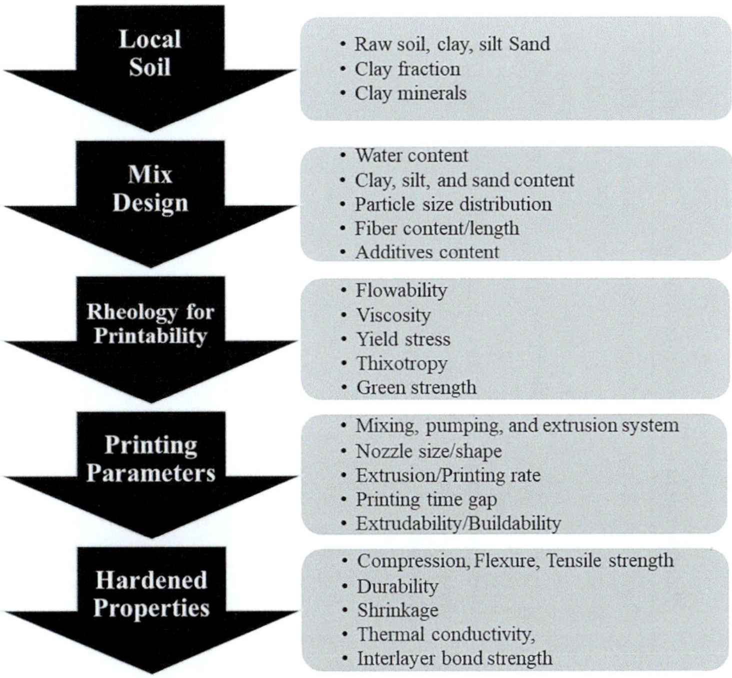

FIGURE 10.3 Critical factors for 3D printing of earth.

Extrudability refers to the ease with which a material can be continuously extruded through a nozzle without interruption, deformation, or clogging. It can be quantified through several measurable attributes, such as surface quality, dimensional consistency, and dimensional conformity of the extruded filament (Zafar, Bakhshi, and Hojati 2023; Zafar et al. 2025). Rheological properties, including plastic viscosity and dynamic yield stress, are key parameters that directly influence extrudability, as they determine the flow behavior of the mix during the pumping and extrusion phases. For instance, Zozaya (Zozaya et al. 2024, 2026; Zozaya 2024) evaluated extrudability by printing a zigzag pattern using clay-rich soils reinforced with rice hull fibers, observing the continuity and uniformity of the extruded path. Similarly, Bhusal (2023) assessed dimension consistency by printing a 50-cm-square filament and measuring the width at regular intervals to check for uniform extrusion and dimensional conformity.

Buildability, on the other hand, is defined as the material's ability to maintain its shape and support additional layers without collapse or excessive deformation. It reflects the green strength of the material and is governed primarily by static yield stress, thixotropy, and the rate of structural build-up (structuration) after deposition (Zafar, Bakhshi, and Hojati 2023; Zafar et al. 2025; Rückrich, Agranati, and Grobman 2023). Buildability tests vary in complexity and scale. Rückrich et al. (Rückrich, Agranati, and Grobman 2023), e.g., evaluated buildability by printing a cylindrical form (15 cm diameter, 10 layers, 1.5 cm thick each, spaced 0.8 cm

TABLE 10.3
Review of 3DP Earthen Materials: Soil Type, Reinforcement, and Additives

Construction Method	Soil	Composition (%)	Additives/Modifiers	Reinforcement	Evaluation Aspects	Ref.
3DP cob	Subsoil	– Clay (19–20) – Sand (80–81)	-	Wheat straw (2% by weight)	– Comp. strength	Gomaa, Vaculik et al. (2021)
3DP earthen wall	Raw earth	-	– Alginate	-	– Rheology – Green strength – Comp. strength	Perrot, Rangeard, and Courteille (2018)
3DP earthen material	Raw earth	– Silt + Clay (68.24) – Sand (28.8) – Gravel (2.96)	– Lime – Rice husk – Marble dust – Cement – Municipal solid waste – Bottom ash	– Jute fiber, coconut fiber, sisal fiber – Goat hair	– Comp. strength – Shrinkage, printability	Faleschini et al. (2023)
3DP earthen wall	Raw earth	– Soil (73) – Water (25)	– -	– Wheat fibers (2%)	– Structural design	Giacomobono et al. (2024)
3D-printed earth-based mortar	Raw earth	– Clay-rich soil	– Sodium alginate – Methylcellulose, xanthan gum – Guar gum – Locust bean gum	– Hemp fiber Wheat straw fiber – Kenaf fiber – Banana fiber	– Comp. strength – Printability – Thermal conductivity	Carcassi et al. 2024)
3DP earthen large-scale structure	Raw earth	– Clay – Silt – Sand	-	– Chopped straw	– Large-scale printing	Curth, Barrak Darweesh, and Rael (2020)

(*Continued*)

TABLE 10.3 *(Continued)*
Review of 3DP Earthen Materials: Soil Type, Reinforcement, and Additives

Construction Method	Soil	Composition (%)	Additives/Modifiers	Reinforcement	Evaluation Aspects	Ref.
3DP earth-based mortar	Raw earth	– Silt (44) – Quarry wash mud (16) – Sea sand (40)	-	– Jute fabric	– Comp. strength – Flexural strength	Tarhan and Perrot (2023)
3DP cob	Clay	– Clay (38.53–52.63) – Sand (11.11–17.73) – Lime (7.37–11.63) – Cement (2.40–2.48) – Straw (1.06–1.50) – Water (18.78)	– Lime – Cement	– Straw	– Comp. strength – Shrinkage – Printability	Alqenaee and Memari (2022)
3DP wall	Subsoil	– Soil (70.42) – Sand (18.78) – Shredded rice husk (1.41) – Hydraulic lime (4.69) – Lime-based binder (4.7)	– Hydraulic lime – Lime-based binder	– Shredded rice husk	– Comp. Strength – Stress–strain – Printability	Ferretti et al. (2022b)

apart), subjecting it to a 2-kg load for 10 seconds, and then measuring deformation. Alqenaee et al. (Alqenaee, Memari, and Hojati 2022) tested cob-based mixes by printing a small 6 × 6-inch beam and were able to stack up to eight layers before failure. A taller print of 18 inches collapsed at 15 inches, demonstrating the limits of unsupported height for that specific mix. Bhusal conducted similar buildability assessments, stacking between 12 and 18 layers depending on the soil formulation, and also fabricated a hollow conical cylinder to explore shape retention and vertical stability, as illustrated in Figure 10.4.

Successful 3D printing of earth-based materials requires an intricate balance between material properties and printing parameters to ensure smooth extrusion and structural stability. Printability encompasses the ability of a material to be extruded through a nozzle and maintain its shape post-deposition. For extrusion-based 3D printing, this involves critical rheological behaviors such as shear-thinning, plasticity, and thixotropy, which govern flow during extrusion and stability after layer placement.

A suitable earth-based mix for 3D printing must exhibit shear-thinning behavior, flowing under applied stress while retaining shape when the stress is removed. This behavior supports layer-by-layer construction by ensuring the material is extrudable yet self-supporting. To achieve this, the mix must display a combination of plastic viscosity, dynamic yield stress, static yield stress, and thixotropy. Plastic viscosity and dynamic yield stress determine pumpability and extrusion ease, while static yield stress and thixotropy govern shape retention and buildability of the deposited layers (Zafar, Bakhshi, and Hojati 2023; Zafar et al. 2025; Bhusal, Sedghi, and Hojati 2023; Li et al. 2020).

To tune these properties, various additives and modifiers are often incorporated. For instance, Bhusal et al. (Bhusal, Sedghi, and Hojati 2023) demonstrated that adding lime to raw earth significantly increased static yield stress from 1,169 Pa to 2,500 Pa and thixotropy from 11,132 Pa/s to 32,766 Pa/s, enhancing the material's structural integrity. Although this treatment also increased plastic viscosity (from 4.1 to 8.3 Pa·s), it lowered the dynamic yield stress (from 354.9 to 84.1 Pa), making the material easier to extrude without compromising buildability. Such rheological

(a) Extrudability test geometry (b) Buildability test geometry

FIGURE 10.4 Extrudability and buildability evaluation. Reproduced from Bhusal, Sedghi, and Hojati (2023), licensed under CC BY 4.0 (Bhusal, Sedghi, and Hojati 2023).

tuning is essential for printing stability, particularly in soil-based formulations where uncontrolled deformation can occur due to inadequate yield stress. Limited studies have linked the rheological properties of earth-based materials to their printability, highlighting the need for more comprehensive and targeted research in this area.

Beyond material formulation, printability also depends on the synchronization of process parameters, including nozzle size, extrusion speed, and printing speed. These variables must be carefully aligned to ensure uniform material deposition and minimize print defects such as over-extrusion or filament discontinuities.

The nozzle diameter and geometry critically control the printability of earth-based materials in large-scale 3D printing. Rectangular nozzles improve layer adhesion and printing speed for structural elements by increasing the interlayer contact area, whereas circular nozzles produce more uniform bead profiles. Larger diameters (≥12 mm) reduce clogging risks with fiber-reinforced or coarse-grained earth mixes but compromise resolution, while smaller nozzles (6–8 mm) enable finer details but require stricter material rheology control. Larger nozzles (20–50 mm) are generally used for large-scale 3D printing of earth mixes with high fiber or aggregate content to reduce clogging. Bhusal et al. (Bhusal, Sedghi, and Hojati 2023) and Ferretti et al. (Ferretti et al. 2022a) used 20-mm nozzles for printing straw-reinforced mixes, while Curth et al. (2024) adopted a 25-mm nozzle for their programmable mud system. Although smaller nozzles offer finer resolution, they are more prone to blockage and require more refined mixtures.

Extrusion speed (i.e., rate) defines how quickly material is pushed through the nozzle and depends on the type of printing system and pumping process. An ideal speed must align with the printing speed to maintain consistent layer dimensions. If extrusion is too fast, it leads to material overflow; if it is too slow, gaps form between layers. Bhusal et al. (Bhusal, Sedghi, and Hojati 2023) found 0.15 rotations per second to be optimal for their screw-based extruder when printing with wheat straw-reinforced earth. Future work should focus on developing specialized pumping systems capable of delivering earth-based mixtures smoothly and continuously to ensure consistent and reliable 3D printing performance.

Printing speed, or nozzle travel rate, is influenced by the scale and configuration of the 3D printing system and plays a critical role in determining final print quality. To achieve uniform and continuous material deposition, the printing speed must be carefully synchronized with the extrusion rate. Optimal printing speeds vary depending on the rheological properties of the earth mix, pump capacity, and hardware configuration. For example, Bhusal et al. (Bhusal, Sedghi, and Hojati 2023) reported a speed of 20 mm/s, Zozaya (Zozaya et al. 2024) used 15 mm/s, Yousaf et al. (Yousaf, Al Rashid, and Koç 2024) employed a range of 25–75 mm/s, and Curth et al. (2024) used speeds between 50 and 300 mm/s, reflecting how print parameters are closely tied to material formulation and printer capabilities.

Together, extrudability and buildability serve as practical and complementary performance indicators to evaluate the feasibility of a given earth mix in 3D printing. By integrating these evaluations with rheological measurements, researchers can better optimize soil formulations for robust and scalable 3D printing applications.

10.5.2 Mechanical Properties of 3D-Printed Earth-Based Mixtures

The structural viability of 3DP earthen materials for full-scale construction is critically governed by their mechanical performance metrics, including compressive strength, flexural strength (modulus of rupture), tensile capacity, and long-term durability under various loading and environmental conditions. These properties dictate the material's ability to serve as a load-bearing component in structural applications.

Among these parameters, *compressive strength* serves as a primary indicator of material suitability. Although earthen materials have lower compressive strength (typically 1–4 MPa) compared to OPC-based materials (>20 MPa), they can still be used for structural applications, particularly in low-rise buildings with thick load-bearing walls and proper design. Modern stabilization methods, such as adding fibers, pozzolans, or small amounts of cement, can enhance their strength and durability. When combined with appropriate engineering and standards (e.g., ASTM and local codes), earthen materials offer viable, sustainable solutions for structural use, especially in regions prioritizing low-carbon construction and local resources. In the absence of stabilizers or fiber reinforcement, studies such as those by Gomaa et al. (Gomaa, Vaculik, et al. 2021) reported compressive strengths ranging from 0.87 for unstabilized raw earth mixtures. A range of studies have documented compressive strengths between 1.1 and 5.04 MPa for stabilized fiber-reinforced 3DP earthen composites (Bhusal, Sedghi, and Hojati 2023; Ferretti et al. 2022a, 2022b). Notably, Ferretti et al. (Ferretti et al. 2022a) achieved compressive strengths of up to 4.5 MPa using shredded rice husk as a natural reinforcement, underscoring the effectiveness of plant-based additives in improving load-bearing behavior.

Further enhancements have been demonstrated in hybrid systems combining both fibers and chemical stabilizers. Bhusal et al. (Bhusal 2023) formulated optimized 3D-printable earthen mixes incorporating lime and wheat straw fibers. Their results showed that raw soil with optimized mix design achieved compressive strengths up to 6.1 MPa and flexural strengths of 4.15 MPa, indicating a substantial improvement over conventional earthen construction benchmarks. Additionally, the same mixes exhibited favorable extrusion consistency and minimal post-deposition deformation, confirming their printability and buildability.

These results are particularly significant when benchmarked against regulatory performance thresholds. The International Building Code (IBC) stipulates minimum compressive and flexural strengths of 2.06 MPa and 0.34 MPa, respectively, for load-bearing earthen masonry (IBC 2018). The mechanical performance achieved through 3D printing not only satisfies but also in many cases exceeds these code requirements. Consequently, this demonstrates that with appropriately engineered mix designs, integrating rheological control, reinforcement strategies, and stabilization, 3DP earthen materials can match or surpass the structural integrity of traditional methods such as adobe and rammed earth. This reinforces their potential as viable, code-compliant alternatives for sustainable and scalable construction.

10.6 CHALLENGES AND FUTURE DIRECTIONS

3D printing of earth-based materials presents a transformative opportunity for sustainable and localized construction, but it remains constrained by several technical, material, and process-related challenges. Addressing these challenges will be essential to unlocking its full potential for both structural and nonstructural applications.

- Shrinkage and Cracking Control: A major limitation of earthen materials is their tendency to shrink and crack during drying, particularly in unstabilized or improperly formulated mixes. Uneven moisture loss, combined with low tensile strength, can lead to visible and structural cracking that compromises both aesthetics and performance. Although reinforcement with natural fibers and the use of shrinkage-reducing additives show promise, comprehensive models linking soil composition, environmental conditions, and drying kinetics are still under development. Further research is required to control volumetric deformation without compromising extrudability or strength.
- Challenges with Swelling Clay and Printability: Certain clay minerals, especially those with high CEC, such as smectite, exhibit excessive swelling and shrinking behavior that impairs printability and dimensional stability. Achieving the necessary rheological profile for successful extrusion and buildability (i.e., the ability to retain shape after deposition) is especially challenging with these clays. Current studies suggest that the appropriate use of innovative stabilizers and careful water content control can mitigate these effects, but robust design frameworks and predictive tools are still lacking.
- Material Variability and Standardization: Earth is a naturally heterogeneous material whose properties, such as particle size distribution, mineral composition, plasticity, and moisture retention, are highly variable and linked to local geological conditions. This makes characterization and standardization particularly difficult when compared to industrial materials like concrete or steel. For 3D printing to become viable at scale, adaptable formulations and real-time testing protocols must be developed. Integrating machine learning for material classification and response prediction could support this need.
- Absence of Unified Codes and Standards: Despite growing interest, there is no unified regulatory framework governing the structural use of 3DP earth materials. This limits their adoption in mainstream construction. Current efforts are scattered across academic and industry initiatives, and harmonization is needed to establish standards for mechanical properties (compressive, flexural, and tensile strengths), durability, and safety. Formalizing testing procedures for properties such as rheology, shrinkage, and thermal conductivity is a necessary next step.
- Low Automation and Digital Integration Gaps: Most 3D printing systems for earth construction still involve substantial manual intervention during material preparation, surface finishing, and quality control. Increasing automation is critical for scalability and quality assurance. Future research should explore integrating robotics, real-time sensors, and digital twins to monitor and adjust process parameters (e.g., moisture, temperature, and extrusion rate) dynamically. AI-driven control systems can optimize layer quality, detect defects, and adapt to varying material inputs.
- Multi-Material Printing and Functional Integration: An emerging direction in construction-scale 3D printing is multi-material 3D printing, where

structural earthen layers are printed alongside insulating, waterproofing, or reinforcing materials. This approach enables wall assemblies with built-in thermal resistance or moisture protection, eliminating the need for secondary treatments. Developing compatible printing heads, synchronized extrusion algorithms, and bonding interfaces between materials is a promising yet complex area of research.

- Prefabrication of Earth Components: While in situ printing has received the most attention, prefabrication of earthen blocks or panels using automated 3D printing methods represents a viable alternative. These prefabricated elements can be printed under controlled conditions and transported for modular assembly. Such approaches combine the architectural freedom of 3D printing with the logistical efficiency of traditional modular construction.
- Application-Driven Innovation: Looking ahead, earth-based 3D printing holds unique potential for use in disaster relief, affordable housing, and extraterrestrial construction. The ability to utilize in situ or locally sourced materials aligns with sustainability goals and logistical constraints in remote or resource-scarce environments. Developing modular, mobile printing platforms and universally adaptable material systems will be key for these applications.

REFERENCES

AFNOR. 1993. *Mesure de La Quantité et de l'activité de La Fraction Argileuse (Norme Française NF*, 68–94. La Défense, Paris, France: Association française de Normalization (ANFOR).

Ahmad, Lina, Wassim Jabi, and Marco Sosa. 2024. "Nonplanar Robotic Printing of Earth-Based Material: A Case Study Using Cob-like Mixture." *Buildings* 14 (8): 2589. https://doi.org/10.3390/buildings14082589.

Al-Atroush, Mohamed Ezzat, Omar Shabbir, Bandar Almeshari, Mohamed Waly, and Tamer A. Sebaey. 2021. "A Novel Application of the Hydrophobic Polyurethane Foam: Expansive Soil Stabilization." *Polymers* 13 (8): 1335. https://doi.org/10.3390/polym13081335.

Alqenaee, Amnah, and Ali M. Memari. 2022. "Experimental Study of 3D Printable Cob Mixtures." *Construction and Building Materials* 324 (March): 126574. https://doi.org/10.1016/j.conbuildmat.2022.126574.

Alqenaee, Amnah, Ali M. Memari, and Maryam Hojati. 2022. "Transition from Traditional Cob Construction to 3D Printing of Clay Homes." *Journal of Green Building* Under Review.

Alrubaye, Ali Jamal, Muzamir Hasan, and Mohammed Y. Fattah. 2017. "Stabilization of Soft Kaolin Clay with Silica Fume and Lime." *International Journal of Geotechnical Engineering* 11 (1): 90–96. https://doi.org/10.1080/19386362.2016.1187884.

Ammari, A., K. Bouassria, M. Cherraj, H. Bouabid, and S. Charif D'ouazzane. 2017. "Combined Effect of Mineralogy and Granular Texture on the Technico-Economic Optimum of the Adobe and Compressed Earth Blocks." *Case Studies in Construction Materials* 7 (December): 240–48. https://doi.org/10.1016/j.cscm.2017.08.004.

Andrade, F. A., H. A. Al-Qureshi, and D. Hotza. 2011. "Measuring the Plasticity of Clays: A Review." *Applied Clay Science* 51 (1–2): 1–7. https://doi.org/10.1016/J.CLAY.2010.10.028.

Arrieta-Escobar, Javier A., Delphine Derrien, Stéphanie Ouvrard, Elnaz Asadollahi-Yazdi, Alaa Hassan, Vincent Boly, Anne-Julie Tinet, and Marie-France Dignac. 2020. "3D

Printing: An Emerging Opportunity for Soil Science." *Geoderma* 378 (November): 114588. https://doi.org/10.1016/j.geoderma.2020.114588.

Artisan-Outdoor. 2025. "Turf vs. Sod – What Makes Sense for Your Lawn?" *Artisan Outdoor*. https://artisanoutdoor.com/turf-vs-sod-what-makes-sense-for-your-lawn/.

ASTM_D4318. 2016. *Standard Test Methods for Liquid Limit, Plastic Limit, and Plasticity Index of Soils*. West Conshohocken, PA: ASTM International. https://www.astm.org/astm-tpt-156.html.

Avila, Fernando, Esther Puertas, and Rafael Gallego. 2021. "Characterization of the Mechanical and Physical Properties of Unstabilized Rammed Earth: A Review." *Construction and Building Materials* 270: 121435. https://doi.org/10.1016/j.conbuildmat.2020.121435

Azizi-Bondarabadi, Hamed, and Neda Haji Sadeghi. 2022. "Local Seismic Culture in Iranian Vernacular Architecture: Evidence from Yazd Earthen Architecture." *Built Heritage* 6 (1): 5. https://doi.org/10.1186/s43238-022-00051-3.

Bailly, Gabo Cyprien, Yassine El Mendili, Athanas Konin, and Eliane Khoury. 2024. "Advancing Earth-Based Construction: A Comprehensive Review of Stabilization and Reinforcement Techniques for Adobe and Compressed Earth Blocks." *Eng* 5 (2): 750–83. https://doi.org/10.3390/eng5020041.

Bajpayee, Aayushi, Mehdi Farahbakhsh, Umme Zakira, Aditi Pandey, Lena Abu Ennab, Zofia Rybkowski, Manish Kumar Dixit, et al. 2020. "In Situ Resource Utilization and Reconfiguration of Soils into Construction Materials for the Additive Manufacturing of Buildings." *Frontiers in Materials* 7 (March). https://doi.org/10.3389/FMATS.2020.00052/FULL.

Balsam, William, Bobby Deaton, and Michael Adler. 2007. "Analysis of Adobe Wall Composition at the Chaves-Hummingbird Site, New Mexico, by Diffuse Reflectance Spectrophotometry." *Geoarchaeology* 22 (8): 825–44. https://doi.org/10.1002/GEA.20197.

Barman, Dharmendra, and Sujit Kumar Dash. 2022. "Stabilization of Expansive Soils Using Chemical Additives: A Review." *Journal of Rock Mechanics and Geotechnical Engineering* 14 (4): 1319–42. https://doi.org/10.1016/j.jrmge.2022.02.011.

Batikha, Mustafa, Rahul Jotangia, Mohamad Yasser Baaj, and Ibrahim Mousleh. 2022. "3D Concrete Printing for Sustainable and Economical Construction: A Comparative Study." *Automation in Construction* 134 (February): 104087. https://doi.org/10.1016/j.autcon.2021.104087.

Bats, Adeline, and Nadia Licitra, eds. 2023. *Storage in Ancient Egypt and Nubia: Earthen Architecture and Building Techniques*. Sidestone Press. https://doi.org/10.59641/o29221ra.

Beckett, C. T. S., C. E. Augarde, D. Easton, and T. Easton. 2018. "Strength Characterisation of Soil-Based Construction Materials." *Géotechnique* 68 (5): 400–9. https://doi.org/10.1680/jgeot.16.P.288.

Beek, Gus W. Van, and Ora Van Beek. 2013. *Glorious Mud! Ancient and Contemporary Earthen Design and Construction in North Africa, Western Europe, the Near East, and Southwest Asia*. Smithsonian Institution.

Beguin, Nestor, Paco Pioline, and Francisco Magnone Rienzi. 2023. "Tension Networks Using Natural Fibers as Support for 3D Printing with Earth." *IAAC Blog*. https://blog.iaac.net/tension-networks-using-natural-fibers-as-support-for-3d-printing-with-earth/.

Bergaya, F., and G. Lagaly. 2006. "Chapter 1 General Introduction: Clays, Clay Minerals, and Clay Science." In *Developments in Clay Science*, 1–18. https://doi.org/10.1016/S1572-4352(05)01001-9.

Bhattacherjee, Shantanu, Anusha S. Basavaraj, A. V. Rahul, Manu Santhanam, Ravindra Gettu, Biranchi Panda, Erik Schlangen, et al. 2021. "Sustainable Materials for 3D Concrete Printing." *Cement and Concrete Composites* 122 (September): 104156. https://doi.org/10.1016/j.cemconcomp.2021.104156.

Bhusal, Shiva. 2023. *3D Printing of Earthen Materials: Toward the Carbon-Zero Construction*. The University of New Mexico.

Bhusal, Shiva, Reza Sedghi, and Maryam Hojati. 2023. "Evaluating the Printability and Rheological and Mechanical Properties of 3D-Printed Earthen Mixes for Carbon-Neutral Buildings." *Sustainability* 15 (21): 15617. https://doi.org/10.3390/su152115617.

Bhusal, Shiva, Francisco Uviña, and Maryam Hojati. 2022. "Preliminary Study on 3D Printing of Locally Available Earthen Materials in New Mexico." In *Earth USA 2022*. https://www.researchgate.net/profile/Shiva-Bhusal-6/publication/365945674_preliminary_study_on_3d_printing_of_locally_available_earthen_materials_in_new_mexico/links/6389be777d9b40514e0a2217/preliminary-study-on-3d-printing-of-locally-available-earthen-ma.

Blundell Jones, Peter. 2012. "History on the Margin: Unravelling the Changing Fortunes of a Devon House and Its Relation to the Locality." *The Journal of Architecture* 17 (2): 191–212. https://doi.org/10.1080/13602365.2012.678637.

Borrero, Monica. 2024. "Nubian Vaults: Sustainable Homes to Earn Carbon Credits for Families in Africa's Sahel." *Climate Change Adaption*. https://www.adaptation-undp.org/nubian-vaults-sustainable-homes-earn-carbon-credits-families-africas-sahel.

Bourgeois, Jean-Louis. 2021. "Adobe: The Tradition and Techniques of Mud Bricks." *Regenerative Skills*. https://regenerativeskills.com/articles-1-2016-5-29-adobe-the-tradition-and-techniques-of-mud-bricks/.

Brindley, G. W., and G. Brown. 1980. *Crystal Structures of Clay Minerals and Their X-Ray Identification*. London: Mineralogical Society. https://nla.gov.au/nla.cat-vn1644537.

Brinson, Linda C. 2023. "How Adobe Construction Works." https://home.howstuffworks.com/home-improvement/construction/materials/adobe-construction2.htm.

Bui, Q.-B., J.-C. Morel, V.-H. Tran, S. Hans, and M. Oggero. 2016. "How to Use In-Situ Soils as Building Materials." *Procedia Engineering* 145: 1119–26. https://doi.org/10.1016/j.proeng.2016.04.145.

Canivell, Jacinto, Juan Jesús Martín-del-Río, Raúl M. Falcón, and Carlos Rubio-Bellido. 2020. "Rammed Earth Construction: A Proposal for a Statistical Quality Control in the Execution Process." *Sustainability* 12 (7): 2830. https://doi.org/10.3390/su12072830.

Caporuscio, Giulia. 2024. "Building Materials Cut From the Earth: A Look at Sod Construction Around the World and in New Mexico." https://digitalrepository.unm.edu/arch_etds/242.

Carcassi, Olga Beatrice, Yierfan Maierdan, Tashania Akemah, Shiho Kawashima, and Lola Ben-Alon. 2024. "Maximizing Fiber Content in 3D-Printed Earth Materials: Printability, Mechanical, Thermal and Environmental Assessments." *Construction and Building Materials* 425 (April): 135891. https://doi.org/10.1016/j.conbuildmat.2024.135891.

Chiappone, A., S. Marello, C. Scavia, and M. Setti. 2004. "Clay Mineral Characterization through the Methylene Blue Test: Comparison with Other Experimental Techniques and Applications of the Method." *Canadian Geotechnical Journal* 41 (6): 1168–78.

Chiusol, Alberto. 2018. "The First 3D Printed House with Earth | Gaia." *WASP*. https://www.3dwasp.com/en/3d-printed-house-gaia/.

Ciancio, Daniela, and Christopher Beckett, eds. 2015. *Rammed Earth Construction*. CRC Press. https://doi.org/10.1201/b18046.

Clarke, G. R. 1951. "The Evaluation of Soils and the Definition of Quality Classes from Studies of the Physical Properties of the Soil Profile in the Field." *Journal of Soil Science* 2 (1): 50–60. https://doi.org/10.1111/j.1365-2389.1951.tb00589.x.

Communities_of_Architects. 2023. "Sun-Dried Bricks: History, Challenges & Modern Applications!" *Swasya Living*. https://www.swasyaliving.com/post/sun-dried-bricks.

Côté, Jean, and Jean-Marie Konrad. 2005. "A Generalized Thermal Conductivity Model for Soils and Construction Materials." *Canadian Geotechnical Journal* 42 (2): 443–58. https://doi.org/10.1139/t04-106.

Crawford, James. 2017. *Fallen Glory: The Lives and Deaths of History's Greatest Buildings.* Macmillan + ORM.

Cruz, Ricardo, José Alexandre Bogas, Andrea Balboa, and Paulina Faria. 2024. "Water Resistance of Compressed Earth Blocks Stabilised with Thermoactivated Recycled Cement." *Materials* 17 (22): 5617. https://doi.org/10.3390/ma17225617.

Curth, Alexander, Barrak Darweesh, and Logman Arja Ronald Rael. 2020. "Advances in 3D Printed Earth Architecture: Onsite Prototyping with Local Materials." In *BE-AM Symposium*. Darmstadt, Germany.

Curth, Alexander, Natalie Pearl, Angelica Castro-Salazar, Caitlin Mueller, and Lawrence Sass. 2024. "3D Printing Earth: Local, Circular Material Processing, Fabrication Methods, and Life Cycle Assessment." *Construction and Building Materials* 421 (March): 135714. https://doi.org/10.1016/j.conbuildmat.2024.135714.

Deng, Fei, Jianguo Lu, Xusheng Wan, Boshi Liu, Binlong Zhang, and Hao Fu. 2025. "Mitigating Frost Heave of a Soil Stabilized with Sisal Fiber Exposed to Freeze-Thaw Cycles." *Geotextiles and Geomembranes* 53 (1): 394–404. https://doi.org/10.1016/j.geotexmem.2024.10.005.

Dente, Anthony, Michael G. Smith, and Massey Burke. 2024. *Essential Cob Construction: A Guide to Design, Engineering, and Building*. New Society Publishers.

Dilberoglu, Ugur M., Bahar Gharehpapagh, Ulas Yaman, and Melik Dolen. 2017. "The Role of Additive Manufacturing in the Era of Industry 4.0." In *Procedia Manufacturing 11*, 545–54. https://doi.org/10.1016/j.promfg.2017.07.148.

Ding, Xinlin, Chris Repka, Zhenghe Xu, and Jacob Masliyah. 2008. "Effect of Illite Clay and Divalent Cations on Bitumen Recovery." *The Canadian Journal of Chemical Engineering* 84 (6): 643–50. https://doi.org/10.1002/cjce.5450840602.

Dixon, J. B. 1989. *Minerals in Soil Environments*. Edited by S. B. Weed. Soil Science Society of American.

Doat, Patrice, Alain Hays, Hugo Houben, Sylvia Matuk, and François Vitoux. 1991. *Building with Earth*. The Mud Village Society.

Downton, Paul, and Dick Clarke. 2013. "Mud Brick." https://www.yourhome.gov.au/materials/mud-brick.

Egenti, C., and J. M. Khatib. 2016. "Sustainability of Compressed Earth as a Construction Material." In *Sustainability of Construction Materials*, 309–41. Elsevier. https://doi.org/10.1016/B978-0-08-100370-1.00013-5.

Eksi Akbulut, Dilek, and Rojda Naz Alibeyoglu. 2025. "Rammed Earth Construction: From Tradition to a Sustainable Future." In *Modern Masonry Construction – The Fusion of Tradition, Technology, and Sustainability [Working Title]*. IntechOpen. https://doi.org/10.5772/intechopen.1010217.

Faleschini, Flora, Daniel Trento, Maryam Masoomi, Carlo Pellegrino, and Mariano Angelo Zanini. 2023. "Sustainable Mixes for 3D Printing of Earth-Based Constructions." *Construction and Building Materials* 398 (September): 132496. https://doi.org/10.1016/j.conbuildmat.2023.132496.

Ferretti, Elena, Massimo Moretti, Alberto Chiusoli, Lapo Naldoni, Francesco de Fabritiis, and Massimo Visonà. 2022a. "Rice-Husk Shredding as a Means of Increasing the Long-Term Mechanical Properties of Earthen Mixtures for 3D Printing." *Materials* 15 (3): 743. https://doi.org/10.3390/ma15030743.

Ferretti, Elena, Massimo Moretti, Alberto Chiusoli, Lapo Naldoni, Francesco de Fabritiis, and Massimo Visonà. 2022b. "Mechanical Properties of a 3D-Printed Wall Segment Made with an Earthen Mixture." *Materials* 15 (2): 438. https://doi.org/10.3390/ma15020438.

Garg, Ankit, He Huang, Vinod Kushvaha, Priyanka Madhushri, Viroon Kamchoom, Insha Wani, Nevin Koshy, and Hong-Hu Zhu. 2020. "Mechanism of Biochar Soil Pore – Gas – Water Interaction: Gas Properties of Biochar-Amended Sandy Soil at Different Degrees of Compaction Using KNN Modeling." *Acta Geophysica* 68 (1): 207–17. https://doi.org/10.1007/s11600-019-00387-y.

Gaudette, H. E. 1964. "The Nature of Illite." *Clays and Clay Minerals* 13 (1): 33–48. https://doi.org/10.1346/CCMN.1964.0130105.

Giacomobono, Leonardo, Maria Argenti, Elena Ferretti, and Giulio Paparella. 2024. "Three-Dimensional Printing with Earthen Materials: A Settlement-Scale Design Experience." *Buildings* 14 (9): 2721. https://doi.org/10.3390/buildings14092721.

Giuffrida, Giada, Laurent Ibos, Abderrahim Boudenne, and Hamza Allam. 2024. "Analysis of the Thermal Performances of Uninsulated and Bio-Based Insulated Compressed Earth Blocks Walls: From the Material to the Wall Scale." *Journal of Building Engineering* 90 (August): 109370. https://doi.org/10.1016/j.jobe.2024.109370.

Gomaa, Mohamed, Wassim Jabi, Alejandro Veliz Reyes, and Veronica Soebarto. 2021. "3D Printing System for Earth-Based Construction: Case Study of Cob." *Automation in Construction* 124 (April): 103577. https://doi.org/10.1016/j.autcon.2021.103577.

Gomaa, Mohamed, Jaroslav Vaculik, Veronica Soebarto, Michael Griffith, and Wassim Jabi. 2021. "Feasibility of 3DP Cob Walls under Compression Loads in Low-Rise Construction." *Construction and Building Materials* 301 (September): 124079. https://doi.org/10.1016/j.conbuildmat.2021.124079.

Gosslar, Joschua. 2023. "With the Rammed Earth Robot on the Construction Site." *Technische Universität Braunschweig*. https://magazin.tu-braunschweig.de/en/pi-post/with-the-rammed-clay-robot-on-the-construction-site/.

Gounni, Ayoub, and Hasna Louahlia. 2020. "Dynamic Behavior and Economic Analysis of Sustainable Building Integrating Cob and Phase Change Materials." *Construction and Building Materials* 262: 120795.

Grim R. E. 1968. *Clay Mineralogy*. Fayetteville, TX: McGraw Hill. https://www.abebooks.com/Clay-Mineralogy-McGraw-Hill-International-Series-Earth/31692831387/bd.

Gualtieri, Alessandro F., Simone Ferrari, Matteo Leoni, Georg Grathoff, Richard Hugo, Mouath Shatnawi, Gianluca Paglia, and Simon Billinge. 2008. "Structural Characterization of the Clay Mineral Illite-1M." *Journal of Applied Crystallography* 41 (2): 402–15. https://doi.org/10.1107/S0021889808004202.

Gupta, Adarsh, and Mukesh Kumar. 2022. "Clayey Soil Stabilization Using Flyash and Jute Fibre." *Materials Today: Proceedings* 48: 1205–10. https://doi.org/10.1016/j.matpr.2021.08.246.

Haddad, Kamal, Simon Lannon, and Eshrar Latif. 2024. "Investigation of Cob Construction: Review of Mix Designs, Structural Characteristics, and Hygrothermal Behaviour." *Journal of Building Engineering* 87 (June): 108959. https://doi.org/10.1016/j.jobe.2024.108959.

Hall, M. R., K. B. Najim, and P. Keikhaei Dehdezi. 2012. "Soil Stabilisation and Earth Construction: Materials, Properties and Techniques." In *Modern Earth Buildings*, 222–55. Elsevier. https://doi.org/10.1533/9780857096166.2.222.

Hamard, Erwan, Bogdan Cazacliu, Andry Razakamanantsoa, and Jean-Claude Morel. 2016. "Cob, a Vernacular Earth Construction Process in the Context of Modern Sustainable Building." *Building and Environment* 106 (September): 103–19. https://doi.org/10.1016/j.buildenv.2016.06.009.

Hohn, Charles M. 2011. "ABCs of Making Adobe Bricks." *Guide G-521*. https://pubs.nmsu.edu/_g/G521/index.html.

Hojati, Maryam, Zhanzhao Li, Ali M. Memari, Keunhyoung Park, Mehrzad Zahabi, Shadi Nazarian, José P. Duarte, and Aleksandra Radlińska. 2022. "3D-Printable Quaternary Cementitious Materials towards Sustainable Development: Mixture Design and Mechanical Properties." *Results in Engineering* 13 (March): 100341. https://doi.org/10.1016/J.RINENG.2022.100341.

Hojati, Maryam, Aleksandra Radlińska, Shadi Nazarian, José P. Duarte, Ali M. Memari, and Sven G. Bilén. 2018. "Synthesis and 3D Printing of One-Part Geopolymer Mortar." In *1st RILEM International Conference on Concrete and Digital Fabrication*. Zurich, Switzerland.

IBC. 2018. "2018 International Building Code Illustrated Handbook." In *International Code Council (ICC)*. IBC.

Ing. Harald Kloft. 2019. "Robotic Rammed Earth." *Technische Universität Braunschweig*. https://www.tu-braunschweig.de/ite/research/roboticrammedearth.

ISO/ASTM 52900. 2015. "Additive Manufacturing – General Principles –Terminology." In *International Organization for Standardization*. Geneva, Switzerland. www.iso.org-www.astm.org.

Jaquin, P. 2012. "History of Earth Building Techniques." In *Modern Earth Buildings*, 307–23. Elsevier. https://doi.org/10.1533/9780857096166.3.307.

Jaquin, P. A., and C. E. Augarde. 2012. *Earth Building: History, Science and Conservation*. https://durham-repository.worktribe.com/output/1124958.

Jarrige, Jean-François, and Richard H. Meadow. 1980. "The Antecedents of Civilization in the Indus Valley." *Scientific American* 243 (2): 122–37. http://www.jstor.org/stable/24966394.

Kita, Yuko, Miguel Domínguez Acosta, Aldo Izaguirre Pompa, Montserrat Hernández Ramírez, Patricia Girón García, Alberto Peña Rodríguez, Rafael Cruz Antillón, and Eduardo Pío Gamboa Carrera. 2024. "Identification of Soil Selection and Construction Techniques of Pre-Columbian Architecture in the Casas Grandes Region." *Journal of Archaeological Science: Reports* 55 (May): 104466. https://doi.org/10.1016/j.jasrep.2024.104466.

Laborel-Préneron, A., J. E. Aubert, C. Magniont, C. Tribout, and A. Bertron. 2016. "Plant Aggregates and Fibers in Earth Construction Materials: A Review." *Construction and Building Materials* 111 (May): 719–34. https://doi.org/10.1016/j.conbuildmat.2016.02.119.

Lemmin-Woolfrey, Ulrike. 2021. "Yemen's Ancient, Soaring Skyscraper Cities." *BBC*. https://www.bbc.com/travel/article/20211004-yemens-ancient-soaring-skyscraper-cities.

Li, Zhanzhao, Maryam Hojati, Zhengyu Wu, Jonathon Piasente, Negar Ashrafi, José P. Duarte, Shadi Nazarian, Sven G. Bilén, Ali M. Memari, and Aleksandra Radlińska. 2020. "Fresh and Hardened Properties of Extrusion-Based 3D-Printed Cementitious Materials: A Review." *Sustainability* 12 (14): 5628. https://doi.org/10.3390/su12145628.

Liuzzi, Stefania, and P. Stefanizzi. 2014. "Soil Based Building Materials for Energy Efficiency." *Key Engineering Materials* 632 (November): 15–38. https://doi.org/10.4028/www.scientific.net/KEM.632.15.

Mai, Inka, Joschua Gosslar, Noor Khader, Dirk Lowke, and Norman Hack. 2024. "Workflow for Earth-Based 3D-Printing." In *Second RILEM International Conference on Earthen Construction, Conference Paper*, 71–80. https://doi.org/10.1007/978-3-031-62690-6_8.

McNally, G. H. 1998. *Soil and Rock Construction Materials*. E & FN Spon. https://www.routledge.com/Soil-and-Rock-Construction-Materials/McNally/p/book/9780419214205.

Meyer, Kirby, and Dean R. Read. 2015. *Soils Earthwork and Foundations: A Practical Approach Based*. International Code Council.

Miller, Sabbie A., Vanderley M. John, Sergio A. Pacca, and Arpad Horvath. 2018. "Carbon Dioxide Reduction Potential in the Global Cement Industry by 2050." *Cement and Concrete Research* 114 (December): 115–24. https://doi.org/10.1016/j.cemconres.2017.08.026.

Minke, Gernot. 2009. *Building with Earth: Design and Technology of a Sustainable Architecture*. 2nd ed. Birkhäuser Architecture.

Moayed, Seyed Amir Mir, Mahyar Arabani, and Hadi Ahmadi. 2025. "Influence of Hemp Fiber on the Mechanical Characteristics and Freeze-Thaw Durability of LKD-Nano Silica-Stabilized Kaolin Soil." *Results in Engineering* 26 (June): 105318. https://doi.org/10.1016/j.rineng.2025.105318.

Moore, D. M., and R. C. Reynolds. 1997. *X-Ray Diffraction and the Identification and Analysis of Clay Minerals*. New York: Oxford university press.

Murthy, V. N. S. 2002. *Geotechnical Engineering: Principles and Practices of Soil Mechanics and Foundation Engineering (Civil and Environmental Engineering)*. CRC Press.

Nagaraj, H. B. 2016. "Influence of Gradation and Proportion of Sand on Stress–Strain Behavior of Clay–Sand Mixtures." *International Journal of Geo-Engineering* 7 (1): 19. https://doi.org/10.1186/s40703-016-0033-8.

Nanda, Radhikesh Prasad, Nijee Priya, and Ashutosh Kumar. 2025. "Enhancement of Strength of Soft Soil Utilizing Construction and Demolition Wastes Reinforced with Recron-3s Fibres." *Journal of Cleaner Production* 505 (May): 145442. https://doi.org/10.1016/j.jclepro.2025.145442.

Narloch, Piotr, Piotr Woyciechowski, Jakub Kotowski, Ireneusz Gawriuczenkow, and Emilia Wójcik. 2020. "The Effect of Soil Mineral Composition on the Compressive Strength of Cement Stabilized Rammed Earth." *Materials* 13 (2): 324. https://doi.org/10.3390/ma13020324.

Natural Resources Conservation Service. 2025. "A Soil Profile." *USDA*. https://www.nrcs.usda.gov/resources/education-and-teaching-materials/a-soil-profile.

Nazarian, Shadi, José P. Duarte, Sven G. Bilén, Ali Memari, Aleksandra Radlinska, Nicholas A. Meisel, and Maryam Hojati. 2021. "Additive Manufacturing of Architectural Structures: An Interplay Between Materials, Systems, and Design." In *Advances in Science, Technology & Innovation*, 111–19. https://doi.org/10.1007/978-3-030-35533-3_15.

Oates, David. 1990. "Innovations in Mud-brick: Decorative and Structural Techniques in Ancient Mesopotamia." *World Archaeology* 21 (3): 388–406. https://doi.org/10.1080/00438243.1990.9980115.

Oudbashi, Omid, Atefeh Shekofteh, and Hamid Fadaei. 2023. "An Exploration of a 2500-Year-Old Monument: A Reappraisal of the Achaemenian Decorative Flooring Plasters in Persepolis World Heritage Site (ca. 550–330 BCE)." *Studies in Conservation* 68 (5): 491–501. https://doi.org/10.1080/00393630.2022.2062562.

Paul, Shantanu, Ashna Tasnim, and Jaya Majumder. 2025. "Enhancement of Micro-Mechanical Characteristics of Expansive Soil through Synergistic Incorporation of Jute and Nylon Fibers with Cement." *Results in Engineering* 26 (June): 104685. https://doi.org/10.1016/j.rineng.2025.104685.

Paul Dana, Leo, and Robert Brent Anderson. 2007. "Taos Pueblo: An Indigenous Community Holding on to Promethean Values." *Journal of Enterprising Communities: People and Places in the Global Economy* 1 (4): 321–36. https://doi.org/10.1108/17506200710833827.

Perrot, A., D. Rangeard, and E. Courteille. 2018. "3D Printing of Earth-Based Materials: Processing Aspects." *Construction and Building Materials* 172 (May): 670–76. https://doi.org/10.1016/j.conbuildmat.2018.04.017.

Puzatova, Anastasia, Pshtiwan Shakor, Vittoria Laghi, and Maria Dmitrieva. 2022. "Large-Scale 3D Printing for Construction Application by Means of Robotic Arm and Gantry 3D Printer: A Review." *Buildings* 12 (11): 2023. https://doi.org/10.3390/buildings12112023.

Ramos Huarachi, Diego Alexis, Giovanna Gonçalves, Antonio Carlos de Francisco, Maria Helene Giovanetti Canteri, and Cassiano Moro Piekarski. 2020. "Life Cycle Assessment of Traditional and Alternative Bricks: A Review." *Environmental Impact Assessment Review* 80 (January): 106335. https://doi.org/10.1016/j.eiar.2019.106335.

Ranjan, Sumit. 2020. "Manhattan of the Desert-Yemen's Ancient Mud Skyscraper City." *Slideshare*. https://www.slideshare.net/slideshow/manhattan-of-the-desertyemens-ancient-mud-skyscraper-city/66773173.

Reynolds, William Earl. 1981. *No the Ethnoarchaeology of Pueblo Architecture*. Arizona State University.

Ricciotti, Laura, Antonio Apicella, Valeria Perrotta, and Raffaella Aversa. 2023. "Geopolymer Materials for Extrusion-Based 3D-Printing: A Review." *Polymers* 15 (24): 4688. https://doi.org/10.3390/polym15244688.

Riza, F. V., and I. A. Rahman. 2015. "The Properties of Compressed Earth-Based (CEB) Masonry Blocks." In *Eco-Efficient Masonry Bricks and Blocks*, 379–92. Elsevier. https://doi.org/10.1016/B978-1-78242-305-8.00017-6.

Rückrich, Stefanie, Galit Agranati, and Yasha Jacob Grobman. 2023. "Earth-Based Additive Manufacturing: A Field-Oriented Methodology for Evaluating Material Printability." *Architectural Science Review* 66 (2): 133–43. https://doi.org/10.1080/00038628.2022.2154739.

Runge, Ryan. 2024. "We Build the World's Best Compressed Earth Block (CEB) Machinery." *AECT*. https://aectearthblock.com/.

Sapriza-Azuri, Gonzalo, Pablo Gamazo, Saman Razavi, and Howard S. Wheater. 2018. "On the Appropriate Definition of Soil Profile Configuration and Initial Conditions for Land Surface–Hydrology Models in Cold Regions." *Hydrology and Earth System Sciences* 22 (6): 3295–309. https://doi.org/10.5194/hess-22-3295-2018.

Samarasinghe, Don. 2024. "Raw-Earth 3D Printing Construction Technology." *CIOB*. https://www.ciob.org/blog/rawearth-3d-printing-construction-technology.

Schmitz, Linus Paul, Joschua Gosslar, Evelien Dorresteijn, Dirk Lowke, and Harald Kloft. 2024. "Experimental Investigations on the Compaction Energy for a Robotic Rammed Earth Process." *Frontiers in Built Environment* 10 (February). https://doi.org/10.3389/fbuil.2024.1363804.

Seed, H. B., Richard J. Woodward Jr., and R. Lundgren. 1962. "Prediction of Swelling Potential for Compacted Clays." *Journal of the Soil Mechanics and Foundations Division* 88 (3). https://doi.org/10.1061/JSFEAQ.000043.

Shilar, Fatheali A., Sharanabasava V. Ganachari, Veerabhadragouda B. Patil, B. E. Bhojaraja, T. M. Yunus Khan, and Naif Almakayeel. 2023. "A Review of 3D Printing of Geopolymer Composites for Structural and Functional Applications." *Construction and Building Materials* 400 (October): 132869. https://doi.org/10.1016/j.conbuildmat.2023.132869.

Shoukat, Rizwan, Marta Cappai, Giorgio Pia, Tadeusz Kubaszek, Roberto Ricciu, Łukasz Kolek, and Luca Pilia. 2025. "Thermal Conductivity of Sustainable Earthen Materials Stabilized by Natural and Bio-Based Polymers: An Experimental and Statistical Analysis." *Energies* 18 (12): 3144. https://doi.org/10.3390/en18123144.

Sir Banister Fletcher, and Dan Cruickshank. 1996. *Sir Banister Fletcher's a History of Architecture*. Architectural Press.

Smith, Edward W., and George S. Austin. 1989. *Adobe, Pressed-Earth, and Rammed-Earth Industries in New Mexico*. Socorro, NM: New Mexico Bureau of Mines & Mineral Resources.

Spencer, A. J. 1979. *Brick Architecture in Ancient Egypt (Oxbow Classics in Egyptology)*: ISBN: 9798888570302. https://studylibfr.com/doc/10119082/spencer--brick-architecture-in-ancient-egypt.1979?utm_source=chatgpt.com.

Środoń, J. 2006. "Identification and Quantitative Analysis of Clay Minerals." In *Developments in Clay Science*, 765–87. https://doi.org/10.1016/S1572-4352(05)01028-7.

Tarhan, Yeşim, and Arnaud Perrot. 2023. "Reinforcement of 3D Printable Earth-Based Mortar with Natural Textile Material." *Materials Today: Proceedings* (August). https://doi.org/10.1016/j.matpr.2023.08.056.

Thompson, David, Charles Augarde, and Juan Pablo Osorio. 2022. "A Review of Current Construction Guidelines to Inform the Design of Rammed Earth Houses in Seismically Active Zones." *Journal of Building Engineering* 54 (August): 104666. https://doi.org/10.1016/j.jobe.2022.104666.

Tinoco, Matheus Pimentel, Érica Martinho de Mendonça, Letícia Ikeda Castrillon Fernandez, Lucas Rosse Caldas, Oscar Aurelio Mendoza Reales, and Romildo Dias Toledo Filho. 2022. "Life Cycle Assessment (LCA) and Environmental Sustainability of Cementitious Materials for 3D Concrete Printing: A Systematic Literature Review." *Journal of Building Engineering* 52 (July): 104456. https://doi.org/10.1016/j.jobe.2022.104456.

Türköz, Murat, and Hasan Tosun. 2011. "The Use of Methylene Blue Test for Predicting Swell Parameters of Natural Clay Soil." *Scientific Research and Essays* 6 (8). https://doi.org/10.5897/SRE10.629.

Umar, Ibrahim Haruna, Sale Abubakar, Abdullahi Balarabe Bello, Hang Lin, Jubril Izge Hassan, and Rihong Cao. 2025. "Stabilization of Expansive Soils Using Cement – Zeolite Mixtures: Experimental Study and Lasso Modeling." *Materials* 18 (10): 2286. https://doi.org/10.3390/ma18102286.

Valenzuela, Marian, Gustavo Ciudad, Juan Pablo Cárdenas, Carlos Medina, Alexis Salas, Angelo Oñate, Gonzalo Pincheira, Shady Attia, and Víctor Tuninetti. 2024. "Towards the Development of Performance-Efficient Compressed Earth Blocks from Industrial and Agro-Industrial by-Products." *Renewable and Sustainable Energy Reviews* 194 (April): 114323. https://doi.org/10.1016/j.rser.2024.114323.

Verhoef, P. 1992. *The Methylene Blue Adsorption Test Applied to Geomaterials*. Vol. 101. Delft University of Technology, Laboratory of Engineering Geology.

Wang, Jingying, and Yan Yung Edward Ng. 2023. "Post-Earthquake Housing Recovery with Traditional Construction: A Preliminary Review." *Progress in Disaster Science* 18 (April): 100283. https://doi.org/10.1016/j.pdisas.2023.100283.

Wangler, Timothy, Ena Lloret, Lex Reiter, Norman Hack, Fabio Gramazio, Matthias Kohler, Mathias Bernhard, et al. 2016. "Digital Concrete: Opportunities and Challenges." *RILEM Technical Letters* 1 (October): 67. https://doi.org/10.21809/rilemtechlett.2016.16.

Wangler, Timothy, Nicolas Roussel, Freek P. Bos, Theo A. M. Salet, and Robert J. Flatt. 2019. "Digital Concrete: A Review." *Cement and Concrete Research* 123 (March). https://doi.org/10.1016/j.cemconres.2019.105780.

Wani, Insha, Vinod Kushvaha, Ankit Garg, Rakesh Kumar, Sambit Naik, and Prabhakar Sharma. 2022. "Review on Effect of Biochar on Soil Strength: Towards Exploring Usage of Biochar in Geo-Engineering Infrastructure." *Biomass Conversion and Biorefinery*, May. https://doi.org/10.1007/s13399-022-02795-5.

Wani, Insha, Shradul R. Narde, Xilong Huang, Neelancherry Remya, Vinod Kushvaha, and Ankit Garg. 2023. "Reviewing Role of Biochar in Controlling Soil Erosion and Considering Future Aspect of Production Using Microwave Pyrolysis Process for the Same." *Biomass Conversion and Biorefinery* 13 (13): 11543–69. https://doi.org/10.1007/s13399-021-02060-1.

Wi, Kwangwoo, Vignesh Suresh, Kejin Wang, Beiwen Li, and Hantang Qin. 2020. "Quantifying Quality of 3D Printed Clay Objects Using a 3D Structured Light Scanning System." *Additive Manufacturing* 32 (March): 100987. https://doi.org/10.1016/j.addma.2019.100987.

Willett, Jason Christopher. 2022. "Clay and Shale (Advance Release)." https://pubs.usgs.gov/myb/vol1/2018/myb1-2018-clay-shale.pdf.

Xie, Liye, Duo Wang, Haitao Zhao, Jiangtao Gao, and Tiziana Gallo. 2021. "Architectural Energetics for Rammed-Earth Compaction in the Context of Neolithic to Early Bronze Age Urban Sites in Middle Yellow River Valley, China." *Journal of Archaeological Science* 126 (February): 105303. https://doi.org/10.1016/j.jas.2020.105303.

Yang, Mengyuan, Dongxian Zhou, Huixian Hang, Shuo Chen, Hua Liu, Jikang Su, Huilin Lv, Huixin Jia, and Gengmao Zhao. 2024. "Effects of Balancing Exchangeable Cations Ca, Mg, and K on the Growth of Tomato Seedlings (Solanum Lycopersicum L.) Based on Increased Soil Cation Exchange Capacity." *Agronomy* 14 (3): 629. https://doi.org/10.3390/agronomy14030629.

Yousaf, Arslan, Ans Al Rashid, and Muammer Koç. 2024. "Parameter Tuning for Sustainable 3D Printing (3DP) of Clay Structures." *Journal of Engineering Research* (May). https://doi.org/10.1016/j.jer.2024.05.027.

Zafar, Muhammad Saeed, Amir Bakhshi, and Maryam Hojati. 2023. "Printability and Shape Fidelity Evaluation of Self-Reinforced Engineered Cementitious Composites."

Construction and Building Materials 408 (December): 133676. https://doi.org/10.1016/j.conbuildmat.2023.133676.

Zafar, Muhammad Saeed, Adnan Shahid, Reza Sedghi, and Maryam Hojati. 2025. "Optimization of Biopolymer Additives for 3D Printable Cementitious Systems: A Design of Experiment Approach." *Case Studies in Construction Materials* 22 (July): e04515. https://doi.org/10.1016/j.cscm.2025.e04515.

Zaini, Muhammad Syamsul Imran, Muzamir Hasan, Sultan Almuaythir, and Masayuki Hyodo. 2024. "Experimental Investigations on Physico-Mechanical Properties of Kaolinite Clay Soil Stabilized at Optimum Silica Fume Content Using Clamshell Ash and Lime." *Scientific Reports* 14 (1): 10995. https://doi.org/10.1038/s41598-024-61854-1.

Zeiher, Laura C. 1996. *The Ecology of Architecture: A Complete Guide to Creating the Environmentally Conscious Building*. Whitney Library of Design. www.buildinggreen.com/newsbrief/ecology-architecture-complete-guide-creating-environmentally-conscious-building.

Zhai, Zhiqiang (John), and Jonathan M. Previtali. 2010. "Ancient Vernacular Architecture: Characteristics Categorization and Energy Performance Evaluation." *Energy and Buildings* 42 (3): 357–65. https://doi.org/10.1016/j.enbuild.2009.10.002.

Zhang, Jiahua, Yunjun Wei, Yuanlei Zhu, Daobing Zhang, and Biao Zhang. 2025. "Stability Analysis of Three-Dimensional Slope Reinforced by Natural Fiber Include Wheat Straw and Bamboo." *Scientific Reports* 15 (1): 4598. https://doi.org/10.1038/s41598-025-89086-x.

Zhou, Xiang, Dong Liu, Hongling Bu, Liangliang Deng, Hongmei Liu, Peng Yuan, Peixin Du, and Hongzhe Song. 2018. "XRD-Based Quantitative Analysis of Clay Minerals Using Reference Intensity Ratios, Mineral Intensity Factors, Rietveld, and Full Pattern Summation Methods: A Critical Review." *Solid Earth Sciences* 3 (1): 16–29. https://doi.org/10.1016/j.sesci.2017.12.002.

Zozaya, Daiquiri Devon. 2024. *3D Printing of Local New Mexican Soil Material: Improving Durability Properties by Assessing Traditional Methods towards Alternative Sustainable Construction*. University of New Mexico.

Zozaya, Daiquiri Devon, Shiva Bhusal, Francisco Uviña-Contreras, and Maryam Hojati. 2024. "Evaluating Grain Size Distribution of Local Soils for 3d Printing in Construction." In *Earth USA 2024*. Santa Fe.

Zozaya, Daiquiri, Hamideh Shojaeian, Francisco Uviña-Contreras, and Maryam Hojati. 2026. "3D Printing of Earth-Based Mixtures: Linking Material Design, Printability, and Structural Performance." *Buildings* 16 (6): 1261. https://doi.org/10.3390/buildings16061261

Zwell, L., and A. W. Danko. 1975. "Applications of X-Ray Diffraction Methods to Quantitative Chemical Analysis." *Applied Spectroscopy Reviews* 9 (1): 167–221. https://doi.org/10.1080/05704927508081490.

Part 3

Modeling and Simulation in 3D Printing of Concrete

11 Numerical and Analytical Modeling of Fresh-State Rheology and Stability in 3D Concrete Printing

Yohan Jacquet and Inka Mai

11.1 INTRODUCTION: TOWARD A DIGITAL MODELING WORKFLOW FOR 3D PRINTING IN CONSTRUCTION

The integration of digital tools in 3D concrete printing (3DCP) is transforming the construction industry by enhancing efficiency, precision, and automation. The digital workflow in 3DCP encompasses a series of computational and simulation-based techniques that facilitate geometric modeling, material behavior prediction, and process optimization. The adoption of building information modeling (BIM) further strengthens these digital workflows; further potential is given in 3DCP by inline process control, which provides not only feedback but also takes action when deviations from as-planned structures is detected in as-built structures.

Digital tools provide a wide range of functionalities that enhance the design of innovative structures in additive manufacturing (AM). Advanced parametric modeling techniques, including generative design and algorithmic modeling, are increasingly being integrated to optimize complex geometries while minimizing material usage and ensuring structural stability (Slepicka et al., 2021).

This parametric method efficiently identifies relationships between printing parameters using the simulated annealing optimization technique. To achieve this, a comprehensive script is generated in Grasshopper, a visual programming plugin for Rhino. This script encapsulates all critical printing parameters, including material properties, geometric configurations, and load conditions. Once generated, the script is executed within computational analysis software controlled via Python, allowing for real-time adjustments during the printing process. Python facilitates the structural analysis of the printed elements, the optimization of printing time, and material consumption. As 3DCP technology continues to evolve, further integration of digital modeling with real-time feedback mechanisms is expected to enhance precision and adaptability.

Several research projects have successfully employed this methodology. For example, Lim et al. (2016) utilized Grasshopper to generate complex, curved print trajectories and seamlessly transfer them to a robotic arm for precise execution of the

DOI: 10.1201/9781003434863-14

designed model. Such integrations of digital tools and automation continue to push the boundaries of 3DCP. Today, it is a standardized workflow used in the 3DCP field, especially for printing systems based on robotic arms.

Beyond geometric optimization, which using topology optimization algorithms provides a huge material saving potential (Dörrie et al., 2023; Jewett & Carstensen, 2019), accurately predicting and controlling material behavior are critical for ensuring high-quality, i.e., geometric precise and smooth, printed structures. Computational techniques help analyze the flow properties of printable concrete mixtures, considering factors like extrusion rate, buildability, interlayer adhesion, but also self-heating (Jacquet et al., 2024). Finite element method (FEM) simulations assess structural stability, deformation, and failure modes (Davtalab et al., 2022), while computational fluid dynamics (CFD) optimizes deposition patterns to mitigate defects such as cold joints and excessive porosity. Besides these numerical approaches, machine learning (ML) models are increasingly used to predict material performance based on composition and environmental conditions (Kazemian et al., 2019; Westphal & Seitz, 2022). Artificial intelligence (AI)-driven real-time monitoring systems are also being developed to dynamically adjust process parameters based on in situ stability assessments (Silva et al., 2024; Yang et al., 2022).

Despite these advancements, accurate material behavior prediction remains one of the major challenges in the field. This chapter explores a hybrid approach that combines experimental rheological measurements with numerical modeling and stability prediction techniques, providing a comprehensive framework for improving material characterization in 3DCP.

11.2 ADVANCEMENTS IN COMPUTATIONAL MODELING RELATING TO 3D CONCRETE PRINTING

11.2.1 Numerical Simulation Techniques and Challenges

Concrete is a highly poly-disperse material due to prevalent particle diameters of concrete components ranging from hundreds of nanometers up to tens of millimeters. Hence, mechanisms determining the disperse and the rheological or mechanical properties of concrete are often considered at different scales (Banfill, 1990; Roussel et al., 2010). Numerical methods that are typically used to describe concrete are divided into three main categories (Roussel et al., 2007; Vasilic et al., 2019):

- Continuous: e.g., CFD including FEM and lattice Boltzmann method (LBM);
- Discrete: e.g., Discrete element method (DEM), smoothed particle hydrodynamics (SPH), and peridynamics (PD);
- Hybrid as a coupling between continuous and discrete methods.

Given the complexity and variability of materials, structures, processes, and their interactions, computational modeling plays a crucial role in enhancing the reliability and performance of 3DCP. By ensuring both flowability and buildability, it contributes to the consistency and stability of printed structures, driving the field toward a more universal approach. Numerous computational techniques have shown promising results in this domain.

Fresh-state construction materials, such as cement paste, mortar, and fresh concrete, exhibit complex rheological behavior, including non-Newtonian flow, yield stress, thixotropy, and particle interactions. Accurately simulating these behaviors requires a combination of continuous, discrete, and hybrid numerical methods.

11.2.1.1 Continuous Methods

These methods treat materials as continuous media, applying fluid dynamics principles to simulate flow behavior.

- CFD: Models fresh concrete as a non-Newtonian fluid with viscosity and yield stress properties. Widely used for simulating casting, pumping, and formwork filling (Choi et al., 2014; Roussel & Gram, 2014).
- FEM for Fluids: Adapts FEM to solve Navier–Stokes equations for fluid-like materials, capturing large deformations and free-surface behavior (Roussel et al., 2007). FEM also provides insights into structural responses under self-weight, mechanical loads, thermal effects from hydration, and deformation stability during printing.
- SPH: A mesh-free method representing fluids as discrete particles, making it effective for highly dynamic flows and large deformations, such as self-compacting concrete (SCC) filling (Kulasegaram et al., 2011; Martys et al., 2010; Tran-Duc et al., 2021).
- LBM: A meso-scale discrete approach bridging microscopic molecular dynamics and macroscopic continuum models. LBM efficiently models flow in porous media and interactions with reinforcement, offering faster computation, higher parallel efficiency, and easier handling of complex boundaries than traditional CFD methods (Li et al., 2021; Martys, 2011).

11.2.1.2 Discrete Methods

These methods represent materials as collections of individual particles, making them ideal for modeling granular behavior and particle–fluid interactions.

- DEM: A computational method that calculates finite particle displacements and rotations while performing contact detection in particle assemblies. Applied to cement-based materials, DEM simulates granular flow, particle collisions, and aggregate distribution in fresh concrete (Cao et al., 2015; Krenzer et al., 2022; Mechtcherine et al., 2014; Mechtcherine & Shyshko, 2015; Shyshko & Mechtcherine, 2013).
- PD for Fluids: A reformulation of classical continuum mechanics, PD models material properties through micro-bond interactions, enabling the simulation of fluid-to-solid transitions. Unlike FEM, DEM, or CFD, PD offers greater discretization flexibility while avoiding FEM's mesh dependency. Although less effective in modeling pure fluid flow than CFD and DEM, PD uniquely integrates both fluid- and solid-like behaviors (Zhu et al., 2023).

11.2.1.3 Hybrid and Coupled Methods

A major shortcoming of pure discrete approaches is their failure to account for the dissipative character of shear flow (Krenzer & Schwabe, 2009; Shyshko & Mechtcherine, 2010). A more accurate description is expected through the coupling of Eulerian and Lagrangian methods that take both the discrete and continuous characteristics of concrete into account. Combining continuous and discrete models enhances the accuracy of multiphase behavior in fresh construction materials.

- CFD–DEM: This method couples CFD (continuous fluid) with DEM (discrete particles) to model aggregate motion, particle segregation, and viscosity evolution in fresh concrete. Such multiscale modeling further enhances predictive capabilities by bridging different levels of material behavior. Cementitious materials transition rapidly between different rheological states during extrusion and deposition, requiring an integrated approach that considers both microscopic particle interactions and macroscopic flow properties.
- Particle Finite Element Method (PFEM): A Lagrangian particle-based extension of FEM, PFEM represents materials as discrete particles while maintaining FEM principles. It accurately captures free-surface flows, fluid-to-solid transitions, and interlayer bonding properties. PFEM is particularly effective for modeling extrusion, mixing, pumping, formwork filling, and reinforcement interactions, ensuring uniform deposition, defect reduction, and improved structural stability.
- Immersed Boundary Method (IBM): This method embeds solid objects (e.g., reinforcement bars) within a CFD-based fluid model, improving simulation efficiency for reinforced concrete casting (Peskin, 2002).

One of the key challenges in numerical simulations these days lies in the absence of well-defined constitutive models for cementitious materials. The time-dependent behaviors of hydration, shrinkage, and creep complicate the accurate prediction of rheological evolution during printing. While finite element (FE) and discrete element (DE) simulations can generate synthetic datasets to compensate for the lack of standardized experimental data, their predictive accuracy remains limited by incomplete material models as, e.g., pure discrete approaches do not take the dissipative character of shear flow into account.

11.2.2 Computational Cost Considerations

The computational demands of FEM and PFEM simulations necessitate efficient strategies to enhance performance while maintaining accuracy. Key considerations include the following:

- Optimized Meshing Strategies: Adaptive mesh refinement techniques reduce computational costs by focusing resolution on critical regions.
- Parallel Computing: Leveraging high-performance computing (HPC) architectures to accelerate large-scale simulations.

- Reduced-Order Models (ROMs): Simplified representations of complex physical processes that retain essential dynamics while significantly reducing computational expense.

Given the large dimensions of 3D-printed concrete (3DPC) structures, traditional FEMs require complex mesh generation and refinement, significantly increasing computational costs. Voxel-based meshing presents a viable alternative, offering a simplified and computationally efficient approach for discretizing the geometric model of printed structures (Vantyghem et al., 2021). The voxelization technique, which transforms 3D geometry into a structured grid of cubic elements (voxels), streamlines the meshing process and facilitates automatic mesh generation without excessive preprocessing efforts. By enabling automated data transfer from CAD models to FE analysis software, voxel-based methods eliminate manual intervention and improve simulation efficiency.

One of the key advantages of voxel-based meshing is its ability to maintain computational efficiency while preserving essential geometric and material details. Unlike conventional meshing techniques, which often involve complex tetrahedral or hexahedral element generation, voxelization ensures a structured and uniform discretization that simplifies numerical processing. This approach has been validated in studies where voxel-based FE simulations have successfully predicted failure mechanisms in 3D-printed structures, demonstrating comparable accuracy to traditional FEMs but with significantly reduced computational overhead (Vantyghem et al., 2021).

Another promising approach to mesh simplification is the CobraPrint method, which segments the print path into discrete elements, allowing for sequential activation during simulation. This technique enables the incorporation of contact properties between layers, improving the representation of interlayer bonding and mechanical interactions (Ooms et al., 2021). Furthermore, combining voxelization with reduced-order modeling techniques enhances computational efficiency by approximating the system's response using lower-dimensional representations, retaining essential dynamics while reducing the numerical burden.

ML combined with numerical simulation holds great potential for predicting the stability and flow behavior of cement-based materials in 3D printing applications. ML can analyze large datasets to identify patterns and optimize process parameters, while numerical simulations provide physics-based insights into material behavior under different conditions. In this way, physics-informed neural networks (PINNs) can improve convergence while lowering the required data amounts (Schøler et al., 2024). This combination enables real-time adjustments, improving printability, structural integrity, and material efficiency. However, significant challenges remain, primarily due to the lack of well-defined constitutive models for cement-based materials. While numerical simulations can generate synthetic training data, the absence of established equations makes it difficult to accurately describe time-dependent behaviors such as hydration, shrinkage, and creep.

A key limitation is the difficulty in generalizing ML models beyond training data. Cement-based materials exhibit complex, evolving rheology influenced by factors such as environmental conditions, printing speed, and layer-by-layer deposition techniques (Carneau et al., 2020). Without underlying physics constraints, ML struggles to adapt to new materials or process variations, reducing its reliability for optimizing real-time 3D printing as ML requires larger learning datasets. Additionally, data scarcity and quality issues hinder the effectiveness of purely data-driven approaches. Unlike bituminous materials (Khadijeh et al., 2025), cementitious systems lack standardized datasets due to variations in mix design, curing conditions, and process parameters. Numerical simulations, such as FEM or DEM, can partially compensate by generating synthetic data, but their accuracy is limited by the absence of robust constitutive models.

Another major challenge is multiscale integration – the transition between the layer scale and the structural scale, both exhibiting a time-dependent evolution. In 3D printing, cementitious materials transition rapidly between different rheological states during extrusion and deposition, making it difficult for ML to capture behavior across scales. Without well-defined constitutive equations, ML-based predictions often lack physical consistency, limiting their ability to optimize printing parameters or predict failure mechanisms. Recent advancements, such as PINNs, attempt to integrate physical principles into ML models, but their application to cement-based materials remains constrained by the incomplete understanding of material behavior, unlike the themes related to hardened concrete for which there is work available, e.g., on the diffusion of chloride and the monitoring of hydration at an early age (Rahman et al., 2024; Shaban et al., 2023).

Despite these challenges, the combination of ML and numerical simulations presents a promising path for enhancing 3D printing of cement-based materials. Future research should focus on developing hybrid models that integrate experimental data, numerical simulations, and physics-based constraints to improve predictive accuracy. By bridging the gap between empirical observations and theoretical modeling, ML-driven approaches can enable more reliable and adaptive solutions, ultimately advancing the efficiency and performance of 3D-printed cementitious structures.

11.3 FROM SINGLE-LAYER TO MULTILAYER 3D CONCRETE PRINTING PROCESS OPTIMIZATION

11.3.1 Relevance of Rheological Parameters to Describe the Complexity of the 3DCP Process

3DCP is a multifaceted process encompassing several interdependent stages, from mixing, pumping to extruding the material (see Figure 11.1).

Each step in the process imposes distinct mechanical and rheological demands on the material. Understanding these requirements is crucial for optimizing the printing process and ensuring the structural integrity of the printed elements. The following is an in-depth exploration of each stage, highlighting the specific loads and the pertinent rheological parameters involved.

11.3.1.1 Dynamic Shear Yield Stress

The dynamic shear yield stress represents the minimum shear stress required to initiate and sustain flow in a material under shear deformation. Yield stress determines

FIGURE 11.1 Process steps in 3DCP. Credit: Inka Mai.

the onset of flow and significantly influences the shape stability of printed layers. A high yield stress prevents excessive deformation and supports the self-weight of deposited layers. However, excessively high yield stress may hinder extrusion, requiring higher pumping pressures and potentially causing nozzle blockages. This parameter is particularly relevant for processes where the material undergoes significant shear, such as mixing, pumping/extrusion, and deposition:

- During mixing, the material is subjected to intense shear forces to ensure homogeneity. A controlled dynamic shear yield stress is necessary to maintain a uniform distribution of particles or additives in the mixture.
- In pumping/extrusion, the material must flow consistently through nozzles or channels, requiring a well-defined dynamic shear yield stress to balance process efficiency and material integrity.
- In deposition, the material must be fluid enough to be extruded and spread onto the previous layer while still maintaining its shape after deposition. The interaction with the previously deposited layer ($N-1$) introduces additional considerations related to both shear and compressive forces.

11.3.1.2 Static Shear Yield Stress

The static shear yield stress is the shear stress required to initiate flow in a material that has been at rest. This parameter is crucial for storage and deposition processes where the material's structuration and thixotropic behavior (time-dependent recovery of structure) play a role:

- In storage, materials often rest for extended periods, allowing structuration to occur due to flocculation, particle interactions, or chemical bonding. The static shear yield stress determines how much stress is needed to restart flow after rest, which is crucial for maintaining consistency in subsequent processes.
- In deposition, the static shear yield stress helps ensure that once the material is placed, it does not excessively spread or lose its form before the next layer is applied.

11.3.1.3 Static Compressive Yield Stress

The static compressive yield stress characterizes the material's ability to withstand compressive forces without collapsing or undergoing significant deformation. It is particularly relevant in deposition and stability processes where gravity or applied loads compress the material.

- In deposition, when a new layer *N* is placed on top of a previously printed layer (*N*–1), the *N*–1 layer must support the weight of the fresh material. This is particularly important in layer-wise AM strategies, where improper compressive strength can lead to deformation or collapse of the structure.
- In stability, the entire printed structure is subjected to gravity-induced compression. The static compressive yield stress determines whether the material can sustain the accumulated weight of upper layers without excessive deformation, ensuring the long-term stability of the printed object. This deformation aspect is highly relevant for the stability prediction, but it's important to note that, at this very moment, the literature usually considers a pseudo-elastic modulus because of the difficulties in capturing the truly elastic strain range.

By meticulously controlling all these rheological and mechanical parameters at each stage of the 3DCP process, practitioners can enhance the quality, durability, and performance of 3DPC structures. All required properties for successful 3DCP are summarized in Table 11.1.

11.3.2 An Overview of Commonly Used Rheological Models

The layer formation in 3DCP is directly influenced by the material's rheology, particularly its ability to capture time-dependent viscosity changes. Various rheological models describe the flow behavior of cementitious materials, affecting their ability to sustain shape stability and interlayer adhesion, including Newtonian,

TABLE 11.1
Relevant Rheological and Mechanical Parameters at Each Printing Process Stage

Process	Type of Load on the Material	Relevant Rheological Value
Mixing	Material significantly sheared	Dynamic shear yield stress
Storage	Material structuration at rest	Static shear yield stress
Pumping/extrusion	Material significantly sheared	Dynamic shear yield stress
Deposition	Material sheared, loading the *N*–1 layer*	Dynamic shear yield stress and static compressive yield stress of the *N*–1 layer
Buildability	Material compressed (gravity)	Static compressive yield stress

* This loading highly depends on the printing strategy: layer-pressing strategy will increase both the shearing of the printable material of the layer *N* and the compressive load locally on the layer *N*–1, compared to the infinite brick strategy (Carneau et al., 2020).

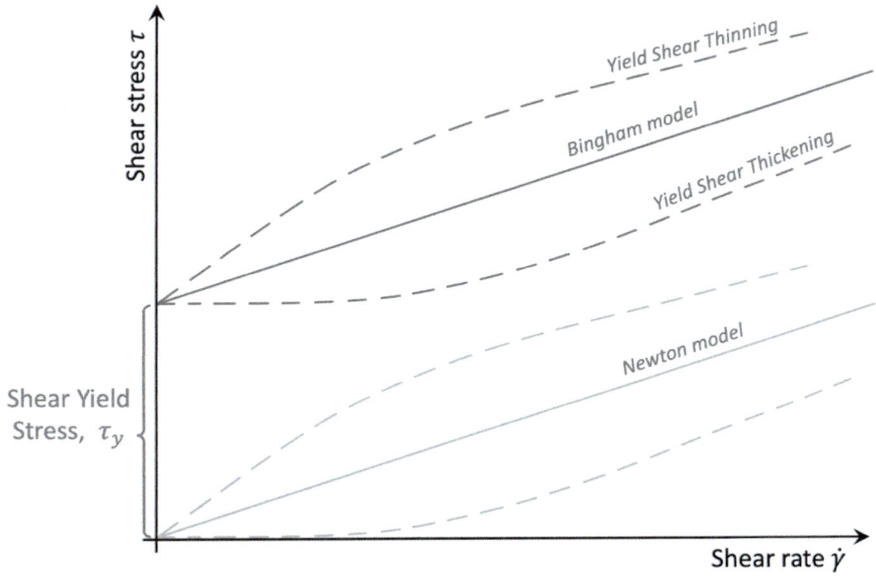

FIGURE 11.2 Typical flow curve of different rheological models commonly used to describe 3D-printable materials. Credit: Yohan Jacquet.

Herschel–Bulkley, power law, and Bingham fluids, each with different implications for viscoelasticity and viscoplasticity. A graphical overview is provided in Figure 11.2.

11.3.2.1 Newtonian Fluids

Newtonian fluids, characterized by a constant viscosity independent of shear rate, do not exhibit significant time-dependent properties. Their inability to capture thixotropy (time-dependent decrease in viscosity) or structuration effects makes them unsuitable for 3DCP, as they lack the necessary yield stress to sustain printed layers. These fluids do not display viscoelastic or viscoplastic behavior, leading to excessive deformation under their own weight. Constitutive equation of a Newtonian fluid can be written as a function of shear rate $\dot{\gamma}$ regarding Eq. (11.1), where μ is the dynamic viscosity of the fluid.

$$\tau = \mu\dot{\gamma} \tag{11.1}$$

11.3.2.2 Bingham Fluids

Bingham fluids possess a yield stress, meaning they behave as a solid until the applied shear stress exceeds this threshold. Once the yield stress is surpassed, they flow with a constant plastic viscosity. While this model captures viscoplasticity, it does not inherently account for time-dependent viscosity changes such as thixotropy or structuration, which are crucial for maintaining layer shape over time. Thus, additional modifications or supplementary models are required to represent structural buildup and workability retention in 3DCP. The constitutive equation of a Bingham fluid can be written as a function of shear rate $\dot{\gamma}$ regarding Eq. (11.2), where τ_0 is the shear yield stress and μ_p is the plastic viscosity (Pa.s).

$$\tau = \tau_0 + \mu_p \dot{\gamma} \tag{11.2}$$

11.3.2.3 Power Law Fluids

Power law fluids describe shear-thinning or shear-thickening behavior without a yield stress. While useful for characterizing flow behavior, they do not incorporate visco-elasticity or viscoplasticity, making them insufficient for accurately predicting layer stability in 3DCP. These fluids fail to represent time-dependent viscosity changes essential for ensuring both extrudability and shape retention. The constitutive equation of a power law fluid can be written as a function of shear rate $\dot{\gamma}$ regarding Eq. (11.3), where K is the consistency index $(Pa.s^n)$ and n is the power law index (-). A power law index less than the unit depicts a shear-thinning behavior, while a power law index higher than the unit depicts a shear-thickening behavior.

$$\tau = K\dot{\gamma}^n \tag{11.3}$$

11.3.2.4 Herschel–Bulkley Fluids

The Herschel–Bulkley model extends the Bingham model by incorporating a shear-thinning or shear-thickening component. It accounts for viscoplastic behavior but does not explicitly include time-dependent viscosity changes unless modified. By tuning its parameters, this model can better describe the complex rheological properties of 3D-printable concrete, although additional terms are required to capture thixotropic effects. The constitutive equation of a Herschel–Bulkley fluid can be written as a function of shear rate $\dot{\gamma}$ regarding Eq. (11.4), where τ_0 is the shear yield stress, K is the consistency index $(Pa.s^n)$, and n is the power law index (-).

$$\tau = \tau_0 + K\dot{\gamma}^n \tag{11.4}$$

Cement-based printable materials exhibit yield stress, primarily influenced by the solid volume fraction and mixture composition. Since Newtonian fluids fail to capture this yield stress, the Bingham model is often more suitable. The 3D printing process involves high shear stress during extrusion or pumping. Due to the nonlinear relationship between shear stress and shear rate in non-Newtonian fluids, both Newtonian and Bingham models fail to fully describe this behavior. In such cases, a power law model is preferred.

Depending on whether the shear yield stress is considered, either the power law model or the Herschel–Bulkley model can be used. The number of rheological parameters increases with the fidelity of rheological behavior description. While the Herschel–Bulkley model provides a highly accurate representation of most 3D-printable materials, the Bingham model is often used as a simpler approximation, despite potential uncertainties at low shear rates.

11.3.3 Layer–Shape and Geometrical Conformity

The precise control of material deposition in 3DCP is crucial for ensuring structural integrity, geometric accuracy, and overall buildability. At the layer scale, the deposition process is influenced by multiple interacting factors, including material rheology, extrusion parameters, and gravitational effects. Numerical simulations play a vital role in predicting and optimizing these interactions to enhance print quality and reduce defects such as filament buckling, tearing, or uncontrolled spreading.

11.3.3.1 Deposition Regimes and Material Flow Behavior

Material extrusion in 3DCP can be categorized into two primary regimes (Roussel, 2018): infinite brick extrusion and free-flow deposition (see Figure 11.3), and an intermediate regime can be referred to as the layer-pressing strategy.

In the former, the extruded material exhibits a high yield strength, maintaining its shape with minimal deformation post-extrusion. This results in a filament cross-section dictated primarily by the nozzle geometry. Conversely, in free-flow deposition, the material undergoes shear deformation before exiting the nozzle, and the final filament shape is governed by a balance between gravity and the material's stress resistance. In cases where the nozzle operates at a reduced height relative to the substrate, the filament can be compressed, influencing its cross-sectional dimensions and potentially affecting interlayer stability.

To model the complex behavior of extruded concrete at the layer scale, CFD methods, particularly the finite volume method (FVM), have been widely employed. A study by Comminal et al. (2020a) utilized these techniques to simulate the impact of nozzle positioning and the velocity ratio between printing speed (V) and extrusion speed (U) on layer formation. Their findings highlight that deviations from an optimal V/U ratio can lead to either excessive spreading or material buildup, affecting

FIGURE 11.3 Material flow behavior after the nozzle shear flow (left), layer-pressing (center), and infinite brick (right). Credit: Yohan Jacquet, inspired from Roussel (2018).

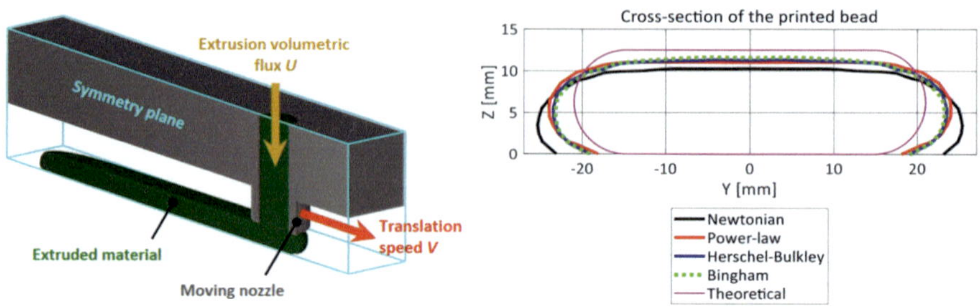

FIGURE 11.4 Numerical modeling of layer deposition based on both printing speed (*V*) and extrusion speed (*U*), and the resulting layer shape based on different rheological models (Comminal et al., 2019).

the dimensional accuracy of printed layers as depicted in Figure 11.4. The effect of velocity was also confirmed by Wolfs et al. in a study where CFD is employed to also consider the effect of tearing or local buckling for estimating the layer geometry (Wolfs et al., 2021).

Using generalized Newtonian and elasto-viscoplastic models, these simulations provide insight into velocity distributions, stress fields, and pressure conditions at the nozzle outlet. The results indicate that reducing the nozzle height leads to increased pressure beneath the nozzle, which may induce local deformations in the underlying layers. Such deformations, if unaccounted for, can compromise the structural stability of the printed object.

The selection of material rheology models is critical for accurately simulating filament formation. Non-Newtonian fluid models such as power law, Herschel–Bulkley, and Bingham plastic models offer realistic representations of concrete flow behavior. Simulations conducted by Comminal et al. demonstrate that filaments formed using non-Newtonian fluids generally exhibit greater thickness and reduced width compared to filaments formed using Newtonian fluids, a discrepancy that slicing software must accommodate (2020b).

Reinold et al. introduced the PFEM to analyze the dynamic interactions between the extruded material and the printing substrate (Reinold & Meschke, 2019). Their 2D simulations indicate that nozzle height, extrusion speed, and print speed significantly influence the stress distribution and layer deformation. Notably, the pressure exerted beneath the nozzle can be up to six times greater than the material's self-weight, underscoring the importance of extrusion force calibration, as presented in Figure 11.5 (left). This study paves the way to a better understanding of the so-called layer-pressing strategy (Carneau et al., 2020).

Closely related to that, Wolfs et al. investigated the effects of extrusion speed mismatches on filament integrity (Wolfs et al., 2021). When $U > V$, the extruded filament exhibits buckling due to accumulated compressive stresses, while $U < V$ results in tensile-induced tearing. Their analytical expressions for buckling and tearing thresholds provide valuable guidelines for optimizing printing parameters to prevent filament defects, according to Figure 11.5 (right).

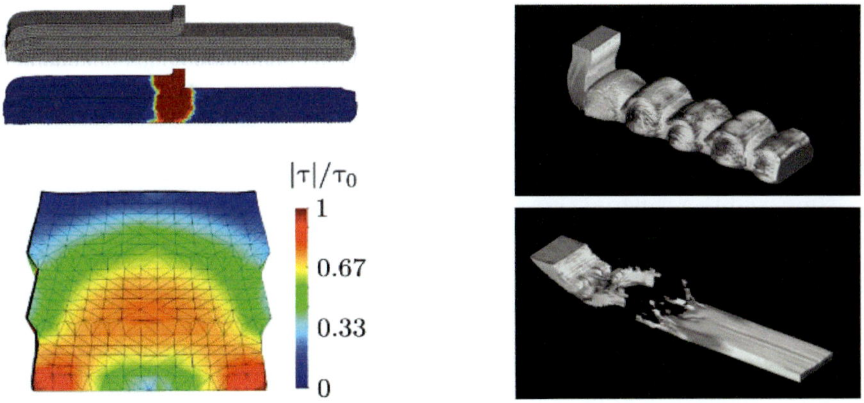

FIGURE 11.5 Stress distribution and layer deformation (Reinold et al., 2019, left); over-extrusion causing buckling and under-extrusion causing tearing (Wolfs et al., 2021, right).

Beyond single-layer deposition, multilayer simulations enable a more comprehensive understanding of buildability and structural evolution. Spangenberg et al. (2021) validated CFD models using experimental data on multilayered walls, demonstrating a high correlation between simulated and actual printed geometries (Spangenberg et al., 2021). Their findings revealed that excessive extrusion pressure can lead to material yielding in lower layers, resulting in progressive deformation and instability.

The simulation of material deposition at the layer scale is instrumental in refining AM processes for cement-based materials. CFD and PFEM approaches provide essential insights into the interplay among extrusion dynamics, material properties, and geometric accuracy. By integrating these simulations with experimental validations, it is possible to optimize process parameters, improve filament consistency, and enhance the overall reliability of 3DCP technologies. Future research should focus on real-time adaptive control strategies to mitigate deviations in layer geometry and further refine predictive models for large-scale applications, as was also presented in Lachmayer et al. (2024).

11.4 FROM MULTILAYER PRINTING TO STABILITY PREDICTION OF 3D-PRINTED STRUCTURES

11.4.1 Constitutive Models: A Review

When analyzing the strength and stability of 3D-printed structures in their fresh state, it is essential to understand how these structures handle different forces. This requires a fundamental understanding of stress components, which we define as follows:

Deviatoric Stress (S): This represents the part of the stress that causes a material to change shape without changing its overall size. Mathematically expressed, the deviatoric stress S can be described as the total stress σ minus the hydrostatic pressure component pI (Eq. (11.5)).

$$S = \sigma - pI \tag{11.5}$$

Mean Stress (p): Also known as hydrostatic pressure, mean stress represents the average stress within a material (Eq. (11.6)).

$$p = \frac{1}{3} trace(\sigma) \tag{11.6}$$

Mises Equivalent Stress (q): This notation is used to predict when a material will undergo permanent deformation. It can be expressed as a function of the deviatoric stress S, according to Eq. (11.7).

$$q = \sqrt{\frac{3}{2} S_{ij} S_{ij}} \tag{11.7}$$

Third Stress Invariant (r): This parameter captures complex stress distribution and refines material behavior predictions (see Eq. (11.8)).

$$r = \frac{1}{3} S_{ij} S_{jk} S_{ki} \tag{11.8}$$

These fundamental stress parameters help define commonly used constitutive models in numerical simulations of 3D-printable material stability. The choice of a model depends on the mechanical properties of the material, expected stress conditions, and available experimental data for model calibration.

11.4.1.1 Von Mises Constitutive Model

The Von Mises model is a classical approach for ductile materials. It states that yielding occurs when the maximum distortion energy equals the distortion energy at yield. The equivalent stress is given by Eq. (11.9):

$$\sigma_v = \frac{1}{\sqrt{2}} \sqrt{(\sigma_1 - \sigma_2)^2 + (\sigma_2 - \sigma_3)^2 + (\sigma_3 - \sigma_1)^2} \tag{11.9}$$

The yield surface F_{V-M} can be written according Equation (11.10):

$$F_{V-M} = q - \sigma_v = 0. \tag{11.10}$$

11.4.1.2 Mohr–Coulomb Constitutive Model

The Mohr–Coulomb criterion is suitable for granular materials under monotonic loading, and it states that the material yields when the ratio of shear stress to normal stress at the shear plane reaches a maximum. It is given by Equation (11.11):

$$\tau_y = C + \sigma_n . \tan\varphi, \tag{11.11}$$

where τ_y and σ_n are the shear yield stress and normal stress, respectively; C is the cohesion yield stress between particles bonded by cement; and φ is the internal friction angle caused by the frictional resistance and interlocking between particles.

The Mohr–Coulomb yield function F_{M-C} is typically written according to Eq. (11.12):

$$F_{M-C} = qf - p \cdot tan\varphi - C = 0, \tag{11.12}$$

where f depends on the Lode angle θ, describing the stress-state shape in the deviatoric plane, regarding Eq. (11.13).

$$f = \left[\frac{1}{\sqrt{3}\, cos\varphi} sin\left(\theta + \frac{\pi}{3}\right) + \frac{1}{3} \tan\varphi cos\left(\theta + \frac{\pi}{3}\right) \right] \tag{11.13}$$

11.4.1.3 Drucker–Prager Constitutive Model

The Drucker–Prager model is another pressure-dependent shear model that accounts for the intermediate principal stress's influence on material strength. Its linear yield surface F_{D-P} is depicted by Eq. (11.14):

$$F_{D-P} = t - p \cdot tan\varphi - C = 0, \tag{11.14}$$

where the deviatoric stress measure t is defined as a function of the ratio of the yield stress in triaxial tension to the yield stress in triaxial compression, K, following Eq. (11.15):

$$t = \frac{1}{2} q \left[1 + \frac{1}{K} - \left(1 - \frac{1}{K}\right) \left(\frac{r}{q}\right)^3 \right]. \tag{11.15}$$

The yield conditions derived above (von Mises, Mohr–Coulomb, and Drucker–Prager) are – in the presented form – purely rate-independent plasticity models, meaning they do not account for the effects of viscosity, strain rate, or time-dependent behavior such as viscoelasticity or viscoplasticity. These models are fundamentally designed for rate-independent plasticity and failure analysis, making the direct applicability not sufficiently precise for materials that experience a significant change over time and that display a high Trouton's ratio, indicating a significant difference between extensional and shear viscosities (Trouton, 1906). In 3DCP, the material undergoes shear flow during extrusion but must also resist deformation in tension and compression once deposited, where classical yield criteria fail to differentiate between these behaviors.

These models assume that yielding occurs instantly once the stress state reaches the yield surface, disregarding the deformation rate. However, in reality, 3DPC exhibits rate-dependent behavior, influenced by the following:

- Early-Age Behavior: Freshly extruded concrete undergoes time-dependent structuration and deformation due to the evolution of its microstructure (i.e., stress relaxation or thixotropy).
- Rate-Dependent Yield Stress: The material gains strength over time due to hydration and structural buildup.
- Shear-Thinning Behavior: The viscosity of fresh concrete decreases under shear but recovers when left at rest.

If the viscoelastic and viscoplastic effects are negligible compared to plastic yielding, they can be ignored, but for highly thixotropic materials like 3DPC, their influence is significant and must be considered in modeling approaches.

11.4.2 Current (Rheology-Based) Buildability Models in 3D Concrete Printing

Various researchers have worked on transferring rheological parameters into analytical or numerical models to predict and calculate the behavior of mortar and concrete structures in 3D printing. The main challenge is that an appropriate time interval between two layers needs to be met.

Several theoretical frameworks have been developed in order to describe *buildability* in models. An overview of the most common models is presented in the following.

(i) Perrot et al.

Perrot et al. developed a theoretical framework to obtain the highest building rate for layer-by-layer construction processes (Perrot et al., 2016). The model is based on the comparison between the mechanical strength of the first deposited strand at the bottom and the load resulting from the weight of subsequently applied strands and its related plastic deformation.

Vertical stress σ_V (N/m²) acting on the first layer is assumed to be described by Eq. (11.16):

$$\sigma_V = \rho g R t, \tag{11.16}$$

with ρ being the specific weight of concrete, t being the time after deposition, and R being the rate of construction describing the vertical building velocity.

Stability of the first layer is given if the vertical stress acting on the bottom layer does not exceed critic failure stress σ_c, according to Eq. (11.17):

$$\sigma_c(t) = \alpha_{geom}\left(\tau_{0,t=0} + A_{thix} t\right), \tag{11.17}$$

where α_{geom} is a geometric factor. $\tau_{0,t=0}$ is the yield stress τ_0 of the deposited strand at the bottom directly after deposition ($t = 0$) and A_{thix} is a linear increase in yield stress over time t at rest.

The time after deposition when a specimen fails (t_{fail}), if a vertical load is increased with rate R and the material's yield stress increases linearly, is as described by Eq. (11.18):

$$t_{fail} = \frac{\tau_{0,t=0}}{\frac{\rho g R}{\alpha_{geom}} - A_{thix}}. \tag{11.18}$$

For longer timespans, depending on the material, for example after approximately 60 minutes, the increase in yield stress may no longer be predominantly linear. In such cases, yield stress evolution can be described by a smooth transition from linear to exponential behavior; see Perrot et al. (2015, 2016) for further details.

The mentioned theoretical framework has been in good agreement with experiments on cement paste for simulated building rates ranging from 1.1 to 6.2 m/h.

(ii) Wangler et al.

Wangler et al. developed a framework in order to derive a time and velocity operation window for material deposition processes (Wangler et al., 2016). In the model, it is assumed that the layer with height h must be able to support itself, which is possible when a minimum yield stress τ_0 of the material is given by Eq. (11.19):

$$\tau_0 = \frac{\rho g h}{\sqrt{3}}, \tag{11.19}$$

where ρ is the density and g is the gravity constant. Accordingly, the same boundary conditions must be fulfilled by the first layer with yield stress $\tau_{0,fin}$ for an object with a total height H_{max}.

A time-dependent yield stress increase, noted as A_{thix}, enables finally to determine a minimum time $t_{h,min}$ needed to print a layer with height h without collapse, following Eq. (11.20):

$$t_{h,min} = \frac{\rho g h}{\sqrt{3}\, A_{thix}}. \tag{11.20}$$

Additionally, a maximum time per layer $t_{h,max}$ in order to prevent cold joints is suggested by Eq. (11.21):

$$t_{h,max} = \frac{\sqrt{\left(\frac{(\rho g h)^2}{12}\right) + \left(\frac{2\mu_p V}{h}\right)^2}}{A_{thix}}, \tag{11.21}$$

where μ_p is the plastic viscosity and V is the horizontal velocity.

From these time necessities, horizontal speed requirements or contour length can be obtained to prevent from cold joints or collapse.

(iii) Wolfs et al.

A numerical investigation to describe the mechanical behavior of 3DPC up to 90 minutes after deposition was performed by Wolfs et al. (2018, 2019). It is based on a time-dependent Mohr–Coulomb failure criterion and linear stress–strain behavior up to failure. The underlying basic equation for failure in the numerical simulation is described by Eq. (11.22):

$$\tau_y = C(t) + \tan(\varphi(t))\sigma_n, \tag{11.22}$$

where τ_y is the shear yield stress, $C(t)$ is the time-dependent cohesion, φ is the time-dependent angle of friction, and σ_n is the acting normal stress.

The time-dependent material properties are needed to allow a structural analysis of the printed object. In Wolfs et al. (2018a), C and φ were experimentally determined by means of a direct shear test. Additionally, Young's modulus $E(t)$ and Poisson's ratio $v(t)$ were determined in an unconfined compression test. Dilatancy angle $\psi(t)$ is assumed to be constant. These latter material parameters are needed to assess the stability of the object during printing.

When printing cylinders with a commercially available mortar, the model was validated, and it was found that the model enables to qualitatively predict the corresponding failure–deformation mode. Quantitatively the model overestimates the strength and stability of the printed object. This is assumed to be due to prevalent material and geometrical imperfections, which are not considered in the model. Additionally, the material parameters were obtained on compacted material, which is not typical in the printing process.

In Wolfs et al. (2019), all parameters were derived from a triaxial compression test on non-compacted samples. This testing procedure is considered to be suitable to assess the structural behavior of concrete components during 3DCP.

Rheology governs the behavior of fresh concrete during pumping, extrusion, and deposition in 3DCP. Proper rheological control is crucial to achieving a balance among extrudability, buildability, and interlayer adhesion. Understanding the key rheological properties and their impact on failure mechanisms helps optimize material formulations and printing parameters.

11.4.3 Self-Buckling and Plastic Collapse

Failures in 3DPC structures are primarily caused by inadequate control of material properties and inefficiencies in the printing process. Understanding these failure mechanisms is essential for optimizing material formulations, printing techniques, and overall structural performance.

While plastic collapse and elastic buckling are the most critical failure modes in the fresh state, other material characteristics also play a significant role in ensuring

FIGURE 11.6 Illustration of plastic collapse (left) and self-buckling (right) failure modes during 3DCP. Credit: Yohan Jacquet.

structural integrity. An illustration of both phenomenon is proposed in Figure 11.6. Strong interlayer adhesion is necessary to prevent delamination and create a cohesive structure. Additionally, the development of compressive strength over time determines the final load-bearing capacity of the printed element. Furthermore, shrinkage behavior and creep resistance impact long-term durability, ensuring the structure remains stable under continuous and environmental loads (Nerella et al., 2019; Wolfs & Suiker, 2019). By refining material composition, optimizing curing conditions, and incorporating reinforcement strategies, it is possible to enhance the performance and reliability of 3DPC structures.

However, the focus of the following sections will be on the two primary failure modes in the fresh state: plastic collapse and self-buckling, as these directly influence the immediate stability of the printed layers.

11.4.3.1 Self-Buckling

Self-buckling is a structural instability that occurs when a printed element deforms out of plane due to insufficient stiffness (Suiker et al., 2020; Wolfs & Suiker, 2019). Sometimes referred to as elastic buckling, this definition is debatable since the rheological properties involved are typically considered pseudo-elastic rather than truly elastic. As the layer height increases, the structure may become unstable and buckle under its own weight or external forces such as wind or vibrations. Factors that contribute to elastic buckling include the following:

- Low Material Stiffness: A high binder-to-aggregate ratio can reduce stiffness, making the structure more prone to deformation.

- Tall and Slender Geometries: Structures with high aspect ratios are inherently more susceptible to buckling.
- Inconsistent Layer Bonding: Weak adhesion between layers reduces overall structural rigidity, increasing instability.

To minimize the risk of elastic buckling, computational modeling can be used to optimize printing parameters. Additional mitigation strategies include reinforcing the material with fibers, adjusting layer geometry for better load distribution, and modifying the printing sequence to enhance stability.

11.4.3.2 Plastic Collapse

Plastic collapse occurs when a printed structure does not have enough static yield stress to support its own weight, leading to excessive deformation or complete failure. This type of failure is most likely to happen in the early stages of printing when the lower layers have not yet gained sufficient strength (Kruger et al., 2020; Roussel, 2018; Suiker et al., 2020). Key factors contributing to plastic collapse include the following:

- Insufficient Static Yield Stress: If the material lacks adequate cohesion and internal strength, it cannot sustain additional layers.
- Excessive Layer Height: Thick layers increase the load on the lower layers, increasing the risk of collapse.
- Inadequate Thixotropic Behavior: The material must develop enough structuration over time to maintain stability after deposition.

To mitigate plastic collapse, it is essential to optimize rheology. This can be achieved by changing the mix design, by adjusting binder content, using viscosity-modifying agents, and carefully controlling the water-to-cement ratio to balance workability and strength development.

By addressing these two primary failure modes – plastic collapse and self-buckling, the structural viability of 3DPC can be significantly improved, ensuring both immediate and long-term performance.

Wolfs et al. (2018) pioneered the use of numerical modeling to analyze structural behavior during 3D printing. They developed a 3D FE model in Abaqus, controlled by a parametric Python script, to simulate layer-by-layer construction, incorporating gravity loads and geometric constraints to ensure layer adhesion. The model employed a Mohr–Coulomb plastic failure criterion and a time-dependent evolution of mechanical properties, derived from shear and compression tests. While the predictions aligned qualitatively with experimental results, the model overestimated the maximum printable layers by 58.6%, likely due to overestimated material properties and the neglect of geometric imperfections. Although valuable for structural analysis, the model does not optimize geometry or printing parameters.

Four studies, conducted by four different authors and summarized in Table 11.2, utilized Abaqus to model the structural behavior of 3DPC, focusing on cylindrical and straight wall geometries. Each study accounted for geometric imperfections through elastic buckling modes and assumed perfect contact between printed layers. However, key differences emerged in their failure models and hardening rules.

TABLE 11.2
Comparison of Different FE Stability Prediction Models of 3DCP

Wolfs et al. (2018)	Software	Abaqus
	Geometry	Cylinder and straight wall
	Failure model	Time-dependent Mohr–Coulomb elastoplastic model
	Imperfections	Defined through elastic buckling mode
	Interfaces	Perfect contact between layers
	Hardening rule	Function of time and independent of plastic strain
Ooms et al. (2021)	Software	Abaqus
	Geometry	Cylinder and straight wall
	Failure model	Time-dependent Mohr–Coulomb elastoplastic model
	Imperfections	Defined through elastic buckling mode
	Hardening rule	Function of time and of plastic strain
Vantyghem et al. (2021)	Software	Abaqus
	Geometry	Cylinder and straight wall
	Failure model	Time-dependent Mohr–Coulomb elastoplastic model
	Imperfections	Defined through elastic buckling mode
	Hardening rule	Function of time and of plastic strain
Liu et al. (2022)	Software	Abaqus
	Geometry	Cylinder and straight wall
	Failure model	Drucker–Prager elastoplastic model
	Imperfections	Defined through elastic buckling mode
	Hardening rule	Function of time and of plastic strain

Wolfs et al. (2018) introduced a time-dependent Mohr–Coulomb elastoplastic model with a hardening rule based solely on time and independent of plastic strain. Later studies by Ooms et al. (2021) and Vantyghem et al. (2021) refined this approach by incorporating plastic strain dependency, allowing for a more accurate representation of material evolution during printing. Liu et al. (2022) diverged further by adopting the Drucker–Prager model instead of the Mohr–Coulomb model, offering a better representation of the pressure-dependent behavior of fresh concrete.

The evolution of these models highlights the increasing effort to enhance predictive reliability by improving material characterization. While the gap between numerical predictions and real printing outcomes has narrowed, significant discrepancies remain. As Wolfs et al. pointed out, a deeper understanding of key rheological parameters is essential to further refine numerical simulations and improve accuracy (Wolfs et al., 2018b).

Alongside FEMs, Chang et al. (2020) utilized the lattice model for numerical simulations of the 3D printing process. This approach, based on a 3D network of Timoshenko beams (Timoshenko & Goodier, 1985), provides a powerful framework for capturing the heterogeneous characteristics of printed concrete structures and the effects of nonuniform gravitational loading (Jacquet et al., 2020). Unlike traditional constitutive models such as Mohr–Coulomb, Drucker–Prager, and Von Mises plasticity criteria, which often struggle to fully describe the intricate mechanical behavior of concrete, the lattice model inherently accounts for its tension–compression asymmetry and high Trouton's ratio. These features are crucial for accurately modeling the material's structural response under complex loading conditions.

The lattice model proposed by Chang et al. successfully reproduced both structural failure – governed by excessive lateral displacement – and material failure, driven by the accumulation of local damage in lower layers (Chang et al., 2021, 2022). Notably, its prediction of critical print height improved with existing models in the literature. When validated against experimental data for cylindrical structures, the lattice model overestimated the critical print height by 41.4%.

Beyond its ability to capture the fundamental mechanical asymmetry of concrete, the lattice model also offers a significant advantage in incorporating viscous contributions, including viscoelastic and viscoplastic effects (Jacquet et al., 2021). Its capacity to model time-dependent behavior was demonstrated in unconfined uniaxial compression tests, where the stiffness and strength of beam elements were calibrated as a function of time. This capability underscores the lattice model's potential as a robust alternative to conventional plasticity-based approaches, particularly for complex, rate-dependent phenomena encountered in concrete materials.

11.4.4 A Case Study: The Hollow Cylinder

A closer examination of a hollow cylinder case study, proposed in Table 11.3, further underscores the gap between numerical predictions and real experimental outcomes. Pioneering work has been conducted by Wolfs et al. (2018) and Liu et al. (2022). Both conducted FE analyses on identical hollow cylinder geometries using Abaqus but differed in their failure criteria and hardening rules. As depicted in Figure 11.7, Wolfs et al. (2018) employed a Mohr–Coulomb model with a hardening rule dependent only on time, while Liu et al. (2022) adopted the Drucker–Prager model, incorporating both time and plastic strain dependency. Despite this refinement, both models still overestimated the experimentally observed maximum printable height – by a mean of 60% and 45%, respectively.

TABLE 11.3
Comparison of the Failure Height Prediction for a Hollow Cylinder Using Two Different Failure Criteria

Publication	Wolfs et al. (2018)	Liu et al. (2022)
Geometry	Hollow cylinder	Hollow cylinder
Software	Abaqus	Abaqus
Failure criterion	Mohr–Coulomb model	Drucker–Prager model
Hardening rule	Function of time and independent of plastic strain (cohesion)	Function of time and of plastic strain

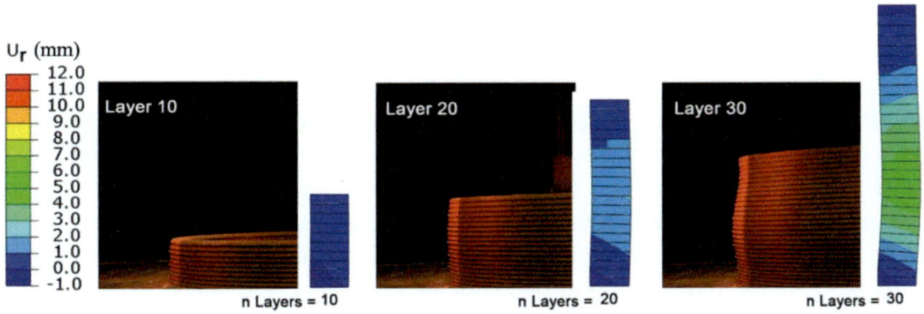

FIGURE 11.7 Comparison of the radial deformation U_r between FE numerical simulation and the real printing of an axisymmetric 3DPC cylinder. Adapted from Wolfs et al. (2018a).

Notably, the overestimation in Wolfs et al. (2018) ranged widely from 37.9% to 82.7%, highlighting the variability in predictive accuracy: this variability, coming from the dispersion of the strength and stiffness parameters, shows that the determination of rheological parameters is of paramount importance. Similarly, Vantyghem et al. (2021) applied a voxel-based approach in Abaqus using Wolfs et al.'s (2018) data, yielding a comparable overestimation of 63%. However, this method proved advantageous in differentiating failure mechanisms, particularly in distinguishing plastic collapse from elastic buckling. A comparison is proposed and summarized in Table 11.3.

These findings emphasize that despite the advancements in modeling approaches, current numerical simulations still struggle not only to accurately capture the complex interplay of rheological properties, material hardening, and failure mechanisms in 3DPC but also in taking into account the dynamic properties of the process, as stated by Reinold and Meschke (2019), in order to quantify maximum buildable heights.

11.5 TRENDS, LIMITS, AND FUTURE DIRECTIONS

The evolution of numerical modeling in 3DCP has significantly improved our ability to predict structural behavior and optimize process parameters, and with this, we are on the way to accurately predict buildability. However, discrepancies between numerical predictions and experimental results highlight the need for further refinement in both modeling techniques and material characterization. ML, particularly PINNs, presents a promising direction for addressing these challenges by integrating physical laws with data-driven approaches, thereby reducing computational costs and improving prediction accuracy.

One of the primary sources of overestimation in numerical modeling is the inaccurate calibration of material properties, particularly elastic parameters. The compaction of material during laboratory tests can lead to inflated mechanical properties, which, when used in numerical simulations, result in an overestimation of structural stability and failure heights. This discrepancy underscores the importance of improving experimental techniques for measuring the elastic and viscoelastic properties of fresh concrete. Current testing methods often fail to capture the full complexity of early-age material behavior, particularly the transition from viscoelasticity to viscoplasticity, but also from the elasto-viscoplastic regime to the hardened state. Novel and more accurate experimental setups and in situ characterization techniques are necessary to refine input parameters and improve the reliability of numerical models (Jacquet et al., 2025).

Another key limitation lies in the reliance on plasticity-based constitutive models, which fail to fully account for the visco-elastoplastic nature of printable concrete. The time-dependent deformation behavior of fresh concrete plays a crucial role in structural stability, especially for multilayer printing, where early-age creep and relaxation effects influence buckling and collapse mechanisms. The development of more advanced constitutive models that integrate viscous contributions alongside plasticity criteria is essential for improving the predictive capabilities of numerical simulations. Such models should also incorporate the influence of process parameters, including extrusion speed, layer deposition time, and environmental factors like temperature and humidity, which significantly affect material behavior.

Furthermore, geometric imperfections and spatial heterogeneity in mechanical properties have often been neglected in deterministic numerical studies. In reality, layer dimensions, deposition accuracy, and material properties vary throughout the printing process, introducing uncertainties that can significantly impact structural stability. Probabilistic modeling approaches, such as stochastic and reliability-based methods, offer a means to quantify and incorporate these uncertainties into numerical simulations. By integrating uncertainty analysis into buildability assessments, researchers can develop more robust and realistic predictive models.

CFD models at the filament scale have shown great potential in characterizing extrusion dynamics and layer formation. However, their application in process optimization remains underexplored. CFD models could be better exploited in the future to understand the interactions between printing parameters, such as nozzle geometry, extrusion velocity, and layer width, which directly influence the quality and stability of printed structures. The coupling of CFD simulations with structural analysis at larger scales could provide a more comprehensive understanding of the entire printing process, from material deposition to final structural performance.

A major challenge in numerical modeling for large-scale 3D printing remains the high computational cost associated with detailed simulations, particularly when considering the diverse range of printable materials such as clay and alternative binders. The variability in rheological and mechanical properties across different materials further complicates the development of universal predictive models. In this context, ML and model reduction techniques offer a powerful solution for optimizing numerical workflows. PINNs stand out as a particularly promising approach, as they integrate physical laws directly into neural network architectures. This allows for more accurate and generalizable predictions across different material types while significantly reducing computational demands. A key challenge, however, lies in continuously expanding the dataset used to train these models. By incorporating a broader range of experimental and numerical data, PINNs can evolve into more universal predictive tools, enabling robust and efficient numerical modeling for various 3D printing applications, regardless of material variability.

Ultimately, the integration of digital tools throughout the 3D printing workflow is transforming the construction industry. From initial design to real-time process monitoring and post-printing structural assessment, numerical modeling and ML algorithms are becoming indispensable for optimizing efficiency and ensuring print quality. Future research should focus on bridging the gap between experimental testing and numerical modeling by improving material characterization techniques, refining constitutive models, and incorporating uncertainty quantification into simulations. By leveraging advancements in AI and computational mechanics, the next generation of 3DCP technologies can achieve greater reliability, scalability, and automation, paving the way for more efficient and resilient construction processes.

REFERENCES

Banfill, P. F. G. (1990). *Rheology of fresh cement and concrete* (P. F. G. Banfill, Ed.; CRC Press). Taylor & Francis. https://doi.org/10.4324/9780203473290

Cao, G., Zhang, H., Tan, Y., Wang, J., Deng, R., Xiao, X., & Wu, B. (2015). Study on the effect of coarse aggregate volume fraction on the flow behavior of fresh concrete via DEM. *Procedia Engineering*, *102*, 1820–1826. https://doi.org/10.1016/J.PROENG.2015.01.319

Carneau, P., Mesnil, R., Ducoulombier, N., Roussel, N., & Baverel, O. (2020). Characterisation of the layer pressing strategy for concrete 3D printing. In F. P. Bos, S. S. Lucas, R. J. M. Wolfs, & T. A. M. Salet (Eds.), *Second RILEM international conference on concrete and digital fabrication* (Vol. 28, pp. 185–195). Springer International Publishing. https://doi.org/10.1007/978-3-030-49916-7_19

Chang, Z., Schlangen, E., & Šavija, B. (2020). Extended lattice model to simulate the printing process of 3D printed cementitious materials. In *RILEM bookseries* (Vol. 28, pp. 814–823). Springer. https://doi.org/10.1007/978-3-030-49916-7_80

Chang, Z., Xu, Y., Chen, Y., Gan, Y., Schlangen, E., & Šavija, B. (2021). A discrete lattice model for assessment of buildability performance of 3D-printed concrete. *Computer-Aided Civil and Infrastructure Engineering*, *36*(5), 638–655. https://doi.org/10.1111/mice.12700

Chang, Z., Zhang, H., Liang, M., Schlangen, E., & Šavija, B. (2022). Numerical simulation of elastic buckling in 3D concrete printing using the lattice model with geometric nonlinearity. *Automation in Construction*, *142*. https://doi.org/10.1016/j.autcon.2022.104485

Choi, M. S., Kim, Y. J., & Kim, J. K. (2014). Prediction of concrete pumping using various rheological models. *International Journal of Concrete Structures and Materials*, *8*(4), 269–278. https://doi.org/10.1007/s40069-014-0084-1

Comminal, R., da Silva, W. R. L., Andersen, T. J., Stang, H., & Spangenberg, J. (2020a). Influence of processing parameters on the layer geometry in 3D concrete printing: Experiments and modelling. *RILEM Bookseries*, *28*, 852–862. https://doi.org/10.1007/978-3-030-49916-7_83

Comminal, R., da Silva, W. R. L., Andersen, T. J., Stang, H., & Spangenberg, J. (2020b). Modelling of 3D concrete printing based on computational fluid dynamics. *Cement and Concrete Research*, *138*, 106256. https://doi.org/10.1016/j.cemconres.2020.106256

Comminal, R., Serdeczny, M. P., Ranjbar, N., Mehrali, M., Pedersen, D. B., Stang, H., & Spangenberg, J. (2019). Modelling of material deposition in big area additive manufacturing and 3D concrete printing. In A. Bernard, R. K. Leach, D. B. Pedersen, & J. S. Taylor (Eds.), Proceedings of the Joint Special Interest Group meeting between euspen and ASPE Advancing Precision in Additive Manufacturing (2019) (pp. 151–154). The European Society for Precision Engineering and Nanotechnology.

Davtalab, O., Kazemian, A., Yuan, X., & Khoshnevis, B. (2022). Automated inspection in robotic additive manufacturing using deep learning for layer deformation

detection. *Journal of Intelligent Manufacturing*, *33*(3), 771–784. https://doi.org/10.1007/S10845-020-01684-W
Dörrie, R., Freund, N., Herrmann, E., Baghdadi, A., Mai, I., Galli, F., David, M., Dröder, K., Lowke, D., & Kloft, H. (2023). Automated force-flow-oriented reinforcement integration for Shotcrete 3D Printing. *Automation in Construction*, *155*. https://doi.org/10.1016/j.autcon.2023.105075
Jacquet, Y., Kawashima, S., & Spangenberg, J. (2025). Unveiling the structural build-up 3D printable cement-based materials: From small amplitude oscillatory shear (SAOS) to extensional (SAOE) rheological workflows. *Cement and Concrete Research*, *197*. https://doi.org/10.1016/j.cemconres.2025.107971
Jacquet, Y., Perrot, A., & Picandet, V. (2021). Assessment of asymmetrical rheological behavior of cementitious material for 3D printing application. *Cement and Concrete Research*, *140*. https://doi.org/10.1016/j.cemconres.2020.106305
Jacquet, Y., Picandet, V., Rangeard, D., & Perrot, A. (2020). Gravity induced flow to characterize rheological properties of printable cement-based materials. *RILEM Technical Letters*, *5*. https://doi.org/10.21809/rilemtechlett.2020.128
Jacquet, Y., Spangenberg, J., & Perrot, A. (2024). Exploring viscosity and friction through temperature: Understanding self-heating dynamics of non-Newtonian 3D printable construction materials via CFD modeling. In D. Lowke, N. Freund, D. Böhler, & F. Herding (Eds.), *Fourth RILEM international conference on concrete and digital fabrication* (pp. 465–473). Springer Nature Switzerland.
Jewett, J. L., & Carstensen, J. V. (2019). Topology-optimized design, construction and experimental evaluation of concrete beams. *Automation in Construction*, *102*, 59–67. https://doi.org/10.1016/j.autcon.2019.02.001
Kazemian, A., Yuan, X., Davtalab, O., & Khoshnevis, B. (2019). Computer vision for real-time extrusion quality monitoring and control in robotic construction. *Automation in Construction*, *101*, 92–98. https://doi.org/10.1016/J.AUTCON.2019.01.022
Khadijeh, M., Kasbergen, C., Erkens, S., & Varveri, A. (2025). Exploring the roles of numerical simulations and machine learning in multiscale paving materials analysis: Applications, challenges, best practices. In *Computer methods in applied mechanics and engineering* (Vol. 433). Elsevier B.V. https://doi.org/10.1016/j.cma.2024.117462
Krenzer, K., Palzer, U., Müller, S., & Mechtcherine, V. (2022). Simulation of 3D concrete printing using discrete element method. *RILEM Bookseries*, *37*, 161–166. https://doi.org/10.1007/978-3-031-06116-5_24
Krenzer, K., & Schwabe, J. H. (2009). Calibration of parameters for particle simulation of building materials, using stochastic optimization procedures. In *Proceedings of the 3rd international RILEM symposium on rheology of cement suspensions such a fresh concrete*. RILEM Publications, Bagneux, Reykjavik, Iceland.
Kruger, J., Cho, S., Zeranka, S., Viljoen, C., & Van Zijl, G. (2020). 3D concrete printer parameter optimisation for high rate digital construction avoiding plastic collapse. *Composites Part B*, *183*, 107660. https://doi.org/10.1016/j.compositesb.2019.107660
Kulasegaram, S., Karihaloo, B. L., & Ghanbari, A. (2011). Modelling the flow of self-compacting concrete. *International Journal for Numerical and Analytical Methods in Geomechanics*, *35*(6), 713–723. https://doi.org/10.1002/NAG.924
Lachmayer, L., Quantz, J., Heeren, H., Recker, T., Dörrie, R., Kloft, H., & Raatz, A. (2024). A spatial multi-layer control concept for strand geometry control in robot-based additive manufacturing processes. *RILEM Bookseries*, *53*, 119–126. https://doi.org/10.1007/978-3-031-70031-6_14/FIGURES/6
Li, Y., Mu, J., Wang, Z., Liu, Y., & Du, H. (2021). Numerical simulation on slump test of fresh concrete based on lattice Boltzmann method. *Cement and Concrete Composites*, *122*, 104136. https://doi.org/10.1016/J.CEMCONCOMP.2021.104136

Lim, S., Buswell, R. A., Valentine, P. J., Piker, D., Austin, S. A., & De Kestelier, X. (2016). Modelling curved-layered printing paths for fabricating large-scale construction components. *Additive Manufacturing*, *12*, 216–230. https://doi.org/10.1016/j.addma.2016.06.004

Liu, H., Ding, T., Xiao, J., & Mechtcherine, V. (2022). Buildability prediction of 3D–printed concrete at early-ages: A numerical study with Drucker–Prager model. *Additive Manufacturing*, *55*. https://doi.org/10.1016/j.addma.2022.102821

Martys, N. S. (2011). A classical kinetic theory approach to lattice Boltzmann simulation. *International Journal of Modern Physics C*, *12*(8), 1169–1178. https://doi.org/10.1142/S0129183101002474

Martys, N. S., George, W. L., Chun, B. W., & Lootens, D. (2010). A smoothed particle hydrodynamics-based fluid model with a spatially dependent viscosity: Application to flow of a suspension with a non-Newtonian fluid matrix. *Rheologica Acta*, *49*(10), 1059–1069. https://doi.org/10.1007/S00397-010-0480-7

Mechtcherine, V., Gram, A., Krenzer, K., Schwabe, J. H., Shyshko, S., & Roussel, N. (2014). Simulation of fresh concrete flow using discrete element method (DEM): Theory and applications. *Materials and Structures/Materiaux et Constructions*, *47*(4), 615–630. https://doi.org/10.1617/S11527-013-0084-7

Mechtcherine, V., & Shyshko, S. (2015). Simulating the behaviour of fresh concrete with the distinct element method – Deriving model parameters related to the yield stress. *Cement and Concrete Composites*, *55*, 81–90. https://doi.org/10.1016/j.cemconcomp.2014.08.004

Nerella, V. N., Hempel, S., & Mechtcherine, V. (2019). Effects of layer-interface properties on mechanical performance of concrete elements produced by extrusion-based 3D-printing. *Construction and Building Materials*, *205*, 586–601. https://doi.org/10.1016/j.conbuildmat.2019.01.235

Ooms, T., Vantyghem, G., Van Coile, R., & De Corte, W. (2021). A parametric modelling strategy for the numerical simulation of 3D concrete printing with complex geometries. *Additive Manufacturing*, *38*. https://doi.org/10.1016/j.addma.2020.101743

Perrot, A., Pierre, A., Vitaloni, S., & Picandet, V. (2015). Prediction of lateral form pressure exerted by concrete at low casting rates. *Materials and Structures/Materiaux et Constructions*, *48*(7), 2315–2322. https://doi.org/10.1617/s11527-014-0313-8

Perrot, A., Rangeard, D., & Pierre, A. (2016). Structural built-up of cement-based materials used for 3D-printing extrusion techniques. *Materials and Structures/Materiaux et Constructions*, *49*(4), 1213–1220. https://doi.org/10.1617/s11527-015-0571-0

Peskin, C. S. (2002). The immersed boundary method. *Acta Numerica*, *11*, 479–517. https://doi.org/10.1017/S0962492902000077

Rahman, M. A., Zhang, T., & Lu, Y. (2024). PINN-CHK: Physics-informed neural network for high-fidelity prediction of early-age cement hydration kinetics. *Neural Computing and Applications*, *36*(22), 13665–13687. https://doi.org/10.1007/s00521-024-09791-y

Reinold, J., & Meschke, G. (2019). Particle finite element simulation of fresh cement paste – Inspired by additive manufacturing techniques. *PAMM*, *19*(1). https://doi.org/10.1002/pamm.201900198

Roussel, N. (2018). Rheological requirements for printable concretes. *Cement and Concrete Research*, *112*, 76–85. https://doi.org/10.1016/j.cemconres.2018.04.005

Roussel, N., Geiker, M. R., Dufour, F., Thrane, L. N., & Szabo, P. (2007). Computational modeling of concrete flow: General overview. *Cement and Concrete Research*, *37*(9), 1298–1307. https://doi.org/10.1016/j.cemconres.2007.06.007

Roussel, N., & Gram, A. (2014). *Simulation of fresh concrete flow: State-of-the art report of the RILEM technical committee 222-SCF* (N. Roussel & A. Gram, Eds.; Vol. 15). Springer Netherlands. https://doi.org/10.1007/978-94-017-8884-7

Roussel, N., Lemaître, A., Flatt, R. J., & Coussot, P. (2010). Steady state flow of cement suspensions: A micromechanical state of the art. *Cement and Concrete Research*, *40*(1), 77–84. https://doi.org/10.1016/j.cemconres.2009.08.026

Schøler, J. P., Rosi, N., Quick, J., Riva, R., Andersen, S. J., Murcia Leon, J. P., Van Der Laan, M. P., & Réthoré, P. E. (2024). RANS wake surrogate: Impact of physics information in neural networks. *Journal of Physics: Conference Series*, *2767*(9). https://doi.org/10.1088/1742-6596/2767/9/092033

Shaban, W. M., Elbaz, K., Zhou, A., & Shen, S. L. (2023). Physics-informed deep neural network for modeling the chloride diffusion in concrete. *Engineering Applications of Artificial Intelligence*, *125*, 106691. https://doi.org/10.1016/J.ENGAPPAI.2023.106691

Shyshko, S., & Mechtcherine, V. (2010). Simulating fresh concrete behaviour – Establishing a link between the Bingham model and parameters of a DEM-based numerical model. *International RILEM Conference on Material Science-MATSCI, II*.

Shyshko, S., & Mechtcherine, V. (2013). Developing a discrete element model for simulating fresh concrete: Experimental investigation and modelling of interactions between discrete aggregate particles with fine mortar between them. *Construction and Building Materials*, *47*, 601–615. https://doi.org/10.1016/J.CONBUILDMAT.2013.05.071

Silva, J. M., Wagner, G., Silva, R., Morais, A., Ribeiro, J., Mould, S., Figueiredo, B., Nóbrega, J. M., & Cruz, P. J. S. (2024). Real-time precision in 3D concrete printing: Controlling layer morphology via machine vision and learning algorithms. *Inventions*, *9*(4), 80. https://doi.org/10.3390/INVENTIONS9040080

Slepicka, M., Vilgertshofer, S., & Borrmann, A. (2021). Fabrication information modeling: Closing the gap between building information modeling and digital fabrication. *38th International Symposiumon Automation and Robotics in Construction (ISARC 2021)*.

Spangenberg, J., da Silva, W. R. L., Comminal, R., Mollah, M. T., Andersen, T. J., & Stang, H. (2021). Numerical simulation of multi-layer 3d concrete printing. *RILEM Technical Letters*, *6*, 119–123. https://doi.org/10.21809/rilemtechlett.2021.142

Suiker, A. S. J., Wolfs, R. J. M., Lucas, S. M., & Salet, T. A. M. (2020). *Elastic buckling and plastic collapse during 3D concrete printing*. https://doi.org/10.1016/j.cemconres.2020.106016

Timoshenko, S., & Goodier, J. N. (1985). *Theory of elasticity* (2nd ed.). McGraw-Hill Book Company.

Tran-Duc, T., Ho, T., & Thamwattana, N. (2021). A smoothed particle hydrodynamics study on effect of coarse aggregate on self-compacting concrete flows. *International Journal of Mechanical Sciences*, *190*. https://doi.org/10.1016/j.ijmecsci.2020.106046

Trouton, F. R. T. (1906). *On the coefficient of viscous traction and its relation to that of viscosity*. https://royalsocietypublishing.org/

Vantyghem, G., Ooms, T., & De Corte, W. (2021). VoxelPrint: A grasshopper plug-in for voxel-based numerical simulation of concrete printing. *Automation in Construction*, *122*. https://doi.org/10.1016/j.autcon.2020.103469

Vasilic, K., Gram, A., & Wallevik, J. E. (2019). Numerical simulation of fresh concrete flow: Insight and challenges. *RILEM Technical Letters*, *4*, 57–66. https://doi.org/10.21809/rilemtechlett.2019.92

Wangler, T., Lloret, E., Reiter, L., Hack, N., Gramazio, F., Kohler, M., Bernhard, M., Dillenburger, B., Buchli, J., Roussel, N., & Flatt, R. (2016). Digital concrete: Opportunities and challenges. *RILEM Technical Letters*, *1*, 67–75. https://doi.org/10.21809/rilemtechlett.2016.16

Westphal, E., & Seitz, H. (2022). Machine learning for the intelligent analysis of 3D printing conditions using environmental sensor data to support quality assurance. *Additive Manufacturing*, *50*, 102535. https://doi.org/10.1016/j.addma.2021.102535

Wolfs, R. J. M., Bos, F. P., & Salet, T. A. M. (2018). Early age mechanical behaviour of 3D printed concrete: Numerical modelling and experimental testing. *Cement and Concrete Research*, *106*, 103–116. https://doi.org/10.1016/j.cemconres.2018.02.001

Wolfs, R. J. M., Bos, F. P., & Salet, T. A. M. (2019). Triaxial compression testing on early age concrete for numerical analysis of 3D concrete printing. *Cement and Concrete Composites, 104*. https://doi.org/10.1016/j.cemconcomp.2019.103344

Wolfs, R. J. M., Salet, T. A. M., & Roussel, N. (2021). Filament geometry control in extrusion-based additive manufacturing of concrete: The good, the bad and the ugly. *Cement and Concrete Research, 150*. https://doi.org/10.1016/j.cemconres.2021.106615

Wolfs, R. J. M., & Suiker, A. S. J. (2019). Structural failure during extrusion-based 3D printing processes. *The International Journal of Advanced Manufacturing Technology*, *104*(1–4), 565–584. https://doi.org/10.1007/s00170-019-03844-6

Yang, X., Lakhal, O., Belarouci, A., & Merzouki, R. (2022). Adaptive deposit compensation of construction materials in a 3D printing process. *2022 IEEE/ASME International Conference on Advanced Intelligent Mechatronics (AIM), 2022-July*, 658–663. https://doi.org/10.1109/AIM52237.2022.9863300

Zhu, J., Ren, X., & Cervera, M. (2023). Peridynamic buildability analysis of 3D-printed concrete including damage, plastic flow and collapse. *Additive Manufacturing, 73*. https://doi.org/10.1016/j.addma.2023.103683

12 Computational Fracture Mechanics of Layered and Additively Manufactured Materials

Numerical Approaches and Constitutive Modeling

Aimane Najmeddine, Shashank Gupta, and Reza Moini

12.1 INTRODUCTION

The advent of additive manufacturing technologies in the construction sector has created interesting challenges for computational modeling and the simulation of fracture in cement-based materials. The discretization of materials into filaments leads to the formation of horizontal and vertical interfaces that significantly alter the size of the fracture process zone (FPZ), associated softening behavior, the crack propagation path, and the overall macroscopic response. These layered microstructural heterogeneities, stemming from capillary pores and macroscopic voids larger than those within the filaments (Moini, Olek, Magee, et al. 2018; Moini, Baghaie, et al. 2021; Gupta et al. 2023; Najmeddine and Moini 2025; Najmeddine, Gupta, and Moini 2025), are absent in the cast counterparts and fundamentally change the fracture properties and crack propagation paths (Gupta et al. 2023; Prihar, Garlock, et al. 2024). While many similarities exist among additively manufactured cement paste, mortar, and concrete with their cast counterparts in terms of crack initiation, the total fracture energy required for propagation and the crack–interface interactions during propagation substantially differ, necessitating new considerations or approaches in computational simulation for these materials.

The mechanisms of energy dissipation in these layered and 3D-printed materials – including microcracking at the interfaces ahead of the crack tip, interface and aggregate bridging and friction between the interfaces behind the crack tip, and crack–microstructure (i.e., interface) interactions – can promote the spread of damage over larger volumes compared to the cast counterparts, effectively enhancing damage dissipation and/or enlarging the FPZ.

 DOI: 10.1201/9781003434863-15

The exploitation of the additional energy mechanisms enabled by engineering the design of materials or the local properties of the interface (relative to the bulk) has recently given rise to engineering tough architected cement-based materials (Najmeddine et al. 2026; Gupta and Moini 2024; Gupta, Esmaeeli, and Moini 2024; Prihar, Gupta, et al. 2024; Rodriguez et al. 2024; Daneshvar et al. 2024; Moini and Makinen 2023; Ralston, Gupta, and Moini 2024; Zhou et al. 2023; Xie, He, et al. 2025; Zhou, Xu, et al. 2025; Zhou, Bol, et al. 2025). The exploration of this class of additively manufactured architected cement-based materials across scales demonstrates significant improvements in damage delocalization and dissipation through engineered crack deflection along interfaces rather than penetration into bulk filaments, among others (Najmeddine et al. 2026; Moini, Olek, Youngblood, et al. 2018; Savija 2020; Chang et al. 2022; Najmeddine, Gupta, and Moini 2025; Najmeddine and Moini 2025). Studies under both quasi-static and impact loading conditions (Liu et al. 2022; Pan et al. 2024; Ramesh et al. 2024; C. Zhang et al. 2025; Xie, Xu, et al. 2025) have also shown enhancements, not attainable through conventional monolithic casting methods or the existing cement-based microstructure. However, simulating these complex fracture phenomena and toughening mechanisms requires computational frameworks that can capture damage across discontinuous interfaces in a unified manner with the damage that takes place in the bulk. It also needs to capture both the size- and scale-dependent aspects (FPZ relative to structure size) that determine the appropriate fracture mechanics theories (i.e., linear and nonlinear).

Recent large-scale deployments of additively manufactured structures in the building and renewable energy sectors (Ma et al. 2022; Bos et al. 2022) have intensified the need for reliable computational methods to analyze the mechanical and structural responses during ultimate loading conditions involving damage and fracture. These structures, often fabricated as thin walls integrated with cast-reinforced concrete for load-bearing systems (Wang et al. 2025; Heever et al. 2022; Gebhard et al. 2022; Gebhard et al. 2021), may require structural analysis through computational approaches in the absence of a design code. The major challenge in computational simulation of these layered structures lies in accounting for the presence, mechanical characteristics, and contributions of interfaces and associated heterogeneous local properties to the overall response. Such characteristics persist across all processing techniques – from extrusion-based (Rabiei and Moini 2025; Rodriguez et al. 2021; Moini et al. 2022; Moini 2020; Moini, Olek, Zavattieri, et al. 2021) to powder-bed methods (Lowke et al. 2021; Lowke et al. 2018) – requiring computational models that incorporate local interfacial properties (cohesive strength and toughness) in both modes I (tension) and II (shear), given the variety of boundary conditions in the resulting structures.

This chapter examines state-of-the-art computational methods for simulating fracture in layered and additively manufactured cement-based materials, building upon foundational work in brittle and quasi-brittle fracture mechanics (Bažant and Planas 2019; Bažant, Le, Salviato, et al. 2021) and layered materials (Salviato et al. 2019; Bažant 2002). More specifically, this chapter focuses on $_{\text{brittle}}$ fracture. Any extensions to the quasi-brittle fracture of additively manufactured concrete require careful applications of linear elastic fracture mechanics (LEFM) and nonlinear elastic fracture mechanics and potentially alternative numerical approaches, depending on the size, scale, and the fracture problem of interest (i.e., macroscopic response, and accurate consideration of crack band).

The chapter evaluates discrete element methods (DEMs), cohesive zone models (CZMs), the extended finite element method (XFEM), variational phase-field (PF) approaches, and recently developed hybrid frameworks that couple PF with CZM to account for interfaces in layered materials. Particular attention is given to the challenge of incorporating interfacial energy dissipation mechanisms into bulk fracture simulations and accounting for local properties in predicting macroscopic structural responses. Throughout, the chapter assesses how these computational approaches can address fundamental questions about accurately simulating cracking phenomena, including crack penetration, deflection, and bridging, and the numerical provisions needed to accurately represent the complex crack–interface interactions that govern failure in 3D-printed cement-based materials.

12.2 LEFM CRITERION FOR CRACK PENETRATION VS. DEFLECTION IN LAYERED MATERIALS

The theoretical foundation for understanding fracture behavior in layered materials with interfaces has been established through decades of work in LEFM. Such classical approaches have provided fundamental insights into the energy balance governing crack propagation and the critical parameters that determine whether cracks will deflect along interfaces or penetrate into bulk materials. Interfacial crack deflection can enable an important toughening mechanism in layered and bi-material systems, where crack propagation into the interface, followed by interfacial debonding, can create additional pathways for energy dissipation. These fundamental approaches to fracture can serve as an initial design guideline for developing materials with tailored mechanical response, such as tough cementitious composite materials and layered structures.

The landmark work by He and Hutchinson (1989) established the theoretical foundation for understanding crack behavior at material interfaces through a purely energy-based approach. In their work, when a crack impinges on a bi-material interface in a normal direction, whether the crack deflects along the interface or penetrates through the bulk substrate (e.g., surrounding filaments in 3D-printed systems) depends on the relative ratio between the energy-release rate at the tip of a small kink extending across (i.e., penetrating through) the interface, G_p, and the energy-release rate at the tip of a small kink deflected along the interface, G_d (Figure 12.1). The condition for crack deflection along the interface can be written as

$$\Gamma_i / \Gamma_s < G_d / G_p, \tag{12.1}$$

where Γ_i is the fracture toughness of the interface under the appropriate mixed-mode conditions, and Γ_s is the fracture toughness of the material (substrate) ahead of the interface. In the general case of a bi-material assembly composed of material A (i.e., cracked material) and material B (uncracked material) with

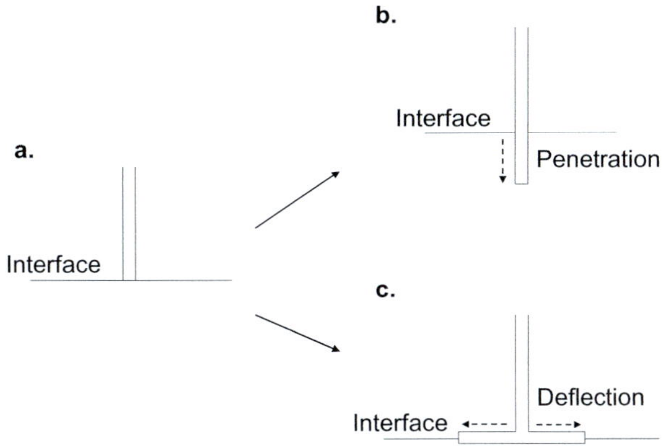

FIGURE 12.1 Schematic representation of penetration vs. deflection crack propagation mechanisms (Parmigiani and Thouless 2006).

elastic mismatch, separated by a *weak* interface normal to pre-existing crack direction (Figure 12.2), the crack propagation mechanism depends not only on the ratio of fracture toughness of the interface to that of bulk material B but also on Dundurs' elastic mismatch parameter, α, quantified as follows (He and Hutchinson 1989):

$$\alpha = \frac{\mu_A(1-\nu_B) - \mu_B(1-\nu_A)}{\mu_A(1-\nu_B) + \mu_B(1-\nu_A)} = \frac{E_A^* - E_B^*}{E_A^* + E_B^*}, \tag{12.2}$$

where μ_i, ν_i, and $E_i^*\,(i = A, B)$ denote the shear modulus, Poisson's ratio, and plane strain Young's modulus, respectively.

When the ratio Γ_i is very small, i.e., when material B has a significantly higher fracture toughness than that of the interface, crack deflection dominates, as illustrated in Figure 12.2. Conversely, crack penetration into material B takes place as the value of the ratio Γ_i to Γ_s increases substantially.

The classical He and Hutchinson framework was later extended by Parmigiani and Thouless (2006), who demonstrated that both cohesive strength and fracture toughness jointly control crack deflection behavior. Their work demonstrated that by controlling interface and bulk properties, three distinct fracture regimes can be identified: toughness-controlled, mixed-mode, and strength-controlled. The transition between these regimes is governed by the relative magnitude of the fracture length scale, defined as $E\Gamma_i / \sigma^2$ (where E is the material modulus and σ is the interface cohesive strength), relative to the appropriate characteristic geometric length that governs the fracture behavior, L, i.e., the relevant physical dimension of the geometry, such as cracked layer thickness or crack/kink length (Suo, Ho, and Gong 1993). When the nondimensional parameter $E\Gamma_i / \sigma^2 L$ is small, toughness controls fracture behavior. When this parameter is large, strength dominates. In the intermediate range, both parameters are important.

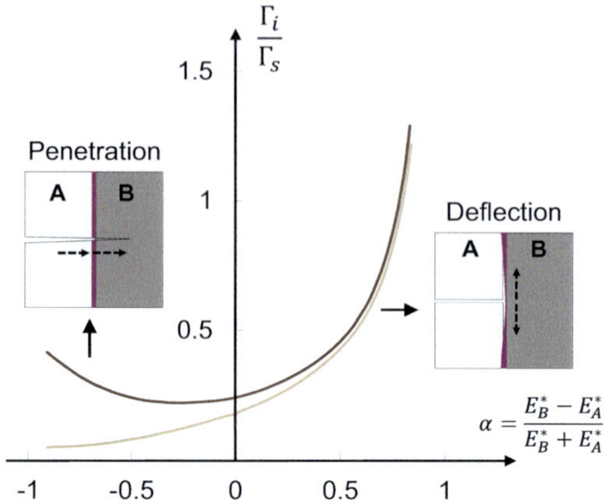

FIGURE 12.2 Schematic illustration of crack propagation mechanisms at a bi-material interface according to He and Hutchinson's criterion (He and Hutchinson 1989) of the layered composite's constituents.

While these analytical frameworks have been essential in providing theoretical insights into the fracture behavior at or due to the presence of interfaces, they have been limited by their assumptions of linear elasticity and simplified (bilayer) geometries. Real layered and composite materials often exhibit complex microstructures, nonlinear material behavior, and 3D fracture and crack evolution that cannot be fully captured by classical LEFM approaches. Furthermore, the probing and prediction of crack initiation, propagation, complex interaction with the microstructure through multiple deflections or branching, and the associated stress and displacement fields during crack propagation all require more sophisticated computational tools. These limitations have recently motivated the development of advanced numerical methods capable of modeling the complexity of fracture in heterogeneous materials (containing interfaces and various bulk constitutive properties) while preserving the fundamental physics established by classical fracture mechanics theory.

Remark. In the case of a bilayer hard–hard assembly containing a crack impinging on an inclined interface, He and Hutchinson (1989) demonstrated that the crack propagation mechanism depends on (i) the ratio of energy release rates for deflection (G_d) and penetration (G_p) and (ii) the interface inclination angle (θ_{int}) for a given elastic mismatch. This is elucidated through the following analytical relationship based on LEFM:

$$\frac{G_d}{G_p} = \frac{1}{16}\left\{\left[3\cos\left(\frac{\theta_{\text{int}}}{2}\right) + \cos\left(\frac{3}{2}\theta_{\text{int}}\right)\right]^2 + \left[\sin\left(\frac{\theta_{\text{int}}}{2}\right) + \sin\left(\frac{3}{2}\theta_{\text{int}}\right)\right]^2\right\}, \quad (12.3)$$

The crack path is determined by comparing the ratio of critical energy release rates for the interface and bulk material, $\frac{G_c^{\text{int}}}{G_c^{\text{bulk}}}$, to the G_d / G_p ratio:

- Deflection occurs if: $\frac{G_c^{\text{int}}}{G_c^{\text{bulk}}} < \frac{G_d}{G_p}$;
- Penetration occurs, otherwise.

12.3 COMPUTATIONAL MODELING OF FRACTURE IN MATERIALS WITH INTERFACES

Computational modeling of fracture in materials with interfaces presents unique challenges that have driven the development of several specialized numerical approaches. Three main approaches include CZMs, the XFEM, and, more recently, PF models. Each method presents a distinct set of opportunities and limitations, addressing different aspects of the fracture process while making distinct assumptions about crack representation and evolution.

CZMs and the XFEM explicitly model crack surfaces as geometric discontinuities, while PF methods represent fracture through continuous field variables that indicate material degradation. The choice of method depends on the specific application requirements, including the need for crack path prediction, probing the crack–microstructure interaction, computational efficiency considerations, and the level of detail required in representing fracture phenomena ahead of the FPZ and behind the crack tip.

The following sections examine three prominent computational fracture mechanics approaches that have proven particularly effective for interfacial fracture problems. CZMs provide a framework for modeling fracture through traction–separation relationships that govern crack surface behavior. CZM approaches are particularly well-suited to fracture problems in layered 3D-printed materials. Unlike traditional finite element methods, where the mesh must conform to crack geometry (requiring crack faces to align with element boundaries and often necessitating remeshing as cracks propagate), the XEFM enables crack representation without mesh conformity by enriching the finite element solution approximation space with additional functions (e.g., Heaviside functions for discontinuous jumps across crack faces and branch functions for singular fields near crack tips), allowing cracks to pass arbitrarily through elements. PF methods model fracture as a continuous process through auxiliary field variables that track material degradation. Each approach offers distinct advantages and limitations, making them suitable for different classes of fracture problems in heterogeneous materials.

In addition to the three main approaches, DEM-based models have recently been used for simulating the fracture and flow behavior of 3D-printed layered concrete. One such model is the lattice discrete particle model (LDPM), which is generally formulated by simulating the concrete mesostructure at the aggregate scale through discrete particles and governs their interaction using meso-scale constitutive laws for tensile fracturing with strain-softening, cohesive and frictional shearing, and nonlinear compressive behavior (Cusatis, Pelessone, and Mencarelli 2011). LDPM has been applied to 3D-printed structures in specific contexts, such as simulating meteorite impact on Martian sulfur concrete (Marscrete) structures additively manufactured in a dome configuration (Troemner et al. 2022). In this study, the layer interfaces were not explicitly modeled, and the material was treated as homogeneous bulk Marscrete in the simulation. While promising, the application of LDPM to simulate interfacial fracture in layered or 3D-printed materials remains under development and subject to future work, and is thus not discussed in detail in this section.

Another DEM-based approach is the discrete fresh concrete (DFC) model, introduced by Ramyara and Cusatis (2021, 2022), which simulates the flow and phase transition of ordinary, self-consolidating, and printable concretes. The DFC model is formulated within the DEM framework, where the interaction between coarse aggregate particles embedded in fine mortar is governed by stress–strain constitutive equations rather than the force displacement laws of conventional DEM. Moreover, a time-dependent extension of the DFC model captures the phase change of concrete from a viscous fluid to a brittle solid, reflecting the evolving rheological and mechanical responses during early-age and throughout setting.

12.3.1 Cohesive Zone Model Approach

12.3.1.1 Theoretical Foundation

The cohesive zone approach was first introduced in the pioneering work of Elliott, Barenblatt, and Dugdale (Elliott 1947; Barenblatt 1959; Dugdale 1960) as an attempt to address the effects of finite stresses near the tip of a crack. This approach introduces a "cohesive zone," a region ahead of the crack tip characterized by a specific traction–separation law (TSL) that describes the process of fracture (Figure 12.3) (Parmigiani and Thouless 2006; Elices et al. 2002). This makes CZMs excellent candidates for capturing fracture phenomena at interfaces, as found in real materials (e.g., layered systems, and composites with dissimilar materials such as fiber–matrix interfaces). Typically, in a TSL, the forces across the crack plane (i.e., cohesive traction) increase with separation (i.e., continuous crack extension) until reaching a peak cohesive strength, after which the traction decreases to zero when a critical separation distance is achieved. At this critical point, the material within the cohesive zone is considered to have completely failed (Figure 12.3). The development of advanced computational methods makes this CZM approach to modeling fracture particularly valuable, enabling crack growth predictions across various geometric configurations (Hillerborg, Modéer, and Petersson 1976; Needleman 1987, 1990; Tvergaard and Hutchinson 1992; Ungsuwarungsri and Knauss 1987). Single-mode (i.e., Mode I or II) fracture behavior is primarily governed by two key TSL parameters: the characteristic fracture toughness Γ (represented by the area under the traction–separation curve) and the characteristic strength σ_{max} (peak cohesive traction in the traction–separation curve).

The mathematical formulation of CZMs has evolved significantly over the past few decades. Several traction–separation constitutive relationships have been developed based on effective displacement ($\bar{\Delta}$) and effective traction ($\bar{T}$) concepts, including cubic polynomial (Tvergaard 1990) (Figure 12.4a), trapezoidal (Tvergaard and Hutchinson 1992) (Figure 12.4b), smoothed trapezoidal (Scheider and Brocks 2003) (Figure 12.4c), exponential (Ortiz and Pandolfi 1999) (Figure 12.4d), linear softening (Camacho and

a. b.

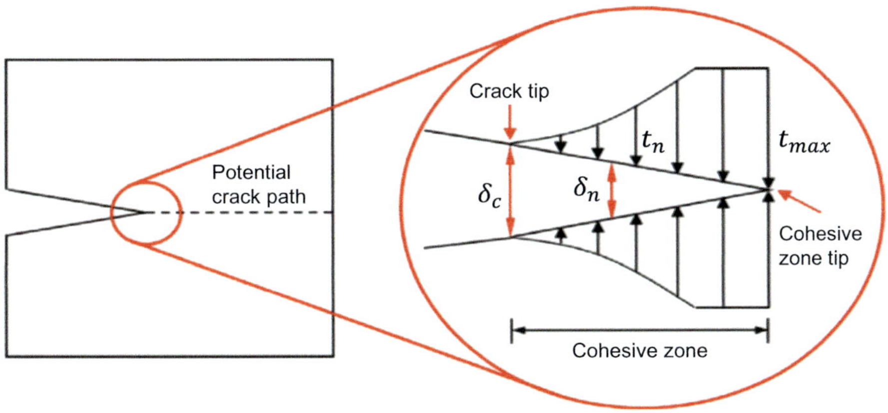

FIGURE 12.3 Schematic of the cohesive zone approach highlighting a generic TSL for single-mode fracture (Mode I) (Dai and Ng 2014).

Ortiz 1996; Geubelle and Baylor 1998; Espinosa and Zavattieri 2003) (Figure 12.4e), and bilinear softening (Petersson 1981; Wittmann et al. 1988) (Figure 12.4f) functions.

The original framework developed by Tvergaard (1990) introduced the effective displacement-based formulation as follows:

$$T_n = \frac{\bar{T}(\bar{\Delta})}{\bar{\Delta}} \frac{\Delta_n}{\delta_n}, \quad T_t = \frac{\bar{T}(\bar{\Delta})}{\bar{\Delta}} \alpha_e \frac{\Delta_t}{\delta_t}, \tag{12.4}$$

where T_n and T_t are the normal and tangential cohesive tractions, respectively; δ_n and δ_t are critical values of the normal and tangential displacements, respectively, at which interfacial failure occurs; and α_e is a nondimensional constant associated with the various formulations shown in Figure 12.4. The nondimensional effective displacement $\bar{\Delta}$ is defined as

$$\bar{\Delta} = \sqrt{(\Delta_n / \delta_n)^2 + (\Delta_t / \delta_t)^2}, \tag{12.5}$$

where Δ_n and Δ_t are the normal and tangential displacement jumps at the interface, respectively. In this formulation, the effective traction $\bar{T}(\bar{\Delta})$ defines the shape of the traction–separation relation. For instance, for the case of the cubic polynomial traction–separation (Figure 12.4a), the effective traction takes the form $\bar{T}(\bar{\Delta}) = \frac{27}{4}\sigma_{\max}\bar{\Delta}(1 - 2\bar{\Delta} + \bar{\Delta}^2)$. Further discussion on the various forms of $\bar{T}(\bar{\Delta})$ and how they relate to the traction–separation plots in Figure 12.4 is discussed in detail in Park and Paulino (2011).

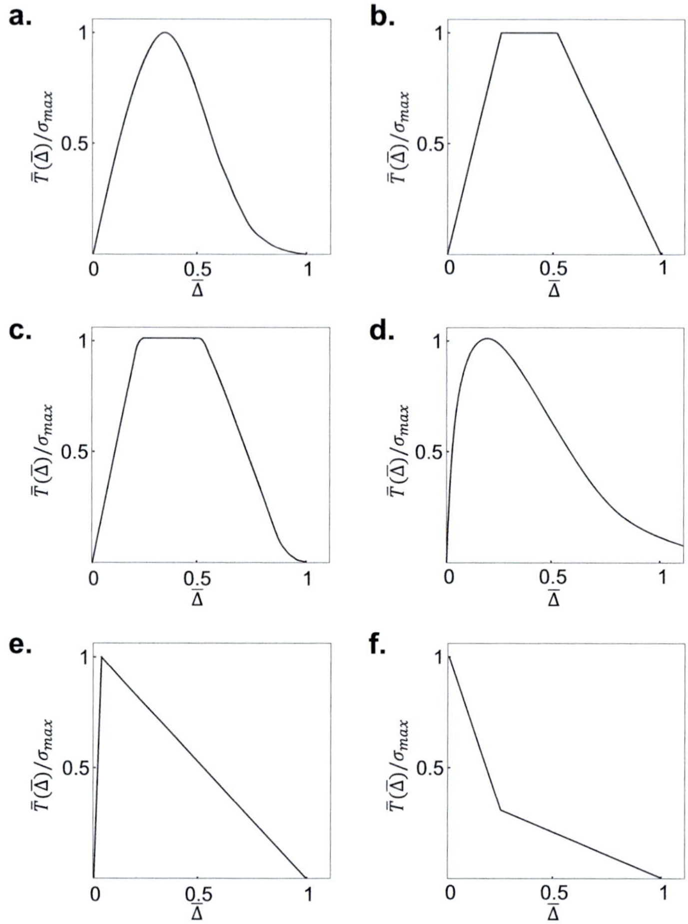

FIGURE 12.4 Effective traction–separation relationships: (a) cubic polynomial, (b) trapezoidal, (c) smoothed trapezoidal, (d) exponential, (e) linear softening, and (f) bilinear softening. Note that the cohesive traction is normalized with respect to the cohesive strength σ_{max} (Park and Paulino 2011).

Effective displacement-based models have certain limitations. Notably, these models can exhibit unrealistic positive stiffness during softening conditions under certain separation paths, where the cohesive traction paradoxically increases as the separation grows (Park and Paulino 2011). Additionally, these formulations apply a constant fracture energy across different fracture modes (i.e., Mode I vs. Mode II), disregarding the specific nature of each mode. Another shortcoming is their inability to differentiate between positive and negative normal separations (Park and Paulino 2011). This is due to the fact that the effective displacement (Eq. (12.5)) is calculated using the square of the normal and tangential separations, which inherently loses the sign information.

As an alternative to effective displacement-based approaches, cohesive traction–separation relationships have been developed on the basis of an energy potential that is typically a function of normal and tangential separations (Δ_n, Δ_t) instead of the effective displacement ($\bar{\Delta}$) (Needleman 1992; Alfano and Crisfield 2001; Park, Paulino, and J. R. Roesler 2009). Potential-based models offer a more nuanced approach by establishing cohesive traction–separation relationships through a potential function that depends on the normal and tangential separations. For a general potential-based model, the normal and tangential cohesive tractions, T_n and T_t, respectively, are derived from a potential function $\varphi^{\text{int}}(\Delta_n, \Delta_t)$ as

$$T_n = \frac{\partial \varphi^{\text{int}}}{\partial \Delta_n}, T_t = \frac{\partial \varphi^{\text{int}}}{\partial \Delta_t}. \tag{12.6}$$

One of the more recent potential-based frameworks that addresses the earlier limitations associated with effective displacement models is the Park–Paulino–Roesler (PPR) model (Park, Paulino, and J. R. Roesler 2009; Park and Paulino 2011). The advantage of using the PPR framework compared to other potential-based approaches is that it is formulated on the basis of consistent fracture boundary conditions and well-defined fracture parameters for each fracture mode: fracture toughness, cohesive strength, initial slope, and shape parameters characterizing brittle, plateau, and quasi-brittle softening responses (Park and Paulino 2011). Furthermore, the PPR model naturally incorporates mode-dependent fracture energies and provides smooth transitions between different loading modes, making it particularly suitable for complex mixed-mode fracture problems encountered in heterogeneous materials and interfaces. The PPR potential is expressed as

$$\begin{aligned}\varphi^{\text{int}}(\Delta_n, \Delta_t) = \min(\mathcal{G}_n, \mathcal{G}_t) + \left[\Gamma_n\left(1 - \frac{\Delta_n}{\delta_n}\right)^{\alpha}\left(\frac{m}{\alpha} + \frac{\Delta_n}{\delta_n}\right)^{m} + \langle \mathcal{G}_n - \mathcal{G}_t \rangle\right] \times \\ \left[\Gamma_t\left(1 - \frac{|\Delta_t|}{\delta_t}\right)^{\beta}\left(\frac{n}{\beta} + \frac{|\Delta_t|}{\delta_t}\right)^{n} + \langle \Delta\mathcal{G}_t - \mathcal{G}_n \rangle\right],\end{aligned} \tag{12.7}$$

where $\mathcal{G}_n$ and $\mathcal{G}_t$ are Mode I and Mode II fracture energies, respectively, and Γ_n and Γ_t are energy constants calculated as

$$\Gamma_n = (-\mathcal{G}_n)^{\frac{(\mathcal{G}_n - \mathcal{G}_t)}{\mathcal{G}_n - \mathcal{G}_t}}\left(\frac{\alpha}{m}\right)^{m}, \Gamma_t = (-\mathcal{G}_t)^{\frac{(\mathcal{G}_t - \mathcal{G}_n)}{\mathcal{G}_t - \mathcal{G}_n}}\left(\frac{\beta}{n}\right)^{n} \text{ for } (\mathcal{G}_n \neq \mathcal{G}_t), \tag{12.8}$$

$$\Gamma_n = -\mathcal{G}_n\left(\frac{\alpha}{m}\right)^{m}, \Gamma_t = \left(\frac{\beta}{n}\right)^{n} \text{ for } (\mathcal{G}_n = \mathcal{G}_t), \tag{12.9}$$

where m and n are exponents of the PPR CZM, α and β are shape parameters that control the softening behavior, as illustrated in Figure 12.5. $\alpha, \beta < 2$ produces concave traction–separation curves, while $\alpha, \beta \gg 2$. yields convex curves typical of quasi-brittle materials; and δ_n and δ_t are the final crack openings in the normal and tangential directions, respectively. Substituting the potential Eq. (12.7) into Eq. (12.6) yields the normal and tangential tractions:

$$T_n = \frac{\Gamma_n}{\delta_n}\left[m\left(1-\frac{\Delta_n}{\delta_n}\right)^{\alpha}\left(\frac{m}{\alpha}+\frac{\Delta_n}{\delta_n}\right)^{m-1} - r\left(1-\frac{\Delta_n}{\delta_n}\right)^{\alpha-1}\left(\frac{m}{\alpha}+\frac{\Delta_n}{\delta_n}\right)^{m}\right]\times \left[\Gamma_t\left(1-\frac{|\Delta_n|}{\delta_n}\right)^{\beta}\left(\frac{n}{\beta}+\frac{|\Delta_t|}{\delta_t}\right)^{n} + \Delta\mathcal{G}_t - \mathcal{G}_n\Delta\right], \tag{12.10}$$

$$T_t = \frac{\Gamma_t}{\delta_t}\left[n\left(1-\frac{|\Delta_t|}{\delta_t}\right)^{s}\left(\frac{n}{s}+\frac{|\Delta_t|}{\delta_t}\right)^{n-1} - s\left(1-\frac{|\Delta_t|}{\delta_t}\right)^{s-1}\left(\frac{n}{s}+\frac{|\Delta_t|}{\delta_t}\right)^{n}\right]\times \left[\Gamma_n\left(1-\frac{\Delta_n}{\delta_n}\right)^{r}\left(\frac{m}{r}+\frac{\Delta_n}{\delta_n}\right)^{m} + \Delta\mathcal{G}_n - \mathcal{G}_t\Delta\right]\frac{\Delta_t}{|\Delta_t|}. \tag{12.11}$$

The normal and tangential traction–separation curves reach their maximum when the normal and tangential separations reach critical crack opening widths δ_{nc} and δ_{tc}, respectively (Figure 12.5):

$$\frac{\partial T_n}{\partial \Delta_n}\Big|_{\Delta_n=\delta_{nc}} = 0, \frac{\partial T_t}{\partial \Delta_t}\Big|_{\Delta_t=\delta_{tc}} = 0. \tag{12.12}$$

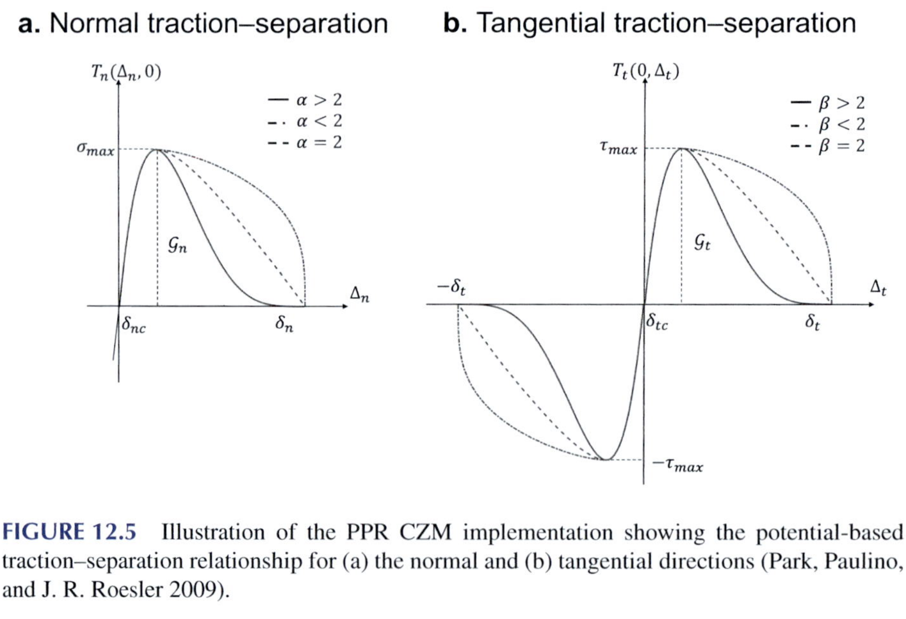

FIGURE 12.5 Illustration of the PPR CZM implementation showing the potential-based traction–separation relationship for (a) the normal and (b) tangential directions (Park, Paulino, and J. R. Roesler 2009).

At these critical points, the maximum tractions correspond to the cohesive strengths (Figure 12.5):

$$T_n\left(\delta_{nc},0\right)=\sigma_{max},\quad T_t\left(0,\delta_{tc}\right)=\tau_{max}. \tag{12.13}$$

12.3.2 Advantages and Limitations

Several researchers have successfully investigated crack propagation in layered cementitious materials using CZMs. Park, Paulino, and J. Roesler (2010) developed a trilinear CZM for functionally graded fiber-reinforced concrete layers, successfully capturing crack propagation through materials with spatially varying fiber content. Dai and Ng (2014) implemented 2D cohesive zone modeling for crack development in layered concrete with distinct microstructural characteristics, demonstrating accurate prediction of crack paths through aggregate–mortar interfaces. Wu and Wriggers (2015) presented a multiscale diffusion–thermal–mechanical CZM for concrete subjected to coupled environmental loading, tracking crack propagation through multiple concrete layers with different aging properties. Lin, Wang, and Xu (2014) presented a CZM-based numerical investigation of steel–concrete composite structures, tracking crack development through multiple material layers. For 3D-printed cementitious materials, L. Kumar et al. (2023) employed CZMs with TSLs to simulate crack propagation in multi directionally loaded specimens, capturing the anisotropic behavior caused by layer interfaces. Warsi, Panda, and Biswas (2024) introduced a composite interface micro-model combining surface-based cohesive zone modeling with a conventional concrete damage plasticity model to predict the structural response of 3D-printed cement-based walls under cyclic loading, successfully capturing interlayer effects and localized damage.

While CZMs have proven valuable for simulating crack propagation in layered cementitious composites and 3D-printed cementitious materials, a few fundamental limitations can restrict their broader application, requiring further investigations per application. The primary challenge is mesh dependency, where cracks are constrained to follow element boundaries, requiring fine discretization that becomes computationally prohibitive for structures with multiple layers or interfaces. However, extrinsic CZMs have been developed to partially address this limitation by adaptively inserting cohesive elements during the simulation (Fukuda et al. 2020). For structures with multiple layers or interfaces, the computational cost becomes particularly prohibitive because cohesive elements must be placed at each interface and potential crack path, multiplying the degrees of freedom by the number of layers – a problem that scales poorly for systems with tens or hundreds of layers, typical in 3D-printed structures. Parameter identification remains a challenge, as cohesive properties must be experimentally determined for each material combination and interface condition. For 3D-printed cement-based materials, the inherent anisotropy from layer-by-layer deposition and complex interaction between interlayer (i.e., interface) and intralayer (i.e., filament) failure modes challenge the typical isotropy and homogeneity assumptions underlying conventional CZM formulations. Additionally, while CZM can model crack propagation effectively once initiated, predicting crack initiation without predefined cohesive surfaces remains challenging. Therefore, capturing the smooth transition between interface debonding and bulk material fracture requires careful calibration that may not generalize across different loading conditions. The computational burden increases significantly with the number of potential crack paths, as cohesive elements must be inserted at all possible failure locations, limiting practical applications to relatively small-scale problems (Najafi Koopas et al. 2024). These constraints indicate that advancing CZM for layered cementitious materials, which can propagate in the bulk filament or interfaces, requires developing adaptive schemes, automated parameter identification methods, and hybrid approaches that can overcome the inherent limitations of the cohesive zone framework (Eckardt and Könke 2008; Marulli et al. 2022).

12.4 EXTENDED FINITE ELEMENT METHOD

The XFEM represents a significant advancement in computational fracture mechanics, enabling the modeling of crack initiation and propagation without the need for mesh conformity to crack surfaces (Moës, Dolbow, and Belytschko 1999; Belytschko and Black 1999). Unlike traditional finite element approaches that require remeshing as cracks evolve, the XFEM enriches the standard finite element approximation space to capture discontinuities independently of the underlying mesh topology (Belytschko and Black 1999).

12.4.1 Theoretical Foundation

The XFEM approach, originally proposed by Belytschko and Black (1999) and Moës, Dolbow, and Belytschko (1999), enriches the standard continuous finite element interpolation of the solution field through the addition of specialized enrichment functions. The enriched displacement field is expressed as

$$u(x,t) = \sum_{I=1}^{n} N_I(x)u_I(t) + \sum_{J=1}^{n} N_J(x)H(x)b_J(t) + \sum_{K=1}^{n} N_K(x)\sum_{L=1}^{4} F_L(x)C_K^L(t), \tag{12.14}$$

where n is the number of nodes per finite element, $N_I(x)$ are the standard finite element shape functions, $u_I(t)$ represent the standard nodal displacements, $H(x)$ is the Heaviside function that captures the jump across crack surfaces, $b_J(t)$ are additional degrees of freedom corresponding to Heaviside enrichment, $F_L(x)$ represent near-tip enrichment functions that capture the asymptotic crack tip displacement fields, and $C_K^L(T)$ are the corresponding near-tip enrichment degrees of freedom.

The Heaviside enrichment allows for the representation of strong discontinuities across crack surfaces, while the near-tip enrichment functions capture the singular stress fields characteristic of crack tips. This formulation requires additional degrees of freedom only for nodes whose influence region is intersected by cracks, making it computationally efficient compared to global mesh refinement approaches.

12.4.2 Phantom Node Implementation Method

The phantom node method, introduced by Hansbo and Hansbo (2004), represents an alternative approach to modeling mesh-independent fracture that offers significant implementation advantages over the original XFEM formulation. This technique provides a more practical way to handle crack propagation in finite element analysis while maintaining the mathematical rigor of XFEMs.

When a crack intersects an element in the phantom node method, the original element is deleted and replaced by two new elements that occupy the same physical location. Each of these "partial elements" contains both a physical portion that represents actual material and a nonphysical portion. The physical portions of both partial elements together cover the entire domain of the original element that was split. The nodes connected to the nonphysical portions of these elements are designated as phantom nodes, giving the method its name. The nonphysical portion exists solely to maintain the standard element topology and connectivity required for finite element formulation but does not contribute to the mechanical response since it represents the void created by the crack. The solution field is then interpolated on these partial elements using standard finite element shape functions, with integration performed only over the physical portions of the elements when computing contributions to the residual.

An important theoretical development came when Areias and Belytschko (2006) demonstrated that the phantom node technique produces exactly the same enrichment as the originally proposed XFEM approach, excluding the near-tip enrichment functions. This equivalence means that both methods add the same number of extra degrees of freedom to the system to represent discontinuities, validating the phantom node method as a mathematically sound alternative to the original XFEM formulation.

The phantom node method offers several compelling advantages from an implementation standpoint. First, it is generally less invasive to existing finite element codes because it does not require the addition of nonstandard degrees of freedom and needs only minor modifications to element integration procedures. The additional degrees of freedom correspond to standard solution fields, which significantly aids in the interpretation of results. The method also simplifies the handling of complex crack patterns, as crack branching and coalescence can be managed through recursive element cutting, whereas the original XFEM requires additional degrees of freedom for elements containing multiple cracks. Finally, for multi-physics applications where different physics share a common mesh, the phantom node technique automatically enriches the degrees of freedom for all solution variables, eliminating the need to explicitly add enrichment degrees of freedom as would be required in the original XFEM approach. These practical benefits make the phantom node method an attractive choice for implementing XFEM capabilities in computational mechanics software. More details on the method can be found in Jiang, Spencer, and Dolbow (2020) and Z. Zhang et al. (2018).

12.4.3 Mesh Cutting Algorithm

The core computational challenge in phantom node XFEM lies in correctly resolving the connectivity of partial elements created when the mesh is cut by evolving cracks. The element fragment algorithm (EFA) addresses this challenge by ensuring that neighboring elements sharing cracked edges maintain proper connectivity on each side of the crack while eliminating connections across crack surfaces (Richardson et al. 2011). Figure 12.6 illustrates the desired connectivity for a representative crack topology.

The EFA operates on a specialized data structure consisting of *fragments* – portions of cut finite elements containing material on one side of cutting planes. When a cutting plane intersects an element, the algorithm creates embedded nodes at intersection points and decomposes the element into fragments, each representing a continuous material region. Each fragment generates a corresponding child element, with temporary nodes created at locations not connected to physical material.

The connectivity establishment phase merges child elements with neighboring elements based on material connections. Two elements are deemed to have a material connection if their fragments share a common edge (2D) or face (3D), indicating physical material continuity across that boundary. The merging process determines which nodes should be shared between adjacent elements to ensure solution continuity in continuous material regions while maintaining discontinuities across cracks. The node merging follows three rules based on node types:

(i) When a temporary node (created outside a fragment) is merged with a permanent node (existing in the mesh), the temporary node is deleted and replaced by the permanent node.
(ii) When two temporary nodes from adjacent elements are merged, both are deleted and replaced by a single newly created permanent node.
(iii) When two permanent nodes with different indices must be merged – a special case occurring at crack tips where nodes are connected to edges both traversed and terminated by cracks – the child node is always deleted in favor of its parent node.

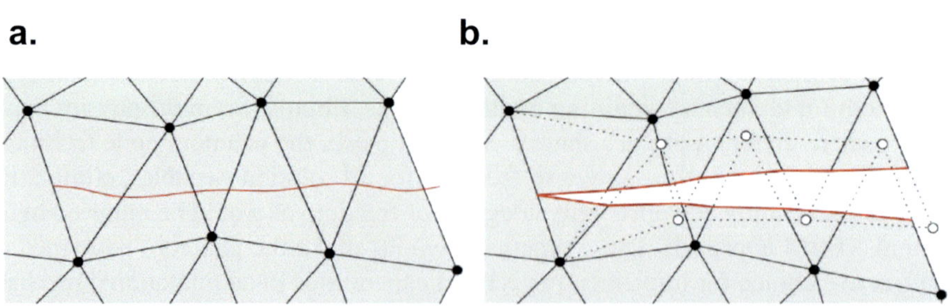

FIGURE 12.6 (a) Original mesh with prescribed cutting line shown in red; (b) cut mesh showing phantom nodes (white circles) and partial elements (dashed lines) (Prince et al. 2025).

This approach enables crack propagation simulation while preserving stateful material data – history-dependent variables such as accumulated plastic strain, damage parameters, and stress states – at integration points. Unlike remeshing strategies that would regenerate the mesh at each time step, the EFA operates incrementally on the previously cut mesh state. This incremental operation prevents data loss that would otherwise occur if the mesh were recreated, thereby maintaining the material history essential for path-dependent constitutive models such as plasticity, damage mechanics, and viscoelastic behavior (Z. Zhang et al. 2018).

Special treatment is required at crack tips to ensure proper topology. When only Heaviside enrichment is used (without near-tip enrichment functions), crack tips must coincide with element boundaries. The algorithm merges the cut edges of partial elements with the uncut element ahead of the crack tip, and when cracks propagate, previously merged edges at former crack tip locations are disconnected to allow continued crack advancement (Jiang, Spencer, and Dolbow 2020).

12.4.4 Integration and Implementation Considerations

The phantom node method requires modifications to standard finite element integration procedures since quadrature integration must be performed only over the physical portions of cut elements to avoid spurious contributions from the nonphysical regions. The standard approach involves decomposing the physical region of each partial element into triangular sub-elements in two dimensions or tetrahedral sub-elements in three dimensions, with integration points placed within these sub-elements.

However, when stateful material data exists at integration points, as is common in nonlinear material models, mapping this data to new integration points during sub-element decomposition can introduce errors and computational overhead. To address this issue, alternative approaches avoid this mapping by scaling the weights of existing integration points. The simplest method (volume scaling) (Dolbow, Moës, and Belytschko 2000; Wells and Sluys 2001) scales weights by the physical volume fraction of the element, while more sophisticated moment fitting approaches (Mousavi and Sukumar 2010; Ventura 2006; Müller, Kummer, and Oberlack 2013) use least squares optimization to determine weights that optimally integrate specified function sets.

The moment fitting approach determines integration weights such that polynomial functions up to a specified order are integrated exactly, and has been shown to provide improved accuracy over volume scaling while maintaining computational efficiency. For typical applications, a four-point integration rule with moment fitting provides adequate accuracy for linear quadrilateral elements commonly used in 2D fracture simulations.

Visualization of phantom node XFEM results requires special consideration due to the overlapping elements created by the method. Standard visualization tools display both physical and nonphysical portions, potentially obscuring meaningful results. Specialized post-processing techniques (Daux et al. 2000; Legrain, Moes, and Verron 2005; Minnebo 2012; Abbas, Alizada, and Fries 2010) clip the nonphysical portions of cut elements to reveal the actual material response. Unlike classical XFEM (Belytschko and Black 1999; Moës, Dolbow, and Belytschko 1999), no modification of computed displacement fields is required since the phantom node method directly computes actual displacements rather than enriched approximations.

12.4.5 Advantages and Limitations

The XFEM approach offers significant advantages for modeling fracture in materials with interfaces. The method enables crack propagation simulations without the geometric constraints imposed by mesh boundaries, allowing cracks to initiate and grow along arbitrary paths dictated by physics-based criteria (e.g., stresses or fracture integrals) that indicate how the cracks should extend, rather than mesh topology (Moës, Dolbow, and Belytschko 1999; Belytschko and Black 1999). This capability is particularly valuable for modeling complex fracture patterns in heterogeneous materials where crack paths cannot be predetermined (Daux et al. 2000; Belytschko et al. 2001).

The phantom node implementation provides additional practical benefits by requiring minimal modifications to existing finite element codes (Song, Areias, and Belytschko 2006). The preservation of stateful material data during crack evolution enables accurate modeling of materials with complex constitutive behavior, including plasticity, damage, and other history-dependent phenomena (Rabczuk et al. 2008; Song and Belytschko 2009).

However, the XFEM also presents certain limitations that must be considered in practical applications. The method requires careful treatment of crack tip regions, particularly when using only Heaviside enrichment, as crack tips must coincide with element boundaries. The integration of partial elements introduces additional computational complexity and potential sources of numerical error. For 3D applications, ensuring continuity of cutting planes between adjacent cracked elements presents additional algorithmic challenges (Duan et al. 2009).

The accuracy of XFEM solutions depends critically on the quality of the underlying mesh and the choice of enrichment functions. While the method eliminates the need for mesh conformity to crack surfaces, adequate mesh resolution is still required to capture stress gradients near crack tips and to ensure accurate representation of crack tip fields through the chosen enrichment functions.

The XFEM has been widely used in the literature to simulate interfacial crack phenomena in layered materials. Early demonstrations, such as Moës et al. (2003), validated the capability of the XFEM to simulate cracks near material interfaces in handling complex microstructures. Natarajan, Song, and Belouettar (2014) improved interfacial crack simulations by introducing scaled-boundary enrichment functions that approximate stress intensity factors and *T*-stresses accurately – without predefining asymptotic fields. Swati et al. (2022) applied the XFEM at the microscale to study fiber–matrix debonding in carbon epoxy composites, validating against experiments and capturing debonding initiation under various fiber orientations. Although the XFEM is a powerful method, it comes with some limitations for use in layered composites: (i) enrichment near interfaces adds integration complexity and more demanding quadrature (e.g., subcells), (ii) mesh-dependent convergence if singular functions are not properly captured, and (iii) inconsistencies in energy release rate calculations (i.e., *J*-integrals) when dealing with nonhomogeneous media (Tafazzolimoghaddam

2017) such as layered and 3D-printed materials. Hybrid approaches combining the XFEM for bulk fracture and cohesive elements for delamination improve modeling the crack path more realistically but tend to increase implementation complexity and computational cost (Stuparu et al. 2016).

12.5 THE PHASE FIELD APPROACH TO FRACTURE

12.5.1 Theoretical Foundation

The PF approach to fracture has recently gained significant traction, given its use to simulate various cracking phenomena in brittle (De Lorenzis and Maurini 2022; Ambati, Gerasimov, and De Lorenzis 2015a; Kumar et al. 2020; Hansen-Dörr et al. 2019; Francfort and Marigo 1998; Miehe, Welschinger, and Hofacker 2010), quasi-brittle (Wu 2017; Yang, Yang, and Zheng 2021; Ulloa et al. 2019; Muñetón-López and Giraldo-Londoño 2024), and ductile materials (Ambati, Gerasimov, and De Lorenzis 2015b; Ambati, Kruse, and De Lorenzis 2016; Miehe, Aldakheel, and Raina 2016; Dittmann et al. 2018).

In PF models, a sharp crack Q_0 contained within a body Ω_0 is regularized through a PF variable which varies both temporally and spatially, as shown in Figure 12.7. A continuous scalar field, $d(X,t)$, is then assigned to every point in the material domain, ranging between 0 (intact material) and 1 (fully fractured material), thus allowing for a gradual transition from intact to damaged regions over a localization band $B \subseteq \Omega_0$, as illustrated in Figure 12.7. The size of the localization band B is given by a characteristic length scale l_c, which, in the limit as l_c approaches 0, recovers the original Griffith theory for fracture (Miehe, Hofacker, and Welschinger 2010).

The PF approach to fracture is based on a variational principle, wherein crack initiation and propagation within a solid body Ω_0 are the direct results of the minimization of the total energy potential of the system. The total energy potential functional, Π, is defined as the sum of the Helmholtz free energy density function integrated over the volume Ω_0, and a crack surface energy component representing the dissipated energy in creating new fractured surface(s) that is diffused over the cracked region of

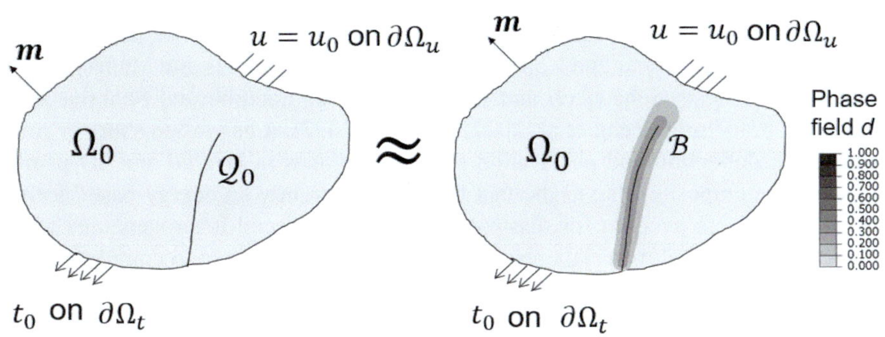

FIGURE 12.7 Substitution of a sharp crack Q_0 within a solid body with a smeared crack representation $\mathcal{B}$ via a continuous scalar PF, d, to describe the fracture in solids.

the body (Miehe, Hofacker, and Welschinger 2010). As such, a solid undergoing fracture can then be completely described by a pair of field quantities representing the displacement field $\boldsymbol{u}(\boldsymbol{x})$ and the PF variable $d(\boldsymbol{x})$, which are determined by solving the following constrained minimization problem:

$$(\boldsymbol{u}(\boldsymbol{x}), d(x)) = \text{Arg }\{\min \prod(u,d)\} \text{ subject to } \dot{d}(x) > 0, d(\boldsymbol{x}) \in [0,1] \quad (12.15)$$
$$\text{with } \boldsymbol{x} \in R^3,$$

Assuming that the solid is subject to a body force per unit volume $\boldsymbol{B}_0$ over Ω_0 and prescribed tractions $\boldsymbol{T}_0$ on the natural boundary $\partial\Omega$, the potential energy functional ∏ can be expressed in terms of the displacement field and the PF variable as follows:

$$\Pi(\boldsymbol{U}, d) = \int_{\Omega_0} \Psi(\boldsymbol{U}, \omega(d))\, \mathrm{dV} + \int_{\Omega_0} G_c \gamma(d, \nabla d) \mathrm{dV} - \int_{\Omega_0} \boldsymbol{B}_0.\boldsymbol{U}\, \mathrm{dV} - \int_{\partial\Omega_t} \boldsymbol{T}_0.\mathrm{U}\, \mathrm{dA},\ \mathrm{U} = \mathrm{U}_0 \text{ on } \partial\Omega_U, \quad (12.16)$$

where $\Psi(\boldsymbol{u}, \omega(d))$ is the degraded strain energy density function defined in terms of $\boldsymbol{u}$ and the PF degradation function $\omega(d)$, G_c is the fracture energy required to propagate a crack per unit area, and $\boldsymbol{u}_0$ is the prescribed displacement on the essential boundary $\partial\Omega_u$ (Figure 12.7). The term $\gamma(d, \nabla d)$ represents the crack surface density function and is defined as $\gamma(d, \nabla d) = \frac{1}{c_\phi}\left[\frac{1}{l_c}\phi(d) + l_c \nabla d.\nabla d\right]$, where $G_c^{(2)} = f_m G_c^{(1)}$ and ∇d is the gradient of the PF variable. The term $\phi(d)$ is a characteristic function for the PF formulation called the crack geometric function. Different versions of $\phi(d)$ have been explored in the literature (Wu 2017); the two most common forms are $\phi(d) = d^2$, representing the so-called AT2 model (Ambrosio and Tortorelli 1990), and $\phi(d) = d$, representing the AT1 model (Pham et al. 2011).

12.5.2 Advantages and Limitations

While the PF approach to fracture has demonstrated remarkable potential in capturing complex crack topologies in single bulk materials that involve nucleation (A. Kumar et al. 2020; Kumar and Lopez-Pamies 2020; De Lorenzis and Maurini 2022), propagation (Miehe, Schaenzel, and Ulmer 2015; Najmeddine and Shakiba 2023, 2024), and branching (Zeng et al. 2022; Xu et al. 2022), it has fallen short in simulating crack deflection and penetration across interfaces in layered and 3D-printed materials or composites due to the fact that it is inherently an energy-based formulation. As such, to account for dissipation across interfacial heterogeneities in the microstructure of real materials, various efforts have been made to complement the PF approach. These effort aim to address the lack of microstructural representation either (i) through an additional order parameter describing interface topology as regularized regions of finite size, as illustrated in Figure 12.8 (Nguyen, Waldmann, and Bui 2019; Tarafder, Dan, and Ghosh 2020; Hansen-Dörr et al. 2020; Hansen-Dörr, Brummund, and Kästner 2021; Schöller et al. 2022), or (ii) by coupling the standard

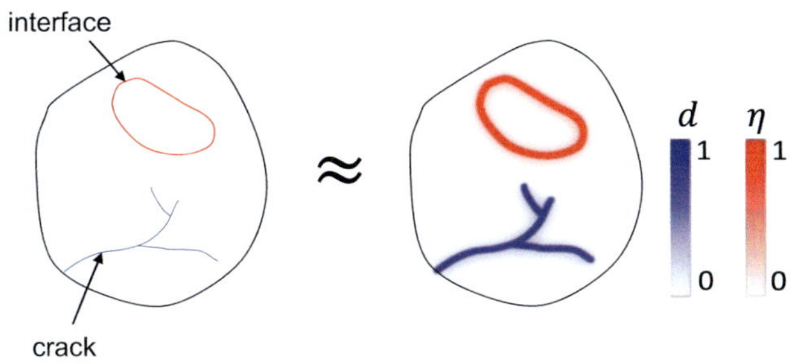

FIGURE 12.8 Schematic of a cracked solid body containing interfaces wherein discrete representations of cracks and interfaces are replaced by regularized representations through scalar PF order parameters d for fracture and η for interface, respectively (Tarafder, Dan, and Ghosh 2020).

PF method with an appropriate cohesive zone model as a means to accurately capture the physics of fracture phenomena across layered hard–hard or hard–soft composites that contain interfaces (Paggi and Reinoso 2017; Carollo, Reinoso, and Paggi 2018; Dusane et al. 2022; Marulli et al. 2022; Najmeddine, Gupta, and Moini 2025; Najmeddine and Moini 2025; Najmeddine et al. 2026).

The complementary interface–order-parameter approach introduces interfaces as diffuse interfacial transition zones of length l_{trans} with an additional PF variable regularizing the sharp interface discontinuity across which a displacement jump is calculated. Figure 12.9 illustrates an example of such a framework developed by Nguyen, Waldmann, and Bui (2019) applied to a bi-material composite, in which various mechanisms of crack propagation are demonstrated across the smeared interface for varying elastic modulus E and fracture energy G_c mismatch factors f_m (where $E^{(2)} = f_m E^{(1)}$ and $G_c^{(2)} = f_m G_c^{(1)}$).

As for the unified PF CZM approach, researchers have coupled the PF method with a CZM to account for crack propagation in composite systems containing interfaces. For example, Paggi and Reinoso (2017) used a linear CZM with a tension cut-off TSL upon failure in combination with PF to study crack propagation in linear elastic composite materials containing interfaces. Their framework was able to capture the competition between crack penetration and deflection in hard–hard material assemblies containing both normal-to-crack and oblique interfaces. Similarly, Carollo, Reinoso, and Paggi (2018) presented a variational framework combining PF and a linear CZM formulation to study complex crack paths in ceramic laminates, considering the effect of tough and quasi-brittle interfaces. The work properly captured crack propagation mechanisms observed experimentally in real laminate structures composed of silicon nitride and boron nitride layers, including deflection, branching, and delamination. Additionally, Dusane et al. (2022) followed a similar approach and presented a coupled PF CZM framework based on a linear TSL formulation to study delamination in matrix fiber and cracking of fiber-reinforced composites. However, these efforts considered only small-deformation kinematics as they focused only on hard–hard material assemblies for which small-deformation

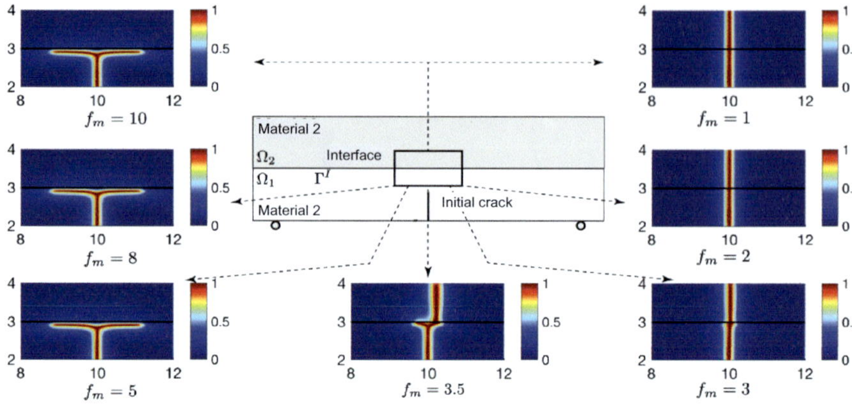

FIGURE 12.9 Crack propagation mechanisms in a bi-material layered structure using the smeared interface model by Nguyen, Waldmann, and Bui (2019) for varying mismatch ratios f_m. The fracture profile across the interface is greatly affected by the mismatch ratio. Adapted from Nguyen, Waldmann, and Bui (2019).

assumption may be appropriate. In contrast, Marulli et al. (2022) developed a coupled PF polynomial-based CZM for large-deformation kinematics for problems such as layered flexible substrates. However, in their work, the CZM was based on the effective displacement approach, and as detailed earlier, these formulations present several limitations when considered in numerical simulations.

12.6 UNIFIED PHASE-FIELD PPR CZM FRAMEWORK

A large-deformation variational formulation capable of modeling crack propagation in layered assemblies composed of hard–soft constituents separated by interfacial zones has recently been developed by Najmeddine, Gupta, and Moini (2025). The formulation provides a comprehensive framework for analyzing crack propagation in layered materials, combining PF modeling of bulk fracture with a potential-based cohesive zone modeling for dissipation along the interface. The framework considers a solid body Ω_0 divided into two regions (Ω_0^1 and Ω_0^2) by an interface Γ_0, with each region potentially having a different constitutive behavior (e.g., hard linear elastic ceramic or cement paste, and soft hyperelastic polymer), as illustrated in Figure 12.10.

The kinematics involve a motion mapping $\boldsymbol{x} = \chi(\boldsymbol{X}, t)$ from the reference to current configuration (Figure 12.10a), with the displacement field defined as $\boldsymbol{u}(\boldsymbol{X}, \boldsymbol{t}) = \boldsymbol{x} - \boldsymbol{X}$, where $\boldsymbol{X}$ is the vector representing the location of an arbitrary material point in the domain of the initial (reference) configuration.

The energy contribution due to the interface dissipation, Π^{int}, is accounted for within the total potential energy functional as follows:

$$\Pi(\boldsymbol{u}, d) = \Pi^{b^k}(\boldsymbol{u}, d) + \Pi^{int}(\boldsymbol{u}) + \mathcal{P}(\boldsymbol{u}), \tag{12.17}$$

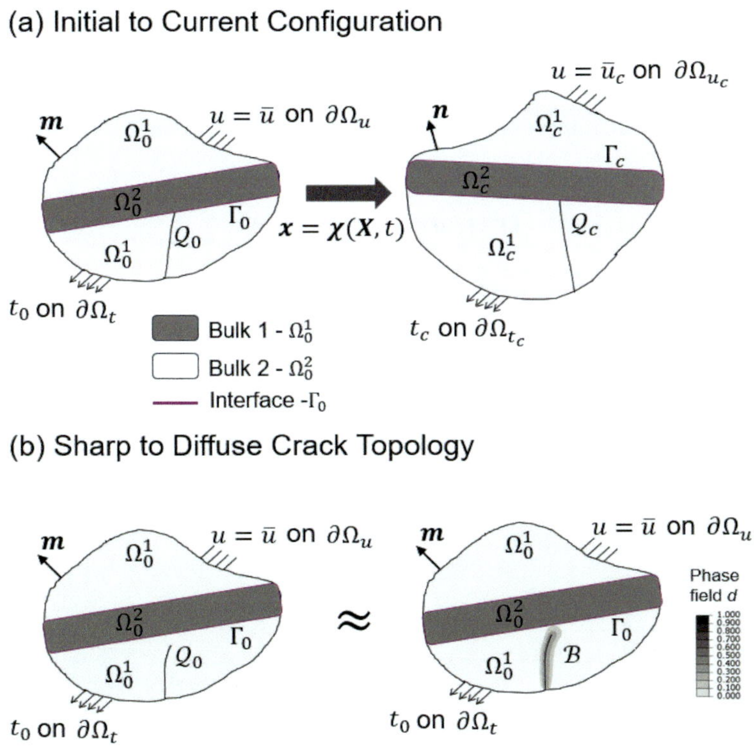

FIGURE 12.10 (a) Kinematics of a solid body containing an initial crack $\mathcal{Q}_0$ and an interface Γ_0 separating two regions Ω_0^1 and Ω_0^2 in the initial, and Ω_c^1 and Ω_0^2 in the current configurations. (b) Substitution of the sharp crack topology $\mathcal{Q}_0$ with a diffuse representation B having a finite width over which a scalar PF variable varies from 0 (intact) to 1 (complete fracture).

where Π^{b^k} represents the bulk energy contributions, with k referring to the bulk material type considered (i.e., hard or soft), and $\mathcal{P}$ represents external work. For the bulk contribution to the total potential in Eq. (12.17), the PF variational formulation described in Section 12.5 is used, wherein $\Pi^{b^k}(\mathrm{u}, \mathrm{d})$ is given by

$$\Pi^{b^k}(\mathrm{u}, \mathrm{d}) = \int_{\Omega_0^k} \Psi^k(\mathbf{F}, d)\mathrm{dV} + \int_{\Omega_0^k} G_c^k \gamma(d, \nabla d)\mathrm{dV}, \tag{12.18}$$

where $\Psi^k(\mathbf{F}, d)$ and G_c^k are the Helmholtz free energy and the fracture toughness of bulk material k, respectively, and the crack surface density function $\gamma(d, \nabla d)$ is given by the following expression:

$$\gamma(d, \nabla d) = \frac{1}{c_\phi}\left[\frac{1}{l_c^k}\phi(d) + l_c^k \nabla d . \nabla d\right], \tag{12.19}$$

where c_ϕ and $\phi(d)$ are described in Section 12.5.

As for the interface, the contribution to the total potential Eq. (12.17) is given by the PPR cohesive potential function $\varphi^{int}(\Delta \boldsymbol{u})$ defined in Eq. (12.7) as a function of the displacement jump across the interface $\Delta \boldsymbol{u} = (\Delta \boldsymbol{u}_t, \Delta \boldsymbol{u}_n)^{\mathrm{T}}$, integrated over the domain Γ_0:

$$\Pi^{int}(\Delta \boldsymbol{u}) = \int_{\Gamma_0} \varphi^{int}(\Delta \boldsymbol{u}) \mathrm{dA}. \tag{12.20}$$

The equilibrium of the system is determined through the principle of stationary (or minimum potential) energy, requiring that the variation in the total potential energy vanishes:

$$\delta \Pi = \delta \Pi^{b^k} + \delta \Pi^{int} + \delta \mathcal{P}. \tag{12.21}$$

The variational expressions with respect to the displacement and PF variable are given by

$$\begin{aligned} \delta \Pi_u^{b^k} &= \int_{\Omega_0^k} \frac{\partial \Psi^k(\mathrm{F,d})}{\partial \mathbf{F}} \frac{\partial \delta \boldsymbol{u}}{\partial \boldsymbol{X}} \mathrm{dV}, \delta \Pi_d^{b^k} \\ &= \int_{\Omega_0^k} \frac{\partial \Psi^k(\mathrm{F,d})}{\partial d} \delta d \mathrm{dV} + \int_{\Omega_0^k} \frac{G_c^k}{c_\phi l_c} \phi'(d) \delta d \mathrm{dV} - \int_{\Omega_0^k} \frac{2G_c^k l_c}{c_\phi} \Delta d \delta d \mathrm{dV}, \end{aligned} \tag{12.22}$$

whereas for the interface, the variation takes the following form:

$$\delta \Pi^{int} = \delta \boldsymbol{u}^{\mathrm{T}} \int_{\Gamma_0} \left(\frac{\delta \Delta \boldsymbol{u}}{\delta \boldsymbol{u}} \right)^{\mathrm{T}} \mathbf{T} \mathrm{dA}. \tag{12.23}$$

where T is the first Piola–Kirchhoff cohesive traction vector, obtained as

$$\mathrm{T} = \frac{\partial \varphi^{\mathrm{int}}(\Delta \boldsymbol{u})}{\partial \Delta \boldsymbol{u}} = (T_t, T_n)^T. \tag{12.24}$$

The coupled large-deformation PF PPR CZM framework has been implemented in Abaqus via a UEL subroutine to examine crack propagation responses in hard–hard (Figure 12.11a) and hard–soft composite structures with interfacial zones (Figure 12.11b,c) (Najmeddine, Gupta, and Moini 2025). Hard–hard layered assemblies can be conceptualized as 3D-printed cementitious layers or filaments that interact across interfacial zones, whereas hard–soft layered and heterogeneous composites can be considered as 3D-printed cement-based composites containing cementitious layers and soft inclusions (e.g., nacre-inspired layered hard–soft 3D-printed cement paste elastomer composites) or fiber-reinforced cementitious composites (Figure 12.12).

Figure 12.13 illustrates the predictions of the coupled PF CZM framework developed by Najmeddine, Gupta, and Moini (2025) for a bi-material composite with a weak interface. The results demonstrate the substantial toughening (energy dissipated

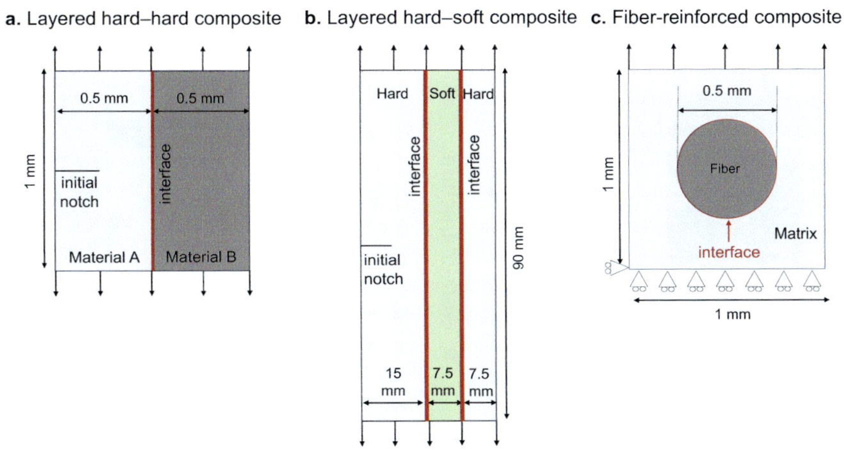

FIGURE 12.11 Benchmark multilayer and heterogeneous composites, which typically exhibit complex crack propagation mechanisms, i.e., crack deflection, crack penetration, crack bridging, and interface delamination, the fracture response of which can be captured by the unified PF PPR CZM framework.

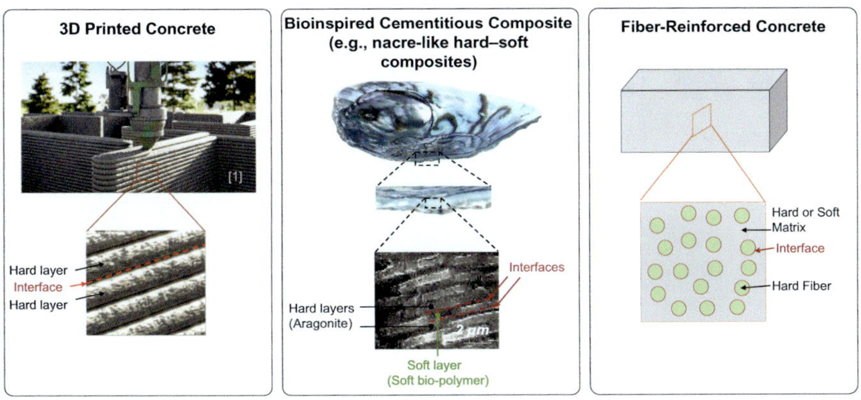

FIGURE 12.12 Application domains related to 3D-printed cement-based materials, fiber-reinforced concrete, and bio-inspired hard–soft composites, where the interface is present and simulating the fracture behavior is crucial for design and engineering across scales.

under the force–displacement curve of the notched simulation) achieved when the crack deflects along the interface as opposed to when it penetrates straight through the interface into material B.

In the case of a layered hard–soft 3D-printed cement paste elastomer composite, Figure 12.14 demonstrates the framework's capability to capture crack propagation across multi-material interfaces, resulting in an emergent crack bridging toughening mechanism due to the inclusion of the soft layer. Such a mechanism can be engineered into layered and 3D-printed materials to substantially enhance the energy dissipated through fracture, while maintaining the load-bearing capacity and using less material compared to its monolithic counterpart (in this case, 33% less material). It

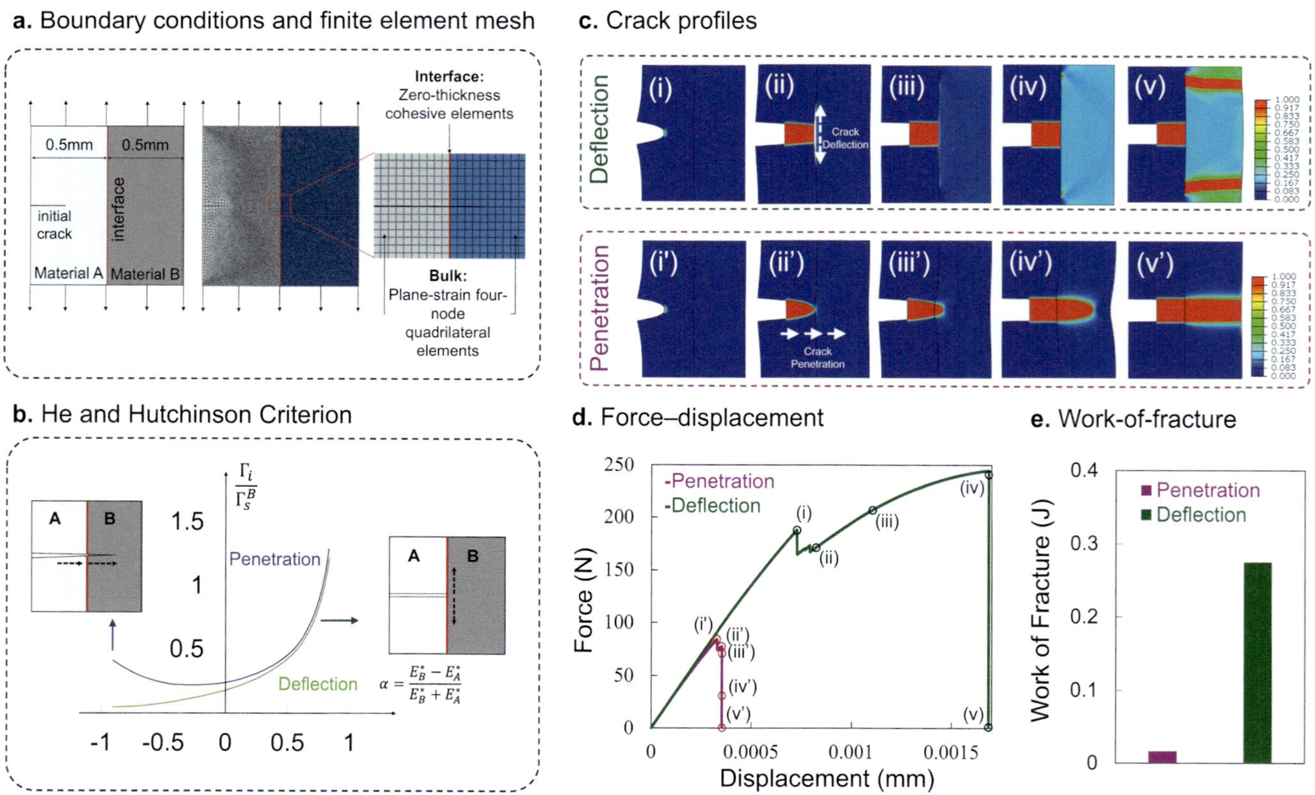

FIGURE 12.13 Crack deflection vs. penetration in a hard–hard bilayer composite simulated using the coupled PF PPR CZM framework (Najmeddine, Gupta, and Moini 2025).

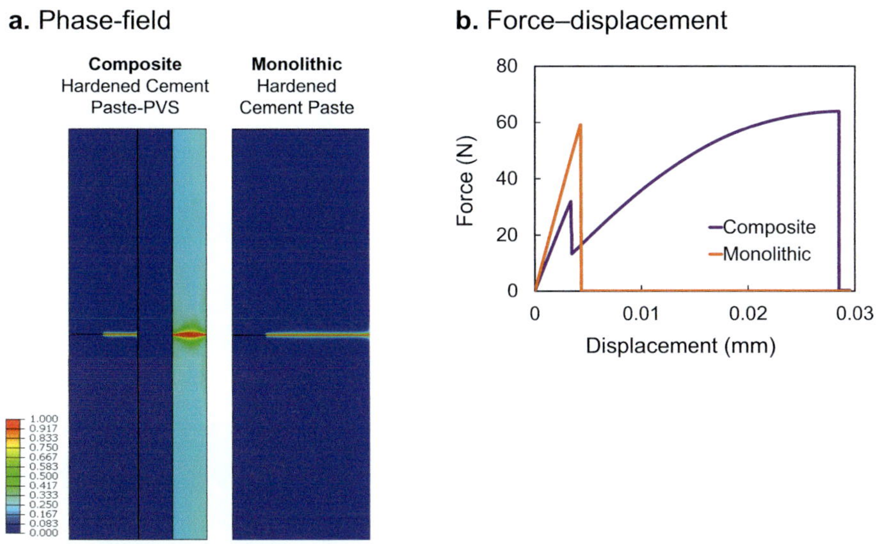

FIGURE 12.14 Crack propagation profile of a tri-layer hardened cement paste elastomer multi-material demonstrating crack bridging by soft elastomeric layer and enhanced fracture toughness, compared to its conventionally brittle monolithic hardened cement paste counterpart. Adapted from Najmeddine, Gupta, and Moini (2025).

has also been demonstrated that thinner soft layers can provide equal or greater fracture toughness in the composite, highlighting the critical role of small mismatches in layered composites in the resulting composite's performance (Najmeddine, Gupta, and Moini 2025).

Besides the utility to simulate 3D-printed single-component cementitious materials or multi-component cementitious polymeric materials, the unified PF PPR CZM framework can also be applied to simulate the response of fiber-reinforced matrix composites. Figure 12.15 illustrates the simulation results pertaining to the case of the fiber–matrix composite, demonstrating remarkable agreement with established experimental observations of failure mechanisms in fiber-reinforced composites (París, Correa, and Mantič 2007). Specifically, fiber–matrix interfacial debonding was observed to initiate at an angle of 64° relative to the normal axis at the center of the composite, in line with the experimentally documented debonding angles of 60°–70° reported by París, Correa, and Mantič (2007). The debonding location corresponds to the region of initial stress concentration along the fiber–matrix interface. The continued debonding leads to increased stress concentration along the 64° diagonal, driving the crack to propagate further into the matrix until complete fracture occurs – a pattern that was also corroborated by previous numerical investigations (Tarafder, Dan, and Ghosh 2020; Labanda, Giusti, and Luccioni 2018; Nguyen et al. 2017). The debonding can be observed as separation at the interface throughout the matrix fracture.

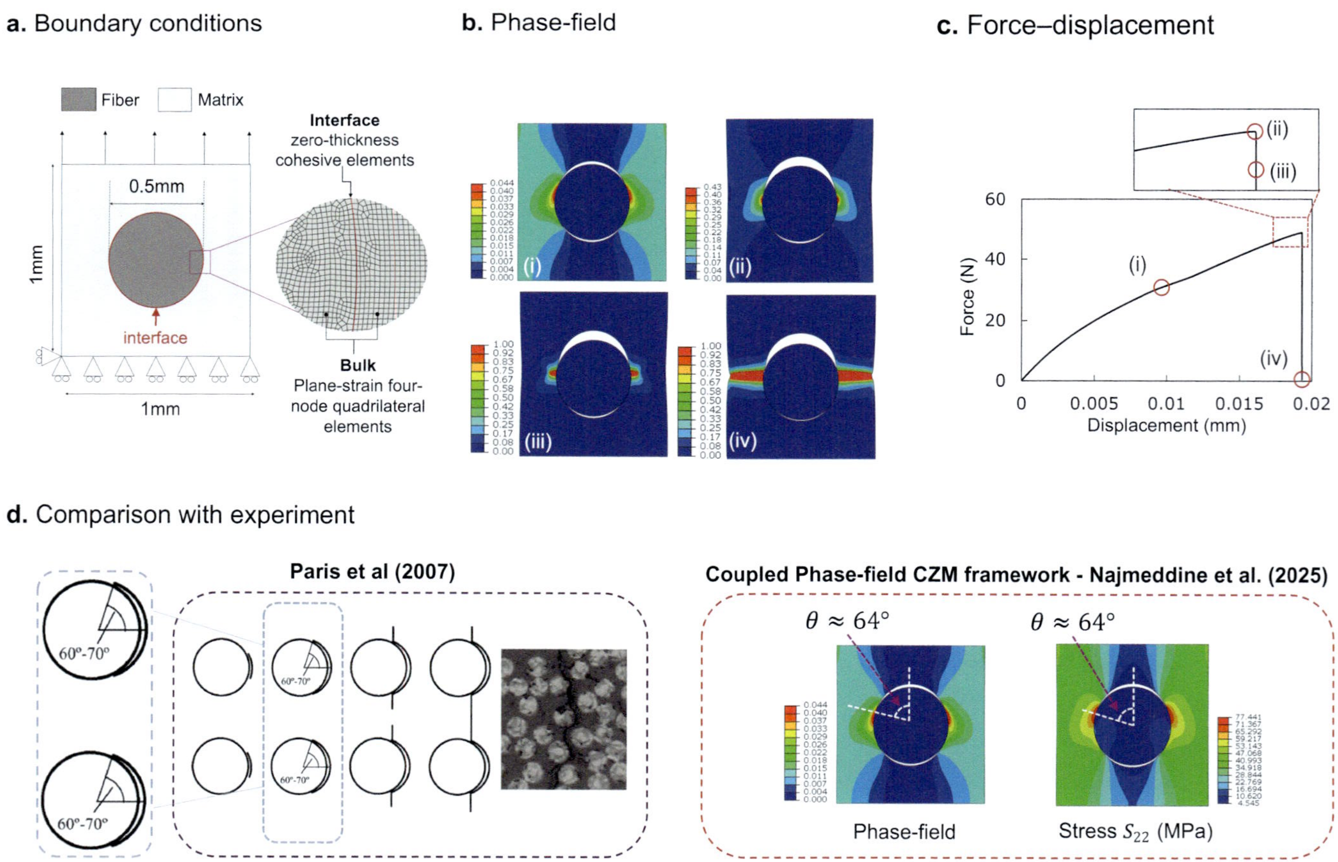

FIGURE 12.15 Fiber-reinforced composite simulation results illustrating fiber–matrix debonding at an angle of approximately 64°, consistent with the experimental finding of París, Correa, and Mantič (2007). This location coincides with the region where stress first begins to localize along the fiber–matrix interface, leading to crack kinking into the matrix slightly adjacent to the 64° diagonal (Najmeddine, Gupta, and Moini 2025).

12.7 CONCLUDING REMARKS

In this chapter, we examined the state-of-the-art computational methods for simulating fracture in brittle layered and additively manufactured cement-based materials. The unique challenges posed by the presence of interfaces in 3D-printed materials are discussed. The review of numerical approaches – including CZMs, the XFEM, PF methods, and hybrid PF CZM frameworks – demonstrates both significant progress and substantial remaining challenges in accurately capturing the complex fracture behavior of these materials.

The computational modeling of 3D-printed cement-based material requires new considerations compared to those developed for cast materials. The presence of interfaces necessitates the use of numerical methods capable of capturing the competition between crack penetration into bulk filaments and deflection along interfaces. While LEFM criteria provide theoretical guidance, extensions to non-LEFM theories and the size effect law are necessary for guiding numerical simulation beyond brittle cement paste materials (i.e., quasi-brittle concrete for which the M7 model discussed in a separate chapter in this book) that offer more advanced constitutive utility. This challenge is exacerbated in computational frameworks for real structures with hundreds of layers (interfaces) and scaling laws for complex loading conditions. Each numerical method offers distinct advantages: CZMs provide direct control over interfacial properties through TSLs, the XFEM enables mesh-independent crack propagation, and the variational PF methods naturally handle complex crack topologies for brittle fracture, including deflection, penetration, and branching.

The recently developed unified PF PPR CZM framework represents a promising advancement, demonstrating the capability to simulate crack propagation in hard–hard (brittle cement paste–cement paste) and hard–soft (cement paste–polymer) layered systems under large deformations (for both interface and bulk). Validation against theoretical solutions and experimental findings for various composite configurations – from simple bilayer systems to complex fiber-reinforced composites – suggests that hybrid approaches may be necessary to capture the full range of fracture mechanisms in 3D-printed cement-based materials.

The computational approaches available for simulating fracture are diverse, each with specific limitations depending on the scale, size, and whether brittle or quasi-brittle response is examined. The constitutive behavior of interfaces, defined by TSLs in normal and tangential directions, significantly alters the overall fracture response of layered additively manufactured materials and structures. The PPR CZM enables the consideration of both Modes I and II, allowing more reliable approaches for complex mixed-mode loading conditions. Interfacial properties must be experimentally characterized and used as input to the PPR model, including cohesive strengths σ and τ (peak load), fracture toughness G_c (area under the traction–separation curve), softening behavior, and ultimately separation. Coupling CZMs with experimentally determined values specific to the interface constitutive behavior at a given size and scale, and for a given failure mode, enables more accurate simulation of interfacial fracture mechanics.

However, predicting crack initiation, propagation, and complex microstructural interactions – including multiple deflections, penetration, or branching – while preserving physical phenomena such as friction, requires additional advancement in computational frameworks. Future development, therefore, must address modeling fracture in heterogeneous materials containing appropriate interfacial physics and more complex microstructure (e.g., concrete) to advance the engineering and design of additively manufactured structures.

As additive manufacturing technologies advance toward large-scale construction applications, reliable computational tools become increasingly critical. The absence of an established design code for 3D-printed cement-based materials and structures places greater emphasis on numerical simulation for ensuring structural safety and performance under ultimate loading conditions. The review of the relevant computational methods for simulating nonlinear behaviors (i.e., fracture and damage) in this chapter provides the foundation for future developments. However, successful future computational development will continue to depend on careful consideration and implementation of the unique characteristics of additively manufactured materials – particularly the dominant role of interfaces in governing the local and global fracture response. Future advancements will also continue to rely on the close integration of the numerical method with experiments and validation against existing theories. Computational mechanics of additively constructed materials and structures will continue to have significant practical use to ensure the simulation capabilities can keep pace with the rapidly evolving field of digital fabrication in construction.

12.8 ACKNOWLEDGMENTS

The authors acknowledge funding from the National Science Foundation, CAREER program (No. 2238993), which partially supported Moini Lab's knowledge-sharing and dissemination efforts leading to this work, to establish the relevant fundamental knowledge for the growing additive technology and to accelerate the translation of knowledge for industry adoption and research community use. The content development in this book chapter is also supported by the educational efforts throughout a new course offered at the Civil and Environmental Engineering Department at Princeton University: CEE 374 – Autonomous Fabrication and Robotics.

REFERENCES

Abbas, Safdar, Alizada, Alaskar, and Fries, Thomas-Peter. 2010. "The XFEM for high-gradient solutions in convection-dominated problems." *International Journal for Numerical Methods in Engineering* 82 (8): 1044–1072.

Alfano, Giulio, and Crisfield, Michael A. 2001. "Finite element interface models for the delamination analysis of laminated composites: Mechanical and computational issues." *International Journal for Numerical Methods in Engineering* 50 (7): 1701–1736. https://doi.org/10.1002/nme.93.

Ambati, Marreddy, Gerasimov, Tymofiy, and De Lorenzis, Laura. 2015a. "A review on phase-field models of brittle fracture and a new fast hybrid formulation." *Computational Mechanics* 55: 383–405.

Ambati, Marreddy, Gerasimov, Tymofiy, and De Lorenzis, Laura. 2015b. "Phase-field modeling of ductile fracture." *Computational Mechanics* 55: 1017–1040.

Ambati, Marreddy, Kruse, Roland, and De Lorenzis, Laura. 2016. "A phase-field model for ductile fracture at finite strains and its experimental verification." *Computational Mechanics* 57: 149–167.

Ambrosio, Luigi, and Tortorelli, Vincenzo Maria. 1990. "Approximation of functional depending on jumps by elliptic functional via t-convergence." *Communications on Pure and Applied Mathematics* 43 (8): 999–1036.

Areias, P. M. A., and Belytschko, T. 2006. "A comment on the article 'A finite element method for simulation of strong and weak discontinuities in solid mechanics' by A. Hansbo and P. Hansbo." *Computer Methods in Applied Mechanics and Engineering* 195 (9–12): 1275–1276.

Barenblatt, Grigory I. 1959. "The formation of equilibrium cracks during brittle fracture. General ideas and hypotheses. Axially-symmetric cracks." *Journal of Applied Mathematics and Mechanics* 23 (3): 622–636.

Bažant, Zdeněk P. 2002. "Concrete fracture models: Testing and practice." *Engineering Fracture Mechanics* 69 (2): 165–205.

Bažant, Zdeněk P., Le, Jia-Liang, Salviato, Marco, et al. 2021. *Quasibrittle Fracture Mechanics and Size Effect: A First Course*. Oxford University Press.

Bažant, Zdeněk P., and Planas, Jaime. 2019. *Fracture and Size Effect in Concrete and Other Quasibrittle Materials*. Routledge.

Belytschko, Ted, and Black, Tom. 1999. "Elastic crack growth in finite elements with minimal remeshing." *International Journal for Numerical Methods in Engineering* 45 (5): 601–620.

Belytschko, Ted, Moës, Nicolas, Usui, Shuji, and Parimi, Chandu. 2001. "Arbitrary discontinuities in finite elements." *International Journal for Numerical Methods in Engineering* 50 (4): 993–1013.

Bos, Freek P., Menna, Costantino, Pradena, Mauricio, et al. 2022. "The realities of additively manufactured concrete structures in practice." *Cement and Concrete Research* 156: 106746.

Camacho, Gerardo Torres, and Ortiz, Michael. 1996. "Computational modelling of impact damage in brittle materials." *International Journal of Solids and Structures* 33 (20–22): 2899–2938. https://doi.org/10. 1016/0020-7683(95)00255-3.

Carollo, Valerio, Reinoso, J., and Paggi, Marco. 2018. "Modeling complex crack paths in ceramic laminates: A novel variational framework combining the phase field method of fracture and the cohesive zone model." *Journal of the European Ceramic Society* 38 (8): 2994–3003.

Chang, Ze, Liang, Minfei, Xu, Yading, Schlangen, Erik, and Šavija, Branko. 2022. "3D concrete printing: Lattice modeling of structural failure considering damage and deformed geometry." *Cement and Concrete Composites* 133: 104719.

Cusatis, Gianluca, Pelessone, Daniele, and Mencarelli, Andrea. 2011. "Lattice discrete particle model (LDPM) for failure behavior of concrete. I: Theory." *Cement and Concrete Composites* 33 (9): 881–890.

Dai, Qingli, and Ng, Kenny. 2014. "2D cohesive zone modeling of crack development in cementitious digital samples with microstructure characterization." *Construction and Building Materials* 54: 584–595.

Daneshvar, Dana, Rabiei, Mahsa, Gupta, Shashank, Najmeddine, Aimane, Prihar, Arjun, and Moini, Reza. 2024. "Geometric fidelity of interlocking bodies in two-component robotic additive manufacturing." In *RILEM International Conference on Concrete and Digital Fabrication*, 134–141. Springer.

Daux, Christophe, Moës, Nicolas, Dolbow, John, Sukumar, Natarajan, and Belytschko, Ted. 2000. "Arbitrary branched and intersecting cracks with the extended finite element method." *International Journal for Numerical Methods in Engineering* 48 (12): 1741–1760.

De Lorenzis, Laura, and Maurini, Corrado. 2022. "Nucleation under multi-axial loading in variational phase-field models of brittle fracture." *International Journal of Fracture* 237 (1): 61–81.

Dittmann, M., Aldakheel, F., Schulte, J., Wriggers, P., and Hesch, C. 2018. "Variational phase-field formulation of non-linear ductile fracture." *Computer Methods in Applied Mechanics and Engineering* 342: 71–94.

Dolbow, John, Moës, Nicolas, and Belytschko, Ted. 2000. "Discontinuous enrichment in finite elements with a partition of unity method." *Finite Elements in Analysis and Design* 36 (3–4): 235–260.

Duan, Qinglin, Song, Jeong-Hoon, Menouillard, Thomas, and Belytschko, Ted. 2009. "Element-local level set method for three-dimensional dynamic crack growth." *International Journal for Numerical Methods in Engineering* 80 (12): 1520–1543.

Dugdale, Donald S. 1960. "Yielding of steel sheets containing slits." *Journal of the Mechanics and Physics of Solids* 8 (2): 100–104.

Dusane, A. R., Budarapu, P. R., Pradhan, A. K., Natarajan, S., Reinoso, J., and Paggi, M. 2022. "Simulation of bridging mechanisms in complex laminates using a hybrid PF-CZM method." *Mechanics of Advanced Materials and Structures* 29 (28): 7743–7771.

Eckardt, Stefan, and Könke, Carsten. 2008. "Adaptive damage simulation of concrete using heterogeneous multiscale models." *Journal of Algorithms & Computational Technology* 2 (2): 275–298.

Elices, Manuel, Guinea, Gustavo V., Gómez, Jaime, and Planas, Jaime. 2002. "The cohesive zone model: Advantages, limitations and challenges." *Engineering Fracture Mechanics* 69 (2): 137–163. https://doi.org/10.1016/S0013-7944(01)00083-2.

Elliott, H. A. 1947. "An analysis of the conditions for rupture due to Griffith cracks." *Proceedings of the Physical Society* 59 (2): 208.

Espinosa, Horacio D., and Zavattieri, Pablo D. 2003. "A grain level model for the study of failure initiation and evolution in polycrystalline brittle materials. Part I: Theory and numerical implementation." *Mechanics of Materials* 35 (3–6): 333–364.

Francfort, Gilles A., and Marigo, J-J. 1998. "Revisiting brittle fracture as an energy minimization problem." *Journal of the Mechanics and Physics of Solids* 46 (8): 1319–1342.

Fukuda, Daisuke, Nihei, Erina, Cho, Sang-Ho, Oh, Sewook, Nara, Yoshitaka, Kodama, Junichi, and Fujii, Yoshiaki. 2020. "Development of a numerical simulator for 3-D dynamic fracture process analysis of rocks based on hybrid FEM-DEM using extrinsic cohesive zone model." *Materials Transactions* 61 (9): 1767–1774.

Gebhard, Lukas, Burger, Joris, Mata-Falcón, Jaime, Lloret Fritschi, Ena, Gramazio, Fabio, Kohler, Matthias, and Kaufmann, Walter. 2022. "Towards efficient concrete structures with ultra-thin 3D printed formwork: Exploring reinforcement strategies and optimisation." *Virtual and Physical Prototyping* 17 (3): 599–616.

Gebhard, Lukas, Mata-Falcón, Jaime, Anton, Ana, Dillenburger, Benjamin, and Kaufmann, Walter. 2021. "Structural behaviour of 3D printed concrete beams with various reinforcement strategies." *Engineering Structures* 240: 112380.

Geubelle, Philippe H., and Baylor, Jeffrey S. 1998. "Impact-induced delamination of composites: A 2D simulation." *Composites Part B: Engineering* 29 (5): 589–602.

Gupta, Shashank, Esmaeeli, Hadi S., and Moini, Reza. 2024. "Tough and ductile architected nacre-like cementitious Composites." *Advanced Functional Materials* 34 (39): 2313516.

Gupta, Shashank, Esmaeeli, Hadi S., Prihar, Arjun, Ghantous, Rita M., Weiss, W. Jason, and Moini, Reza. 2023. "Fracture and transport analysis of heterogeneous 3D-Printed lamellar cementitious materials." *Cement and Concrete Composites* 140: 105034.

Gupta, Shashank, and Moini, Reza. 2024. "Tough cortical bone-inspired tubular architected cement-based material with disorder." *Advanced Materials* 36 (52): 2313904.

Gupta, Shashank, Weiss, W. Jason, Ghantous, Rita M., and Moini, Reza. 2025. "Pore connectivity and arrangement in additively manufactured cementitious material." Available at *SSRN 5269422*.

Hansbo, A., and Hansbo, P. 2004. "A finite element method for the simulation of strong and weak discontinuities in solid mechanics." *Computer Methods in Applied Mechanics and Engineering* 193 (33–35): 3523–3540.

Hansen-Dörr, Arne Claus, Borst, René de, Hennig, Paul, and Kästner, Markus. 2019. "Phase-field modelling of interface failure in brittle materials." *Computer Methods in Applied Mechanics and Engineering* 346: 25–42.

Hansen-Dörr, Arne Claus, Brummund, Jörg, and Kästner, Markus. 2021. "Phase-field modeling of fracture in heterogeneous materials: Jump conditions, convergence and crack propagation." *Archive of Applied Mechanics* 91 (2): 579–596.

Hansen-Dörr, Arne Claus, Dammaß, Franz, Borst, René de, and Kästner, Markus. 2020. "Phase-field modeling of crack branching and deflection in heterogeneous media." *Engineering Fracture Mechanics* 232: 107004.

He, Ming-Yuan, and Hutchinson, John W. 1989. "Crack deflection at an interface between dissimilar elastic materials." *International Journal of Solids and Structures* 25 (9): 1053–1067.

Heever, Marchant van den, Bester, Frederick, Kruger, Jacques, and Zijl, Gideon van. 2022. "Numerical modelling strategies for reinforced 3D concrete printed elements." *Additive Manufacturing* 50: 102569.

Hillerborg, Arne, Modéer, Matz, and Petersson, Per-Erik. 1976. "Analysis of crack formation and crack growth in concrete by means of fracture mechanics and finite elements." *Cement and Concrete Research* 6 (6): 773–781.

Jiang, Wen, Spencer, Benjamin W., and Dolbow, John E. 2020. "Ceramic nuclear fuel fracture modeling with the extended finite element method." *Engineering Fracture Mechanics* 223 (January): 106713.

Kumar, Aditya, Bourdin, Blaise, Francfort, Gilles A., and Lopez-Pamies, Oscar. 2020. "Revisiting nucleation in the phase-field approach to brittle fracture." *Journal of the Mechanics and Physics of Solids* 142: 104027.

Kumar, Aditya, and Lopez-Pamies, Oscar. 2020. "The phase-field approach to self-healable fracture of elastomers: A model accounting for fracture nucleation at large, with application to a class of conspicuous experiments." *Theoretical and Applied Fracture Mechanics* 107: 102550.

Kumar, Lalit, Dey, Dhrutiman, Panda, Biranchi, and Muthu, Nelson. 2023. "Experimental and numerical evaluation of multi-directional compressive and flexure behavior of three-dimensional printed concrete." *Frontiers of Structural and Civil Engineering* 17 (11): 1643–1661.

Labanda, Nicolás A., Giusti, Sebastián M., and Luccioni, Bibiana M. 2018. "Meso-scale fracture simulation using an augmented Lagrangian approach." *International Journal of Damage Mechanics* 27 (1): 138–175.

Legrain, Grégory, Moes, Nicolas, and Verron, Erwan. 2005. "Stress analysis around crack tips in finite strain problems using the extended finite element method." *International Journal for Numerical Methods in Engineering* 63 (2): 290–314.

Lin, J. P., Wang, J. F., and Xu, R. Q. 2014. "Cohesive zone model based numerical analysis of steel-concrete composite structure push-out tests." *Mathematical Problems in Engineering* 2014 (1): 175483.

Liu, Junli, Li, Shuai, Fox, Kate, and Tran, Phuong. 2022. "3D concrete printing of bioinspired Bouligand structure: A study on impact resistance." *Additive Manufacturing* 50: 102544.

Lowke, Dirk, Dini, Enrico, Perrot, Arnaud, Weger, Daniel, Gehlen, Christoph, and Dillenburger, Benjamin. 2018. "Particle-bed 3D printing in concrete construction – Possibilities and challenges." *Cement and Concrete Research* 112: 50–65.

Lowke, Dirk, Vandenberg, Aileen, Pierre, Alexandre, Thomas, Amaury, Kloft, Harald, and Hack, Norman. 2021. "Injection 3D concrete printing in a carrier liquid – Underlying physics and applications to lightweight space frame structures." *Cement and Concrete Composites* 124: 104169.

Ma, Guowei, Buswell, Richard, Silva, Wilson Ricardo Leal da, Wang, Li, Xu, Jie, and Jones, Scott Z. 2022. "Technology readiness: A global snapshot of 3D concrete printing and the frontiers for development." *Cement and Concrete Research* 156: 106774.

Marulli, M. R., Valverde-González, A., Quintanas-Corominas, A., Paggi, M., and Reinoso, J. 2022. "A combined phase-field and cohesive zone model approach for crack propagation in layered structures made of nonlinear rubber-like materials." *Computer Methods in Applied Mechanics and Engineering* 395: 115007.

Miehe, C., Welschinger, F., and Hofacker, M. 2010. "Thermodynamically consistent phase-field models of fracture: Variational principles and multi-field FE implementations." *International Journal for Numerical Methods in Engineering* 83 (10): 1273–1311.

Miehe, Christian, Aldakheel, Fadi, and Raina, Arun. 2016. "Phase field modeling of ductile fracture at finite strains: A variational gradient-extended plasticity-damage theory." *International Journal of Plasticity* 84: 1–32.

Miehe, Christian, Hofacker, Martina, and Welschinger, Fabian. 2010. "A phase field model for rate independent crack propagation: Robust algorithmic implementation based on operator splits." *Computer Methods in Applied Mechanics and Engineering* 199 (45–48): 2765–2778.

Miehe, Christian, Schaenzel, Lisa-Marie, and Ulmer, Heike. 2015. "Phase field modeling of fracture in multi-physics problems. Part I. Balance of crack surface and failure criteria for brittle crack propagation in thermo-elastic solids." *Computer Methods in Applied Mechanics and Engineering* 294: 449–485.

Minnebo, Hans. 2012. "Three-dimensional integration strategies of singular functions introduced by the XFEM in the LEFM." *International Journal for Numerical Methods in Engineering* 92 (13): 1117–1138.

Moës, Nicolas, Cloirec, Mathieu, Cartraud, Patrice, and Remacle, J-F. 2003. "A computational approach to handle complex microstructure geometries." *Computer Methods in Applied Mechanics and Engineering* 192 (28–30): 3163–3177.

Moës, Nicolas, Dolbow, John, and Belytschko, Ted. 1999. "A finite element method for crack growth without remeshing." *International Journal for Numerical Methods in Engineering* 46 (1): 131–150.

Moini, Mohamadreza. 2020. "Buildability and mechanical performance of architectured cement-based materials fabricated using a direct-ink-writing process." PhD diss., Purdue University.

Moini, Mohamadreza, Olek, Jan, Magee, Bryan, Zavattieri, Pablo, and Youngblood, Jeffrey. 2018. "Additive manufacturing and characterization of architectured cement-based materials via X-ray micro-computed tomography." In *RILEM International Conference on Concrete and Digital Fabrication*, 176–189. Springer.

Moini, Mohamadreza, Olek, Jan, Youngblood, Jeffrey P., Magee, Bryan, and Zavattieri, Pablo D. 2018. "Additive manufacturing and performance of architectured cement-based materials." *Advanced Materials* 30 (43): 1802123.

Moini, Reza, Baghaie, Ahmadreza, Rodriguez, Fabian B., Zavattieri, Pablo D., Youngblood, Jeffrey P., and Olek, Jan. 2021. "Quantitative microstructural investigation of 3D-printed and cast cement pastes using micro-computed tomography and image analysis." *Cement and Concrete Research* 147: 106493.

Moini, Reza, and Makinen, William. 2023. *Multi-material Concrete 3D Printing with Thermoplastic and Elastomeric Polymers*. US Patent App. 18/122,432, September.

Moini, Reza, Olek, Jan, Zavattieri, Pablo D., and Youngblood, Jeffrey P. 2021. "Open-span printing method for assessment of early-age deformations of additively manufactured cement-based materials using an isosceles triangle." In *ASTM Symposium on Standards Development for Cement and Concrete Used in Additive Construction*, edited by S. Z. Jones and E. L. Kreiger, 1–12.

Moini, Reza, Olek, Jan, Zavattieri, Pablo D., and Youngblood, Jeffrey P. 2022. "Early-age buildability – Rheological properties relationship in additively manufactured cement paste hollow cylinders." *Cement and Concrete Composites* 131: 104538.

Mousavi, S. E., and Sukumar, N. 2010. "Generalized Duffy transformation for integrating vertex singularities." *Computational Mechanics* 45 (2): 127–140.

Müller, Björn, Kummer, Florian, and Oberlack, Martin. 2013. "Highly accurate surface and volume integration on implicit domains by means of moment-fitting." *International Journal for Numerical Methods in Engineering* 96 (8): 512–528.

Muñetón-López, Rogelio A., and Giraldo-Londoño, Oliver. 2024. "A phase-field formulation for cohesive fracture based on the Park–Paulino–Roesler (PPR) cohesive fracture model." *Journal of the Mechanics and Physics of Solids* 182: 105460.

Najafi Koopas, Rasoul, Rezaei, Shahed, Rauter, Natalie, Ostwald, Richard, and Lammering, Rolf. 2024. "Comparative analysis of phase-field and intrinsic cohesive zone models for fracture simulations in multiphase materials with interfaces: Investigation of the influence of the microstructure on the fracture properties." *Applied Sciences* 15 (1): 160.

Najmeddine, Aimane, Gupta, Shashank, and Moini, Reza. 2025. "Coupled large deformation phase-field and cohesive zone model for crack propagation in hard-soft multi-materials." *Journal of the Mechanics and Physics of Solids* 196: 106016.

Najmeddine, Aimane, Gupta, Shashank, Makinen, William, Lin, Zayvinn, Delnoce, Krystal, and Moini, Reza. 2026. "Tough, ductile, and strong hard-soft cementitious composite enabled by multi-material additive manufacturing." *Advanced Materials* e1546. https://advanced.onlinelibrary.wiley.com/doi/full/10.1002/adma.202515461.

Najmeddine, Aimane, and Moini, Reza. 2025. "Role of the weak interface on toughening in layered brittle materials: A coupled phase field-cohesive zone model approach to fracture." In *Proceedings of the 12th International Conference on Fracture Mechanics for Concrete and Concrete Structures,* edited by B. L. A. Pichler, P. Preinstorfer, and Ch. Hellmich. Vienna, Austria.

Najmeddine, Aimane, and Shakiba, Maryam. 2023. "Physics and chemistry-based phase-field constitutive framework for thermo-chemically aged elastomer." *International Journal of Mechanical Sciences,* 108721. ISSN: 0020-7403.

Najmeddine, Aimane, and Shakiba, Maryam. 2024. "Efficient BFGS quasi-Newton method for large deformation phase-field modeling of fracture in hyperelastic materials." *Engineering Fracture Mechanics* 310: 110463.

Natarajan, Sundararajan, Song, Chongmin, and Belouettar, Salim. 2014. "Numerical evaluation of stress intensity factors and T-stress for interfacial cracks and cracks terminating at the interface without asymptotic enrichment." *Computer Methods in Applied Mechanics and Engineering* 279: 86–112.

Needleman, A. 1987. "A continuum model for void nucleation by inclusion debonding." *Journal of Applied Mechanics* 54 (3) (September): 525–531. ISSN: 0021-8936.

Needleman, A. 1990. "An analysis of decohesion along an imperfect interface." *International Journal of Fracture* 42: 21–40.

Needleman, A. 1992. "Micromechanical modelling of interfacial decohesion." *Ultramicroscopy* 40 (3): 203–214.

Nguyen, Thanh-Tung, Waldmann, Danièle, and Bui, Tinh Quoc. 2019. "Role of interfacial transition zone in phase field modeling of fracture in layered heterogeneous structures." *Journal of Computational Physics* 386: 585–610.

Nguyen, Vinh Phu, Nguyen, Giang Dinh, Nguyen, Chi Thanh, Shen, Luming, Dias-da-Costa, Daniel, El-Zein, Abbas, and Maggi, Federico. 2017. "Modelling complex cracks with finite elements: A kinematically enriched constitutive model." *International Journal of Fracture* 203: 21–39.

Ortiz, Michael, and Pandolfi, Anna. 1999. "Finite-deformation irreversible cohesive elements for three dimensional crack-propagation analysis." *International Journal for Numerical Methods in Engineering* 44 (9): 1267–1282. https://doi.org/10.1002/(SICI)1097-0207(19990330)44:9<1267::AID-NME486> 3.0.CO;2-7.

Paggi, Marco, and Reinoso, José. 2017. "Revisiting the problem of a crack impinging on an interface: A modeling framework for the interaction between the phase field approach for brittle fracture and the interface cohesive zone model." *Computer Methods in Applied Mechanics and Engineering* 321: 145–172.

Pan, Jinlong, Ping, Pengxin, Ding, Boyin, Zhu, Binrong, Lin, Yuanzheng, Ukrainczyk, Neven, Zhang, Hong, and Cai, Jingming. 2024. "Impact behaviour of 3D printed fiber reinforced cementitious composite beams." *Composites Part A: Applied Science and Manufacturing* 182: 108175.

París, Federico, Correa, Elena, and Mantič, Vladislav. 2007. "Kinking of transversal interface cracks between fiber and matrix." *Journal of Applied Mechanics* 74(4): 703–716.

Park, Kyoungsoo, and Paulino, Glaucio H. 2011. "Cohesive zone models: A critical review of traction separation relationships across fracture surfaces." *Applied Mechanics Reviews* 64 (6): 060802.

Park, Kyoungsoo, Paulino, Glaucio H., and Roesler, Jeffery R. 2009. "A unified potential-based cohesive model of mixed-mode fracture." *Journal of the Mechanics and Physics of Solids* 57 (6): 891–908.

Park, Kyoungsoo, Paulino, Glaucio H., and Roesler, Jeffery R. 2010. "Cohesive fracture model for functionally graded fiber reinforced concrete." *Cement and Concrete Research* 40 (6): 956–965.

Parmigiani, J. P., and Thouless, M. D. 2006. "The roles of toughness and cohesive strength on crack deflection at interfaces." *Journal of the Mechanics and Physics of Solids* 54 (2): 266–287.

Petersson, Per-Erik. 1981. "Crack growth and development of fracture zones in plain concrete and similar materials." [Doctoral Thesis (monograph), Division of Building Materials]. Division of Building Materials, LTH, Lund University.

Pham, Kim, Amor, Hanen, Marigo, Jean-Jacques, and Maurini, Corrado. 2011. "Gradient damage models and their use to approximate brittle fracture." *International Journal of Damage Mechanics* 20 (4): 618–652.

Prihar, Arjun, Garlock, Maria E. M., Najmeddine, Aimane, and Moini, Reza. 2024. "Mechanical performance of sinusoidally architected concrete enabled by robotic additive manufacturing." *Materials & Design* 238: 112671.

Prihar, Arjun, Gupta, Shashank, Esmaeeli, Hadi S., and Moini, Reza. 2024. "Tough double-bouligand architected concrete enabled by robotic additive manufacturing." *Nature Communications* 15 (1): 7498.

Prince, Zachary M., Munday, Lynn, Yushu, Dewen, Nezdyur, Max, Guddati, Murthy. 2025. "MOOSE optimization module: Physics-constrained optimization." *SoftwareX* 26: 101754.

Rabczuk, Timon, Zi, Goangseup, Gerstenberger, Axel, and Wall, Wolfgang A. 2008. "A new crack tip element for the phantom-node method with arbitrary cohesive cracks." *International Journal for Numerical Methods in Engineering* 75 (5): 577–599.

Rabiei, Mahsa, and Moini, Reza. 2025. "Extrusion under material uncertainty with pressure-based closed-loop feedback control in robotic concrete additive manufacturing." *Automation in Construction* 180: 106494.

Ralston, Nadia, Gupta, Shashank, and Moini, Reza. 2024. "3D-printing of architected calcium silicate binders with enhanced and in-situ carbonation." *Virtual and Physical Prototyping* 19 (1): e2350768.

Ramesh, Akilesh, Rajeev, Pathmanathan, Xu, Shanqing, Sanjayan, Jay, and Lu, Guoxing. 2024. "Impact response of textile-reinforced 3D printed concrete panels." *Engineering Structures* 315: 118489.

Ramyara, Elham, and Cusatis, Gianluca. 2021. "Time-dependent discrete model for concrete 3D printing simulations." In *The Biot-Bazant Conference*. Proceedings of the Biot-Bažant Conference on Engineering Mechanics and Physics of Porous Materials.

Ramyar, Elham, and Cusatis, Gianluca. 2022. "Discrete fresh concrete model for simulation of ordinary, self-consolidating, and printable concrete flow." *Journal of Engineering Mechanics* 148 (2): 04021142.

Richardson, Casey L., Hegemann, Jan, Sifakis, Eftychios, Hellrung, Jeffrey, and Teran, Joseph M. 2011. "An XFEM method for modeling geometrically elaborate crack propagation in brittle materials." *International Journal for Numerical Methods in Engineering* 88 (10): 1042–1065.

Rodriguez, Fabian B., Moini, Reza, Agrawal, Shubham, Williams, Christopher S., Zavattieri, Pablo D., Olek, Jan, Youngblood, Jeffrey P., and Varma, Amit H. 2024. "Mechanical response of small-scale 3D-printed steel-mortar composite beams." *Cement and Concrete Composites* 154: 105795.

Rodriguez, Fabian B., Olek, Jan, Moini, Reza, Zavattieri, Pablo D., and Youngblood, Jeffrey P. 2021. "Linking solids content and flow properties of mortars to their three-dimensional printing characteristics." *ACI Materials Journal* 118 (6): 371–382.

Salviato, Marco, Kirane, Kedar, Bažant, Zdeněk P., and Cusatis, Gianluca. 2019. "Mode I and II interlaminar fracture in laminated composites: A size effect study." *Journal of Applied Mechanics* 86 (9): 091008.

Savija, Branko. 2020. "Use of 3D printing to create multifunctional cementitious composites: Review, challenges and opportunities." *RILEM Technical Letters* 5: 16–25.

Scheider, I., and Brocks, W. 2003. "Simulation of cup – Cone fracture using the cohesive model." *Engineering Fracture Mechanics* 70 (14): 1943–1961.

Schöller, Lukas, Schneider, Daniel, Herrmann, Christoph, Prahs, Andreas, and Nestler, Britta. 2022. "Phase-field modeling of crack propagation in heterogeneous materials with multiple crack order parameters." *Computer Methods in Applied Mechanics and Engineering* 395: 114965.

Song, Jeong-Hoon, Areias, Pedro M. A., and Belytschko, Ted. 2006. "A method for dynamic crack and shear band propagation with phantom nodes." *International Journal for Numerical Methods in Engineering* 67 (6): 868–893.

Song, Jeong-Hoon, and Belytschko, Ted. 2009. "Cracking node method for dynamic fracture with finite elements." *International Journal for Numerical Methods in Engineering* 77 (3): 360–385.

Stuparu, F., Constantinescu, D. M., Apostol, D. A., and Sandu, M. 2016. "A combined cohesive elements – XFEM approach for analyzing crack propagation in bonded joints." *The Journal of Adhesion* 92 (7–9): 535–552.

Suo, Z., Ho, S., and Gong, X. 1993. "Notch ductile-to-brittle transition due to localized inelastic band." *Journal of Engineering Materials and Technology* 115 (3) (July): 319–326. ISSN: 0094-4289.

Swati, Raees Fida, Qureshi, Saad Riffat, Sohail, Muhammad Umer, Munir, Adnan, Qureshi, Omer Masood, and Khan, Abid Ali. 2022. "Micro-mechanical investigation of interfacial debonding in carbon fiber reinforced composites using extended finite element method (XFEM) approach." *Micromachines* 13 (8): 1226.

Tafazzolimoghaddam, Behrooz. 2017. "Computational mechanics of fracture and fatigue in composite laminates by means of XFEM and CZM." PhD diss., University of Sheffield.

Tarafder, Preetam, Dan, Saikat, and Ghosh, Somnath. 2020. "Finite deformation cohesive zone phase field model for crack propagation in multi-phase microstructures." *Computational Mechanics* 66 (3): 723–743.

Troemner, Matthew, Ramyar, Elham, Meehan, Jonathan, Johnson, Benton, Goudarzi, Nima, and Cusatis, Gianluca. 2022. "A3D-printing centered approach to Mars habitat architecture and fabrication." *Journal of Aerospace Engineering* 35 (1): 04021109.

Tvergaard, Viggo. 1990. "Effect of fibre debonding in a whisker-reinforced metal." *Materials Science and Engineering: A* 125 (2): 203–213.

Tvergaard, Viggo, and Hutchinson, John W. 1992. "The relation between crack growth resistance and fracture process parameters in elastic-plastic solids." *Journal of the Mechanics and Physics of Solids* 40 (6): 1377–1397.

Ulloa, Jacinto, Rodríguez, Patricio, Samaniego, Cristobal, and Samaniego, Esteban. 2019. "Phase-field modeling of fracture for quasi-brittle materials." *Underground Space* 4 (1): 10–21.

Ungsuwarungsri, T., and Knauss, W. G. 1987. "The role of damage-softened material behavior in the fracture of composites and adhesives." *International Journal of Fracture* 35: 221–241.

Ventura, Giulio. 2006. "On the elimination of quadrature subcells for discontinuous functions in the extended finite-element method." *International Journal for Numerical Methods in Engineering* 66 (5): 761–795.

Wang, Qiang, Yang, Wenwei, Wang, Li, Bai, Gang, and Ma, Guowei. 2025. "Reinforcement design and structural performance for the topology optimized 3D printed concrete truss beams." *Engineering Structures* 332: 120064.

Warsi, Syed Bustan Fatima, Panda, Biranchi, and Biswas, Pankaj. 2024. "Structural analysis of 3D-printed concrete walls under quasi-static cyclic loading using composite micro-model." *Progress in Additive Manufacturing*, 1–21.

Wells, Garth N., and Sluys, L. J. 2001. "A new method for modelling cohesive cracks using finite elements." *International Journal for Numerical Methods in Engineering* 50 (12): 2667–2682.

Wittmann, F. H., Rokugo, K., Brühwiler, E., Mihashi, H., and Simonin, Ph. 1988. "Fracture energy and strain softening of concrete as determined by means of compact tension specimens." *Materials and Structures* 21: 21–32.

Wu, Jian-Ying. 2017. "A unified phase-field theory for the mechanics of damage and quasi-brittle failure." *Journal of the Mechanics and Physics of Solids* 103: 72–99.

Wu, T., and Wriggers, P. 2015. "Multiscale diffusion – Thermal–mechanical cohesive zone model for concrete." *Computational Mechanics* 55 (5): 999–1016.

Xie, Jinbao, He, Shan, Xu, Yading, Meng, Zhao Zheng, Zhou, Wen, Schlangen, Erik, and Šavija, Branko. 2025. "Enhanced elastomer-like auxetic cementitious materials through strain-hardening cementitious composites (SHCC) with extended softening properties." *Cement and Concrete Composites* 161: 106069.

Xie, Jinbao, Xu, Yading, Meng, Zhaozheng, Liang, Minfei, Zhou, Yubao, and Šavija, Branko. 2025. "Impact behavior of auxetic cementitious cellular composites (ACCCs) architected through additive manufacturing (AM) assisted casting: Experiment and modelling." *Construction and Building Materials* 471: 140692.

Xu, Wen Long, Yu, Hao, Zhang, Jia Ning, Lyu, ChengSi, Wang, Quan, Micheal, Marembo, and Wu, Heng An. 2022. "Phase-field method of crack branching during SC-CO2 fracturing: A new energy release rate criterion coupling pore pressure gradient." *Computer Methods in Applied Mechanics and Engineering* 399: 115366.

Yang, Liang, Yang, Yongtao, and Zheng, Hong. 2021. "A phase field numerical manifold method for crack propagation in quasi-brittle materials." *Engineering Fracture Mechanics* 241: 107427.

Zeng, Jun, Zhang, Mengnan, Yang, Erjie, Tian, Fucheng, and Li, Liangbin. 2022. "A tracking strategy for multi-branched crack tips in phase-field modeling of dynamic fractures." *International Journal for Numerical Methods in Engineering* 123 (3): 844–865.

Zhang, Chao, Ren, Juanjuan, Zhang, Shuhao, Guo, Yipu, Li, Ning, Li, Wengui, and Qu, Fulin. 2025. "Advanced impact resistance design through 3D-printed concrete technology: Unleashing the potential of additive manufacturing for protective structures." *Journal of Building Engineering* 111: 113533.

Zhang, Ziyu, Jiang, Wen, Dolbow, John E., and Spencer, Benjamin W. 2018. "A modified moment-fitted integration scheme for X-FEM applications with history-dependent material data." Computational Mechanics 62 (2) (August): 233–252.

Zhou, Wen, Bol, Rowin JM, Zhou, Yubao, Meng, Zhaozheng, Xu, Yading, Xie, Jinbao, and Šavija, Branko. 2025. "Reinforcing mechanism of lattice-reinforced cementitious composites: Insight into flexural performance and material interactions." *Materials & Design* 256: 114332.

Zhou, Wen, Mcgee, Wes, Gökçe, H Süleyman, and Li, Victor C. 2023. "A bio-inspired solution to alleviate anisotropy of 3D printed engineered cementitious composites (3DP-ECC): Knitting/tilting filaments." *Automation in Construction* 155: 105051.

Zhou, Wen, Xu, Yading, Meng, Zhaozheng, Xie, Jinbao, Zhou, Yubao, Schlangen, Erik, and Šavija, Branko. 2025. "Filament stitching: An architected printing strategy to mitigate anisotropy in 3D-printed engineered cementitious composites (ECC)." *Cement and Concrete Composites* 160: 106044.

13 Artificial Intelligence and Machine Learning in 3D Printing of Concrete

Ali Kazemian, Kasra Banijamali, Zhanzhao Li, and José P. Duarte

13.1 INTRODUCTION

Construction is classified as a "low digitization" sector currently in the early stages of digital transformation, yet it has the potential to become a significant driver of economic growth in the future (Manyika et al. 2015; Siebel 2019). Productivity in the construction sector has remained almost the same during the past few decades, while other industries such as manufacturing have achieved significant improvements in productivity (Delgado et al. 2019). Infusing digital technology, new materials, and advanced automation are key factors in realizing a much-needed productivity boost and adopting a manufacturing-inspired production approach in the construction industry.

There are various ongoing efforts and technological advancements toward the digitization of the construction industry. Building information modeling (BIM), e.g., provides a digital platform to streamline design and data collection and analysis by different stakeholders and teams (Tang et al. 2019; Davtalab et al. 2018). In addition, integrating cloud computing technology with BIM allows project stakeholders to work and collaborate in real time from different locations (Birje et al. 2017). Cloud-BIM data can be accessed by mobile devices, enabling timely access to updated information, thereby improving the decision-making process (Duarte, Brown et al. 2024a; Matthews et al. 2015). Internet of Things (IoT) is another area with a potential to contribute to the digital transformation in construction. IoT is related to a system where objects are connected to the Internet, via wireless or wired network connections, and cloud cyber infrastructure (Chatziantoniou et al. 2011), through integrated or attached sensors. IoT solutions can be applied to monitor construction progress, track real-time locations of materials and equipment, monitor workers' health, and measure vibration, deformation, stress, and temperature in different structures and equipment (Adepoju et al. 2022; Craveiro et al. 2019).

While technologies such as BIM and IoT have led to some improvements in the construction industry, the actual construction processes are still predominantly

DOI: 10.1201/9781003434863-16

manual and heavily reliant on skilled construction workers. Low construction productivity, poor construction quality, high accident rates, and cost and schedule overruns are common consequences of these traditional construction methods. Construction-scale additive manufacturing through extrusion, also known as 3D printing of concrete (3DPC), has received significant attention in recent years as a highly automated construction technology with the potential to revolutionize the industry. 3DPC robots can efficiently build complex structures without the use of formwork, resulting in significant cost savings, accelerated construction processes, increased job site safety, minimized waste, and reduced environmental impact (Kazemian et al. 2023). The possibility of using colors and manufacturing complex textures also increases its appeal from a design perspective. Common applications of 3DPC include, but are not limited to, the construction of low-income housing units, emergency shelters, bridges, highway walls and barriers, custom architectural elements, and modular prefabricated units. Depending on the application requirements and project location, different types of customized robotic platforms based on articulated robots, gantry, delta, and crane type robots have been developed and used for 3DPC, with gantries and articulated robots being the most commonly used configurations.

Another interesting application of the 3DPC technology is space construction. To promote successful exploration efforts and enable a long-term presence on the Moon and Mars, supporting infrastructure such as landing and launch pads, habitats, hangars, research labs, and protective shields are necessary to safeguard humans and exploratory technological assets in the harsh lunar and Martian environments (Naser and Chehab 2018; Giwa et al. 2022, 2024a, 2024b). Given these inhospitable conditions and the limited available human workforce, conventional construction methods that rely on manual efforts cannot be adopted for extraterrestrial construction. Furthermore, the limited motion capabilities of suited astronauts on the Moon highlight the necessity of lunar construction automation. Given these constraints, a "fully autonomous" 3DPC process is considered a viable solution for construction in such extreme conditions and has been explored since 2004 by NASA (Khoshnevis 2004; Schuldt et al. 2021; Khoshnevis et al. 2004).

Despite the numerous advantages of 3DPC and the significant number of successful demonstration projects worldwide, several technical barriers hinder its widespread adoption. The most critical issues are related to the robustness, reliability, and level of automation of existing 3DPC systems. To fully harness the potential of the 3DPC technology, higher levels of automation and robustness are needed. Achieving this requires utilizing digital technologies like reality capture systems and advanced sensory systems (Wolfs et al. 2024), combined with artificial intelligence (AI) algorithms. These solutions can minimize the need for manual intervention and enhance the efficiency of the 3DPC process. This chapter provides an overview of how data-driven approaches, such as machine learning (ML), and reality capture technologies can enhance 3DPC as a robotic construction method capable of reliable operations under various conditions, and guide the design of 3D-printed materials and structures.

13.2 SMART CONCRETE 3D PRINTERS WITH AI AND SENSOR INTEGRATION

3DPC has not yet reached its full potential in large-scale construction due to several technical and regulatory challenges. Given the more complex nature of 3DPC compared to conventional construction, multiple factors related to the printing material and the printing process affect both the deposition process and the final 3D-printed structure. The mixture proportions must be carefully optimized to achieve the required properties both during construction and in the long term. Moreover, printing parameters such as extrusion rate and printing speed further contribute to the complexity, as they must be precisely controlled to achieve the desired geometrical accuracy and structural integrity. Models capable of providing accurate and reliable information about both the fresh and hardened state properties of 3D-printed concrete are highly desired (Giwa et al. 2023). Data-driven approaches, such as ML and other AI tools, offer promising solutions to address these limitations and can provide insights into the influence of different ingredients and proportions on the performance of the printing materials (Ghasemi and Naser 2023). The following sections provide an overview of the existing literature on leveraging AI and ML to advance different aspects of the 3DPC process.

13.2.1 AI-Driven Design of Printing Materials

Finding the optimal printing material design is an important step in 3DPC, often demanding extensive experimentation and resources (Li et al. 2022). The printing material should satisfy two main performance requirements in the fresh state: extrudability and shape stability. Extrudability refers to the ability of the material to pass through a nozzle opening for continuous deposition while maintaining dimensional accuracy and layer quality (Li et al. 2020), while shape stability (sometimes described in terms of buildability) refers to the ability of the material to withstand the potential deformations caused by the extrusion pressure, self-weight, and weight of the following layers (Mechtcherine et al. 2022a). While numerous researchers have studied printing material design optimization to meet both extrudability and shape stability requirements (Ahmed and Yehia 2022; Liu et al. 2021; Cho et al. 2020; Rahul and Santhanam 2020), the variety of existing raw materials, ambient conditions, and extrusion systems makes it challenging to develop universal guidelines and mixture design approaches. Therefore, extensive trials are commonly conducted in each case to optimize the printing material design.

In the absence of standards for extrudability and shape stability assessments, the traditional workability and rheological parameters of the printing materials are often used to provide information on their extrudability and shape stability. Rheological parameters such as static and dynamic yield stress, and plastic viscosity offer valuable insights into the fresh-state properties of the printing material. The static yield stress is linked to the microstructural build-up and is commonly considered the primary rheological parameter governing shape stability and buildability, while plastic viscosity plays a crucial role in determining the extrudability of a printing material.

Thixotropy, the build-up and breakdown of the internal 3D structure within cementitious paste, is another important rheological property that affects the time-dependent shape stability of deposited cementitious layers (Kazemian et al. 2017; Mechtcherine et al. 2022b). In some studies, estimating the rheological properties is targeted to reduce the number of experimental trials and achieve cost and time savings (Jeong et al. 2019). Gao et al. (2024) developed an ML algorithm to predict the rheological properties of printing materials. This study focuses on predicting plastic viscosity and yield stress, primarily based on the mixture composition and the time elapsed after water addition. A database consisting of 153 data points was utilized to evaluate the effectiveness of different ML algorithms by these researchers. Based on the results, the extreme gradient boosting (XGBoost) model provided the highest accuracy with a coefficient of determination of higher than 90% for both cases. They also defined a printable zone based on the plastic viscosity and yield stress of mixtures to minimize trial-and-error efforts.

In a study by Nazar et al. (2023), gene expression programming (GEP) was selected as the ML algorithm to predict the plastic viscosity and yield stress of the 3D-printable material. Portland cement, sand, water, small- and medium-size coarse aggregate, and superplasticizers were selected as the input parameters of the GEP model due to their high impact on the rheological properties of the fresh material. The developed GEP model was able to predict the plastic viscosity and yield stress with a coefficient of determination of 97.8% and 99.8%, respectively. Amin et al. (2022) developed Artificial Neural Network (ANN) and Random Forest (RF) models to predict the rheological properties, namely, yield stress and plastic viscosity, of the printable material. The results revealed that the RF model outperforms the ANN model in predicting both properties, with the coefficient of determination of 92% and 96% for plastic viscosity and yield stress, respectively. Dynamic yield stress is another rheological property that was predicted by Uddin et al. (2023) using ML algorithms. This study focused on the properties of 3D-printed polypropylene fiber-reinforced concrete, and several ML algorithms were evaluated to determine the most effective approach. A total of 40 mixtures were tested, and the light gradient boosting machine (LightGBM) algorithm demonstrated the best performance, with a coefficient of determination of 98%. Table 13.1 presents a summary of the ML models discussed in this section, aiming at printing material design optimization.

13.2.2 Performance Evaluation and Strength Prediction

Another challenge in implementing 3DPC in the construction industry is the uncertainty surrounding the mechanical strength evolution of 3D-printed structures (Yoo et al. 2013; Banijamali et al. 2024a). To achieve the promised rapid construction with 3DPC, the 3D-printed elements must attain a minimum load-bearing capacity to be able to withstand the service loads within a relatively short period after finishing the printing process. Multiple approaches have been implemented to estimate the mechanical strength of 3D-printed concrete (Banijamali et al. 2024b; Wolfs et al. 2018a, 2018b; Duarte, Duarte et al. 2024b). Although these studies provided valuable information regarding the mechanical strength estimation of 3D-printed concrete, they are often limited to specific concrete grades, ingredients, and curing conditions,

TABLE 13.1
Summary of the Described ML Models for Material Optimization of 3D-Printed Structures

Tested ML Algorithms	Best ML Algorithm	Model Inputs	Model Output	Database	Performance	Study
ANN	ANN	Set accelerator, hardening accelerator, nano clays, and viscosity-modifying agent (VMA)	Yield stress (YS) and mini-slump test	16	YS: $R^2 = 99.9\%$ Slump: $R^2 = 96\%$	Charrier and Ouellet-Plamondon (2022)
Linear regression, SVR, RF, XGBoost, and MLP (neural network (NN))	XGBoost	Cement, sulfoaluminate cement (SAC), fly ash (FA), silica fume (SF), superplasticizer (SP), and water, and time after water addition	Plastic viscosity (PV) and YS	153	$R^2 > 90\%$ for both	Gao et al. (2024)
GEP	GEP	Cement, sand, water, small-size and medium-size coarse aggregate, and SP	PV and YS	PV: 142 YS: 137	PV: $R^2 = 97.8\%$ YS: $R^2 = 99.8\%$	Nazar et al. (2023)
NN and RF	RF	Sand, cement, water, small and medium gravel, and SP	PV and YS	139	PV: $R^2 = 92\%$ YS: $R^2 = 96\%$	Amin et al. (2022)
RF, Support vector machine (SVM), XGBoost, LightGBM	LightGBM	Water-to-binder ratio, sand-to-binder ratio, polypropylene (PP) fiber, and SP	Dynamic YS	40	$R^2 = 98\%$	Uddin et al. (2023)

and a broader and generalized predictive model still remains elusive. ML algorithms provide a potential solution to address these issues and provide a more in-depth analysis of the influential parameters on the mechanical strength evolution of 3D-printed structures (Geng et al. 2023).

Duarte, Duarte et al. (2024b) developed a predictive model for assessing the structural performance of printed parts, focusing on early-age failure modes: plastic collapse, elastic buckling, and flexural collapse. This model, grounded in experimental data on the time-dependent strength of fresh concrete, powers a structural behavior simulator. The simulator slices any shell or linear model into layers, performing incremental structural analysis as each layer is added, with material properties adjusted according to the age of the concrete. Finite element analysis at each printing stage assesses collapse risks based on evolving rheological and mechanical properties. Validation through printing experiments confirmed the model's accuracy for predicting collapse in various structures, such as a solid cylinder, thin wall, and overhang forms. Integrated into a broader design workflow that includes form and toolpath optimization, this model – although developed for a specific concrete mix – can be adapted to other mixtures (Duarte, Duarte et al. 2024b).

Accurate prediction of concrete's compressive strength, a basic but key property of concrete, is essential for structural design and ensuring safety in construction projects. Izadgoshtasb et al. (2021) employed a combination of ANN and multi-objective grasshopper optimization algorithm (MOGOA) to predict the compressive strength of 3D-printed mortars based on the mixture design. Seven different models with varying levels of complexity (in terms of the number of hidden layers) were developed and evaluated. The results revealed that the more complex models resulted in more accurate predictions. Specifically, the most complex model, featuring three layers with 5, 15, and 1 neuron, demonstrated the highest accuracy. In another study by Alyami et al. (2024), ML algorithms were proposed to predict the compressive strength of 3D-printed fiber-reinforced concrete. To identify the most accurate model, seven different ML algorithms were evaluated, including support vector regression (SVR), decision tree (DT), SVR-bagging, SVR-boosting, RF, gradient boosting (GB), and GEP. The models were trained and tested using 299 data points. The DT, RF, GB, and GEP models produced the most accurate results, with coefficients of determination values higher than 98%.

A recent study (Ali et al. 2023) focused on predicting the flexural and tensile strength, evaluating the performance of various ML algorithms for estimating the 28-day strength values for fiber-reinforced 3D-printed concrete. The DT, SVM, XGBoost regression, and gaussian process regression are the selected approaches in this study. These researchers created a database from the existing literature consisting of 77 mixtures for flexural strength and 49 mixtures for tensile strength. Eleven inputs were selected for training purposes, classified into three categories: mixture design (seven inputs), printing parameters (two inputs), and fiber properties (two inputs). The SVM model provided the most accurate results, with an accuracy of 88.9% and 89.3% for tensile strength and flexural strength, respectively.

Reducing the carbon footprint is a major concern in the concrete construction industry since portland cement production is both energy-intensive and carbon-intensive (Sizirici et al. 2021). To address this issue, researchers aim to partially or completely replace portland cement with eco-friendly alternatives and industrial by-products in cast and 3D-printed concrete (Hojati et al. 2022). Estimating the mechanical strength of eco-friendly 3D-printed concrete using ML algorithm was studied by Zhu et al. (2024). The compressive and flexural strength tests were carried out after 28 days on 3D-printed materials made with fine aggregates. These researchers developed printing materials with portland cement contents in the range of 550–600 kg/m^3. To predict the mechanical strength, the SVR algorithm was employed, and a coefficient of determination of higher than 90% was achieved for predicting both parameters.

An important consideration in strength estimation, which has received minimal attention from researchers, is the curing conditions of 3D-printed concrete. In conventional construction, concrete specimens are covered and protected by formwork, aiding the cement hydration process. However, in the absence of formwork in 3DPC, the 3D-printed elements are exposed to ambient conditions, which can significantly impact the cement hydration reactions (Kazemian et al. 2017). For instance, when a 3D-printed structure is subjected to hot and dry curing conditions, excessive water evaporation can impede the cement hydration progress and severely affect its early-age and long-term structural performance.

To address this knowledge gap, Banijamali et al. 2024b proposed a prediction model leveraging the electrical properties and internal temperature of 3D-printed concrete, to estimate the early-age strength of 3D-printed specimens under different curing conditions. An embedded sensor was used to continuously collect the electrical resistivity (ER), permittivity (P), and internal temperature (T) of the 3D-printed concrete over time (Figure 13.1). Two different empirical models based on the ER (ERT model) and permittivity (PT model) were developed and evaluated for estimating the early-age flexural strength of 3D-printed concrete under different curing conditions. Based on the results, the permittivity-based model showed a consistently higher accuracy compared to the ER-based model, with an R^2 value of more than

FIGURE 13.1 (a) Final 3D-printed specimen with the embedded sensor in between the second and third layers; (b) placement of the embedded sensor on top of the second layer during the printing process (Banijamali et al. 2025).

TABLE 13.2
Accuracy of the Proposed ML Models for Estimating the Flexural Strength of 3D-Printed Concrete (Banijamali et al. 2025)

ML Model	Accuracy
LR	84.5%
SVR	93.5%
ANN	95.1%

83% for all cases across three different concrete grades. A follow-up study by these researchers (Banijamali et al. 2025) focused on leveraging ML to improve the accuracy of early-age strength estimations for 3D-printed concrete under various curing conditions, based on the same sensory data. Six input parameters were selected for training the ML algorithms: water-to-cement ratio, curing condition, testing age, ER, average internal temperature, and dielectric permittivity. Linear regression (LR), SVR, and ANN models were designed and evaluated for estimating the flexural strength of the 3D-printed concrete specimens, using 144 data points. A summary of the accuracy of the proposed models is provided in Table 13.2, and based on that, the ANN model showed a considerably higher accuracy compared to the other ML and empirical models (ERT and PT), achieving a coefficient of determination of 95.1%. A summary of the discussed ML models used for strength evaluation and prediction of the 3D-printed structures is provided in Table 13.3.

13.2.3 Automated Quality Control and Inspection

Quality inspection is a critical aspect of any construction and fabrication process. In conventional construction, manual inspection methods are typically employed, which are susceptible to human errors, technical challenges, and inaccuracies. Additionally, the manual inspections are time-consuming and not well-suited for fast and automated building processes such as 3DPC. To reduce the required time for inspection and to increase the 3DPC automation level, inline and automated quality inspection methods are required (Mechtcherine et al. 2022b; Buswell et al. 2018). Different methods, such as 2D computer vision and LiDAR sensors, have been implemented to enable automated quality monitoring and inspection of 3D-printed concrete (Martin et al. 2024b; Lachmayer et al. 2023; Kazemian et al. 2019; Nair et al. 2022; Maqsoodi and Irizarry 2024). Martin et al. (2024a) introduced a near-real-time methodology and associated algorithms for performing geometrical inspection of 3D-printed structures using 2D LiDAR data. In their study, the proposed variable standard deviation (VSD) algorithm for continuous measurement of extrudate width and height was evaluated using a comprehensive validation plan, achieving an average error of less than 1.2 mm and a maximum error of less than 1.33 mm when data collection cycles exceeded 1 second. The proposed system also lays the groundwork for future

TABLE 13.3
Summary of the Described ML Models for Strength Prediction of 3D-Printed Structures

Tested ML Algorithm	Best ML Algorithm	Model Input	Model Output	Database	Performance	Study
ANN-MOGOA	ANN-MOGOA1	Water-to-cement ratio, fine and coarse aggregate, and SP	Compressive strength	65	$R^2 = 98\%$	Izadgoshasb et al. (2021)
SVR, DT, SVR-bagging, SVR-boosting, RF, GB, and GEP	DT	Fiber type, diameter of fiber, length of fiber, curing age, loading direction, fiber volume fraction, water, hydroxypropyl methylcellulose, SP, silica fume (SF), ground slag, sand, fly ash (FA), and cement	Compressive strength	299	$R^2 = 98.7\%$	Alyami et al. (2024)
DT, SVM, XGBoost regression, and Gaussian process regression	SVM	Water, cement, SF, FA, coarse aggregate, fine aggregate, viscosity modifying agent (VMA), fibers, fiber diameter and strength, print speed, and nozzle area	Flexural strength (FS) and tensile strength (TS)	77 49	FS: $R^2 = 89.3\%$ TS: $R^2 = 88.9\%$	Ali et al. (2023)
SVR	SVR	Material properties (i.e., sand, rubber, epoxy, SF, and glass fiber), geometry properties (i.e., length, width, and cross-sectional area), and dry weight and wetness weight of the 3D-printed concrete specimens	Compressive and FS	25 mixtures' designs (number of data points not reported)	$R^2 > 90\%$ for both	Zhu et al. (2024)
LR, SVR, and ANN	ANN	Internal temperature, electrical resistivity, permittivity, water-to-cement ratio, curing condition, and testing age	FS	144	$R^2 = 95.1\%$	Banijamali et al. (2025)

research on remote process monitoring by enabling inline geometric data acquisition and extrudate visualization (Figure 13.2). In a recent study by Banijamali et al. (2026), a novel and automated geometrical inspection system was proposed to enable inline dimensional measurements during the 3D printing process. The system leverages structured laser light (SLL) machine vision, where a 2D camera captures images of two parallel laser lines projected onto the 3D-printed specimens (Figure 13.3). By analyzing the shifts and deformations in the captured laser line patterns, the system can accurately extract the main geometric features such as width and height measurements and detect the potential deformations of the extrudates in real time. Experimental results showed that the system can deliver measurements with a submillimeter accuracy (less than 1 mm error) across different extrudate types, with a processing time of 200 milliseconds. Lindemann et al. (2019) used shotcrete 3D printing to fabricate a doubly curved, fully reinforced concrete wall. After printing the vertical wall, the as-built structure was scanned using a 3D laser scanner to create an accurate digital representation of the existing surface and then to generate the toolpath for the next process step. Additionally, 3D laser scanning and photogrammetry were employed at various stages to create a 3D reconstruction of the structure.

ML algorithms can be leveraged in combination with noncontact measurement and reality capture techniques to design and deploy integrated automated inspection systems, accounting for the complex interplay among various environment, material, and process parameters involved in 3DPC.

Defects and deformations in 3D-printed elements are critical because they can lead to process failures, such as wall collapse during layer deposition, or compromise the post-construction structural integrity. In the prefabrication scenario, deformations and geometrical deviations from the initial design can lead to major challenges and delays during the assembly phase at the job site. Davtalab et al. (2022) developed

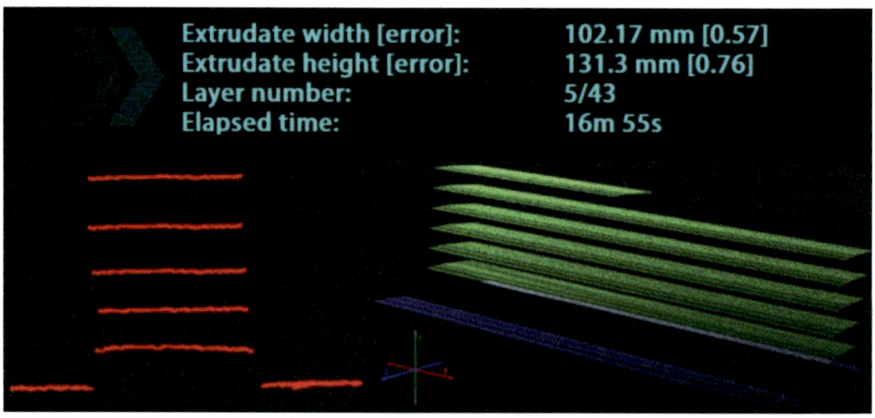

FIGURE 13.2 Data acquisition and extrudate visualization based on raw LiDAR data (bottom left), cumulative 2D cross-sectional profile of the extrudate in real time, (bottom right) 3D reconstruction and visualization of the layers (Martin et al. 2024a).

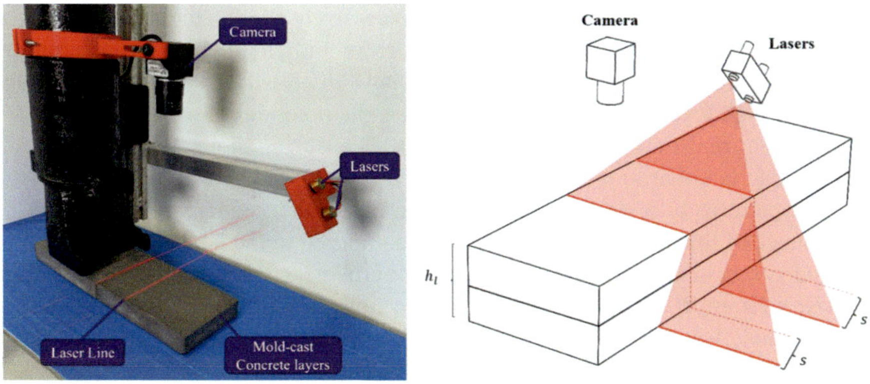

FIGURE 13.3 (a) SLL module, including a 2D camera and two parallel laser emitters, both of which mounted on the extruder; b) measurement configuration of the proposed method: s indicates the laser line shifts when projected on extrudates and hl demonstrates the total extrudate height.

a system to detect the layer surface defects and deformations of the 3D-printed layers. The system consists of a deep convolutional neural network (D-CNN) model inspired by SegNet (Badrinarayanan et al. 2015). The proposed algorithm comprises three steps: image data preprocessing, layer edge extraction, and defect detection using an outlier detection technique. The proposed algorithm achieved an accuracy of 97.5% for the classification of defective and non-defective layers.

To improve the surface quality of the deposited layers, Lao et al. (2020) developed an ANN model to directly control extrudate formation by determining the appropriate nozzle shape. The model was trained using images from the middle layer of three-layer 3D-printed elements using 13 different nozzle shapes. The optimized nozzle shape obtained from the ANN model showed a significant improvement in the surface quality of the 3D-printed objects compared to the rectangular nozzle shape, which consisted of a vertical wall, an inclined wall, and a curved part. Based on these findings, Lao et al. (2021) later developed a variable-geometry nozzle capable of controlling the extrudate geometry during the printing process.

In another study, Garfo et al. (2020) proposed an algorithm to automate defect detection of 3D-printed concrete. For this purpose, image processing and deep learning-based object detection adopted by Google Cloud Machine Learning Engine were utilized. A combination of MobileNet – a convolutional neural network (CNN) model used for image feature extraction – and the Single Shot MultiBox Detector (SSD) as the object detector, called MobileNet-SSD, was utilized as the deep-learning model. The results demonstrated the superiority of the proposed model compared to the conventional CNN models for defect detection. In 2022, Rill-García et al. (2022) studied a 2D computer vision-based system for the inline quality control of 3D-printed concrete structures. This system leverages a CNN model, initially trained for crack segmentation. The CNN model is trained at classifying the textural features of the printed layers into four distinct categories: excessively fluid, excessively dry, good quality, and tearing. By analyzing the texture of individual layers, the proposed system achieved a macro-average *F*-score of 94% in classifying the fresh extrudate layer. This high level of accuracy demonstrates the potential of using advanced ML techniques for real-time quality control in 3DPC, ensuring more consistent and reliable construction operations. Chang et al. (2022) proposed a CNN model to predict the crack patterns and stress–crack curves in 3D-printed concrete. To validate the performance of this model, the outputs were compared to the lattice numerical analysis, and the results revealed a quantitative agreement with the lattice numerical analysis, with an R^2 value of 75%. In another study, a real-time quality monitoring method coupled with a closed-loop feedback system was developed (Zhang et al. 2024). In this study, a secondary mixer was used to add the liquid accelerator to the previously mixed mixture to adjust the rheology of the printing material before deposition. After deposition, images were captured to train two different ML models: VGG-16 network and Modified Inception-ResNet. To evaluate the performance of the developed and trained ML models, 3 different concrete layers with different widths were printed by adjusting the extrusion rate and the addition rate of the accelerator to the second mixing process. The Modified Inception-ResNet showed higher accuracy compared to the VGG-16 network with a minimum accuracy of 93.51%. In a two-part study, Versteege et al. (2025a) developed a data-driven framework for real-time quality monitoring of 3D-printed concrete by converting raw sensory data into meaningful features through knowledge-driven feature engineering (KDFE). The first part focused on extracting geometric features, including layer width, ridge height, and surface texture, from in-line laser triangulation scans to assess deposition accuracy and process stability. The second part of the study (Versteege et al. 2025b) expanded the framework to include multi-modal sensory data, such as environmental conditions, process parameters, and material properties, structured within a layered digital architecture. These studies lay the groundwork for future research focused on process optimization and control. The extracted features can be integrated into adaptive feedback loops for continuously adjusting printing parameters and can also support predictive modeling of properties such as interlayer bond strength, ultimately advancing the automation and reliability of concrete 3D printing. Table 13.4 presents a summary of the ML models discussed in this section, limited to the studies that applied ML models for automated inspection of 3D-printed concrete and reported the accuracy of the models.

13.3 LIMITATIONS AND EXISTING CHALLENGES

Greater robustness of the 3DPC process is an essential prerequisite for the widespread industrial adoption of this technology, particularly for onsite construction. 3DPC systems must be designed to withstand variations in the ambient conditions and material properties such that these anticipated variations do not lead to process failure or undesired deformations and geometrical deviations. For instance, some variation levels in the printing materials seem unavoidable due to inconsistencies in the moisture content and grading of aggregates, which are commonly obtained from natural sources such as river beds and quarries. Such variations highlight the need

TABLE 13.4
Summary of the Described ML Models for Quality Inspection

Tested ML Algorithm	Best ML Algorithm	Model Input	Model Output	Database	Performance	Study
D-CNN	D-CNN	Images	Defect detection	1,000,000	$R^2 = 97.5\%$	Davtalab et al. (2022)
ANN	ANN	Images	Optimizing the nozzle geometry	10,000,000	Arithmetic averaged roughness (*Ra*) reduced by 0.313 mm	Lao et al. (2020, 2021)
Deep learning-based object detection	Deep learning-based object detection	Images	Defect detection	> 20,000	Quicker than the conventional surface defect detection methods	Garfo et al. (2020)
CNN	CNN	Images	Texture analysis	> 1,000 per iteration	Macro-averaged *F*-score = 94%	Rill-García et al. (2022)
CNN	CNN	Images	Crack patterns	132,883	$R^2 = 75\%$	Chang et al. (2022)
VGG-16 network and Modified Inception-ResNet	Modified Inception-ResNet	Images	Adjust the extrusion rate and pump rate of the accelerator	3,519	VGG-16 network: $R^2 > 86.44\%$ Modified Inception-ResNet: $R^2 > 93.51\%$	Zhang et al. (2024)

for quality control during the 3DPC process. In traditional concrete construction, after a mixture is designed, quality monitoring of its fresh properties on the construction site is often limited to slump measurements, sometimes complemented by unit weight and air content measurements. These traditional quality control methods are manual, are off-line, need sample preparation, and are time-consuming. While sufficient for conventional construction, these tests cannot be used for quality control of 3DPC (Kazemian et al. 2019). A major distinction between the two construction approaches is that in traditional construction, large volumes of concrete are poured into the formwork at once, and as long as performance requirements are later satisfied by the hardened concrete, some variations in the fresh properties of the material are acceptable. In the formwork-free 3DPC process, however, thin layers are continuously extruded, and variations in the printing mixture or the layers can lead to process failure, such as the collapse of freshly printed layers (Kazemian et al. 2023). During the printing process, the stability of the deposited layers depends on two factors: the increasing strength and stiffness caused by thixotropic structural build-up and setting of the 3D-printed materials vs. the gradually increasing load as more layers are deposited on each other (Kazemian et al. 2017; Wolfs et al. 2018a). An imbalance between the loading condition and layer stiffness, or the extrusion of layers with unacceptable properties as a result of variations in the proportions, could lead to extensive deformations and progressive collapse of the freshly printed structure. Other than variations in the material properties and the hardening rate, deposition inaccuracies and variations in the layer dimensions can also contribute to the instability of a freshly printed structure. Therefore, real-time and automated customized quality control systems are needed to enable reliable 3DPC systems.

A review of existing studies highlights the feasibility of a combined use of advanced sensory systems and data-driven approaches such as ML to enhance the robustness of the 3DPC process and enable automated process adjustments. ML algorithms have shown their great potential for the design of printing materials and structures, by minimizing the number of trials and exploring unconventional and optimized architectural designs. However, the majority of existing ML models are based on a limited number of data points and do not consider all the relevant process, environment, and material parameters. For instance, nozzle design has been shown to affect the properties of 3D-printed layers, yet it is rarely accounted for in existing ML models. Similarly, variations in material delivery and extrusion mechanisms, ambient (curing) conditions, and the type of printing materials are typically overlooked. These factors warrant more comprehensive investigation to support the development of ML models with broader scope and greater applicability across diverse 3DPC scenarios. Recent efforts, such as those by RILEM TC 303-PFC and TC 304-ADC: Testing and Assessment of Printable and Printed Concrete (Springer Nature n.d.), can be utilized to address this gap by establishing interlaboratory databases for additive manufacturing in construction. Nonetheless, extracting meaningful insights from rheological and fresh property data of 3D-printed concrete remains a significant and unresolved challenge in the field.

13.4 FUTURE PROSPECTS

The integration of AI and advanced sensory systems in the design of 3DPC systems holds significant promise for advancing the field. By leveraging these technologies, it is possible to design more robust and reliable 3DPC systems, with enhanced predictive maintenance and adjustment capabilities to prevent system breakdowns and optimize performance. AI/ML algorithms can analyze large amounts of data to improve the precision and consistency of printed structures. Additionally, AI can enable adaptive control systems that adjust printing parameters in real time to accommodate varying ambient conditions, thereby expanding the operational envelope of 3D printers. Furthermore, AI algorithms can be leveraged to optimize 3DPC reinforcement strategies and toolpath design, enabling intelligent placement and distribution of structural and reinforcement materials to enhance the structural integrity and performance of the printed structures. These advancements pave the way for higher levels of automation, reducing the need for human intervention and increasing the efficiency and scalability of 3DPC in construction. In addition, these enhancements will further promote the use of 3DPC robots in extreme environments, such as post-disaster sites and on the surfaces of celestial bodies. In such extreme environments, very high levels of automation and robustness are strictly required, which can be achieved through the use of advanced sensory systems and adaptive AI-powered process control algorithms.

13.5 SUMMARY AND CONCLUSIONS

The successful implementation of 3DPC in the construction industry requires overcoming technical challenges related to process robustness, material variability, and quality control. The integration of AI with advanced embedded sensors offers promising solutions to these challenges by enabling more accurate material design, predictive maintenance, and automated defect detection and process parameter adjustments. This chapter highlighted several knowledge gaps and limitations, particularly in understanding the complex interactions among process parameters, environmental conditions, and material properties. For a more extensive investigation of data-driven approaches, future research focusing on developing ML models with broader applicability, considering a wider range of parameters, and incorporating more extensive datasets is needed. As the 3DPC technology continues to evolve, the role of AI-powered advanced sensory systems will become increasingly critical in achieving fully autonomous and robust robotic construction systems, capable of operating in diverse and extreme environments. These advancements will not only enhance the efficiency and scalability of 3DPC but also pave the way for its deployment in challenging scenarios such as post-disaster recovery and extraterrestrial construction on the Moon and Mars.

13.6 CONFLICTS OF INTEREST

One of the co-authors, José P. Duarte, and The Pennsylvania State University own equity in X-Hab 3D, whose activity is related to concrete printing. Their ownership in this company has been reviewed by the University's Individual and Institutional Conflict of Interest Committees and is currently being managed by the University.

REFERENCES

Adepoju, Omoseni, Clinton Aigbavboa, Nnamdi Nwulu, Michael Onyia, and Omoseni Adepoju. 2022. "Internet of Things (IoT)." *Re-skilling Human Resources for Construction 4.0: Implications for Industry, Academia and Government*:171–194.

Ahmed, Sara, and Sherif Yehia. 2022. "Evaluation of workability and structuration rate of locally developed 3D printing concrete using conventional methods." *Materials* 15 (3):1243.

Ali, Ammar, Raja Dilawar Riaz, Umair Jalil Malik, Syed Baqar Abbas, Muhammad Usman, Mati Ullah Shah, In-Ho Kim, Asad Hanif, and Muhammad Faizan. 2023. "Machine learning-based predictive model for tensile and flexural strength of 3D-printed concrete." *Materials* 16 (11):4149.

Alyami, Mana, Majid Khan, Muhammad Fawad, R Nawaz, Ahmed WA Hammad, Taoufik Najeh, and Yaser Gamil. 2024. "Predictive modeling for compressive strength of 3D printed fiber-reinforced concrete using machine learning algorithms." *Case Studies in Construction Materials* 20:e02728.

Amin, Muhammad Nasir, Ayaz Ahmad, Kaffayatullah Khan, Waqas Ahmad, Saqib Ehsan, and Anas Abdulalim Alabdullah. 2022. "Predicting the rheological properties of super-plasticized concrete using modeling techniques." *Materials* 15 (15):5208.

Badrinarayanan, Vijay, Alex Kendall, and Roberto Cipolla SegNet. 2015. "A deep convolutional encoder-decoder architecture for image segmentation." *arXiv preprint* arXiv:1511.005615.

Banijamali, Kasra, Mary Dempsey, Jianhua Chen, and Ali Kazemian. 2025. "Machine learning approach to predict the early-age flexural strength of sensor-embedded 3D-printed structures." *Progress in Additive Manufacturing*:1–13.

Banijamali, Kasra, Michael Martin, David Mascarenas, and Ali Kazemian. 2026. "3D geometric measurements of 3D printed concrete via structured light machine vision." *Journal of Computing in Civil Engineering* 40 (2): 04025136.

Banijamali, Kasra, Payam Vosoughi, Gabriel Arce, Hassan Noorvand, Marwa Hassan, and Ali Kazemian. 2024a. "Early-age strength monitoring of sensor-embedded 3D printed structures." *Construction Research Congress 2024*:137–147. https://doi.org/10.1061/9780784485262.015.

Banijamali, Kasra, Payam Vosoughi, Gabriel Arce, Hassan Noorvand, Joseph Lamendola, Marwa Hassan, and Ali Kazemian. 2024b. "Automated strength monitoring of 3D printed structures via embedded sensors." *Automation in Construction* 166:105681. https://doi.org/10.1016/j.autcon.2024.105681.

Birje, Mahantesh N., Praveen S. Challagidad, R. H. Goudar, and Manisha T. Tapale. 2017. "Cloud computing review: Concepts, technology, challenges and security." *International Journal of Cloud Computing* 6 (1):32–57.

Buswell, R. A., W. R. Leal de Silva, S. Z. Jones, and J. Dirrenberger. 2018. "3D printing using concrete extrusion: A roadmap for research." *Cement and Concrete Research* 112:37–49. https://doi.org/10.1016/j.cemconres.2018.05.006.

Chang, Ze, Zhi Wan, Yading Xu, Erik Schlangen, and Branko Šavija. 2022. "Convolutional neural network for predicting crack pattern and stress-crack width curve of air-void structure in 3D printed concrete." *Engineering Fracture Mechanics* 271:108624. https://doi.org/10.1016/j.engfracmech.2022.108624.

Charrier, Malo, and Claudiane M. Ouellet-Plamondon. 2022. "Artificial neural network for the prediction of the fresh properties of cementitious materials." *Cement and Concrete Research* 156:106761. https://doi.org/10.1016/j.cemconres.2022.106761.

Chatziantoniou, Damianos, Katerina Pramatari, and Yannis Sotiropoulos. 2011. "Supporting real-time supply chain decisions based on RFID data streams." *Journal of Systems and Software* 84 (4):700–710.

Cho, Seung, Jacques Kruger, Frederick Bester, Marchant van den Heever, Algurnon van Rooyen, and Gideon van Zijl. 2020. "A compendious rheo-mechanical test for printability assessment of 3D printable concrete." *Second RILEM International Conference on Concrete and Digital Fabrication*, Cham.

Craveiro, Flávio, José P. Duarte, Helena Bartolo, and Paulo Jorge Bartolo. 2019. "Additive manufacturing as an enabling technology for digital construction: A perspective on Construction 4.0." *Automation in Construction* 103:251–267.

Davtalab, Omid, Ali Kazemian, and Behrokh Khoshnevis. 2018. "Perspectives on a BIM-integrated software platform for robotic construction through contour crafting." *Automation in Construction* 89:13–23.

Davtalab, Omid, Ali Kazemian, Xiao Yuan, and Behrokh Khoshnevis. 2022. "Automated inspection in robotic additive manufacturing using deep learning for layer deformation detection." *Journal of Intelligent Manufacturing* 33 (3):771–784.

Delgado, Juan Manuel Davila, Lukumon Oyedele, Anuoluwapo Ajayi, Lukman Akanbi, Olugbenga Akinade, Muhammad Bilal, and Hakeem Owolabi. 2019. "Robotics and automated systems in construction: Understanding industry-specific challenges for adoption." *Journal of Building Engineering* 26:100868.

Duarte, Gonçalo, Nathan C. Brown, and Jose P. Duarte. 2024a. "Workflow for generating, simulating, and optimizing form and toolpath in 3D concrete printing of vaults." *Automation in Construction* 166:105590.

Duarte, Gonçalo, José P. Duarte, Nathan Brown, Ali Memari, and Juan Pablo Gevaudan. 2024b. "Design for early-age structural performance of 3D printed concrete structures: A parametric numerical modeling approach." *Journal of Building Engineering* 94:109986. https://doi.org/10.1016/j.jobe.2024.109986.

Gao, Huaxing, Lang Jin, Yuxuan Chen, Qian Chen, Xiaopeng Liu, and Qingliang Yu. 2024. "Rheological behavior of 3D printed concrete: Influential factors and printability prediction scheme." *Journal of Building Engineering* 91:109626. https://doi.org/10.1016/j.jobe.2024.109626.

Garfo, Selorm, M. A. Muktadir, and Sun Yi. 2020. "Defect detection on 3D print products and in concrete structures using image processing and convolution neural network." *Journal of Mechatronics and Robotics* 4 (1):74–84.

Geng, Songyuan, Qiling Luo, Kun Liu, Yunchao Li, Yuchen Hou, and Wujian Long. 2023. "Research status and prospect of machine learning in construction 3D printing." *Case Studies in Construction Materials* 18:e01952.

Ghasemi, Alireza, and M. Z. Naser. 2023. "Tailoring 3D printed concrete through explainable artificial intelligence." *Structures* 56:104850. https://doi.org/10.1016/j.istruc.2023.07.040.

Giwa, Ilerioluwa, Mary Dempsey, Michael R. Fiske, and Ali Kazemian. 2024a. "3D printed sulfur-regolith concrete performance evaluation for waterless extraterrestrial robotic construction." *Automation in Construction* 165:105571. https://doi.org/10.1016/j.autcon.2024.105571.

Giwa, Ilerioluwa, Daniel Game, Hassan Ahmed, Hassan Noorvand, Gabriel Arce, Marwa Hassan, and Ali Kazemian. 2023. "Performance and macrostructural characterization of 3D printed steel fiber reinforced cementitious materials." *Construction and Building Materials* 369. https://doi.org/10.1016/j.conbuildmat.2023.130593.

Giwa, Ilerioluwa, Marc Hebert, Joseph Lamendola, Michael R. Fiske, and Ali Kazemian. 2024b. "Planetary robotic construction using large-scale 3D printing with sulfur concrete." *Construction Research Congress 2024.*

Giwa, Ilerioluwa, Dharla Moore, Michael R. Fiske, and Ali Kazemian. 2022. "Planetary construction 3D printing using lunar and Martian in situ materials." *Earth and Space 2022*: 817–831.

Hojati, Maryam, Zhanzhao Li, Ali M. Memari, Keunhyoung Park, Mehrzad Zahabi, Shadi Nazarian, José P. Duarte, and Aleksandra Radlińska. 2022. "3D-printable quaternary cementitious materials towards sustainable development: Mixture design and mechanical properties." *Results in Engineering* 13:100341. https://doi.org/10.1016/j.rineng.2022.100341.

Izadgoshasb, Hamed, Amirreza Kandiri, Pshtiwan Shakor, Vittoria Laghi, and Giada Gasparini. 2021. "Predicting compressive strength of 3D printed mortar in structural members using machine learning." *Applied Sciences* 11 (22):10826.

Jeong, Hoseong, Sun-Jin Han, Seung-Ho Choi, Yoon Jung Lee, Seong Tae Yi, and Kang Su Kim. 2019. "Rheological property criteria for buildable 3D printing concrete." *Materials* 12 (4):657.

Kazemian, Ali, Ilerioluwa Giwa, and Mahmut Ekenel. 2023. "Large-scale additive manufacturing for automated construction – An overview." In *Additive Manufacturing Design and Applications*, edited by Mohsen Seifi, et al., 0. ASM International.

Kazemian, Ali, Xiao Yuan, Evan Cochran, and Behrokh Khoshnevis. 2017. "Cementitious materials for construction-scale 3D printing: Laboratory testing of fresh printing mixture." *Construction and Building Materials* 145:639–647. https://doi.org/10.1016/j.conbuildmat.2017.04.015.

Kazemian, Ali, Xiao Yuan, Omid Davtalab, and Behrokh Khoshnevis. 2019. "Computer vision for real-time extrusion quality monitoring and control in robotic construction." *Automation in Construction* 101:92–98. https://doi.org/10.1016/j.autcon.2019.01.022.

Khoshnevis, Behrokh. 2004. "Automated construction by contour crafting – related robotics and information technologies." *Automation in Construction* 13 (1):5–19. https://doi.org/10.1016/j.autcon.2003.08.012.

Khoshnevis, Behrokh, Melanie P. Bodiford, Kevin H. Burks, Ed Ethridge, Dennis Tucker, Won Kim, Houssam Toutanji, and Michael R. Fiske. 2004. "Lunar contour crafting: A novel technique for ISRU-based habitat development." *AIAA 43rd Aerospace Sciences Meeting and Exhibit*.

Lachmayer, Lukas, David Böhler, Niklas Freund, Inka Mai, Dirk Lowke, and Annika Raatz. 2023. "Modelling the influence of material and process parameters on shotcrete 3D printed strands-cross-section adjustment for automatic robotic manufacturing." *Automation in Construction* 145:104626.

Lao, Wenxin, Mingyang Li, and Tegoeh Tjahjowidodo. 2021. "Variable-geometry nozzle for surface quality enhancement in 3D concrete printing." *Additive Manufacturing* 37:101638.

Lao, Wenxin, Mingyang Li, Teck Neng Wong, Ming Jen Tan, and Tegoeh Tjahjowidodo. 2020. "Improving surface finish quality in extrusion-based 3D concrete printing using machine learning-based extrudate geometry control." *Virtual and Physical Prototyping* 15 (2):178–193.

Li, Zhanzhao, Maryam Hojati, Zhengyu Wu, Jonathon Piasente, Negar Ashrafi, José P. Duarte, Shadi Nazarian, Sven G. Bilén, Ali M. Memari, and Aleksandra Radlińska. 2020. "Fresh and hardened properties of extrusion-based 3D-printed cementitious materials: A review." *Sustainability* 12 (14):5628.

Li, Zhanzhao, Jinyoung Yoon, Rui Zhang, Farshad Rajabipour, Wil V Srubar III, Ismaila Dabo, and Aleksandra Radlińska. 2022. "Machine learning in concrete science: Applications, challenges, and best practices." *npj Computational Materials* 8 (1):127.

Lindemann, H., R. Gerbers, S. Ibrahim, F. Dietrich, E. Herrmann, K. Dröder, A. Raatz, and H. Kloft. 2019. "Development of a shotcrete 3D-printing (SC3DP) technology for additive manufacturing of reinforced freeform concrete structures." *First RILEM International Conference on Concrete and Digital Fabrication – Digital Concrete 2018*.

Liu, Chao, Xianggang Wang, Yuning Chen, Chao Zhang, Lei Ma, Zhicong Deng, Chun Chen, Yamei Zhang, Jinlong Pan, and Nemkumar Banthia. 2021. "Influence of hydroxypropyl methylcellulose and silica fume on stability, rheological properties, and printability of 3D printing foam concrete." *Cement and Concrete Composites* 122:104158. https://doi.org/10.1016/j.cemconcomp.2021.104158.

Manyika, James, Sree Ramaswamy, Somesh Khanna, Hugo Sarrazin, Gary Pinkus, Guru Sethupathy, and Andrew Yaffe. 2015. "Digital America: A tale of the haves and have-mores." *McKinsey Global Institute*:1–120.

Maqsoodi, Aras, and Javier Irizarry. 2024. "A framework and cyber-physical system architecture for cloud-based construction monitoring with autonomous quadrupeds." *ISARC. Proceedings of the International Symposium on Automation and Robotics in Construction.*

Martin, Michael, Kasra Banijamali, Hunter Gilbert, David Mascarenas, and Ali Kazemian. 2024a. "LiDAR-based real-time geometrical inspection for large-scale additive manufacturing." *Progress in Additive Manufacturing*:1–17. https://doi.org/10.1007/s40964-024-00817-4.

Martin, Michael, Kasra Banijamali, and Ali Kazemian. 2024b. "Reality capture technologies for automated quality control during construction 3D printing." *Computing in Civil Engineering 2023*:361–370. https://doi.org/10.1061/9780784485224.044.

Matthews, Jane, Peter E. D. Love, Sam Heinemann, Robert Chandler, Chris Rumsey, and Oluwole Olatunj. 2015. "Real time progress management: Re-engineering processes for cloud-based BIM in construction." *Automation in Construction* 58:38–47.

Mechtcherine, Viktor, Shirin Fataei, Freek P. Bos, Richard A. Buswell, Wilson Ricardo Leal da Silva, Emmanuel Keita, Hans-Werner Krauss, Dirk Lowke, Arnaud Perrot, and Venkatesh Naidu Nerella. 2022a. "Digital fabrication with cement-based materials: Underlying physics." *Digital Fabrication with Cement-based Materials: State-of-the-art Report of the RILEM TC 276-DFC*:49–98.

Mechtcherine, Viktor, Kim van Tittelboom, Ali Kazemian, Eric Kreiger, Behzad Nematollahi, Venkatesh Naidu Nerella, Manu Santhanam, Geert de Schutter, Gideon Van Zijl, Dirk Lowke, Egor Ivaniuk, Markus Taubert, and Freek Bos. 2022b. "A roadmap for quality control of hardening and hardened printed concrete." *Cement and Concrete Research* 157. https://doi.org/10.1016/j.cemconres.2022.106800.

Nair, Sooraj A. O., Gaurav Sant, and Narayanan Neithalath. 2022. "Mathematical morphology-based point cloud analysis techniques for geometry assessment of 3D printed concrete elements." *Additive Manufacturing* 49:102499. https://doi.org/10.1016/j.addma.2021.102499.

Naser, Mohannad Z., and Alaa I. Chehab. 2018. "Materials and design concepts for space-resilient structures." *Progress in Aerospace Sciences* 98:74–90.

Nazar, Sohaib, Jian Yang, Muhammad Faisal Javed, Kaffayatullah Khan, Lihui Li, and Qingfeng Liu. 2023. "An evolutionary machine learning-based model to estimate the rheological parameters of fresh concrete." *Structures* 48:1670–1683. https://doi.org/10.1016/j.istruc.2023.01.019.

Rahul, A. V., and Manu Santhanam. 2020. "Evaluating the printability of concretes containing lightweight coarse aggregates." *Cement and Concrete Composites* 109:103570. https://doi.org/10.1016/j.cemconcomp.2020.103570.

Rill-García, Rodrigo, Eva Dokladalova, Petr Dokládal, Jean-François Caron, Romain Mesnil, Pierre Margerit, and Malo Charrier. 2022. "Inline monitoring of 3D concrete printing using computer vision." *Additive Manufacturing* 60:103175. https://doi.org/10.1016/j.addma.2022.103175.

Schuldt, Steven J., Jeneé A. Jagoda, Andrew J. Hoisington, and Justin D. Delorit. 2021. "A systematic review and analysis of the viability of 3D-printed construction in remote environments." *Automation in Construction* 125:103642.

Siebel, Thomas M. 2019. *Digital Transformation: Survive and Thrive in an Era of Mass Extinction*: RosettaBooks.

Sizirici, Banu, Yohanna Fseha, Chung-Suk Cho, Ibrahim Yildiz, and Young-Ji Byon. 2021. "A review of carbon footprint reduction in construction industry, from design to operation." *Materials* 14 (20):6094.

Springer Nature. n.d. "RILEM TC 303-PFC and TC 304-ADC: Testing and Assessment of Printable and Printed Concrete." *Materials and Structures*. Accessed August 6, 2026. https://link.springer.com/collections/ggicbhafff.

Tang, Shu, Dennis R. Shelden, Charles M. Eastman, Pardis Pishdad-Bozorgi, and Xinghua Gao. 2019. "A review of building information modeling (BIM) and the internet of things (IoT) devices integration: Present status and future trends." *Automation in Construction* 101:127–139.

Uddin, Md Nasir, Faharidine Mahamoudou, Bo-Yu Deng, Moneef Mohamed Elobaid Musa, and Landry Wilfried Tim Sob. 2023. "Prediction of rheological parameters of 3D printed polypropylene fiber-reinforced concrete (3DP-PPRC) by machine learning." *Materials Today: Proceedings*.

Versteege, J., R. J. M. Wolfs, and T. A. M. Salet. 2025a. "Data-driven additive manufacturing with concrete: Enhancing in-line sensory data with domain knowledge, Part I: Geometry." *Automation in Construction* 172:106020.

Versteege, J., R. J. M. Wolfs, and T. A. M. Salet. 2025b. "Data-driven additive manufacturing with concrete: Enhancing in-line sensory data with domain knowledge, Part II: Moisture and heat." *Automation in Construction* 177:106327.

Wolfs, R. J. M., F. P. Bos, and T. A. M. Salet. 2018a. "Early age mechanical behaviour of 3D printed concrete: Numerical modelling and experimental testing." *Cement and Concrete Research* 106:103–116. https://doi.org/10.1016/j.cemconres.2018.02.001.

Wolfs, R. J. M., F. P. Bos, and T. A. M. Salet. 2018b. "Correlation between destructive compression tests and non-destructive ultrasonic measurements on early age 3D printed concrete." *Construction and Building Materials* 181:447–454. https://doi.org/10.1016/j.conbuildmat.2018.06.060.

Wolfs, Rob, Derk Bos, Jean-François Caron, Markus Gerke, Romain Mesnil, Richard Buswell, Nicolas Ducoulombier, Norman Hack, Emmanuel Keita, and Peter Kinnell. 2024. "On-line and in-line quality assessment across all scale levels of 3D concrete printing." *Cement and Concrete Research* 185:107646.

Yoo, Doo-Yeol, Jung-Jun Park, Sung-Wook Kim, and Young-Soo Yoon. 2013. "Early age setting, shrinkage and tensile characteristics of ultra high performance fiber reinforced concrete." *Construction and Building Materials* 41:427–438. https://doi.org/10.1016/j.conbuildmat.2012.12.015.

Zhang, Hanghua, Yanke Tan, Lucen Hao, Shipeng Zhang, Jianzhuang Xiao, and Chi Sun Poon. 2024. "Intelligent real-time quality control for 3D-printed concrete with near-nozzle secondary mixing." *Automation in Construction* 160:105325. https://doi.org/10.1016/j.autcon.2024.105325.

Zhu, Ronghua, King-James I. Egbe, Hadi Salehi, Zhongtian Shi, and Pengcheng Jiao. 2024. "Eco-friendly 3D printed concrete with fine aggregate replacements: Fabrication, characterization and machine learning prediction." *Construction and Building Materials* 413:134905.

Part 4

Construction Applications and Future Opportunities

14 Case Studies in 3D-Printed Concrete

Xiqiang Lin, Hailong Wang, Guanfei Tian, and Xiaoyan Sun

14.1 INTRODUCTION

In the past decade, concrete 3D printing technology, as an innovative approach, has made significant progress in the construction industry (Lu et al. 2024; De Schutter et al. 2018). The development of this technology has gone through three main stages: basic laboratory research, technical demonstration in practical engineering applications, and commercialization and promotion of projects. Initially applied to the printing of residential buildings, 3D printing has now expanded to include public facilities, landscape design, and pedestrian bridges (Zuo et al. 2024; Zhang et al. 2025). In the Middle East, traditional construction methods struggle to meet the housing needs brought about by rapid urbanization and population growth due to the hot and humid climate and a shortage of construction workers. The application of concrete 3D printing technology provides a new solution for the construction industry in the region. Leveraging the advantages of this technology, Dubai has set an ambitious goal for 25% of new buildings to be constructed using 3D printing technology by 2030 (Dubai Municipality 2025). This goal not only reflects Dubai's high regard for this technology but also demonstrates the region's determination to drive change in the construction industry.

In the United States and Europe, the construction industry faces multiple challenges, including high material costs, labor shortages, and growing housing demands (Nasdaq 2024; NAIOP 2023). This has prompted continuous exploration of new construction methods. As a leader in concrete 3D printing for residential projects, the United States has achieved significant results in Texas, demonstrating the potential of this technology and providing valuable experience for the global construction industry (3DPrinting.com 2023; PERI USA 2025). Europe, leading in digital design and robotics, has also achieved many exemplary results in residential and public buildings, greatly promoting the application and development of 3D printing technology in the construction industry (BUILD UP 2025; European Commission 2025). Meanwhile, in Asia, countries like China regard 3D-printed buildings as an important direction for the intelligent development of the construction industry, investing significant resources in R&D and project implementation, and achieving many successful cases (Ma et al. 2018).

There are two main construction methods for concrete 3D printing: on-site printing and prefabricated component printing followed by on-site assembly (Raza et al.

DOI: 10.1201/9781003434863-18

2024; Li et al. 2024). Both methods show great potential for development and technological advantages. Compared with traditional construction methods, concrete 3D printing not only improves construction efficiency and reduces costs but also offers significant environmental and design benefits (Hassan et al. 2024; de Rubeis et al. 2024). In recent years, breakthroughs in 3D printer technology, the development of printing materials, the optimization of printing processes, and the advancement of application technologies have laid a solid foundation for the rapid development of concrete 3D printing (Kopitha et al. 2025; -Sovetova and Calautit 2024). An increasing number of concrete 3D printing cases are being reported worldwide (Xiao et al. 2021). This chapter will select some representative cases for in-depth analysis, examining various aspects such as printer types, material properties, printing process details, and printing efficiency, aiming to provide readers with comprehensive technical information and deepen their understanding of this construction technology.

14.2 3D PRINTING EQUIPMENT FOR CONSTRUCTION

3D printers are crucial in concrete 3D printing. They use concrete as “ink” to turn digital design models into physical structures, bridging design, materials, and construction. Concrete 3D printers are now moving toward being larger, modular, flexible, and smarter.

In hardware terms, construction 3D printers are mainly divided into gantry, frame, and robotic-arm types. Their key parts cover the frame, motor drive system, material mixing and pumping system, print head, and 3D printing slicing and path planning software. A construction 3D printer’s development potential depends not only on reliable hardware but also on user friendly software and control systems. A good printer should make its software and control systems easy for engineers at the construction site to master and use.

Like fused deposition modeling (FDM; a 3D printing technology that builds objects by heating and extruding thermoplastic material layer by layer) 3D printers, frame 3D printers have a rigid frame holding the *XYZ* three-axis motion system, with the print head on the *X*-axis. When used with concrete mixing and pumping equipment, concrete is delivered to the print head via flexible pipes, forming a complete 3D printing and placement system. The size of frame 3D printers can be adjusted by increasing the height and span of the frame columns, enabling the printing of larger two- or three-story buildings. However, frame 3D printers face challenges in modular design and rapid installation for on-site deployment. As shown in Figures 14.1 and 14.2, these examples illustrate the typical structural form of frame-type 3D construction printers.

Gantry-style 3D printers are structurally similar to frame-type ones but feature a unique retractable *Y*-axis track, allowing theoretically unlimited length extension. This gives them greater flexibility in handling various building size printing tasks, especially in large-scale printing operations, where their mobilization advantage is evident. With a modular design, these printers are easy to assemble and disassemble, greatly improving transportation efficiency. As shown in Figures 14.3 and 14.4, representative gantry-style 3D printers demonstrate the structural characteristics and practical applications of this technology. Currently, they are widely used

FIGURE 14.1 COBOD BOD2 3D construction printer. Image © COBOD International, PERI 3D Construction.

FIGURE 14.2 Construction frame 3D printer. Image © CSCEC.

in the construction of large communities and multistory buildings. However, there is still room for optimization in lightweight design, height extension, and stability. To meet the demands of printing two- to three-story buildings, it is necessary to make lightweight improvements to the printer while ensuring structural stability and enhancing its vertical direction extension capability. Meanwhile, the overall design

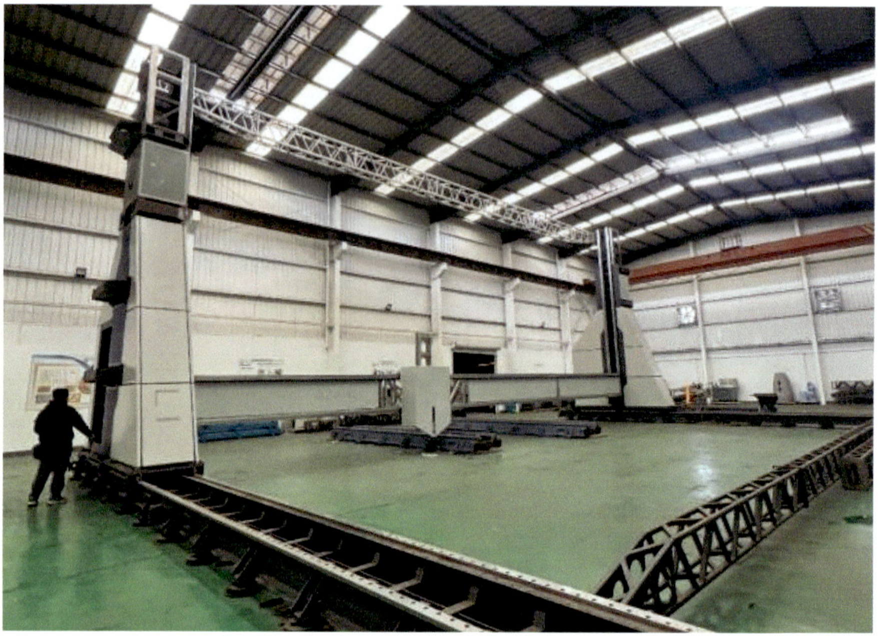

FIGURE 14.3 Construction gantry 3D printer. Image © GUANLI.

FIGURE 14.4 COBOD BOD3 gantry 3D printer. Image © COBOD.

of the printer needs to be refined to boost its stability and reliability in complex construction environments, thereby promoting the wider application of gantry-style 3D printers in multistory building printing.

Robotic-arm 3D printers are highly regarded for their operational flexibility in space and multidimensional directions. With extensive industrial application experience, they can be easily integrated into construction 3D printing systems, making them popular among companies and research institutions. By mounting industrial robotic arms on mobile bases, these printers can move flexibly around construction sites. Some robotic-arm printers can also operate on tracks, further expanding their application scope. As shown in Figures 14.5 and 14.6, vehicle-mounted robotic-arm printers exemplify this mobility and adaptability in construction applications.

Despite their flexibility, mobile robotic-arm 3D printers have a relatively small printing range for construction tasks. Printing an entire building in situ usually requires multiple moves and repositionings to complete different areas. Consequently, they are mainly used for printing components and parts. In addition, 3D printing slicing and path planning for robotic arms require professionals to use specialized software to generate complex printing paths, which places high demands on the expertise of on-site personnel. It should be noted that slicing and path planning are not unique to robotic-arm printers; gantry-style and frame-type 3D printers also require these processes. However, because robotic arms have more degrees of freedom, their slicing and path planning are generally more complex and technically demanding, making this challenge especially prominent in robotic-arm systems.

FIGURE 14.5 Vehicle-mounted robotic arm printer. Image © GUANLI.

FIGURE 14.6 Vehicle-mounted robotic arm printer. Image © Vertico.

To overcome the limitations of traditional industrial robotic arms in printing size, ultra-large-reach robotic-arm printers have been developed specifically for construction 3D printing. These advanced printers can be rapidly deployed, reducing installation time and costs, and are capable of printing approximately 100 sq. m of building structure in one go. Figures 14.7 and 14.8 illustrate two representative types of such large-scale robotic-arm systems: a foldable version developed by Guanli and a telescopic version introduced by ICON. In the future, such large-scale robotic-arm printers are expected to be used in more construction projects.

Many representative companies have emerged in the field of R&D and production of construction 3D printers. Their products have been successfully applied in construction projects. These companies include ICON, Mighty Buildings, SQ4D, Apis Cor, Diamond Age, and Branch in the United States; COBOD, CyBe, Vertico, XtreeE, Construtions-3D, and WASP in Europe; and Guanli, WinSun, Jiayi, and Huachuang in China. Table 14.1 presents the major global construction 3D printer types and their application fields, providing a comparative overview of the technological landscape (Puzatova et al. 2022).

Currently, robotic-arm 3D printers are primarily used for printing building components and small-scale houses. However, with the increasing size of buildings and the growing complexity of architectural designs, there is a rapidly growing demand for large-scale construction 3D printers capable of on-site holistic printing. These printers need to be designed with modularity to facilitate expansion, disassembly, and installation. From the user's perspective, it is important to select the appropriate printer based on the characteristics and scope of application of different types of printers, considering the actual engineering and on-site conditions.

FIGURE 14.7 Construction gantry 3D printer. Image © GUANLI.

FIGURE 14.8 Telescopic construction large-reach robotic-arm printer. Image © 3DXB.

TABLE 14.1
Major Global Construction 3D Printer Types and Their Application Fields

No.	3D Printer	Developer	Technology Form	In Situ Printing	Application Field
1	Contour crafting (Illumin 2025)	University of Southern California	Gantry	Yes	Components
2	CONPrint3D (Mechtcherine et al. 2024)	Technical University of Dresden	Concrete placing boom arm	Yes	House
4	CyBe (CyBe Construction 2025)	CyBe	Robotic arm	Yes	Components
5	Vulcan (ICON 2025)	ICON	Gantry	Yes	House
6	Phoenix (ICON 2025)		Robotic arm	Yes	House
7	BigDelta/Crane (WASP 2025)	WASP	Cantilever	Yes	House
8	Office building printer (Lin et al. 2024)	CSCEC & Guanli	Frame type	Yes	House
9	Apis Cor (Apis Cor 2025)	Apis Cor	Tower style	Yes	House
10	XtreeE (XtreeE 2025)	XtreeE	Robotic arm	No	Components
11	Constructions-3D (Construction 3D 2025)	Constructions-3D	Robotic arm	No	Components
12	3D concrete printing (Eindhoven University of Technology 2025)	Eindhoven University	Gantry	No	Components
13	Robotic fabrication (ETH Zurich 2025)	ETH Zurich	Robotic arm	No	Components
14	BOD2 (COBOD International 2024)	COBOD International	Frame type	Yes	House
15	BOD3 (COBOD International 2024)		Gantry	Yes	House
16	GL-3DPRT-C4 (Hangzhou Guanli Intelligent Technology Co., Ltd. 2025)	Guanli	Gantry	Yes	House
17	GL-3DPRT-TJA (Hangzhou Guanli Intelligent Technology Co., Ltd. 2025)		Robotic arm	Yes	House
18	Mighty Buildings (Mighty Buildings 2025)	Mighty Buildings	Gantry	Yes	Components
19	ATLAS Robot (Vertico 2025)	Vertico		No	
20	Team of mobile robots (Ackerman 2018)	Nanyang Technological University	Robotic arm	No	Components
21	Berlim-1/Tilikum-1 (Twente Additive Manufacturing 2025)	Twente Additive Manufacturing	Robotic arm	No	Components
22	ARCS (SQ4D 2025)	S Square SQ4D	Gantry	Yes	House

14.3 CASE STUDIES OF 3D-PRINTED BUILDINGS

14.3.1 Representative Cases in the Middle East

Case 1: Dubai "Office of the Future"

In May 2016, the world's first office building constructed using 3D printing technology, known as the "Office of the Future", was completed at the Dubai International Financial Centre. This was not only the world's first operational concrete 3D-printed building but also the first project to combine 3D-printed architectural structural design with the concrete 3D printing technology. The architectural design was completed by Gensler, the Chinese company WinSun Building Technology was responsible for 3D printing the building components, and the China State Construction Engineering Corporation (CSCEC) completed the overall construction. The total construction area is 250 sq. m, and the building features a streamlined, curved structure that integrates futuristic elements with Middle Eastern cultural aesthetics (Dubai Future Foundation 2020). As shown in Figure 14.9, the Office of the Future demonstrates a landmark achievement in the application of 3D printing in construction.

Technical features of the project:

(i) Targeted architectural and structural design

The reinforcement and assembly process design of the printed components are highly integrated with the 3D printing process. A truss steel mesh reinforcement method is used, with the steel bars aligned with the printing path. The vertical steel bars and lifting components are designed rationally. Prestressing, wall component connection, and anchoring, as well as the integrated design of roof and floor structures, are comprehensively applied in the assembly construction. This effectively enhances the performance of the components and the structural safety. As illustrated in Figure 14.10,

FIGURE 14.9 Real-world view of the 3D-printed Office of the Future. Image © CSCEC.

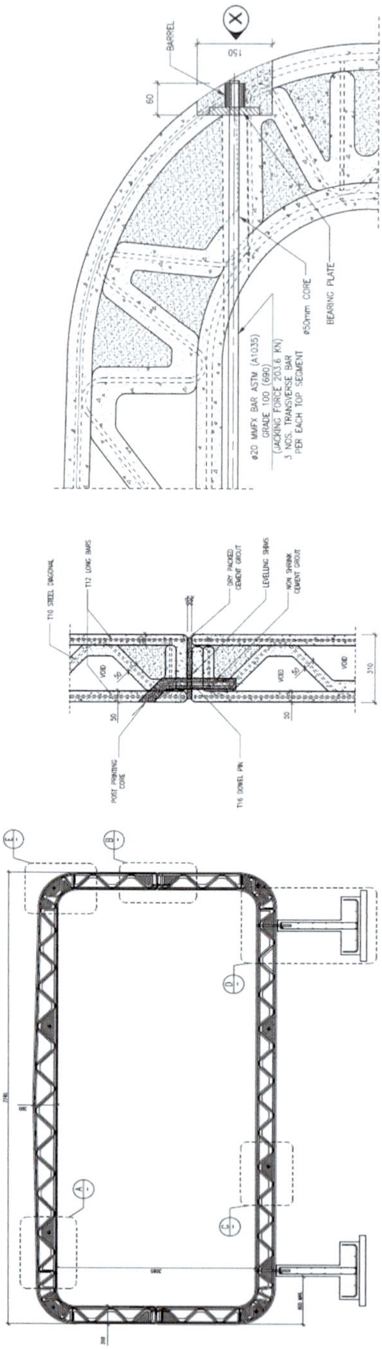

FIGURE 14.10 Reinforcement and connection design of structural components. Image © CSCEC.

FIGURE 14.11 Hoisting and installation process of 3D-printed structural components. Image © CSCEC.

the reinforcement and connection design of structural components demonstrates the integration of steel reinforcement with the 3D printing process.

(ii) Eco-friendly printing materials

The 3D printing mortar used is developed by WinSun. It is primarily composed of high-grade portland cement, industrial solid waste admixtures, fine tailing aggregates, glass fibers, and additives. This composition enhances the material strength and printing stability. The compressive strength of the concrete specimens reaches 49 MPa, and the flexural strength reaches 6 MPa. The use of recyclable materials helps reduce construction waste and carbon emissions. The hoisting and installation process of 3D-printed structural components is clearly illustrated in Figure 14.11.

(iii) Fast construction speed

By using 3D-printed components and prefabricated construction methods, half-ring-shaped concrete components were printed in the factory and then transported to the site for assembly. The main structure was printed in just 17 days, and the component assembly took 2 days. The subsequent decoration and finishing took about 1 month, demonstrating the high efficiency of the 3D printing construction process. Figure 14.12 presents the assembly and construction process of the 3D-printed office, highlighting the speed and effectiveness of this building method.

The Dubai "Office of the Future" is a milestone project in global 3D-printed construction technology, marking the transition of 3D-printed architecture from "proof of concept" to "functional application." Its modular construction model, rapid construction period, and cross-border collaborative efforts provide a replicable case study and design solution for global public buildings, showcase architecture, and digital construction strategies.

Case 2: Saudi Arabia 3D-Printed Mosque

In April 2024, the world's first 3D-printed mosque was completed and inaugurated in Saudi Arabia. The project was jointly constructed by Fursan Real Estate Company of Saudi Arabia and Guanli Technology Company. It aims to implement the Saudi "Vision 2030" and promote innovation and sustainable development in

FIGURE 14.12 Assembly and construction process of the 3D-printed office. Image © CSCEC.

FIGURE 14.13 Architectural design and completed view of the mosque. Image © Guanli, Fursan.

the construction sector. The project covers an area of 5,600 sq. m and employs a framework structure, including traditional columns, beams, and roofing. All vertical walls are installed using 3D-printed wall modules, which not only ensure structural safety but also enhance construction flexibility and efficiency (Fakharany 2024). Figure 14.13 presents the architectural design and completed view of the mosque.

Technical Features and Implementation Process:

(i) 3D-printed concrete material
The 3D-printed concrete used in the project is formulated with local portland cement grade 32.5 from Saudi Arabia, fine sand, chemical admixtures, and water. Over 96% of the raw materials are sourced locally. The material exhibits excellent compressive strength: 14.3 MPa after 1 day, 25.7 MPa after 3 days, and 47.6 MPa after 28 days.

(ii) Printing equipment and process
Three large gantry-style concrete 3D printers were employed for the on-site printing of wall components. The printers are equipped with Moli slicing and control software, which enables automatic slicing of the wall 3D model and optimized design of the printing path. The print head extrudes mortar layer by layer to complete the formwork-free construction of wall components. As shown in Figure 14.14, the wall components were printed directly on site using these gantry-style printers.

FIGURE 14.14 On-site printing of mosque wall components. Image © Guanli, Fursan.

(iii) Construction efficiency
The project went from the start of construction to completion in just 6 months. The total printing time for the wall components was only 400 hours. The 3D printing and assembly processes were carried out concurrently, with all wall assemblies completed within 2 months. Figure 14.15 illustrates the assembly process of the 3D-printed mosque wall components, demonstrating the efficiency of the integrated workflow. This highlights the high efficiency advantages of 3D printing technology in the field of construction.

(iv) Controlling the costs of 3D printing construction
Although this project utilized a prefabricated assembly process, it innovatively employed three large 3D printers on site. With the use of 3D digital models, these printers can directly fabricate building components and simultaneously carry out the installation of these components. This approach significantly reduces costs associated with various aspects of traditional prefabricated construction, such as the extensive use of formwork, the need for component manufacturing plants and storage facilities, and the transportation of components. Moreover, it conserves a substantial amount of labor, minimizes the use of lifting equipment, and shortens the construction time. As presented in Figure 14.16, the completed view of the mosque demonstrates the success of this innovative approach.

FIGURE 14.15 Assembly of 3D-printed mosque wall components. Image © Guanli, Fursan.

FIGURE 14.16 Completed view of the 3D-printed mosque. Image © Guanli.

The project successfully integrated traditional structural forms with 3D-printed components in construction, fully demonstrating the feasibility and efficiency advantages of 3D-printed modular building structures. It provides new ideas and case references for the future application and development of 3D printing technology and also brings more possibilities to the design and construction of modular prefabricated buildings. As a typical public building in the Middle East region with extensive construction demand, the mosque is expected to inspire more similar 3D-printed mosque projects implemented in countries like Saudi Arabia and Dubai in the future, further promoting the widespread application of 3D printing technology in the field of construction.

14.3.1.1 Other Cases: In Situ 3D-Printed Villas

In the Middle East, several in situ 3D-printed construction projects have been completed, demonstrating their potential in the region's construction sector.

In 2019, Apis Cor completed a government office in Dubai with a construction area of 640 sq. m and a height of 9.5 m. The project utilized a rotating telescopic arm 3D printer and a gypsum-based mixture as the printing material, which had a density of only 50% of that of conventional concrete, effectively reducing the structural self-weight. The printer performed in situ printing on site, enabling rapid construction of the building (Apis Cor 2019). As shown in Figure 14.17, this project illustrates a government office in Dubai constructed using 3D printing technology.

In July 2024, Saudi Arabian developer Dar Al Arkan, in collaboration with COBOD, completed a 3D-printed villa in Riyadh with a construction area of 330 sq. m and a height of 9.9 m. During the printing process, D.fab admixture, co-developed

FIGURE 14.17 3D-printed government office in Dubai. Image © Apis Cor.

FIGURE 14.18 3D-printed villa in Riyadh. Image © COBOD International, Dar Al Arkan.

FIGURE 14.19 Two-story 3D-printed private villa. Image © Guanli, BTI.

by COBOD and CEMEX, was used in the 3D-printed concrete material, accounting for 1% of the material composition, with the remainder sourced from local raw materials. The total printing time for the entire building was 26 days, showcasing the efficiency and feasibility of 3D printing technology in residential construction. As illustrated in Figure 14.18, the completed villa demonstrates the practical application of 3D printing in large-scale residential projects.

In May 2025, Dubai-based 3D printing company BTI and Guanli completed a private villa in Dubai with a construction area of 420 sq. m and a height of 9.5 m. A large gantry-style printer, the GL-3DPRT-C4 with dimensions of 20 × 15 × 10 m, was used for the construction. The first floor's walls were printed continuously for 3.5 days, and the second floor was printed continuously for 4 days. The 3D-printed concrete achieved a compressive strength of 23.5 MPa after 1 day and 52.0 MPa after 28 days. Over 96% of the materials in the mix were sourced locally, supplemented by a small amount of additives. The use of local materials not only reduced costs but also minimized carbon emissions associated with transportation. Figure 14.19 shows the two-story 3D-printed private villa, and Figure 14.20 shows the single-story 3D-printed house delivered in Dubai.

In December 2023, 3DXB Group used the MaxiPrinter to complete a 3D-printed villa in Dubai with a construction area of 400 sq. m. The project demonstrated the application potential of the new MaxiPrinter building 3D printer in large-scale

FIGURE 14.20 Single-story 3D-printed private villa. Image © Guanli, BTI.

FIGURE 14.21 3D-printed 400-m^2 villa project. Image © 3DXB.

construction projects. The new robotic-arm printer is efficient and flexible, capable of rapid on-site deployment and achieving a high printing speed of up to 250 mm per second while maintaining high-precision printing. The project utilized Sikacrete733 material, which not only has excellent printing performance but also significantly reduces carbon emissions. Figure 14.21 shows the 400-m^2 3D-printed villa project.

14.3.2 Representative Cases in North America

Case 1: 3D-Printed Communities in the United States

The United States is at the forefront of the global 3D-printed construction sector and is the first country to successfully commercialize 3D-printed building real estate. Among the leading companies, ICON has become a globally renowned 3D-printed construction firm, thanks to its advanced printers, materials, and design technologies. ICON, in partnership with Lennar, a residential construction company, has developed the Wolf Ranch community in Austin, Texas. This community is the world's largest 3D-printed residential community, consisting of 100 single-story homes. Each home has a construction area ranging from 140 to 190 sq. m, with three to four bedrooms and a garage, meeting the housing needs of middle income American families. Figure 14.22 shows the 3D-printed Wolf Ranch community.

FIGURE 14.22 3D-printed Wolf Ranch community. Image © ICON.

Technical Features:

(i) Advanced 3D Printer
The project utilized the Vulcan 3D printer developed by ICON. This printer has a large print size, capable of constructing single-story homes with an area of up to 3,000 sq. ft. The Vulcan printer operates on a tablet based system and uses ICON's proprietary BuildOS printing software, which converts architectural drawings into 3D-printed lines, facilitating rapid deployment on the project site. The printer is equipped with a dedicated material mixing and pumping system, ensuring efficient and stable printing processes. As shown in Figure 14.23, multiple Vulcan 3D printers can operate simultaneously, significantly improving construction efficiency.

(ii) Eco-Friendly 3D Printing Material
The 3D printing material used in the project, Lavacrete, is formulated from locally sourced natural raw materials, including a precise blend of cement, sand, water, and additives. This mixture creates an environmentally friendly and mechanically robust 3D-printed concrete. The use of Lavacrete reduces construction waste and carbon emissions. It has excellent strength properties, with compressive strength greater than 10 MPa after 1 day and greater than 45 MPa after 28 days. Additionally, Lavacrete is waterproof, mold-resistant, termite-resistant, and capable of withstanding extreme weather conditions, providing a solid structure and good durability for the homes. Figure 14.24 shows the performance status of Lavacrete in 3D-printed concrete applications.

(iii) Environmental and Cost Advantages
3D printing allows for precise layer-by-layer material deposition, eliminating waste and aligning with sustainable development principles. The construction efficiency is high, with a three- to four-bedroom home taking approximately 3 weeks to print. Compared to traditional construction methods, this significantly reduces the construction period and increases building efficiency. Overall, the 3D-printed community project demonstrates significant advantages in terms of speed, cost, labor requirements, and material waste. As the technology advances and scales up, costs are expected to decrease further.

The 3D-printed residential communities mark a milestone in global construction technology by pioneering the application of concrete 3D printing in large-scale

FIGURE 14.23 Multiple 3D printers constructing simultaneously. Image © ICON.

FIGURE 14.24 Performance status of 3D-printed concrete. Image © ICON.

residential projects. They have successfully demonstrated the technical feasibility, cost-effectiveness, and market potential of 3D printing technology in mass residential construction. This innovation offers new solutions to address the housing shortage faced by middle- and low-income families in the United States. Moreover, the completion of these communities has spurred the development of relevant supporting standards, including 3D-printed residential structural codes, streamlined approval processes, and adapted building insurance systems. These advancements have had a positive impact on the formulation of construction regulations and policies. In terms of design collaboration, construction efficiency, and industrial cooperation, the project has provided valuable experience and serves as a reference for the construction of replicable 3D-printed residential communities worldwide.

Other Cases: 3D-Printed Houses in the United States

(i) The World's First 3D-Printed Hotel
The world's first 3D-printed hotel is being constructed in the West Texas desert, led by American hotelier Liz Lambert and ICON, a cutting-edge 3D construction company. The project, which spans 60 acres, aims to redefine the renowned "El Cosmico" campsite. Utilizing the 3D printing technology of the architectural firm BIG, the project integrates Lambert's community-oriented vision with the state-of-the-art construction technology. ICON's 3D printing technology can create various shapes, including arches, domes,

FIGURE 14.25 World's first 3D-printed hotel. Image © ICON.

FIGURE 14.26 Walmart 3DCP project. Image © Alquist.

and curved structures, which harmonize with the desert environment and utilize locally sourced materials. Figure 14.25 shows the architectural concept and design of the 3D-printed hotel.

(ii) Walmart 3D Concrete Printing Project
In May 2025, Walmart and Alquist announced the completion of the second groundbreaking 3D concrete printing (3DCP) project at the Walmart Supercenter in Irondale, Alabama. The 5,000-sq. ft pickup area's load-bearing walls in the expansion zone were constructed within 1 week. The technical team utilized two 3DCP systems to print 16-ft-high walls in just 75 hours over 7 working days, a speed far exceeding that of traditional construction methods. Sika USA provided a specially formulated concrete mixture that met the requirements for speed and durability under actual construction conditions. Figure 14.26 presents the Walmart 3DCP project.

(iii) The First 3D-Printed Starbucks
In December 2024, a brand new Starbucks opened in Brownsville, Texas, constructed using 3D printing technology. With a total construction area of approximately 130 sq. m, the project was commissioned by Starbucks Corporation and completed by PERI 3D Construction using a COBOD gantry-style 3D printer. The store began operations on April 28, 2025. Figure 14.27 presents the 3D-printed Starbucks in Brownsville.

FIGURE 14.27 3D-printed Starbucks in Brownsville. Image © COBOD International, PERI 3D Construction.

FIGURE 14.28 3D-printed home projects by SQ4D in New York. Image © SQ4D.

(iv) 3D-Printed Home Project in New York
SQ4D's 3D-printed home project in Riverhead, New York, is the first 3D-printed house listed for sale in the United States. The home features three bedrooms, two bathrooms, and an open-concept floor plan. It is constructed using high-strength mortar through 3D printing, which is approximately 33% stronger than existing concrete block houses on the market. Compared to similar recently constructed homes in the area, the price is reduced by 50%. Figure 14.28 presents the 3D-printed home project by SQ4D in New York.

(v) 3D-Printed Barracks at Fort Bliss
In March 2025, the US Army completed its first 3D-printed barracks project at Fort Bliss, Texas. The project was constructed in collaboration between the US Army Engineer Research and Development Center and ICON. Each barracks building has a construction area of approximately 529 sq. m. The project utilized ICON's Vulcan 3D printer and its proprietary concrete-based material, Lavacrete. The completion of the barracks marks a successful application of 3D printing technology in military infrastructure construction, demonstrating its advantages in rapidly and economically building high-quality structures. Figure 14.29 presents the 3D-printed barracks at Fort Bliss.

FIGURE 14.29 3D-printed barracks at Fort Bliss. Image © ICON.

14.3.3 Representative Cases in Europe

In recent years, Europe has faced significant housing shortages in some cities and regions due to population growth, urbanization, and immigration. Meanwhile, the European construction industry has encountered labor shortages and rising costs. Building 3D printing technology, with its advantages of improving construction efficiency, quickly printing wall structures, and shortening construction periods, effectively reduces costs: it reduces the use of formwork, lowers the demand for labor and skilled personnel, and minimizes material waste, thereby alleviating the pressure on housing costs and helping ease the housing demand in Europe. In 2021, the first two-story 3D-printed residential building in Europe was completed in Beckum, Germany. Since then, 3D-printed construction cases in Europe have continued to emerge, covering both residential and public buildings. The following are typical 3D-printed cases in Europe.

(i) Europe's First Multistory Residential Project

Europe's first multistory residential project was jointly constructed by the German construction technology company PERI Group and the Danish 3D printing company COBOD. The building consists of two stories, with a total construction area of 160 sq. m. The project utilized the BOD2 printing system, and the 3D-printed concrete mixture was custom developed by Heidelberg Material to meet the requirements for structural load-bearing performance and stable printing construction.

The evoBuild® 3D-printed concrete material features a maximum aggregate size of 4 mm and is designed for easy pumping with excellent print extrusion performance. It offers a printing open time of 20 minutes at ambient temperatures ranging from 10 to 30°C, making it suitable for printing speeds between 150 and 250 mm/s. The material achieves a compressive strength greater than 10 MPa after just 1 day, and this strength exceeds 45 MPa after 28 days. Additionally, it boasts outstanding environmental performance, with a 55% reduction in CO_2 emissions compared to traditional portland cement systems.

The project successfully passed the German building code review and obtained a construction permit. The walls feature a layered insulation structure design, employing an integrated three-layer wall structure (inner wall, insulation layer, and outer wall) to enhance the building's energy-saving performance. Figure 14.30 presents Europe's first multistory 3D-printed residential project.

(ii) Heidelberg "Wave House" Data Center
In 2024, the "Wave House" data center was completed in Heidelberg, Germany. It is the largest 3D-printed building by area in Europe. The project was designed by SSV and Mense Korte and constructed by Peri 3D Construction. The exterior walls of the building measure 54 m in length, 10.7 m in width, and 9 m in height, and were printed using the COBOD BOD2 3D construction printer. The printing concrete was made from the evoBuild® 3D-printed dry mix developed and produced by Heidelberg Materials. By adopting a wavy wall design, the "Wave House" not only offers greater visual aesthetics and uniqueness but also demonstrates the advantages of 3D printing technology in architectural shaping. Figure 14.31 presents the Heidelberg "Wave House" data center.

(iii) Cable Car Upper Station in Kopřivnice, Czech Republic
In January 2025, the upper station of the cable car in Kopřivnice, Czech Republic, was completed. The building was designed by ATELIER 3M and constructed using 3DCP technology by the Czech industrial automation company ICE. The station is divided into two main areas: the cable car

FIGURE 14.30 Europe's first multistory residential project. Image © Peri, COBOD.

FIGURE 14.31 Heidelberg "Wave House" data center. Image © Peri, COBOD.

operating area and a visitor daytime lounge. The project integrates prefabricated elements with on-site production, showcasing the versatility of 3DCP technology. The 3D-printed walls are designed to integrate insulation, internal functionality, and external detailing into a single process. Figure 14.32 presents the cable car upper station in Kopřivnice.

(iv) 3D-Printed Apartment Building in Germany

RUPP and COBOD constructed a three-story, 380-sq. m 3D-printed apartment building in Wallenhausen, Bavaria, which includes a basement and five apartments. This project is one of the largest 3D-printed residential buildings in Europe. The building envelope was entirely constructed using the COBOD BOD2 gantry-type printer, with each floor taking only 25 hours to print. The project utilized a new 3D-printed concrete formula developed by Heidelberg Materials, with a high cement content. The exterior walls were finished with special adhesives and plaster materials to ensure comfort and durability, demonstrating the scalability and potential of 3DCP technology in multiunit residential construction. Figure 14.33 presents the 3D-printed apartment building in Germany.

FIGURE 14.32 Cable car upper station in Kopřivnice. Image © ATELIER 3M.

FIGURE 14.33 3D-printed apartment building in Germany. Image © COBOD International, PERI 3D Construction.

(v) Tor Alva
On May 28, 2025, commissioned by the cultural foundation Nova Fundaziun Origen, researchers from ETH Zurich constructed Tor Alva using 3D-printed concrete components. Located in Mulegns, Switzerland, this tower is the world's tallest 3D-printed concrete tower, standing at 30-m high with four stories. The design of Tor Alva was entirely created using parametric scripting, which means the digital model was generated by computer algorithms and then adjusted, visualized, and simulated for production within the constraints of the robotic 3D printer. Figure 14.34 presents the White Tower (Tor Alva), the world's tallest 3D-printed concrete structure.

(vi) Two-Story Terraced Houses in Northern Ireland, UK
The two-story terraced houses in Northern Ireland, UK, are among the first 3D-printed residential homes to be put into use in the country. Each house has an area of 110 sq. m, divided into two floors, and fully complies with relevant standard requirements, having passed government approval. The construction cost of each residence is approximately £205,000, which covers the traditional roof structure and standard internal and external finishes. Figure 14.35 presents the 3D-printed two-story terraced houses in Northern Ireland.

FIGURE 14.34 World's tallest 3D-printed structures: the White Tower in Switzerland. Image © Hasnseyer/Dillenburger Architects.

FIGURE 14.35 3D-printed two-story terraced houses. Image © COBOD International, HTL Tech.

FIGURE 14.36 3D-printed defensive structures. Image © ATELIER 3M.

(vii) Defensive Structures at the NATO Air Force Base
The defensive structures at the NATO Air Force Base have adopted 3D printing technology, which is an innovative application in the field of military construction. The Czech military is implementing a modular defensive system at the Čáslav Military Base in the Czech Republic, demonstrating the potential of 3D printing technology in military engineering and providing solutions for the rapid deployment and construction of military infrastructure. Figure 14.36 presents the 3D-printed defensive structures.

14.3.4 Representative Cases in Asia

Asia has also made many breakthroughs in the research and application of 3D-printed construction technology. China, e.g., has achieved significant progress in the research and application of 3D-printed construction, ranging from equipment development to material innovation, and from residential construction to bridge engineering, cultural and tourism landscapes, and municipal facilities. Chinese companies have developed and manufactured various types of large-scale 3D construction printers, including frame-type, gantry-type, and robotic-arm models. Meanwhile, many research institutions and enterprises have achieved numerous research results in 3D-printed concrete and have applied them in projects. The following are some typical engineering projects:

Case 1: China's 3D-Printed "Mars Nest Hotel" Project

The "Mars Base 1" is located in the Gobi Desert in the Jinchang area, Gansu Province. It is a Mars-simulating habitat cultural tourism project built on the basis of the local Mars-like terrain, covering children's space technology experience and educational themes. The base includes five egg-shaped nest hotel buildings constructed by 3DCP, each with a height of 9 m, a diameter of 6 m, and a two-story interior covering a construction area of 46 sq. m. This project was jointly implemented by Zhejiang University in China and Guanli 3D printing company. It integrates architectural design, modular printing, structural assembly, and decoration, exploring the potential of digital construction in the harsh Gobi environment. Figures 14.37 and 14.38 present the design and interior views of the 3D-printed Mars Nest Hotel.

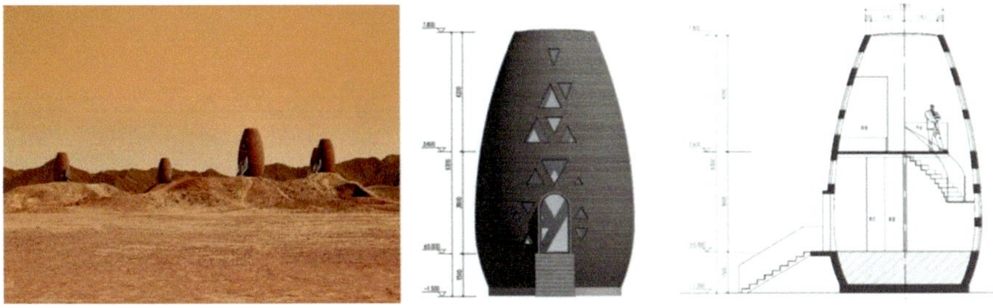

FIGURE 14.37 Design of the 3D-printed Mars Nest Hotel. Image © Guanli.

FIGURE 14.38 Interior views of the 3D-printed Mars Nest Hotel. Image © Guanli.

The project adopts a "factory-printed components + on-site modular assembly" model. Its main features are as follows:

(i) Structural Design and Reinforcement
During printing, circumferential horizontal reinforcement meshes are first placed and welded into a single unit, with one layer set every 25 cm to enhance the structural strength and stability. After the exterior shell is printed, vertical rebars of diameter 12 mm are inserted and tied to the horizontal reinforcement. C40 self-consolidating concrete is then poured to form a composite reinforcement structure. The foundation structure is made of traditionally reinforced concrete. Figure 14.39 illustrates the model segmentation and structural system of the Mars Nest Hotel.

(ii) Large-Scale 3D Printing Equipment and Customized Materials
Two gantry-style concrete 3D printers are used, capable of printing large-sized components and achieving precise shaping of complex components. The specially developed 3D-printed concrete has high flowability, pumpability, and early strength characteristics. Combined with reinforcement bars and poured concrete, it constructs a highly durable and high-strength structure. The printing parameters are as follows: layer height of 15 mm, speed of 120 mm per second, extrusion width of 40 mm, setting time of 41 minutes, and compressive strength reaching 47.9 MPa after 28 days. Figure 14.40 presents the printing and manufacturing of the modular components.

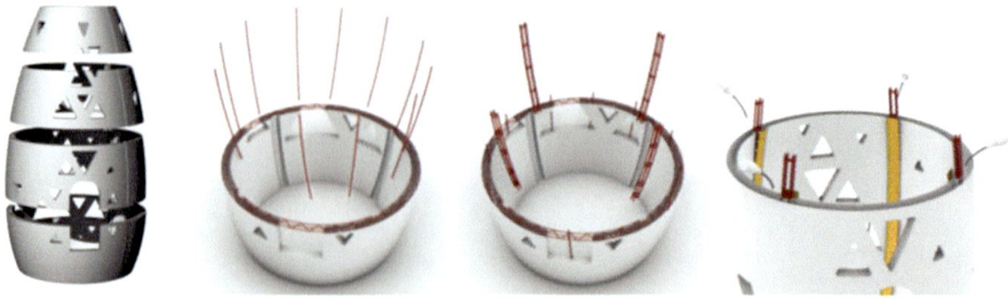

FIGURE 14.39 Model segmentation and structure of the 3D-printed Mars Nest Hotel. Image © Guanli.

FIGURE 14.40 Printing and manufacturing of modules. Image © Guanli.

(iii) On-Site Assembly and Construction Process

After the printed wall components are transported to the site, they are assembled into a circular structure in sequence using a crane, followed by welding the rebar cages and pouring the mortar. The installation of the four circular structures took only 1 day, with a safe and rapid construction process. Figures 14.41–14.45 show the assembly process, including component installation, rebar welding, and concrete pouring. Figure 14.44 presents the final assembly view of all five houses. This project is one of the largest oval 3D-printed structures both domestically and internationally, and the first 3D-printed hotel project to achieve commercial operation. Its construction provides technical experience for 3D-printed structures and promotes the development of the additive intelligent construction industry.

Case 2: Two-Story Visitor Service Center

The project is located in Anhui Province, China. It is a visitor service center with a construction area of 350 sq. m and a total building height of 10.45 m. The architectural design is both innovative and challenging, featuring inclined and twisted exterior walls, with some external columns having varying cross sections and tilt angles. The functional areas of the building include an exhibition hall, café, sales area, and restrooms. Figure 14.46 shows the architectural design of the two-story visitor service center.

FIGURE 14.41 On-site installation of components for the "Mars Nest Hotel." Image © Guanli.

FIGURE 14.42 Reinforcement bar welding of columns for the "Mars Nest Hotel." Image © Guanli.

FIGURE 14.43 Overall assembly of the "Mars Nest Hotel." Image © Guanli.

FIGURE 14.44 Reinforcement bar welding and concrete pouring of columns. Image © Guanli.

FIGURE 14.45 Assembled view of all five houses. Image © Guanli.

FIGURE 14.46 Architectural design of the two-story visitor service center. Image © Guanli.

Technical Features:

(i) Architectural and Structural Design
The architectural design of the project features multifaceted walls that tilt outward, with irregularly shaped columns at angles connecting from the ground to the roof. Using traditional methods to fabricate concrete formwork for such a design poses significant challenges. The structural design employs a traditional pile foundation, and the two-story building above the ground adopts a reinforced concrete frame structure with beams, slabs, and columns. Figure 14.47 illustrates the foundation and beam–column structure of the building.

(ii) 3D Printing Equipment
The project utilized the GL-3DPRT-TJA, an ultra large robotic-arm 3D printer developed by Guanli Technology in China. This printer has an arm span of 8 m and a track length of 18 m and can reach a height of up to 6 m. It enables flexible and rapid deployment of the robot on-site, enhancing the efficiency of the printing construction and reducing time costs. The project completed the printing of the first-floor building walls on-site within just 7 days. Figure 14.47 shows on-site printing by the TJA 3D printing robot.

(iii) Construction Process
The inclined walls and other vertical wall structures of the first floor were printed on site by the robotic-arm 3D printer. The floor slabs were cast in place using formwork and poured concrete. The second floor features a framework structure composed of structural columns, beams, and a concrete

FIGURE 14.47 Foundation and beam–column structure of the building. Image © Guanli.

FIGURE 14.48 On-site printing by the TJA 3D printing robot. Image © Guanli.

roof, with the walls made of glass curtain walls. The column form shells in the building were prefabricated by 3D printing in the factory and then transported to the site for assembly. The printed concrete column form shells were individually fitted over the already tied rebar cages, and after securing and supporting them, the core concrete was poured to complete the construction of the structural columns. The roof structure was then built using traditional reinforced concrete techniques, integrating with the 3D-printed walls and columns to form the complete building structure. Figures 14.49–14.51 show the model breakdown, in situ printing, and prefabricated column form shells used in the project.

(iv) Material and Technical Features

The project uses 3D-printed concrete material with a compressive strength greater than 50 MPa. During the printing of the inclined walls, a hollow structure is adopted to reduce self-weight, and longitudinal columns are set within the walls to connect with the foundation. Additionally, 0.6 m of concrete is poured at the bottom of the hollow walls to strengthen the anchorage between the inclined walls and the foundation. Custom-made wooden supports are used during the printing process to ensure wall stability, and these supports are removed only after the beams and roof are cast to form an integrated structure. Figure 14.52 presents the completed construction of the two-story building.

FIGURE 14.49 Model breakdown of on-site printing and prefabricated component printing. Image © Guanli.

FIGURE 14.50 In situ printing of a portion of the building. Image © Guanli.

FIGURE 14.51 Formwork shell surrounding the prefabricated printed column. Image © Guanli.

FIGURE 14.52 Completed construction of the two-story building. Image © Guanli.

The project employs a robotic-arm 3D printer for construction, which can be flexibly deployed. Through automated construction, it significantly reduces labor requirements, cutting labor costs by 30%~50%. Additionally, it eliminates the need for custom formwork for irregular walls and columns, resulting in faster construction and a shortened project duration. 3DCP is highly efficient in material utilization. The printer can precisely control the amount of material used based on the design, achieving accurate distribution and minimizing waste to the greatest extent. In traditional construction, formwork is an essential part. However, custom formwork for irregular shapes not only has high production costs but also requires a large amount of storage space and transportation costs. 3D-printed concrete technology does not use formwork. The printer directly relies on the digital model design to control the layer-by-layer addition of materials to form the required structure, without the need for additional formwork components. The project saves about 30%~40% on formwork costs.

14.3.5 Other Cases: Other Typical Projects in Asia

(i) Japan's 3DPOD Pavilion
The project was completed in 2023 by Obayashi Corporation in Japan. It is the first 3D-printed seismic-resistant building to receive certification and construction permits from the Minister of Land, Infrastructure, Transport, and Tourism. The building was printed using a special mortar made from ultra-high-strength fiber-reinforced concrete "SLIM-Crete®." This material serves both as the interior and exterior finish and as the load-bearing structural framework, eliminating the need for additional steel reinforcement or frames. The curved shape is designed to maximize the proportion between the internal space and the wall surface area, achieving the largest internal volume with the least amount of material under the constraints of printing technology and site conditions. The walls feature a multilayer structure that integrates load-bearing elements, insulating material cavities, utility wiring, and radiant heating and cooling systems. The floor design employs a ribbed structure strengthened along the main stress lines, offering an economical material solution that enhances both aesthetics and spatial quality. A robotic printer was installed on site to directly print the mortar templates onto the foundation. The roof slab templates, printed as prefabricated components in a nearby facility, were placed on top of the walls, and the 3D printer was then positioned on the roof slab to complete the parapet for the walkable deck, demonstrating the technical feasibility of multistory printed construction.

(ii) Singapore Pilot Project – A 3D-Printed Children's Center
In August 2025, the pilot project – a 3D-printed children's center in Singapore – showcased on-site 3DCP using a commercially available mobile construction printer. The initiative was led by Woh Hup (Private) Limited, with the Department of Civil and Environmental Engineering, National University of Singapore (NUS) providing technical support in material characterization, structural validation, and compliance with the Building and Construction Authority (BCA) requirements. Hangzhou Guanli Intelligent

FIGURE 14.53 3D printing process of the children's center. Image © Guanli, Woh Hup.

Technology Co., Ltd. (Guanli) supplied the construction 3D-printing robot (mobile construction printing platform) and on-site printing support. Compared with gantry-based systems, the mobile platform enabled direct in situ deposition at the construction site, reducing logistical constraints and illustrating a viable pathway for field deployment. The collaboration emphasized structural applications, integrating mix design and rheology control with performance and safety verification. This case demonstrates the feasibility of engineering-scale adoption of 3DCP in high-density urban environments and highlights how contractor–university–equipment-supplier collaboration can accelerate the transfer of technology from laboratory to practice. Figure 14.53 illustrates the 3D printing process of the children's center.

(iii) Thailand's First 3D-Printed Medical Center

In 2023, Siam Cement Group (SCG) of Thailand, in collaboration with COBOD, completed the country's first two-story 3D-printed medical center with a construction area of 345 sq. m. The medical center features a unique wavy wall design, achieved through 3D printing technology, offering high design flexibility. The 3D-printed mortar used has strength grades of C75 for load-bearing walls and C30 for non-load-bearing walls, with the building structure capable of withstanding seismic loads. Figure 14.54 shows Thailand's first 3D-printed medical center.

(iv) Other 3D-Printed Houses in China

In addition to the cases mentioned earlier, other companies in China have also successfully completed various 3D-printed construction projects. For example, traditional Chinese architectural styles have been combined with modern 3D printing technology to build Chinese-style villas. Figure 14.55 shows a 3D-printed Chinese-style villa.

Large-scale in situ 3D printers have also been used to directly print two-story residential buildings, demonstrating the potential of 3D printing for mid-scale housing

FIGURE 14.54 Thailand's first 3D-printed medical center. Image © COBOD, SCG.

FIGURE 14.55 3D-printed Chinese villa. Image © WinSun.

projects. Figure 14.56 presents the completed construction of a two-story 3D-printed building. Furthermore, in rural areas of China, where there is a need for renovation and reconstruction of single-story houses, Chinese companies have piloted several rural housing projects using 3D printing, focusing on cost control and construction efficiency to improve living conditions in the countryside. Figures 14.57 and 14.58 show examples of 3D-printed rural houses.

These cases demonstrate the wide application and diverse practices of 3D printing technology in China's construction sector, covering both residential and public facilities, and highlight its potential to drive innovation in the construction industry while meeting various social needs.

FIGURE 14.56 Completed construction of the two-story building. Image © CSCEC.

FIGURE 14.57 3D-printed rural house 1. Image © XWG Studio.

14.4 CASE STUDIES OF 3D-PRINTED CONCRETE BRIDGES

Concrete 3D printing technology is rapidly developing in the field of infrastructure construction, particularly in bridge building. It has demonstrated significant advantages in design innovation, construction efficiency, optimized material usage, cost effectiveness, and environmental sustainability. This technology allows for more flexible and complex bridge designs, enabling the creation of structures that are

FIGURE 14.58 3D-printed rural house 2. Image © Guanli.

FIGURE 14.59 3D-printed concrete bridge in the Netherlands. Image © Witteveen+Bos.

difficult to achieve with traditional methods. Several notable cases around the world have showcased its potential in practical engineering applications.

In 2017, the world's first 3D-printed concrete bridge was built in a small town in the Netherlands. The pedestrian bridge components were entirely 3D-printed with concrete and then assembled on-site with post-tensioning to create a bicycle bridge. The printing quality was excellent, and the design was aesthetically pleasing (Figure 14.59).

In September 2021, the Netherlands completed another 3D-printed concrete pedestrian bridge with a span of 29 m. It is currently the longest 3D-printed concrete bicycle bridge. The project utilized topology optimization and parametric design, and its structural components were manufactured using concrete 3D printing. Compared to traditional methods, this approach saved up to 50% of the materials (Figure 14.60). Figure 14.61 shows Striatus, a 3D-printed concrete unreinforced arch bridge built in Venice in 2021, combining historic masonry compression principles with modern 3D printing technology.

FIGURE 14.60 3D-printed concrete pedestrian bridge. Image © Reprodução/Eindhoven University of Technology.

FIGURE 14.61 3D-printed Striatus bridge. Image © Naaro.

In China, concrete 3D printing technology has also been applied to bridge construction. In 2019, a pedestrian bridge with a total length of 26.3 m and a width of 3.6 m was built in Shanghai. It features a single-arch structure for load bearing and was entirely assembled from components printed by a concrete 3D printing robot (Figure 14.62). In 2021, Hebei University of Technology completed the 3D-printed construction of the assembled Zhaozhou Bridge, later used as a pedestrian bridge over the school's river (Figure 14.63).

3D printing technology is increasingly valued in the field of bridge construction, as it can provide faster construction speeds, lower costs, and greater design flexibility. As the technology continues to develop and mature, more innovative projects of this kind are expected to emerge. A notable example is the 3D-printed bridge in Ningbo, China (August 2024), which distinguishes itself by incorporating reinforcement close to that of traditional reinforced concrete bridges, effectively demonstrating the feasibility of reinforced concrete 3D-printed bridges.

Case: Reinforced Concrete 3D-Printed Pedestrian Bridge Project in Ningbo, China

The 3D-printed bridge is located within the Zhejiang University campus in Ningbo. The bridge features a crescent arch shape, with a total length of 16 m, a width of 3.3 m, and a height of 3.3 m. It spans across a river on the campus. The bridge

FIGURE 14.62 Shanghai concrete 3D-printed footbridge. Image © XWG Studio.

FIGURE 14.63 Hebei concrete 3D-printed Zhaozhou bridge. Image © HEBUT.

design includes an elliptical window and an arched sunshade on the top, presenting a visually appealing appearance. The project timeline began in July 2024, with installation and construction carried out in early August 2024. The total printing time for the components was 15 working days, with the five bridge components taking a total of 120 hours to print. The installation was completed at the end of August 2024, and

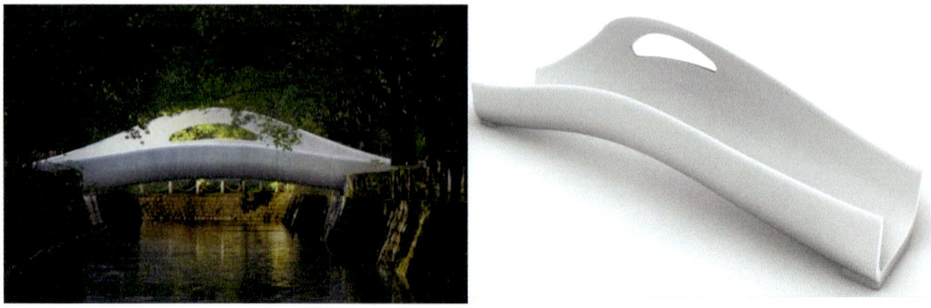

FIGURE 14.64 Design of the 3D-printed pedestrian bridge project in Ningbo. Image © Guanli.

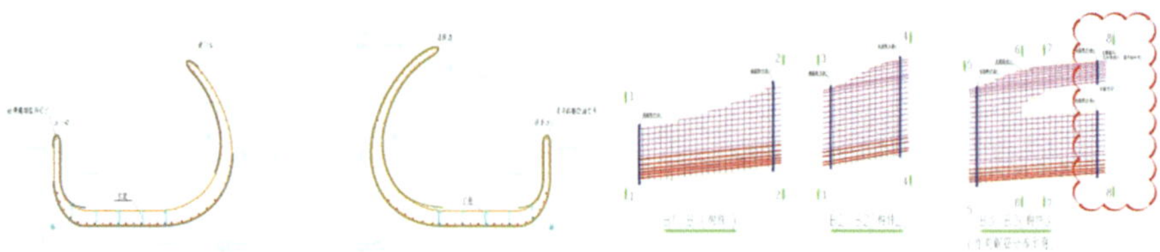

FIGURE 14.65 Structural design of the 3D-printed bridge. Image © Guanli.

the bridge installation and construction took 7 working days. Figure 14.64 shows the design of the 3D-printed pedestrian bridge in Ningbo.

The key technical processes of the project are as follows:

(i) Structural Design
Based on the functional requirements of the bridge and relevant standards, the load-bearing criteria and reinforcement design of the bridge were established. Professional structural analysis software was utilized to conduct load simulation analyses on the bridge structure. The results of these analyses were then assessed to determine the reinforcement requirements. Considering the structural load characteristics and printing requirements of the bridge, the overall bridge model was segmented into several sections, each approximately 3 m in length. A certain width of post-pour strip was reserved for subsequent reinforcement welding and concrete pouring of the components; regarding the bridge body, a post-pour strip was also reserved for its entire length for continuous reinforcement and concrete pouring after the assembly of all components, ensuring the integrity and structural continuity of the bridge. The bridge architectural design drawings were then converted into a 3D geometric model suitable for 3D printing, such as STL or Gcode format printing files. Figures 14.65–14.67 illustrate the structural design, block segmentation, and installation sequence of the 3D-printed bridge.

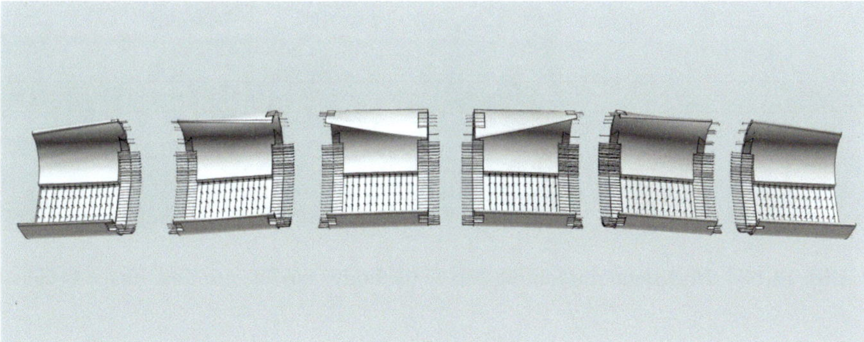

FIGURE 14.66 Structural block of the 3D-printed bridge. Image © Guanli.

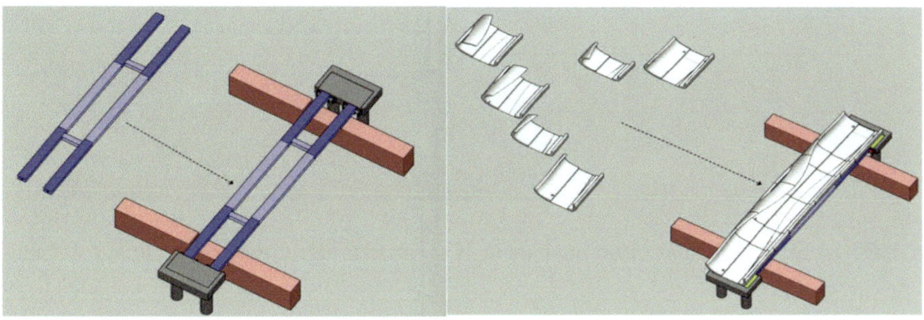

FIGURE 14.67 Schematic of the bridge installation sequence. Image © Guanli.

(ii) Printer and Materials

3D-printed concrete was prepared according to the design mix ratio, and its properties such as fluidity, setting time, and strength development were tested to meet the requirements of the bridge. A concrete printing robot with printing dimensions suitable for the component size and a large gantry-style construction 3D printer were selected. The 3D model was imported into the printer control software for slicing and printing path planning. Based on the component dimensions, printing path, and material properties, the equipment parameters were set, such as printing width, single-layer thickness, material flow rate, and printing speed. Figures 14.68 and 14.69 show the placement of horizontal and vertical reinforcements during the printing of bridge components.

(iii) Placement of Horizontal and Longitudinal Reinforcements in the Bridge Shell

Horizontal reinforcement mesh is laid inside the shell, with prefabricated horizontal reinforcement mesh panels positioned and arranged according to design requirements during the printing process. Longitudinal main bars are inserted inside the shell at the structural design locations. Concrete of the required design strength is poured in layers inside the printed component shell.

FIGURE 14.68 Horizontal reinforcing bars in the bridge module printing. Image © Guanli.

FIGURE 14.69 Vertical reinforcing bars in 3D-printed bridge components. Image © Guanli.

(iv) Bridge Position Installation

Components are transported to the site and lifted into place. A steel frame is erected at the bridge position, and the modules are hoisted onto the frame for assembly. The reinforcement bars between modules are adjusted manually to ensure accurate positioning and welded together. After the modules are assembled, continuous reinforcement bars are tied in the post-pour strip. Concrete is poured and vibrated to ensure density in the bridge post-pour strip. Finally, a protective and decorative coating is sprayed on the 3D-printed bridge to enhance its overall aesthetics and durability. Figures 14.70 and 14.71 illustrate the reinforcement connection design and the completed bridge structure after the post-pour joint was finished.

The concrete 3D-printed pedestrian bridge project is based on a composite construction mechanism of "additive manufacturing and reinforcement and concrete pouring." 3D printing technology is used to print the shell, in which reinforcement is placed and concrete is poured. The bridge structure is then assembled on-site to form an integrated construction system that combines structural design, printing construction, reinforcement placement, and bridge assembly for small- and medium-sized concrete bridges. This approach is also applicable to the 3D printing construction of projects such as garden walkways, municipal landscape bridges, campus footbridges, and pedestrian passages in scenic areas with small and medium spans.

FIGURE 14.70 Reinforcing bar connection design for the assembly of the 3D-printed bridge. Image © Guanli.

FIGURE 14.71 Overall bridge structure after the completion of the post-pour joint. Image © Guanli.

14.5 OTHER APPLICATIONS OF 3D-PRINTED CONCRETE

The core advantages of 3DCP technology stem from the deep integration of digital design and material science, which have opened up entirely new dimensions of freedom and feasibility for complex shapes in spatial design and construction. This innovative process sculpts unique interlayer textures, and its key technical feature – the monolithic forming of irregular curved surfaces – significantly enhances the precision of spatial construction, creating a smooth and natural experience. Relying on digitally controlled shaping printing technology, complex irregular curved surfaces can be seamlessly and integrally formed. This not only greatly improves the accuracy of the shape but also effectively avoids the error problems that are difficult to overcome in traditional processes. It significantly reduces the dependence on manual skills and becomes a perfect example of the integration of technology and art.

In application scenarios such as landscape structures, urban furniture, and outdoor landscape sculptures, these projects are usually small in size, have low load requirements, and have relatively simple functional needs, making 3DCP technology more convenient to apply. Compared with the construction field, landscape projects, with their diverse functional needs and flexible design requirements, may provide a broader space for the application of 3DCP technology. Figures 14.72–14.74 show application cases of concrete 3D-printed flower pots, urban furniture, and landscape sculptures, respectively.

FIGURE 14.72 Application of concrete 3D-printed flower pots. Image © Guanli.

FIGURE 14.73 Concrete 3D-printed urban furniture. Image © Guanli.

FIGURE 14.74 Concrete 3D-printed outdoor sculptures and landscaping. Image © Guanli.

Although the current 3D concrete material technology still faces some technical bottlenecks, such as the need to further improve the uniformity of stress and the limited means of reinforcement, the widely used 3D-printed concrete materials generally have a compressive strength of more than 40 MPa. This key performance indicator is sufficient to meet the structural stability requirements of most landscape facilities and structures.

14.6 FUTURE PROSPECTS FOR CONCRETE 3D PRINTING TECHNOLOGY

In the future, the construction costs of concrete 3D printing are expected to continue to decrease, making the overall on-site printing of buildings and customized houses more economically competitive. Through the joint efforts of research institutions and application enterprises worldwide, the establishment of a standardization system for

3D printing construction will help regulate the design, construction, and acceptance of 3D-printed buildings, promoting the orderly development of future technology. The application of concrete 3D printing technology will primarily focus on several key directions.

Material research and development will place greater emphasis on localized production, making full use of local cement, sand, and gravel to reduce transportation costs and minimize environmental impact. At the same time, efforts will be made to increase the utilization of industrial solid waste in 3D-printed concrete, achieving effective recycling of resources. The promotion of green concrete materials will also be vigorously pursued to reduce carbon emissions and mitigate the environmental burden throughout the entire life cycle of buildings. Establishing a comprehensive material system for 3D printing construction is crucial. This will ensure that the performance and quality of materials meet printing requirements and construction standards, thereby providing solid material support for 3D-printed buildings.

The development of concrete 3D printers will follow two major trends. On the one hand, large gantry-style or frame-type printers with larger construction surfaces will be developed to meet the printing needs of large-scale building components and entire buildings. On the other hand, multi-module collaborative construction and robotically mobile printer forms will be explored to achieve efficient coordination and intelligent control in construction. In the future, 3D printing construction equipment will be integrated with existing construction machinery, adopting modular design to achieve more flexible construction methods. Additionally, the conveying and extrusion processes of printing concrete need to be further optimized to accommodate the requirements of multi-material or dual-component printing.

It is necessary to establish a design and structural system specifically for 3D-printed buildings, fully considering the characteristics and advantages of 3D printing technology, and developing adaptive design methods and structural forms. Leveraging the flexibility of 3D printing technology, building structures should meet the needs of personalized development by designing irregular structures to reduce structural self-weight and optimize load-bearing performance. At the same time, the close integration of structural design and material design should be realized. Structural designers should propose design requirements based on the load-bearing performance of materials, while material engineers should design 3D-printed building materials specifically.

REFERENCES

3DPrinting.com. 2023. "ICON Nears Completion of 100 Home Neighborhood in Texas." Last modified October 12, 2023. https://3dprinting.com/news/icon-nears-completion-of-100-home-neighborhood-in-texas/.

Ackerman, Evan. 2018. "Mobile Robots Cooperate to 3D Print Large Structures." *IEEE Spectrum*, August 28, 2018. https://spectrum.ieee.org/mobile-robots-cooperate-to-3d-print-structures.

Apis Cor. 2019. "Dubai Municipality Office: The World's Largest 3D Printed Building." Accessed June 20, 2025. https://www.apis cor.com.

Apis Cor. 2025. "Home – Apis Cor Construction with Robotic Precision." Accessed June 20, 2025. https://apis-cor.com/.

BUILD UP. 2025. "Digitalisation and Artificial Intelligence: Reshaping the Competitiveness of the Construction Sector." Accessed June 20, 2025. https://build-up.ec.europa.eu/en/resources-and-tools/articles/digitalisation-and-AI-reshaping-competitiveness-construction-sector.

COBOD International. 2024. "The Latest Advancement in 3D Construction Printing Technology." Last modified October 23, 2024. https://cobod.com/.

Construction 3D. 2025. "Our 3D Printers." Accessed June 20, 2025. https://www.constructions-3d.com/en/our-3d-printers/.

CyBe Construction. 2025. "CyBe Robot Track (RT)." Accessed June 20, 2025. https://cybe.eu/3d concrete printing/printers/cybe robot track/.

de Rubeis, Tullio, Annamaria Ciccozzi, Letizia Giusti, and Dario Ambrosini. 2024. "On the Use of 3D Printing to Enhance the Thermal Performance of Building Envelope – A Review." *Journal of Building Engineering* 95: 110284. https://doi.org/10.1016/j.jobe.2024.110284.

De Schutter, G., et al. 2018. "Vision of 3D Printing with Concrete – Technical, Economic and Environmental Potentials." *Cement and Concrete Research* 112: 25–36. https://doi.org/10.1016/j.cemconres.2018.06.001.

Dubai Future Foundation. 2020. "Dubai's 3D Printed Office of the Future Sets New World Record." Last modified February 26, 2020. https://www.dubaifuture.ae/latest-news/dubais-3d-printed-office-future-sets-new-world-record/.

Dubai Municipality. 2025. "3D Printing in the Construction Sector." Accessed June 20, 2025. https://www.dm.gov.ae/municipality-business/3d-printing-in-the-construction-sector/.

Eindhoven University of Technology. 2025. "3D Concrete Printing." Accessed June 20, 2025. https://www.tue.nl/en/research/research-groups/structural-engineering-and-design/3d-concrete-printing/.

ETH Zurich. 2025. "Robotic Fabrication Lab (RFL)." Accessed June 20, 2025. https://systems.arch.ethz.ch/rfl/robotic-fabrication-lab.

European Commission. 2025. "Evo Constructor – Fully Autonomous Robotic and 3D Printing Solution for Large Scale Buildings (Project ID: 868218)." Accessed June 20, 2025. https://cordis.europa.eu/project/id/868218.

Fakharany, N. 2024. "World's First Ever 3D Printed Mosque Opens in Jeddah, Saudi Arabia." Translated by July Shao. *ArchDaily*, April 2, 2024. https://www.archdaily.com/.

Hangzhou Guanli Intelligent Technology Co., Ltd. 2025. "In Situ 3D Construction Printer." Accessed June 20, 2025. https://guanli3d.net/3d-construction printers.

Hassan, Habibelrahman, Edwin Rodriguez Ubinas, Adil Al Tamimi, Esra Trepci, Abraham Mansouri, and Khalfan Almehairbi. 2024. "Towards Innovative and Sustainable Buildings: A Comprehensive Review of 3D Printing in Construction." *Automation in Construction* 163: 105417. https://doi.org/10.1016/j.autcon.2024.105417.

ICON. 2025. "Vulcan Construction System – Technology." Accessed June 20, 2025. https://www.iconbuild.com/technology.

Illumin. 2025. "Contour Crafting: A Future Method of Building." University of Southern California. Accessed June 20, 2025. http://illumin.usc.edu/assets/submissions/755/Contour%20Crafting%20revision.pdf.

Kopitha, K. Rajeev, P. Sanjayan, and Y. Elakneswaran. 2025. "CO_2 Sequestration and Low Carbon Strategies in 3D Printed Concrete." *Journal of Building Engineering* 99: 111653.

Li, Y., H. Wu, X. Xie, L. Zhang, P. F. Yuan, and Y. M. Xie. 2024. "FloatArch: A Cable--Supported, Unreinforced, and Re-Assemblable 3D---Printed Concrete Structure Designed Using Multi-Material Topology Optimization-." *Additive Manufacturing* 81: 104012. https://doi.org/10.1016/j.addma.2024.104012.

Lin, Xiqiang, Hailong Wang, Xiaoyan Sun, Wael Kikhia, Zhennan Wu, Lu Lan, and Jiawei Zhao. 2024. "Construction Technology of the Curved 'Mars House' Digital Hotel Project in China Using Concrete 3D Printing." In *Construction 3D Printing (3DcP 2023)*, 241–250. Cham: Springer. https://doi.org/10.1007/978-3-031-64269-2_30.

Lu, W., et al. 2024. "Digital Technologies for Construction Sustainability: Status Quo, Challenges, and Future Prospects." *NPJ Materials Sustainability* 2: 10. https://doi.org/10.1038/s44296-024-00010-2.

Ma, Guo Wei, Li Wang, and Yang Ju. 2018. "State-of-the-Art of 3D Printing Technology of Cementitious Material – An Emerging Technique for Construction." *Science China Technological Sciences* 61, no. 4: 475–495. https://doi.org/10.1007/s11431-016-9077-7.

Mechtcherine, Viktor, Markus Taubert, Steffen Müller, Marko Butler, Frank Will, Florian Storch, Paul Plaschnick, Jens Otto, and Patrick Maiwald. 2024. "Integration of Steel Reinforcement into Extrusion Based 3D Concrete Printing of Monolithic Concrete Elements." In *Proceedings of the Fourth RILEM International Conference on Concrete and Digital Fabrication (DC 2024)*, 311–318. Cham: Springer. https://doi.org/10.1007/978-3-031-68691-7_35.

Mighty Buildings. 2025. "Mighty Kit System™ – 3D Printed Wall Systems." Accessed June 20, 2025. https://www.mightybuildings.com/.

NAIOP. 2023. "Construction Cost Challenges Shift from Materials to Labor." *Development Magazine*, Fall 2023. https://www.naiop.org/research-and-publications/magazine/2023/fall-2023/business-trends/construction-cost-challenges-shift-from-materials-to-labor/.

Nasdaq. 2024. "Rising Construction Costs and Labor Shortages Challenge US Housing Rebound." Last modified December 15, 2024. https://www.nasdaq.com/articles/rising-construction-costs-and-labor-shortages-challenge-us-housing-rebound.

PERI USA. 2025. "Nation's Largest 3D Printed Home, Houston, TX." Accessed June 20, 2025. https://www.peri-usa.com/projects/houston-3d.html.

Puzatova, Anastasia, Pshtiwan Shakor, Vittoria Laghi, and Maria Dmitrieva. 2022. "Large-Scale 3D Printing for Construction Application by Means of Robotic Arm and Gantry 3D Printer: A Review." *Buildings* 12, no. 11: 2023. https://doi.org/10.3390/buildings12112023.

Raza, Muhammad Huzaifa, Svetlana Besklubova, and Ray Y. Zhong. 2024. "Economic Analysis of Offsite and Onsite 3D Construction Printing Techniques for Low-Rise Buildings: A Comparative Value Stream Assessment." *Additive Manufacturing* 89 (June): 104292. https://doi.org/10.1016/j.addma.2024.104292.

Sovetova, Mariia, and Jillian Kaiser Calautit. 2024. "Thermal and Energy Efficiency in 3D--Printed Buildings: Review of Geometric Design, Materials and Printing Processes-." *Energy and Buildings* 323: 114731.

SQ4D. 2025. "3D Printed Houses – Changing the Way the World Is Built™." Accessed June 20, 2025. https://www.sq4d.com/.

Twente Additive Manufacturing. 2025. "Product Range 2021." Accessed June 20, 2025. https://www.twente am.com/app/uploads/2021/05/200825_TAM-Product-Range-Document-2021.pdf.

Vertico. 2025. "3D Concrete Printer – High Precision Robotic Arm." Accessed June 20, 2025. https://www.vertico.xyz/3d-concrete-printer.

WASP. 2025. "BigDelta WASP 12MT: The Largest Delta 3D Printer in the World." Accessed June 20, 2025. https://www.3dwasp.com/en/giant-3d-printer-bigdelta-wasp-12mt/.

Xiao, Jiabei, et al. 2021. "Large--Scale 3D Printing Concrete Technology: Current Status and Future Opportunities-." *Cement and Concrete Composites* 122: 104115.

XtreeE. 2025. "Robots." Accessed June 20, 2025. https://www.xtreee.com/printers/robots.

Zhang, Yuying, et al. 2025. "3D Printing Technology in Concrete Construction." *Nature Reviews Clean Technology* 1, no. 4. https://doi.org/10.1038/s44359-025-00047-z.

Zuo, Zibo, Wouter De Corte, Yulin Huang, Xiaoming Chen, Yamei Zhang, Jin Li, Longlong Zhang, Jianzhuang Xiao, Yong Yuan, Ketao Zhang, Lulu Zhang, and Viktor Mechtcherine. 2024. "Propelling the Widespread Adoption of Large-Scale 3D Printing." *Nature Reviews Materials* 9, no. 11: 754–756. https://doi.org/10.1038/s41578-023-00626-1.

15 Design and Additive Manufacturing of a Martian Habitat Using Geopolymer Concrete with In Situ Martian Material

José P. Duarte, Sven G. Bilén, Shadi Nazarian, Ali M. Memari, Aleksandra Radlińska, Nicholas A. Meisel, and Nathan D. Watson

15.1 INTRODUCTION: BACKGROUND AND DRIVING FORCES

Human exploration and the innate drive for discovery have fueled advancements in technology and innovation. As humanity sets its sights on increasingly ambitious goals of interplanetary exploration, technological progress must keep pace to ensure the safety of explorers and the feasibility of these missions. The drive to colonize Mars stems from a combination of motivations: the pursuit of scientific discovery, the development of cutting-edge technologies, the exploration of untapped resources, and the potential to foster new industries such as space tourism. At its core, this ambition is also fueled by a profound desire to secure humanity's long-term survival beyond Earth. Mars presents a compelling case – it shares geological features with our planet, shows evidence of ancient oceans and possibly life, and orbits the Sun at a distance that could support habitable conditions. The vision of sending humans to Mars – as outlined by NASA, SpaceX, and other entities – has sparked significant public interest and enthusiasm. NASA's strategy to employ lunar missions as a preparatory step for crewed Mars missions, alongside several ambitious proposals for manned Mars expeditions, has been working to make this once-distant aspiration achievable (Dunbar, 2019; Musk, 2024). As plans take shape, however, questions arise regarding whether a mission to Mars is worth the associated risks and expense. Some critics argue that the potential costs and hazards are unjustifiable, given the unprecedented nature of such an endeavor (Szocik, 2019). Nevertheless, many contend that the profound scientific, educational, and societal benefits of a human mission to Mars outweigh these concerns and provide sufficient justification (Ehlmann

DOI: 10.1201/9781003434863-19

et al., 2005). Progressing toward this goal requires significant advancements, new technologies, and comprehensive research.

Ensuring human safety in extraterrestrial environments presents unique challenges that vary based on mission parameters. From Yuri Gagarin's historic flight into low Earth orbit, to multi-day lunar missions, to long-term habitation on the International Space Station, each milestone has required substantial advances to protect human life (Ocampo and Klaus, 2013). Missions to the Moon and Mars, however, represent the most prolonged and distant expeditions attempted thus far, necessitating preparation for extended periods of stay without the option of a quick return and the challenges associated with delayed communications due to vast distances in space. These factors demand thorough planning, including autonomous precursor missions to construct and test habitats before the arrival of a human crew. Such measures reduce risks by ensuring that astronauts are met with safe, reliable living conditions, thus safeguarding their lives and the success of the mission (Bodiford et al., 2005).

The chapter details the Penn State team's participation in NASA's 3D-Printed Habitat Challenge, a Centennial Challenge designed to advance autonomous construction technology for Martian habitats. The challenge unfolded across three phases: (i) the Design Competition, focused on conceptual 3D-printed habitats (completed in 2015); (ii) the Structural Member Competition, centered on developing materials and technologies for structural components (2017); and (3) the On-Site Habitat Competition, requiring the autonomous construction of a one-third-scale habitat (2019). Phases 2 and 3 involved multiple physical construction tasks, with Phase 3 divided into virtual construction and physical construction levels. In the virtual construction level, teams developed full-scale habitat designs and demonstrated construction processes, while the physical construction level required autonomous printing of foundations, complex geometries, overhangs, embedded objects, and large structures under Mars-like conditions. This chapter presents the Penn State team's sustainable approach to Martian habitats, integrating digital design with autonomous additive construction (AAC) and utilizing geopolymer concrete in a parametric building information modeling (BIM) platform to adapt designs for both Mars and Earth environments. The chapter also describes the results from the physical construction level.

15.2 MARTIAN CONDITIONS

The Martian surface presents numerous hazards that pose significant risks to human life. These challenges stem primarily from Mars' smaller size relative to Earth, resulting in a gravitational force that is approximately one-third that of Earth's. Due to this reduced gravity and the lack of a magnetic field, the atmosphere is only about 1% of that on Earth. The sparse atmosphere and lack of a magnetic field allow solar radiation to penetrate the Martian atmosphere, resulting in surface radiation levels that are significantly higher than those experienced on Earth (Saganti et al., 2004). Additionally, Martian soil is toxic due to its high concentration of perchlorates, necessitating further protective measures for human safety (Dartnell et al., 2007). As a result, a sealed and pressurized habitat with sufficiently thick walls for radiation shielding is essential for survival.

These conditions raise critical questions: "What requirements must these habitats meet?" and "What materials will they be constructed from?" A key challenge is the logistical complexity and high cost of transporting materials to Mars, alongside the need to fully understand available resources on the Martian surface. Even with conservative estimates, the expense of transporting construction materials is prohibitively high (Bodiford et al., 2005). Leveraging Martian resources for habitat construction could drastically reduce the required transport payload. Additionally, to ensure human safety, habitats must be constructed autonomously. Exposure to Mars' extreme conditions during resource gathering and construction could pose serious risks to human life. Therefore, deploying autonomous construction methods is essential to establish safe, functional habitats ahead of human arrival.

15.3 DESIGN REQUIREMENTS

Bodiford et al. (2005) outlined essential requirements for Martian habitats, which include supporting a pressurized, shirt-sleeve environment for the crew and ensuring protection from worst-case radiation exposure, such as galactic cosmic radiation and solar particle events. The habitats must also shield occupants from micrometeorites and exhaust plumes, be capable of semi-autonomous fabrication before the crew's arrival to offer immediate safety, and adhere to early, achievable, and visible milestones for mission success. Additionally, the development of these habitats should be evolutionary and scalable to adapt to future needs.

15.4 STRATEGIES FOR MARS HABITATS

Numerous strategies have been proposed to support human life on Mars, each offering distinct approaches. Broadly, these strategies can be categorized into two main groups: those employing in situ resource utilization (ISRU) and those that do not. An example of a non-ISRU method is presented by Cadogan et al. (1999) that proposes transporting the entire habitat and inflating it on-site after landing. This method relies on an inflatable, lightweight, thin-walled habitat that is compact for transport. The design uses a multilayered membrane capable of maintaining the necessary internal pressure with a safety factor of four, addressing key concerns related to cosmic radiation shielding. A safety factor of four means that the membrane is designed to withstand four times the expected operational pressure. For instance, if the expected internal pressure is 1 unit, the membrane is built to safely handle 4 units. This provides a margin of safety to account for unexpected stresses, material imperfections, degradation over time, or other uncertainties. Polyethylene is utilized as the primary material for radiation protection, as demonstrated by Nagamatsu et al. (2015), who showed that it reduces radiation transmission. However, the exact thickness needed to ensure safety on Mars remains uncertain. While the thin membrane may mitigate radiation exposure, it is unlikely to reduce it to levels considered safe for long-term human habitation, nor would it provide sufficient protection against meteorite impacts.

Non-ISRU approaches offer the advantage of quicker assembly due to the absence of complex material processing, potentially allowing a crewed mission to arrive with a prefabricated habitat to deploy for their immediate use. However, constructing a habitat that meets safety requirements for extended stays, including adequate radiation shielding and micrometeorite protection, would entail significant expense and potentially exceed practical payload limits. One major challenge for habitats transported from Earth without ISRU is the substantial weight and cost associated with providing sufficient radiation shielding and impact protection. Achieving the necessary safety standards would require heavier, more expensive launch vehicles. Additionally, scalability is a significant constraint, making it difficult to create habitats that meet ergonomic and psychological requirements for extended human presence, and future need for additional units would require greater number of deployments.

In contrast, ISRU offers greater flexibility by utilizing the resources available on Mars, allowing for the construction of larger and more robust structures than what could be transported from Earth. This approach can significantly reduce the payload required for transport and enable more sustainable and scalable habitat construction. Although initial construction and furnishing of ISRU-based habitats may still require some Earth-sourced resources, future advancements, such as in-orbit manufacturing facilities, could overcome these limitations and enhance self-sufficiency.

The concept of terraforming Mars, aimed at making the planet habitable, addresses the significant challenges posed by its harsh environment, particularly solar wind and radiation that have stripped Mars of its atmosphere. In the 1980s, research into low-pressure bioregenerative systems indicated that hypobaric plant growth modules could effectively function at around 10 kPa (0.1 bar), where photosynthesis remains viable (Richards et al., 2006). Thus, achieving this minimum surface pressure would be essential for supporting plant life on Mars.

In the 1980s, NASA's "Case for Mars" report proposed strategies for recycling and repurposing spacecraft from previous missions and using Martian regolith to shield habitats against cosmic rays (Richards et al., 2006). A 1985 study outlined necessary infrastructural elements for a Martian base, which included water wells, segmented vehicles, greenhouses, and robotic equipment. The design suggested that many components, such as Quonset huts, would be transported from Earth due to their lightweight and easy assembly, originally developed for military use (NASA, 2007). In 1989, NASA introduced a concept for a self-sustaining Martian outpost capable of supporting up to seven astronauts. This outpost was envisioned to evolve through phases, starting with the Emplacement Phase, which involved establishing essential modules like living spaces, a rover dock, and an unpressurized storage area. The Consolidation Phase would follow, expected to take about 8 years for full operational capability (NASA, 1989). The 2004 Mars Homestead Project at MIT proposed advanced homes constructed using native materials and mission remnants, incorporating features such as greenhouses and reinforced airtight structures to prevent pressure leaks (Than, 2005). This project highlighted the need for building materials that could be locally produced, such as bricks formed from Martian regolith. The Mars Foundation expanded on these ideas in 2005, presenting designs that included skylight-friendly domes and vaulted roofs made from Martian bricks. They suggested that Mars could eventually produce essential materials like plastics and metals, aiming for a self-sustaining settlement (Digital City Briefs, 2012).

In contrast, in 2013 the German firm ZA Architects proposed an underground habitat design featuring expansive caves carved into Martian basalt. This approach aimed to provide natural protection from the surface's harsh conditions while incorporating an automated system for oxygen and water supply (ZA Architects, 2013). Spanish architect Alberto Villanueva's 2015 proposal suggested using mycelium alongside Martian materials to create resilient structures. This innovative approach would also incorporate bioluminescent bacteria as a natural light source (Lynch, 2016). The Mars Case, developed by OPEN Architecture in collaboration with Xiaomi, proposed a minimalist habitat that integrates advanced mechanical engineering and electronic systems to create a closed-loop ecosystem for energy, water, and air recycling (OPEN Architecture, 2018). The "Martian Seed of Life" concept by Malaysian architects Warith Zaki and Amir Amzar explored using bamboo as a primary building material. Bamboo's resilience and rapid growth potential in a CO_2-rich atmosphere make it a promising resource for sustainable human habitation on Mars (Harrouk, 2019). The Nüwa project, designed by ABIBOO Studio in collaboration with SONet, envisioned a scalable habitat on Martian cliffsides, accommodating up to one million residents. This design includes automated, CO_2-enriched farming systems and dedicated shelter zones for emergencies (ABIBOO Studio, 2024).

These studies collectively underscore the need for constructing Martian habitats using 3D-printed concrete mixes derived from local materials. To facilitate this, it is crucial that autonomous robotic systems be deployed to build fully enclosed structures before human arrival, ensuring that these habitats are self-sustaining and capable of withstanding the extreme conditions on Mars.

15.5 AUTONOMOUS ADDITIVE CONSTRUCTION

Utilizing in situ resources for construction can be achieved through automated extraction and processing of regolith, followed by automated building techniques using large-scale additive manufacturing (AM), which can also be referred to as AAC. This approach deposits or bonds material layer by layer to create a 3D structure without the need for human intervention (Camacho et al., 2018). Current technologies include large-scale versions of powder bed fusion, extrusion methods, and binder jetting. Large-format robotic AM, such as 3D concrete printing (3DCP), is particularly appealing for terrestrial construction due to its design flexibility and autonomous operation in hazardous environments. Moreover, the process can be adapted to interplanetary construction, as Martian regolith provides the necessary constituents for producing workable binders (Labonnote et al., 2016).

Cesaretti et al. (2014) proposed a hybrid strategy that involves the use of lunar regolith and an inflatable interior shell. This method envisions a pressurized structure brought from Earth in a compact form, which is inflated and then covered with lunar regolith bonded using the 3D-printing technology developed by D-Shape. This technology selectively jets droplets onto the surface of each powder layer, reinforcing the regolith that covers the inflatable structure. The additional bonded layers provide protection against radiation, micro-meteorite impacts, and extreme thermal conditions. Although importing materials from Earth remains necessary, minimizing consumable imports is crucial (Bodiford et al., 2005).

Other proposals, such as those by Kading and Straub (2015), involve bringing reusable large-scale construction equipment from the Earth for modular construction. Their plan uses a fused deposition modeling (FDM) method, in which Martian basalt is melted and extruded layer by layer to build the habitat. Initially, triangular sections are printed within the spacecraft to form a large dome, after which a gantry system is assembled inside the dome to construct habitat modules. While this approach offers reusability, it demands substantial energy for basalt melting and uses single-purpose equipment, limiting scalability.

Khoshnevis et al. (2005) suggested a gantry-based approach for lunar implementation. This system utilizes in situ resources to produce a cement-like material from processed lunar regolith, with water sourced through chemical extraction from the Moon or based on recent discoveries, possibly from polar ice deposits. The gantry system is track-mounted, facilitating automated construction of large-scale structures. Although material processing consumes significant energy, the construction itself requires relatively little energy compared to thermal extrusion, which demands substantial energy storage.

Buchner et al. (2018) explored a different binding method using phosphoric acid. Although this binder would need to be transported from Earth, minimal regolith processing is needed. This method also requires small amounts of water. While promising, further research is necessary to optimize the printing process and fully understand how the material would behave in extraterrestrial environments.

As with most early-stage research aimed at ambitious goals such as autonomous habitat construction on other planets, numerous challenges and uncertainties remain. This chapter seeks to present a strategy that advances beyond current technology readiness levels (TRLs) from approximately TRL 1–3 to concepts nearing TRL 5 (NASA, 2012).

15.6 PRINTING SYSTEM CONCEPTS

Deploying a construction system on the surface of Mars or the Moon involves transporting the system to the celestial body, setting it up, and operating it autonomously to produce construction material and, subsequently, build the habitat. This requires a compact, lightweight system capable of self-unpacking and self-assembly. Large-scale printing systems typically fall into several categories: gantry systems, cable-suspended systems, swarm robotics, and multipurpose robotic-arm solutions (Labonnote et al., 2016).

Gantry systems are essentially scaled-up versions of standard desktop plastic extrusion printers. Their straightforward design enables simpler control systems, but they lack flexibility. Additionally, the print volume is constrained by the physical dimensions of the gantry system itself. This implies that constructing large structures requires a system proportionate in size to the habitats being printed, which must be transported to Mars. The size of the gantry system and the limitations of rocket cargo capacity necessitate assembly on-site, presenting further challenges. On Earth, assembling such a system typically requires several people and large cranes over several days.

Cable-suspended systems share similar challenges with gantry-based systems. While they do not require structural elements that span the entire length and width of the build area, they do need tall anchor points for cable connections. These anchor points would require assembly and secure placement (Chessera et al., 2018). Furthermore, the capabilities of cable-suspended systems are limited, making them suitable primarily for habitat construction and of limited use once assembly is complete.

Swarm robotic solutions involve multiple mobile robots capable of printing simultaneously by maintaining awareness of their positions relative to each other and the structure being built (Oxman et al., 2014). This approach has significant advantages for off-world construction due to the limited cargo space and weight constraints for transportation. Swarm robots can vary in size and work within a theoretically unlimited build volume. Depending on their size, these robots may need to climb previously printed sections to continue construction at greater heights (Urhal et al., 2019). This method offers scalability, as the number of robots can determine the overall speed of construction. Given the extended window for construction on Mars due to its orbital cycle relative to Earth, a smaller number of robots could be used effectively. However, this approach also comes with limitations in terms of the features it can build, and transporting construction materials to the printing site poses an additional challenge not addressed here.

Multipurpose robotic arms provide a versatile approach to AM. These arms are compact when folded, eliminating the need for on-site assembly and simplifying transport and autonomous setup. Robotic arms can support various tool attachments for different construction tasks and post-construction operations needed by astronauts. Their reusability allows for the consolidation of necessary hardware, ultimately reducing the launch weight (Bodiford et al., 2005).

The adaptability of robotic arms was the reason the Penn State team selected this approach for their proposed system. This included a six-axis robotic arm paired with an integrated mixer–pump system that mixed a limited amount of water required for the material and pump it to the robot's end-effector. The material was supplied from a silo containing a processed dry Martian-based composition. The extreme temperature fluctuations and low atmospheric pressure on Mars make it challenging to use liquid water, as water sublimates directly from ice to vapor below the triple point. To address this, the printing process would occur within a pressurized, insulated environment maintained above the triple point and below the boiling point of water (Haberle et al., 2001). This pressurized environment would be created by an unfolding dome, which would also protect the curing concrete (Shoukry et al., 2011). Figure 15.1 illustrates the printing system along with the unfolding dome. Figure 15.2 demonstrates how a habitat with multiple units could be constructed by relocating the printing system to different site locations to print each unit individually.

15.7 MATERIALS

Smaller robots would complement the primary six-axis robotic arm, collecting and processing raw materials needed to produce the dry mix for 3D printing. The Penn State team developed a material called MarsCrete™, designed using analogs to Martian resources for both the aggregate and binder components. This material comprises fine basalt as the aggregate and metakaolin as the primary binder,

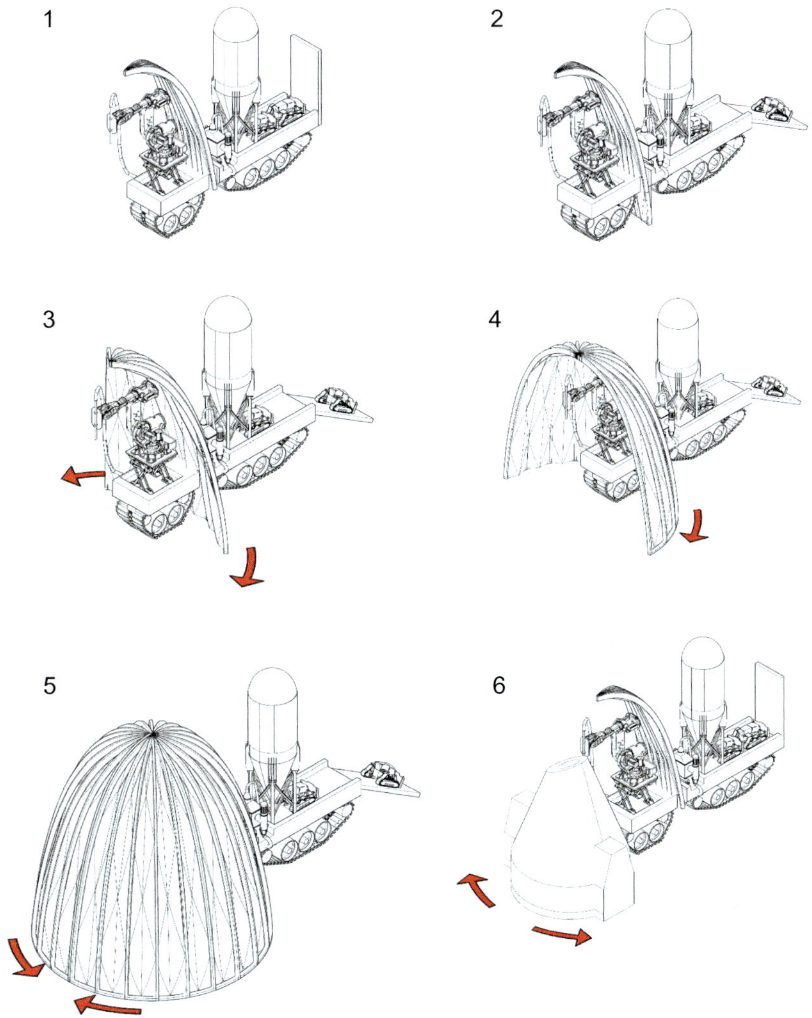

FIGURE 15.1 Printing system with unfolding dome for pressurization and insulation.

combined with an activating solution primarily consisting of water with Na_2O and SiO_2. Notably, over 98% of MarsCrete's components can be sourced directly from Mars (Hojati et al., 2018).

Developing a material that can be produced on Mars greatly reduces launch weight and enhances the potential scale and scalability of Martian habitats. This stands in contrast to options that require transporting entire habitats from Earth, which limits construction size to what can be launched and assembled, offering no scalability. Although no method can rely solely on 100% Martian materials, the choice of material impacts how much needs to be imported from Earth and the equipment required

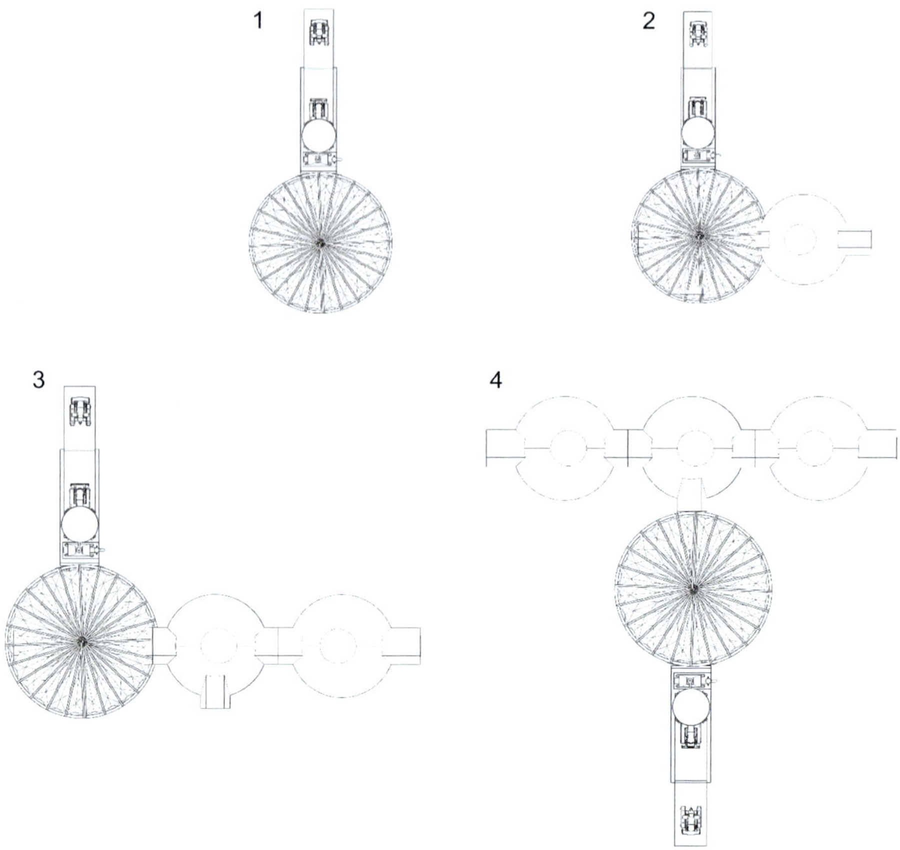

FIGURE 15.2 Top view of the printing sequence of a habitat with multiple units.

for processing. For example, while the D-Shape method requires minimal equipment to process Martian regolith, it still depends on binders brought from Earth (Kading and Straub, 2015), which significantly limits scalability. Investing in equipment to produce MarsCrete allows for a greater volume of construction, reducing design constraints.

Another key factor when selecting a material is its resistance to cosmic rays and solar radiation. Certain polymers, such as polyethylene, are known to be effective radiation barriers due to their hydrogen content, which can absorb the energy from cosmic ray particles because of their comparable size (Nagamatsu et al., 2015). Concrete can also block cosmic rays if used in sufficient thickness, especially against alpha particles and protons, although it is less effective at blocking muons (Zigler and Watson, 1981). Given Mars' weak magnetic field, alpha particles and protons from cosmic rays are the dominant sources of the radiation reaching the surface, suggesting that concrete could be a viable radiation shield when thick enough (Molina-Cuberos et al., 2001). Additionally, the radiation-shielding properties of concrete could potentially be enhanced with specific additives (Kanagaraj et al., 2023). Thus, using locally sourced concrete may prove more practical than relying on polyethylene.

These trade-offs are also evident when considering other approaches, such as sintering regolith or using a phosphoric acid binder. Sintering, while requiring minimal material processing, demands significant energy and supporting infrastructure (Watson et al., 2019). Further research is necessary to identify the specific equipment needs for each method.

15.8 DESIGN CONCEPT

The shelter design was informed by material behavior analysis and the specifications of the printing system. A Mars habitat must effectively protect inhabitants from extreme environmental conditions, offer a comfortable living space for extended periods, and be autonomously constructed. Once the system design and construction material are established, the habitat can be tailored to meet the design constraints posed by these elements. To maximize habitat size relative to the printing system's capabilities, a modular approach was adopted. Each module is sized based on the printing system's limitations, but multiple modules can be assembled to form a scalable habitat that expands as the Martian settlement grows, as shown in Figure 15.3. Interconnecting these modules reduces the need for external exits and entries, thereby limiting exposure to the harsh Martian conditions, including its significant radiation levels.

For a robotic-arm printing with concrete, the maximum module size is constrained by the reach of the arm. Testing revealed that, given the printing technology available, concrete structures required a minimum overhang angle between 60 and 70° from the horizontal plane to remain stable under the gravity conditions of Earth. Although this constraint could be relaxed under the lower gravity found on Mars, it was retained for the competition, which involved constructing a full-scale replica of the habitat on Earth. To avoid the need for removable supports – which would complicate the construction process – the habitat was designed so that all walls adhere to this minimum overhang angle. This constraint led to a modular structure composed of conical segments. The height of the cone is determined by the cone's diameter and the arm's height limitations, resulting in a truncated cone design.

As with any architectural problem, the design solution needed to address a variety of contextual factors, including environmental conditions, programmatic

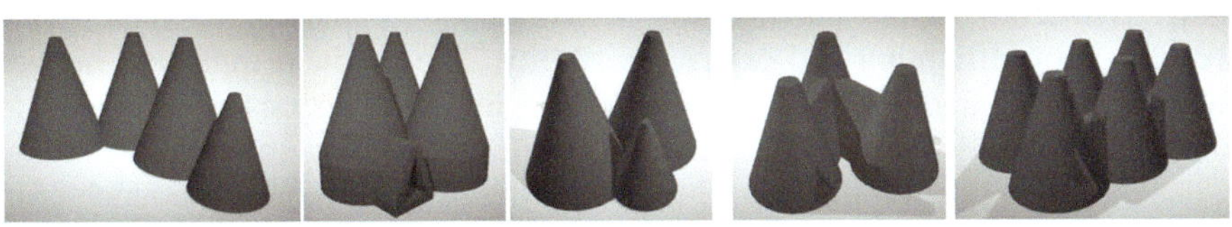

FIGURE 15.3 Different design configurations achieved by combining modular units.

requirements, and the construction methods and materials available. The primary objective was to create an environment that supports both the physical and psychological well-being of the astronauts. Designing a habitat for Mars, however, presents an extreme case of this problem, demanding innovative solutions to meet the unique and harsh conditions of the Martian landscape.

15.9 DESIGN BRIEF AND PROBLEM FORMULATION

A key initial design decision was selecting the location for the habitat to optimize for both safety and functionality. The aim was to minimize the adverse effects of Mars' harsh environmental conditions – cosmic radiation, extreme temperature fluctuations, low gravity, and a thin atmosphere – while maximizing factors such as ease of landing, availability of construction materials, and scientific interest in the surrounding area. The chosen site was a flat region near the equator, adjacent to the Endurance Crater, known for its scientific value based on previous Mars missions. The equatorial location offers temperatures ranging from −55 to +21°C throughout the year, with significant daily variations. Due to Mars' thin atmosphere, a pressurized interior was essential for human habitation.

Programmatic requirements were dictated by the NASA 3D-Printed Habitat Challenge, which specified a habitat for four astronauts with a minimum habitable area of 93 m. This space needed to encompass work, eating, sleeping, sanitation, and leisure areas. The design also had to include mechanical, electrical, and plumbing (MEP) systems and an environmental control and life support system (ECLSS). Furthermore, it required two ports – one for astronaut entry/exit and another for rovers – and a viewport for overseeing external operations. The project required the use of BIM, with the MEP system and ECLSS modeled at a schematic level of development (LOD 100) and the envelope and partitions at LOD 300, suitable for virtual construction phases.

The capacity of the newly developed MarsCrete material to shield against cosmic radiation set the minimum wall thickness at 60 cm. The homogeneity and opacity of concrete presented limitations in terms of external views and thermal performance. To address these constraints, functionally graded concrete was proposed, allowing for strategic adjustments in material properties. By partially replacing sand aggregates with cork and borate glass granules, the design leveraged the unique benefits of these materials. Cork was chosen for its acoustic and thermal insulation properties, compressive strength, and availability due to its use as a spacecraft ablative heat shield (required for landing on Mars). Its inclusion increased toward the exterior of the walls, protecting against low temperatures and freeze–thaw cycles while lightening the structure. Borate glass, known for its radiation shielding, was incorporated to enable natural lighting and external visibility. The gradient transition between materials avoided traditional joint issues, simplifying construction and enhancing the habitat's airtightness, thus reducing the risk of life-threatening air leaks.

The printing system introduced three major design constraints. The reach of the robotic arm defined the maximum printable dimensions. Autonomy necessitated the printing of unsupported structures (i.e., without formwork or temporary supports), eliminating the need for post-construction removal of formwork. Preliminary tests under Earth gravity indicated that the maximum print angle was 60° from the horizontal plane for a concrete-like material.

To optimize the habitat's size relative to the printing system, a modular design was proposed. Each module was dimensioned within the limits of the printing system, allowing scalable construction as the settlement expanded. Interconnecting modules created a continuous living space, reducing the need for frequent exterior exposure and thus limiting radiation risks. The module shape was influenced by structural and environmental constraints. Unlike on Earth, the low Martian gravity and significant pressure difference between the interior and exterior placed greater emphasis on the lateral forces over vertical loads, favoring rounded structures. A fully spherical form was impractical due to print angle constraints, so a cylindrical module with a conical top was selected. The module's dimensions, finalized to ensure stability and functionality, included two floors with a smooth transition between shapes to prevent abrupt directional changes.

Figure 15.4 illustrates different configurations achieved by combining these modular units. To support future scalability, three types of connections were designed: intersecting modules, tangentially adjacent modules, and modules linked by a connecting element.

15.10 PARAMETRIC PROBLEM FORMULATION

The shelter's design emerged from a complex response to environmental and programmatic constraints, alongside limitations dictated by selected materials and the printing system. To explore a wide range of design solutions suitable for both Mars and Earth, these constraints were integrated into a parametric design model in Revit,

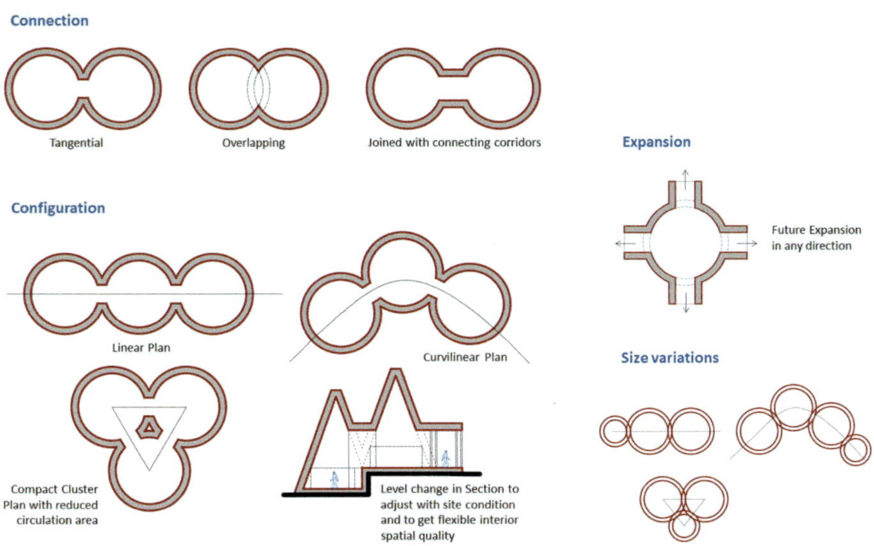

FIGURE 15.4 Design concept featuring adaptable configurations.

enhanced with Dynamo scripting. This BIM-based platform comprised five primary components: formulation, generation, analysis, optimization, and construction simulation.

The formulation component focused on encoding the design concept parametrically – cylindrical and conical units that could be interconnected in various configurations. Key design variables, including wall thickness, taper angle, module radius and height, and smoothing factor, were identified and encoded using Dynamo for Revit. The generation component facilitated the automated creation of parametric configurations by assigning values to these design variables. This enabled rapid iterations and modifications during the design process. To ensure comprehensive evaluation, the generation component was connected to multiple performance analysis tools through custom application programming interfaces (APIs). These tools assessed various aspects of the design, including structural integrity, environmental performance, constructability, and metrics related to area, volume, and cost. The resulting data from geometric modeling and performance analyses were stored in a cloud-based relational database, allowing seamless integration with the analysis and optimization process. The optimization component was designed to run concurrently with performance analysis, using an interface into which users could input specific design objectives and constraints. This facilitated the identification of optimal variable combinations that met project requirements. The optimizer also enabled real-time updates to building geometry based on the results of performance evaluations, streamlining the iterative design process.

An additional component of the framework was the construction simulation component, which supported constructability analysis, including robotic toolpath simulation, logistical planning, setup configuration, and sequencing. These simulations were critical to evaluating construction feasibility and were incorporated into the broader optimization workflow. To prepare for actual construction, a pre-construction simulation of the printing setup and sequence was developed within this BIM platform, ensuring practical implementation strategies were accounted for. Further details about the BIM platform's structure and operation can be found in the work by Muthumanickam et al. (2021).

Using this BIM-based multi-objective optimization framework, two distinct design solutions were proposed – one tailored for virtual construction and another prepared for physical construction in the context of the NASA 3D-Printed Habitat Challenge, as discussed in the following sections.

15.11 SELECTED DESIGN: VIRTUAL CONSTRUCTION LEVEL

At the virtual construction level, we selected a single design from multiple options to best illustrate our concept. Each habitat unit featured a protective MarsCrete shell and a lighter interior of MarsCrete and recycled plastics, arranged across two floors. The selective use of functionally graded materials optimizes the performance of the habitat from the structural, optical, and thermal viewpoints, as illustrated in Figure 15.5. A gradual MarsCrete-to-glass transition at the top provides ambient light through skylights, while sharper transitions create windows with Martian landscape views

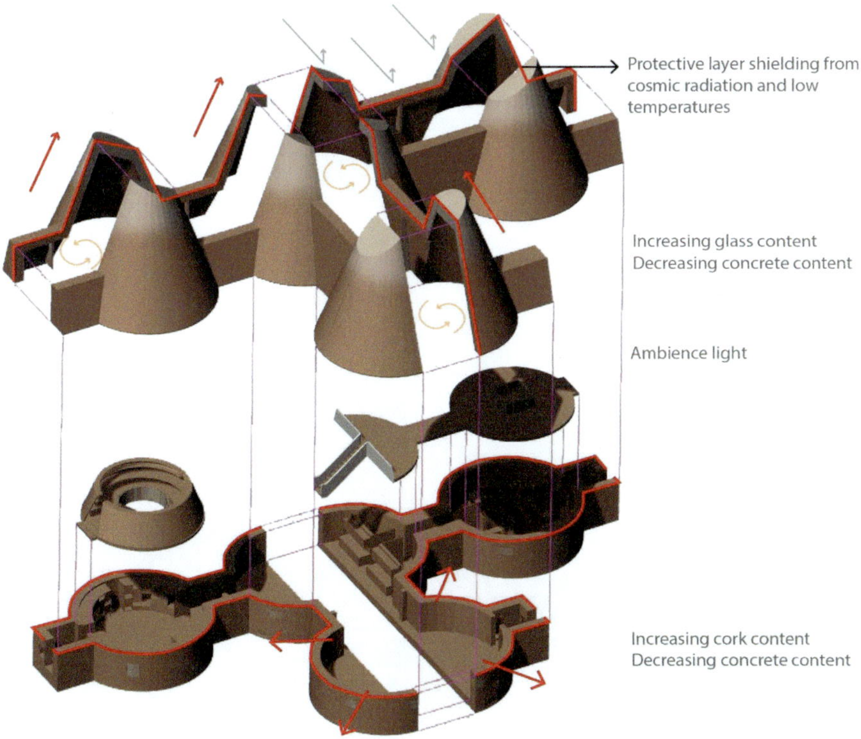

FIGURE 15.5 Protective shell incorporating functionally graded materials for optimized structural, optical, and thermal performance.

(Nazarian et al., 2022). A cork gradient toward the exterior surface forms a protective layer against cosmic radiation and low temperatures (Craveiro et al., 2017).

The habitat was composed of four primary units designated for work, living, sleeping, and gardening. Figure 15.6 displays the plans, sections, and elevations of the habitat, while Figure 15.7 presents an exterior rendering of the proposed modular habitat. Entry chambers at each end of the habitat facilitated both exterior access and emergency exits. The first unit housed a wet lab, meeting area, and storage on the ground floor, with individual workspaces located on the second floor, connected by a central staircase. The second unit included the dining area, kitchen, and bathrooms on the first floor, with a leisure area occupying the second floor, accessible via a separate staircase. The third unit provided sleeping quarters on the first floor, with workout and leisure areas on the second. This arrangement allowed for visual connectivity between the different units and floors, creating a spatially dynamic and engaging environment. The fourth unit housed life support and hydroponic systems, equipped with provisions for emergency situations. Each unit could be individually sealed and supported by its life support system during repairs. The hydroponic

FIGURE 15.6 Plans of the proposed habitat: lower floor (top) and upper floor (bottom). From left to right: work, living, sleeping, and garden units.

system in the fourth unit supported food production, oxygen generation, and wastewater treatment. Figure 15.8 provides interior views, highlighting the design's goal of achieving spatial complexity and openness. By incorporating generous areas with views of the Martian landscape, the design aims to foster privacy and reduce feelings of claustrophobia.

Efficient system distribution was achieved through floor ducts that supplied clean, pressurized air, water, and electricity. Gray water was directed back through sanitary piping for treatment via the hydroponic and mechanical filtration systems. A centralized wet wall distributed clean and wastewater piping to kitchen and bathroom

FIGURE 15.7 Exterior renderings of the proposed modular habitat.

fixtures, with additional piping extending to the work unit, which was equipped with fume hoods to safely vent noxious fumes from chemical experiments. Solar films and nuclear thermoelectric generators provided electricity, while two uninterrupted power supply units ensured consistent power availability during periods of low energy production.

15.12 SELECTED DESIGN: PHYSICAL CONSTRUCTION LEVEL

For the physical construction level of the NASA Challenge, the habitat's envelope was to be printed at one-third-scale. Given the significant differences in environmental

FIGURE 15.8 Interior renderings. From left to right and from top to bottom: lab space, office space, kitchen, living, bathroom, corridor, bedroom, workout area, and hydroponic garden.

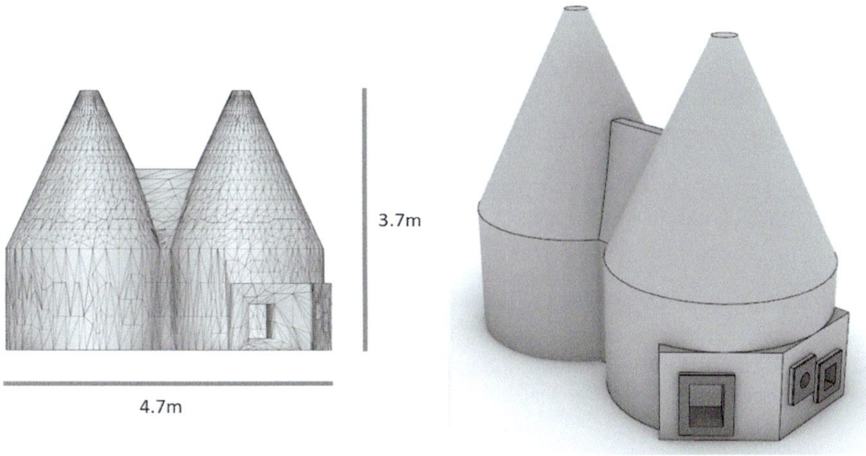

FIGURE 15.9 Design adapted to be printed on Earth.

conditions between Mars and Earth, as well as the need to complete the printing within 3 days, with 10 hours of printing per day allowed during the on-site head-to-head competition, a modified solution was developed using the BIM platform. Figure 15.9 shows the adapted design.

In this solution, the floor area was slightly above the minimum required, resulting in a more compact design that excluded the hydroponic garden. The design was compressed into just two modular units to fit within the allowable printing area of 4.5 × 4.5 m, while also ensuring that the printed structure remained within the reach of the robotic arm. Additionally, the entire setup, including all printing equipment, had to fit within a 10 × 10-m area.

The printing process involved the use of two robotic arms: one for printing the habitat's structure and another to insert the view, suit, and rover hatch-frames for the three ports during printing. The positioning of both robotic arms had an impact on the design, as it was necessary to carefully position the viewports to avoid collisions between the robots during their operations.

15.13 DEMONSTRATION

Phase 3, Level 3 of the NASA 3D-Printed Habitat Challenge involved printing a one-third-scale Martian habitat, which included a rover hatch, a suit hatch, and a circular window hatch near the suit hatch. The full-scale habitat required a minimum floor area of 93 m^2, with a ceiling height of at least 2.25 m. The design was based on the same parametric approach used in the virtual design phase, but instead of printing each cone modularly, the structure was printed using a static robot, positioned just outside where two cones would meet. The limited reach of the robot also influenced the diameter and height ratio needed to meet the required internal area.

The material used for this construction was a portland cement–based mix, chosen as a functional simulant for MarsCrete due to manufacturing challenges to produce

FIGURE 15.10 BIM-based virtual model of the construction setup and the actual setup at the competition venue during the finals.

significant quantities of the MarsCrete material. Tests were conducted to determine the minimum printable overhang angle, resulting in a cone with a 65° angle from horizontal for the cone sections, and straight walls supporting angles of up to 70°. The cone's angle and internal diameter determined its height, while cylindrical vertical walls were added beneath the cones to increase the usable internal space. The resulting internal volume was calculated to meet the minimum required area, ensuring that the habitat's floor space was at least 93 m.

A critical feature of the printed habitat was the inclusion of penetrations for the rover hatch, suit hatch, and circular window hatch. These had to be autonomously integrated into the structure during the printing process. A second robotic arm was employed to place wooden frame inserts into the printed voids for these features. The frames were placed flush with the top of the concrete voids, allowing the printing system to print over them smoothly. The printing setup used for the physical construction level is shown in Figure 15.10. It included a delivery truck, trailer, silo for feed stock material storage, water tank, mixer, pumps, two ABB IRB 6640 robots, computer stations, safety barricades, window fixtures, scissor jack for build space monitoring, and camera for visual feedback. More information about the printing setup can be found in Muthumanickam et al. (2021).

The final structure (Figure 15.11), standing 3.7 m tall with a maximum width of 4.7 m, represented the first fully enclosed, concrete-printed structure at the architectural scale, without the need for support. The final structure demonstrated several important features: maximizing the printable volume, the ability to embed components during the print, and the autonomous handling of support structures.

Several issues emerged while printing the habitat. Uneven ground caused over-extrusion and gravel displacement on the shelter floor, highlighting the need for improved leveling or autonomous tactics for adjusting the position of the print nozzle. The nozzle support arm experienced entanglement, causing the hose to twist and pull back, resulting in damage to a nozzle screw and a misalignment in the structure. The robot shook when using all five servos simultaneously, leading to zig-zag patterns around window frames, which affected fit and required recalibration through remote intervention, which meant that the operator was not allowed to directly view

FIGURE 15.11 One-third-scale habitat printed in the physical construction level.

the operation. Window frames presented sizing challenges; additional layers were needed to match the height of the first frame, while others did not align due to vibrations and excessive material deformation. Frequent power and conveyor belt malfunctions disrupted material flow, with one outage reversing the belt's motion, causing water-only extrusion. Mixer sensors also failed to detect low material levels, requiring manual intervention to reset sensors and clear drying concrete. Finally, software license issues caused difficulties in G-code generation, temporarily halting printing.

The print would have taken approximately 12.5 hours to complete nonstop, but time constraints and troubleshooting extended the process. Nevertheless, the resulting structure met all essential requirements, showing the potential for autonomous, scalable habitat construction for Mars and beyond.

15.14 CONCLUSION

As humanity prepares to explore new planets, ensuring astronauts' safety from space hazards is paramount. Mars presents a hostile environment: its thin, oxygen-poor atmosphere and lack of a magnetic field make its surface lifeless and exposed to radiation. By leveraging in situ resource utilization, scalable and modular habitats can be constructed on Mars using local materials. This minimizes the launch mass, which in turn lowers mission costs and maximizes the materials available for creating a safe and comfortable living environment.

AAC methods further enhance mission viability, allowing habitats to be built before astronauts arrive, which reduces risk. For the NASA 3D-Printed Habitat Challenge, a one-third-scale habitat was autonomously printed to demonstrate these principles. Using a six-axis robotic arm, the project showed that a multipurpose robot could serve various functions on Mars, reducing the hardware required for construction. A modular design also allows for a compact printing system, with the flexibility to scale up as needed to create larger habitats. By incorporating in situ resource utilization (ISRU), the habitat structure can effectively shield inhabitants from cosmic rays and solar radiation, providing a secure base for human exploration.

The development of modular, 3D-printed habitats for Mars represents an essential step in enabling sustained human exploration of extraterrestrial environments. By leveraging ISRU, we can minimize reliance on Earth-supplied materials, making Martian habitats more sustainable, cost-effective, and adaptable. This chapter has detailed our approach to designing and testing a scalable habitat system for Mars, combining autonomous construction with a parametric design model to adapt to diverse conditions on both Mars and Earth.

Our approach illustrates that modular habitat designs, when optimized with simulation and performance analysis, can provide safe and comfortable environments for astronauts. In the context of NASA's 3D-Printed Habitat Challenge Competition, we demonstrated the feasibility of constructing robust, layered shells from MarsCrete, an ISRU-derived material, potentially capable of shielding inhabitants from radiation and the hostile Martian environment. The dual configuration of virtual and physical designs exemplified the adaptability of our system to varying gravity conditions and construction constraints, underscoring the versatility of autonomous, robotic construction for future Mars missions.

Our findings also suggest that a modular, robotics-driven approach to habitat construction has broader implications for planetary exploration. By physically demonstrating that autonomous systems can execute complex building tasks with minimal human intervention, this work supports the vision of pre-deployed construction missions that establish foundational infrastructure in advance of crewed landings. Additionally, the high level of control and customization afforded by a parametric BIM platform ensures that these habitats can evolve to meet new requirements as mission objectives expand.

Future research will focus on advancing materials and robotic systems to improve scalability and autonomy, enabling increasingly complex architectural solutions that address the unique challenges of off-world construction. As humanity looks beyond Earth for exploration and long-term habitation, the methods and technologies outlined here bring us closer to building resilient, sustainable habitats that will support a human presence on Mars and beyond. To achieve this goal, it is necessary to make an interdisciplinary effort, integrating architecture, engineering, and science, and it requires global collaboration.

Several research institutions and space agencies worldwide are conducting Earth-based simulations – such as Biosphere 2 in the United States and Lunar Palace 1 in China – to study closed-loop life support systems and human resilience in isolated environments. These missions help test environmental controls, food production, and crew dynamics under conditions mimicking Mars. Technologies like ISRU, which harness Martian soil, ice, and solar energy, are being refined to support sustainable living. NASA's NextSTEP program is pioneering inflatable and composite habitat designs for Mars missions, while international partnerships explore modular, multipurpose structures. Architectural firms have also contributed visionary concepts, including Foster + Partners' 3D-printed regolith dwellings and ABIBOO Studio's Nüwa cliffside city, emphasizing automation, modularity, and the use of native materials to enable self-sufficient and adaptable living on the Red Planet.

15.15 ACKNOWLEDGMENTS

This research was supported by prize money from NASA's 3D-Printed Habitat Centennial Challenge and by grants from the College of Arts and Architecture and the College of Engineering and developed at the Stuckeman Center for Design

Computing (SCDC) and at the Additive Construction Laboratory, the Pennsylvania State University. The authors acknowledge the valuable contribution to the research and express their gratitude to the following members of the Penn State team: Negar Ashrafi, Randall Bock, Eduardo Castro e Costa, Flávio Craveiro, Samuel Dzwill, Maryam Hojati, Zhanzhao Li, Naveen Muthumanickam, Keunhyang Park, Andrew Przyjemski, Aiden Smith, and Joe Straka.

15.16 CONFLICTS OF INTEREST

Three of the co-authors, José P. Duarte, Sven G. Bilén, Shadi Nazarian, and The Pennsylvania State University own equity in X-Hab 3D, whose activity is related to concrete printing. Their ownership in this company has been reviewed by their respective University's Individual and Institutional Conflict of Interest Committees and is currently being managed by the Universities.

REFERENCES

ABIBOO Studio. 2024. "Nüwa, the First Sustainable City on Mars." ABIBOO Studio Website. Accessed November 11, 2025. https://abiboo.com/projects/nuwa/.

Bodiford, Melanie, Michael R. Fiske, Walter McGregor, and Regina Pope. 2005. "In-Situ Resource-Based Lunar and Martian Habitat Structures Development at NASA/MSFC." *1st Space Exploration Conference: Continuing the Voyage of Discovery* 2: 974–980. https://doi.org/10.2514/6.2005-2704.

Buchner, Christoph, Roland H. Pawelke, Thomas Schlauf, Alexander Reissner, and Advenit Makaya. 2018. "A New Planetary Structure Fabrication Process Using Phosphoric Acid." *Acta Astronautica* 143 (February): 272–284. https://doi.org/10.1016/j.actaastro.2017.11.045.

Cadogan, D., J. Stein, and M. Grahne. 1999. "Inflatable Composite Habitat Structures for Lunar and Mars Exploration." *Acta Astronautica* 44 (7–12): 399–406. https://doi.org/10.1016/S0094-5765(99)00103-4.

Camacho, Daniel D., Patricia Clayton, William J. O'Brien, Carolyn Seepersad, Maria Juenger, Raissa Ferron, and Salvatore Salamone. 2018. "Applications of Additive Manufacturing in the Construction Industry – A Forward-Looking Review." *Automation in Construction* 89 (May): 110–119. https://doi.org/10.1016/j.autcon.2017.12.031.

Cesaretti, Giovanni, Enrico Dini, Xavier De Kestelier, Valentina Colla, and Laurent Pambaguian. 2014. "Building Components for an Outpost on the Lunar Soil by Means of a Novel 3D Printing Technology." *Acta Astronautica* 93 (January): 430–450. https://doi.org/10.1016/j.actaastro.2013.07.034.

Chessera, Phillip C., Brian K. Posta, Alex Roschlia, Randall F. Linda, Alex M. Boulgera, Lonnie J. Lovea, and Katherine T. Gaula. 2018. "Fieldable Platform for Large-Scale Deposition of Concrete Structures." *29th Annual International Solid Freeform Fabrication Symposium*: 2020–2032.

Craveiro, Flávio, Helena M. Bártolo, A. Gale, Jose P. Duarte, and Paulo J. Bártolo. 2017. "A Design Tool for Resource-Efficient Fabrication of 3D-Graded Structural Building Components Using Additive Manufacturing." *Automation in Construction* 82 (October): 75–83. https://doi.org/10.1016/j.autcon.2017.05.006.

Dartnell, L.R., L. Desorgher, J.M. Ward, and A.J. Coates. 2007. "Modelling the Surface and Subsurface Martian Radiation Environment: Implications for Astrobiology." *Geophysical Research Letters* 34 (2): 4–9. https://doi.org/10.1029/2006GL027494.

Digital City Briefs. 2012. "Future Habitats: MIT's Mars Homestead Project." *Digital City Briefs: Future Habitats,* June 24. http://future-habitats.blogspot.com/2012/06/mits-mars-homestead-project.html.

Dunbar, Brian. 2019. "Apollo's Legacy Is NASA's Future." NASA. https://www.nasa.gov/specials/apollo50th/back.html.

Ehlmann, Bethany L., Jeeshan Chowdhury, Timothy C. Marzullo, R. Eric Collins, Julie Litzenberger, Stuart Ibsen, Wendy R. Krauser, Brandon DeKock, Michael Hannon, Jessica Kinnevan, Rebekah Shepard, and F. Douglas Grant. 2005. "Humans to Mars: A Feasibility and Cost-Benefit Analysis." *Acta Astronautica* 56 (9–12): 851–858. https://doi.org/10.1016/j.actaastro.2005.01.010.

Haberle, Robert M., Christopher P. McKay, James Schaeffer, Nathalie A. Cabrol, Edmon A. Grin, Aaron P. Zent, and Richard Quinn. 2001. "On the Possibility of Liquid Water on Present-Day Mars." *Journal of Geophysical Research* 106 (E10): 317–326.

Harrouk, Christele. 2019. "Designers Imagine Bamboo Colony on Mars." *Martian Seed of Life. ArchDaily,* August 30. https://www.archdaily.com/924023/designers-imagine-bamboo-colony-on-mars.

Hojati, Maryam, Shadi Nazarian, Jose P. Duarte, Aleksandra Radlińska, Negar Ashrafi, and Flávio Craveiro. 2018. "3D Printing of Concrete: A Continuous Exploration of Mix Design and Printing Process." *42nd IAHS World Congress on the Housing for the Dignity of Mankind* (April): 1–14.

Kading, Benjamin, and Jeremy Straub. 2015. "Utilizing In-Situ Resources and 3D Printing Structures for a Manned Mars Mission." *Acta Astronautica* 107 (February–March): 317–326. https://doi.org/10.1016/j.actaastro.2014.11.036.

Kanagaraj, Balamurali, N. Anand, A. Diana Andrushia, M.Z. Nase. 2023. "Recent Developments of Radiation Shielding Concrete in Nuclear and Radioactive Waste Storage Facilities – A State-of-the-Art Review." *Construction and Building Materials* 404 (November): 133260. https://doi.org/10.1016/j.conbuildmat.2023.133260.

Khoshnevis, Behrokh, Melanie Bodiford, Kevin Burks, Ed Ethridge, Dennis Tucker, Won Kim, Houssam Toutanji, and Michael R. Fiske. 2005. "Lunar Contour Crafting – A Novel Technique for ISRU-Based Habitat Development." *43rd AIAA Aerospace Sciences Meeting and Exhibit, Session: MSSP-3: In-Situ Resource Utilization and Space Manufacturing* (January): 7397–7409. https://doi.org/10.2514/6.2005-538.

Labonnote, Nathalie, Anders Rønnquist, Bendik Manum, and Petra Rüther. 2016. "Additive Construction: State-of-the-Art, Challenges and Opportunities." *Automation in Construction* 72, Part 3 (December): 347–366. https://doi.org/10.1016/j.autcon.2016.08.026.

Lynch, Patrick. 2016. "Esta propuesta de colonización en Marte imprimiría en 3D torres de hongos biodegradables que no dejan rastro." Translated by Isadora Stockins. *ArchDaily Colombia*, March 9. https://www.archdaily.co/co/783717/esta-propuesta-de-colonizacion-en-marte-imprimiria-en-3d-torres-de-hongos-biodegradables-que-no-dejan-rastro.

Molina-Cuberos, G.J., W. Stumptner, H. Lammer, N.I. Kömle, and K. O'Brien. 2001. "Cosmic Ray and UV Radiation Models on the Ancient Martian Surface." *Icarus* 154 (1): 216–222. https://doi.org/10.1006/icar.2001.6658.

Musk, Elon. 2024. "Mars & Beyond: The Road to Making Humanity Multiplanetary." *SpaceX*. Accessed September 16, 2025. https://www.spacex.com/mars.

Muthumanickam, Naveen K., Jose P. Duarte, Shadi Nazarian, Ali Memari, and Sven G. Bilén. 2021. "Combining AI and BIM in the Design and Construction of a Mars Habitat." In *Routledge Companion for AI in Architecture*, edited by Imdat As and Prithwish Basu. Abingdon, UK: Routledge. https://doi.org/10.4324/9780367824259-17.

Nagamatsu, Aiko, Marco Casolino, Oscar Larsson, Tsuyoshi Ito, Nakahiro Yasuda, Keiichi Kitajo, Ken Shimada, Kazuo Takeda, Shuichi Tsuda, and Tatsuhiko Sato. 2015. "Space Radiation Dosimetry to Evaluate the Effect of Polyethylene Shielding in the Russian Segment of the International Space Station." *Physics Procedia* 80: 25–35. https://doi.org/10.1016/j.phpro.2015.11.083.

NASA. 1989. "Outpost Consolidation Illustration." FY-89 Office of Exploration Annual Report, NASA Archive, Photo-S89_51054. https://web.archive.org/web/20000304055928/http://spaceflight.nasa.gov/gallery/images/mars/marsbases/html/s89_51054.html.

NASA. 2007. "Artist's Concept of Possible Exploration Programs." NASA Archive, Photo-s85_31484, August 2. https://web.archive.org/web/20070802140112/http://spaceflight.nasa.gov/gallery/images/exploration/marsexploration/html/s85_31484.html.

NASA. 2012. "Technology Readiness Level." *NASA*. https://www.nasa.gov/directorates/heo/scan/engineering/technology/txt_accordion1.html.

Nazarian, Shadi, Carlo Pantano, Paolo Colombo, and Mauro Marangoni. 2022. "Seamless and Impermeable Joints Resulting in a Functionally Graded Material: Transitioning from Glass to Geopolymer Mortar." U.S. Patent No. 11414345B2, August 16.

Ocampo, Robert P., and David M. Klaus. 2013. "A Review of Spacecraft Safety: From Vostok to the International Space Station." *New Space* 1 (2): 73–80. https://doi.org/10.1089/space.2013.0015.

OPEN Architecture. 2018. "MARS Case / OPEN Architecture." *ArchDaily*, October 9. https://www.archdaily.com/903434/mars-case-open-architecture.

Oxman, Neri, Jorge Duro-Royo, Steven Keating, Ben Peters, and Elizabeth Tsai. 2014. "Towards Robotic Swarm Printing." *Architectural Design* 84 (3): 108–115.

Richards, Jeffrey T., Kenneth A. Corey, Anna-Lisa Paul, Robert J. Ferl, Raymond M. Wheeler, and Andrew C. Schuerger. 2006. "Exposure of *Arabidopsis thaliana* to Hypobaric Environments: Implications for Low-Pressure Bioregenerative Life Support Systems for Human Exploration Missions and Terraforming on Mars." *Astrobiology* 6 (6): 851–866. https://doi.org/10.1089/ast.2006.6.851.

Saganti, Premkumar B., Francis. A. Cucinotta, John W. Wilson, Lisa C. Simonsen, and Cary Zeitlin. 2004. "Radiation Climate Map for Analyzing Risks to Astronauts on the Mars Surface from Galactic Cosmic Rays." *Space Science Reviews* 110: 143–156. https://doi.org/10.1023/B:SPAC.0000021010.20082.1a.

Shoukry, Samir N., Gergis W. William, Brian Downie, and Mourad Y. Riad. 2011. "Effect of Moisture and Temperature on the Mechanical Properties of Concrete." *Construction and Building Materials* 25 (2): 688–696. https://doi.org/10.1016/j.conbuildmat.2010.07.020.

Szocik, Konrad. 2019. "Should and Could Humans Go to Mars? Yes, but Not Now and Not in the Near Future." *Futures* 105: 54–66. https://doi.org/10.1016/j.futures.2018.08.004.

Than, Ker. 2005. "The Homestead Project: Making a Mars Settlement a Reality." *Space.com*, August 10. Accessed September 16, 2025. https://www.space.com/1419-homestead-project-making-mars-settlement-reality.html.

Urhal, Pinar, Andrew Weightman, Carl Diver, and Paulo Bartolo. 2019. "Robot Assisted Additive Manufacturing: A Review." *Robotics and Computer-Integrated Manufacturing* 59 (October): 335–345. https://doi.org/10.1016/j.rcim.2019.05.005.

Watson, Nate D., Nicholas A. Meisel, Sven G. Bilén, Jose P. Duarte, Shadi Nazarian. 2019. "Large-Scale Additive Manufacturing of Concrete Using a 6-Axis Robotic Arm for Autonomous Habitat Construction." *30th Annual International Solid Freeform Fabrication Symposium – An Additive Manufacturing Conference*: 1583–1595.

ZA Architects. 2013. "Mars Colonization." *ZAarchitects.com*. https://zaarchitects.com/en/other/103-mars-colonization.html.

Ziegler, J.F., and T.J. Watson. 1981. "The Effect of Concrete Shielding on Cosmic Ray Induced Soft Fails in Electronic Systems." *IEEE Transactions on Electron Devices* 28 (5): 560–565. https://doi.org/10.1109/T-ED.1981.20383.

16 On the Development of Planetary Applications for Cementitious Materials Using Additive Construction

Michael R. Fiske

16.1 INTRODUCTION

As the reader understands by now, additive construction is a process of joining layered building materials to create building entities/structural elements from 3D model data. It is closely linked to the type of material that is being printed. The method used to create the building is partly dependent on the material used, with concrete, metals, or polymers (plastics) being the most common.

As man moves beyond Earth, it is imperative to develop the required infrastructure in advance to support this migration. As distances increase, it becomes increasingly obvious that, like our ancestors in the great migration to the US West in the 1800's, we must learn to "live off the land." Rockets are an inherently expensive method for the transportation of raw materials, requiring the development of in situ resource utilization (ISRU)-based materials, technologies, and specialized equipment to support them. It is understood that deployment of these infrastructure elements will be time-phased in terms of their level of ISRU-based materials, as shown in Figure 16.1 (Kennedy 2002).

To this end, a number of ISRU-based materials and processes have been developed in support of infrastructure development on the Lunar surface. These include thermal processes such as sintering or melting of lunar regolith using lasers, solar energy, or microwaves. This chapter will focus on the development of cementitious materials for these applications. Perhaps unsurprisingly, many of these lunar "concretes" are based on concretes commonly used on Earth.

16.2 ADDITIVE CONSTRUCTION DEVELOPMENT HISTORY

Charles Hull is the inventor of stereolithography (SLA), the first commercial rapid prototyping technology commonly known as 3D printing (NIHF 2005). Invented in

 DOI: 10.1201/9781003434863-20

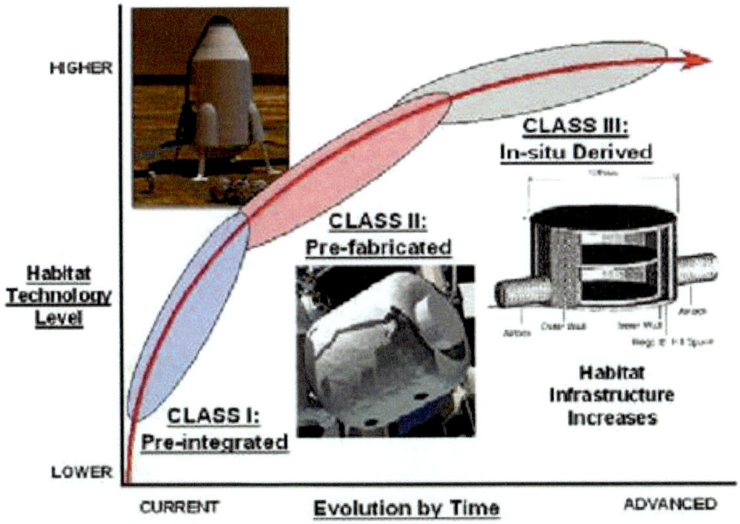

FIGURE 16.1 Habitat development vernacular. Image Credit: Kriss Kennedy.

1986, the earliest applications were in research and development for labs and tool rooms, but 3D printing applications and technologies have been identified almost continuously since then, including printing of various polymers, metals, ceramics, and, most recently, construction materials for both terrestrial and planetary applications.

Unbeknownst to many, the world's first 3D-printed building (shown in Figure 16.2) was fabricated behind a small warehouse in Valparaiso, Indiana, in 1939, by inventor William E. Urschel (Curth 2022). His machine, shown in Figures 16.3 and 16.4, would be used to fabricate multistory structures with integrated reinforcement printed with concrete and no formwork. While lacking the automation element of today's efforts, this work predated the seminal paper by Pegna (1997) and was 60 years before the first modern examples of construction 3D printing.

If Urschel was the "grandfather" of additive construction, Dr. Behrokh Khoshnevis of the University of Southern California (USC) and now also Contour Crafting Corporation (CCC) was the "father." He defined an automated concrete construction process as early as 1998 (Khoshnevis 1998), and patents on the first large-scale 3D printing of concrete were published by Khoshnevis in the early 2000s (Khoshnevis 2004; Khoshnevis and Bekey 2002). In these works, Khoshnevis defined Contour Crafting (CC) as "an additive fabrication technology that uses computer control to exploit the superior surface-forming capability of troweling to create smooth and accurate planar and free-form surface." The idea of CC was to use two trowels to form a solid planar surface for external edges. Filler material such as concrete can then be poured to fill the extruded area. This automated, large-scale additive construction technique could use a number of materials including clay, gypsum, ceramic, and other cementitious materials. Fixed or movable trowels on the sides of the extrusion

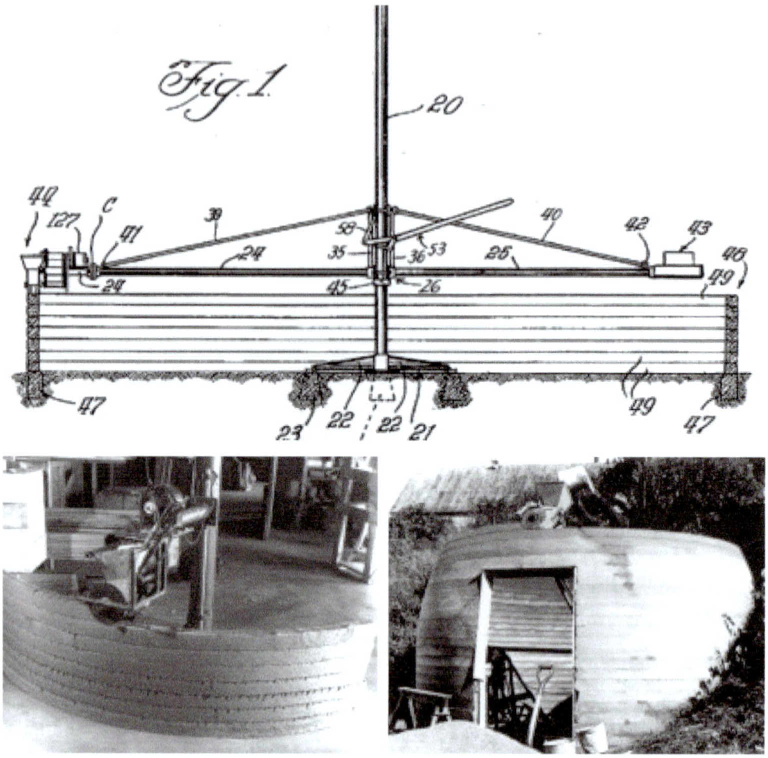

FIGURE 16.2 Urschel's wall-building machine in 1939. Image Credit: Urschel History – Urschel Laboratories.

FIGURE 16.3 Molding elements added to the extruder, creating interlocking channels between layers. Image Credit: Urschel History – Urschel Laboratories.

FIGURE 16.4 Urschel completing a dome structure with minimal overhang support. Image Credit: Urschel History – Urschel Laboratories.

nozzle helped control the angle of bead sides and provide layer-to-layer smoothing, as shown in Figure 16.5.

This nozzle was capable of full six-axis positioning and could extrude both sides and filler material. The CC nozzle could also be used for forming paint-ready surface, placing reinforcement before pouring concrete, plastering and tiling, plumbing, and installing electrical modules and communication line wiring. Khoshnevis also developed an optimized method for CC machines to efficiently construct complicated large-scale structures. Extensive research was carried out to avoid collisions between multiple nozzles. They concluded that using the CC method is significantly faster than traditional construction methods.

However, the presence of the side trowels drove the need for a rotatable nozzle, somewhat constraining the ability to make turns of 90° or larger. At some point, researchers began extruding through straight vertical tubes with no trowels, resulting in the stacked layer look prevalent in most of today's additive construction efforts, as shown in Figure 16.6, and in a variety of wall design patterns, as shown in Figure 16.7.

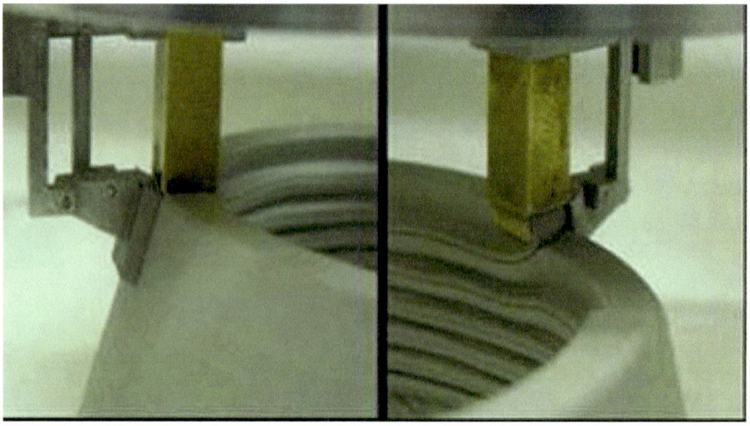

FIGURE 16.5 Smoothed layers resulting from CC. Image Credit: B. Khoshnevis.

FIGURE 16.6 Rounded beads resulting from vertical extrusion with no trowels. Image Credit: ICON.

Khoshnevis also presented design concepts for various mobility systems, including open and closed gantries and robot-arm designs, and for the integration of mechanical, electrical, and plumbing (MEP) concepts into 3D-printed walls, as shown in Figure 16.8. This early work inspired many of the efforts that followed, especially as his early patents expired. As interest in additive construction grew through the early 2000s, entities around the world began to understand the implications of these technologies and develop their own printers, print materials, and mobility system configurations. Figure 16.9 shows the exponential growth of additive construction activities between 1996 and 2018.

In 2012, Khoshnevis, Carlson, and Thangavelu were awarded a Phase II NIAC (Khoshnevis et al. 2020). Between 2012 and 2017, they significantly advanced the state-of-the-art in printing of lunar- and Martian regolith-based sulfur concrete, design of habitats and landing pads, and the incorporation of reinforcement and MEP

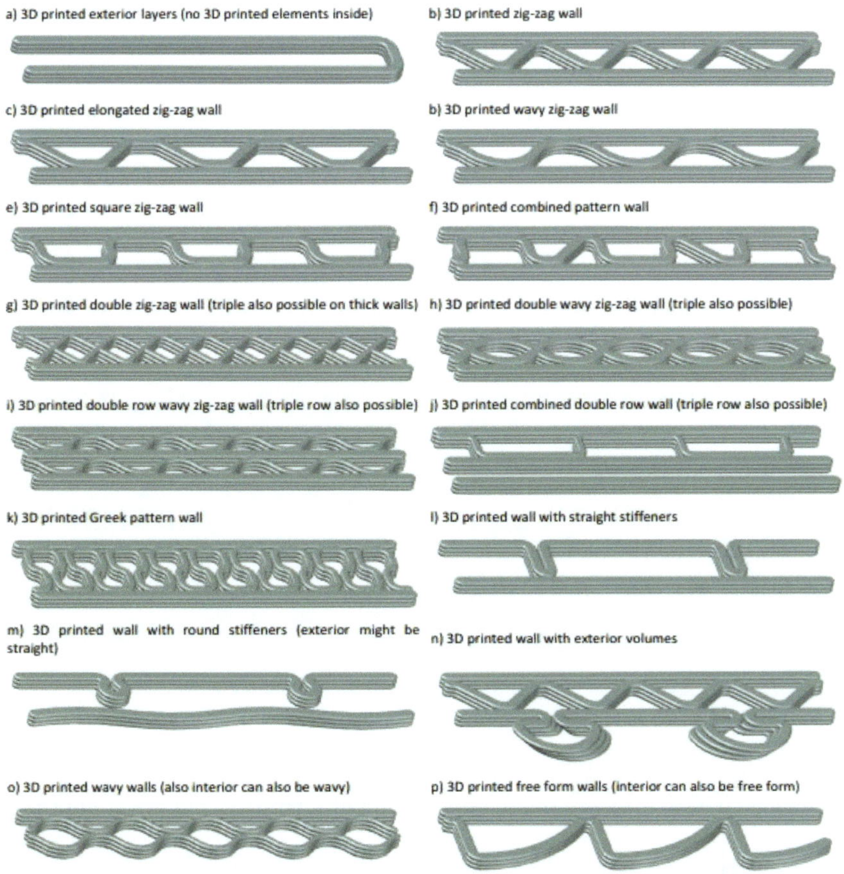

FIGURE 16.7 Wall design pattern for 3D-printed concrete structure walls. Image Credit: Mansouri et al. (2022).

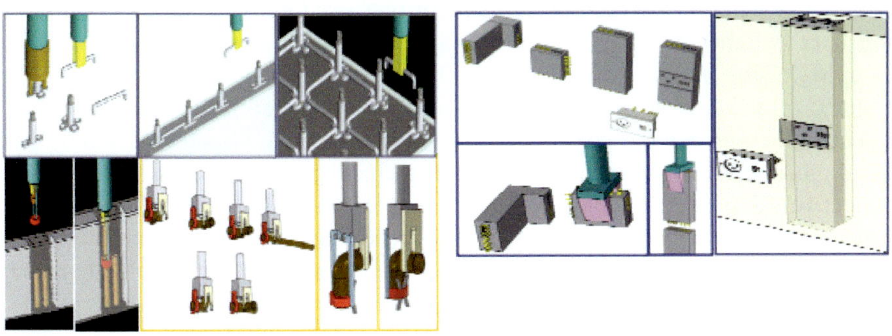

FIGURE 16.8 Generic design concepts for the integration of MEP components. Image Credit: Khoshnevis et al. (2005).

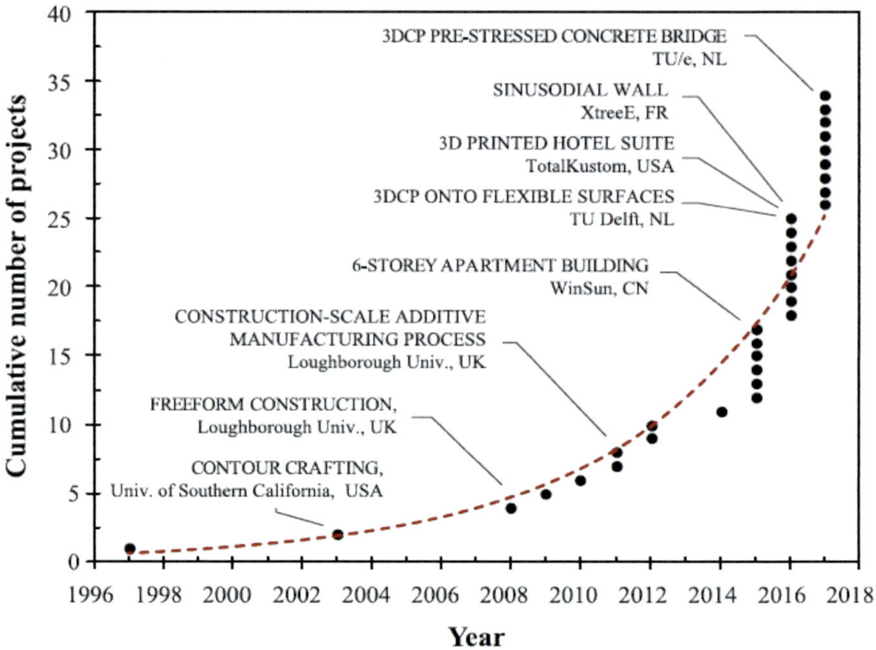

FIGURE 16.9 Rise in large-scale additive construction applications since concept inception in 1997 (Buswell et al. 2018). Image Credit: R. Buswell et al.

components. These efforts significantly benefited terrestrial 3D printing efforts as well.

A few years after Khoshnevis developed CC, Enrico Dini developed D-Shape printing, a large-scale 3D printing technology, primarily known for its ability to print stone-like objects by binding sand with a magnesium-based binder (Dini). After years developing industrial automation, he filed the first D-Shape patents in 2005 and founded industry-pioneering Monolite UK in 2006. The following year, Enrico printed the iconic Radiolaria Pavilion – the world's first full-scale 3D-printed architectural structure. The first D-Shape printer used epoxy resin as a binder, but Dini later switched to a more suitable magnesium-based binder due to issues with the epoxy. Dini patented the design with the magnesium-based binder in 2008. D-Shape gained significant attention due to its potential for large-scale construction projects, allowing for the creation of complex, stone-like structures directly from digital models. The D-Shape printer utilizes a binder-jetting process, where a binder is selectively deposited layer by layer onto a bed of sand, effectively "gluing" the sand particles together to form a solid object. The technology quickly garnered significant media coverage, establishing itself as one of the first construction-scale 3D printers in the world.

The disadvantage of these types of process is an additional deposition device requiring more maintenance, cleaning, and control instructions, and the secondary structure must be cleaned away in a post-processing operation. The simplicity of powder-based approaches such as D-Shape must be balanced against the large volumes of material that have to be deposited in the build area and then removed to reveal the part (Lim et al. 2012). Even with this constraint, the D-Shape technology is being developed by the European Space Agency for consideration for lunar habitat fabrication (Dini 2020).

With an eye toward planetary construction, NASA/MSFC identified large-scale 3D printing of concrete as a potential construction technique that could be significantly automated. As mentioned, early mobility system configurations were many and varied. The following list identifies a number of mobility system configurations and identifies government, commercial, or academic entities that currently use or have used this type of mobility system for large-scale additive construction.

(i) SCARA platform (CyBE and XTreeE)
(ii) Cylindrical coordinate platform (Apis Cor)
(iii) Articulated platform (MIT-DCP, Construction-3D, and ICON Phoenix)
(iv) Stewart/Delta platforms (WASP)
(v) Sky-Cam (derivative)
(vi) Cable-suspended platform (NIST Robo-crane)
(vii) Cartesian platforms (Gantry, open or closed) (Contour Crafting, ICON Vulcan, NASA ACME-2, USACE ACES-3, Bet Abram P1, COBOD, Mudbots, Total Kustom Stroybot, and S-Squared ARCS)
(viii) Deployable gantry platforms (Contour Crafting)
(ix) Structure-riding platform (Mindbuilder – Barcelona)
(x) Hybrid systems (AI Space Factory and X-Hab 3D)

FIGURE 16.10 3D-printed concrete dome built at NASA/MSFC in 2005. Image Credit: NASA/Marshall Space Flight Center.

A number of other design concepts have been developed over the years, but this author is not aware of their actual use in large-scale construction.

In late 2005, using a closed gantry-style printer developed by Behrokh Khoshnevis and procured from the University of Southern California with a CC nozzle, MSFC fabricated the iconic 3D-printed concrete dome that still graces the entrance to the NASA/MSFC Additive Construction Laboratory, shown in Figure 16.2. This dome is a testament to the use of portland cement and 3D printing for long-term sustainability. Using a mix developed in-house at MSFC (with chemical additives for rheology and crack control, and still used today), the dome uses interior walls as overhang support.

16.3 HISTORY OF 3D PRINTING IN LOW GRAVITY

As human spaceflight has become more mainstream, and especially with the advent of additive manufacturing technologies in the late 1980s and early 1990s, it was a natural progression to the idea of 3D printing in low gravity.

With Johnson Space Center (JSC) and Milwaukee School of Engineering as partners, Ken Cooper at NASA/MSFC began researching fused filament fabrication in the late 1990s in the lab environment as a viable in-space manufacturing process. In the summer of 1999, the technology advanced further when Cooper successfully conducted a total of four flights on NASA's KC-135 Low-Gravity Research Aircraft (aka, the Vomit Comet) in the first ever demonstration of 3D printing in a low-gravity environment. Shown in Figure 16.11, a modified Stratasys fused filament fabrication

FIGURE 16.11 Stratasys 1600 during KC-135 reduced-gravity testing, 1999. Image Credit: Ken Cooper.

system experienced a weightless environment due to the KC-135's parabolic arc flight path. A total of seven different part designs were manufactured using acrylonitrile butadiene styrene (ABS) filament. The total flight time was 10–12 hours with a total time of 1 hour and 20 minutes spent in a weightless environment. Due to the nature of the parabolic arc, continuous operation time in a weightless environment is limited to around 25 seconds (Cooper and Griffin 2003).

This work paved the way for the Exploration Science and Technology Division at MSFC to establish an In Situ Fabrication and Repair (ISFR) Program Element, which housed the later development of MSFC's In-Space Manufacturing (ISM) group and planetary additive construction activities.

The ISM group at MSFC funded a Small Business Innovative Research (SBIR) grant with Made in Space (MIS) that resulted in the launch of 3D printing in Zero-G technology demonstration mission in 2014 aboard the ISS. This was the first long-duration application of 3D printing in low gravity and resulted in 55 parts being printed of ABS plastic (Prater 2019). In 2016, the Additive Manufacturing Facility (AMF), a second-generation printer (shown in Figure 16.12) also built by MIS (acquired by Redwire in 2020), was flown to the ISS. The AMF was used to conduct additional materials characterization studies using the flight printer and a ground-based equivalent unit. The AMF studies derive their testing methodology from composites, in that mechanical test specimens are machined from printed panels rather than specimens that are printed directly (Prater 2018).

In the same time frame, Owens and de Weck at MIT began evaluating the idea of using additive manufacturing as a component of reliability and maintainability, especially on longer duration missions (Owens et al. 2015). This work led to the idea of designing for maintainability, now a first principle for long military operations

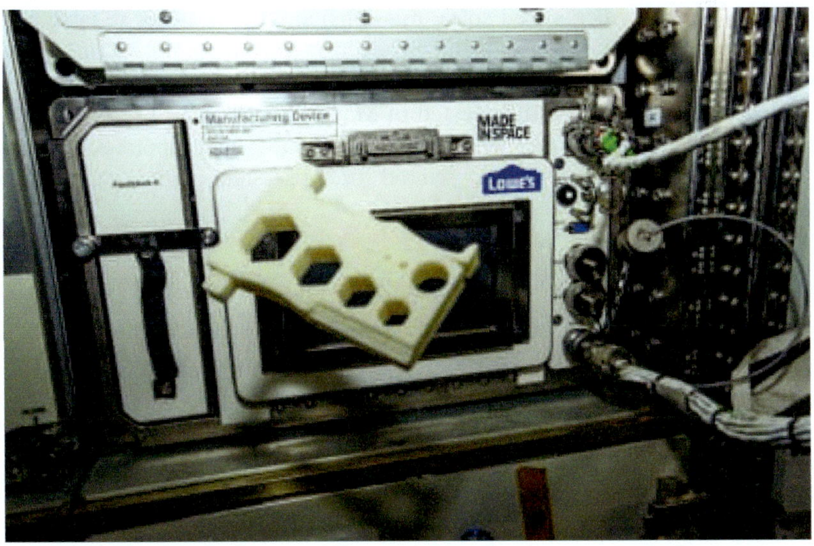

FIGURE 16.12 MIS AMF aboard the ISS, 2016. Image Credit: T. Prater.

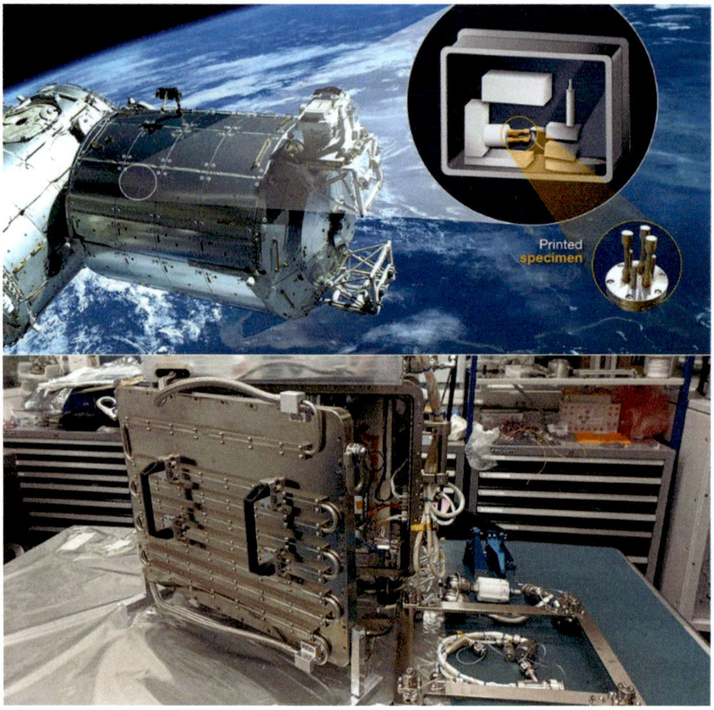

FIGURE 16.13 ESA/Airbus metal 3D printer for ISS, 2024. Imager Credit: T. Prater.

(subs, overseas deployments, and space missions) and the need for reliable additive manufacturing capabilities for real-time production of spare parts.

In January 2024, the first metal 3D printer was shipped to the ISS by an ESA/Airbus team and was successfully tested in May 2024. Shown in Figure 16.13, this system utilized a stainless steel wire that was fed into a melt pool being heated by a high-power laser. Four parts were printed for subsequent comparison to ground-based reference prints.

16.4 ISRU-BASED PLANETARY ADDITIVE CONSTRUCTION

The basic premise of this section is that rockets are an inherently inefficient method of transportation, resulting in the need to maximize the use of in situ materials for the construction of lunar surface infrastructure. The first mention of ISRU in the literature appears to be in the manufacturing of propellant on the surface of Mars to support return to Earth (Lim 2017). In 1989, NASA issued a report identifying the need for ISRU-based materials for a lunar outpost (Cohen 1989). One of the identified applications in this report for ISRU materials was planetary surface infrastructure construction. An excellent review of lunar resources was performed in Crawford

(2015), and the characterization there identified lunar regolith as an excellent resource for metals and volatiles.

Payload costs to low earth orbit (LEO) are still in the \$10,000/kg range, estimates from Earth to the lunar surface are in the \$1 M/kg range, and costs from Earth to Mars are probably an order of magnitude larger. Therefore, every kilogram of payload has to buy its way on to a mission, and the idea of transporting construction raw materials from Earth to the Moon becomes prohibitively expensive, even at low mass percentages.

However, there are many material analogs for both lunar regolith and planetary cements that can be used on Earth to develop and refine planetary construction processes, technologies, and infrastructure design. Polymer-based concretes and portland cement-based concretes come to mind. For example, polymer concretes require mixing of the regolith with aggregate and subsequent heating to liquefy the polymer prior to pumping/extrusion and is therefore an excellent analog material for planetary-based sulfur concrete construction technology development.

The use of in situ resources is made more challenging by key aspects of the lunar environment that all construction equipment (excavators, beneficiation equipment, extraction systems, construction, and test equipment) and built structures (landing pads, roads, blast shields, habitats, labs, and shelters) will be exposed to both during and after construction. These environmental concerns are summarized in Table 16.1.

TABLE 16.1
Challenging Aspects of the Lunar Surface Environment

Factor	Description	Implications for Exploration/ Construction
Atmosphere and vacuum	The Moon has an extremely thin atmosphere (exosphere) with almost no pressure (less than 1 picobar during the day), which is considered a hard vacuum	No protection from solar radiation or micrometeoroids; requires pressure suits and specialized systems for human and robotic missions
Gravity	The Moon's surface gravity is about 1.62 m/s^2 or approximately one-sixth of Earth's gravity	Makes handling heavy equipment easier but can complicate mobility and locomotion for crews
Temperature	Temperatures fluctuate dramatically between day and night, especially at the lunar equator	Requires robust thermal management systems to keep equipment and habitats within operational temperature ranges • Equatorial day: up to 123°C (253°F) • Equatorial night: down to −153°C (−243°F) • Polar regions: as low as −233°C (−387°F) in permanently shadowed regions

(Continued)

TABLE 16.1 (*Continued*)
Challenging Aspects of the Lunar Surface Environment

Factor	Description	Implications for Exploration/ Construction
Radiation	Without a protective atmosphere or global magnetic field, the Moon is exposed to two primary sources of radiation	Presents a major health risk to crews, requiring shielded habitats and special precautions • Galactic cosmic rays (GCRs): high-energy particles from outside the solar system • Solar energetic particles (SEPs): Bursts of high-energy particles during solar flares and other events
Illumination	A lunar day and night each last about 14–15 Earth days. The incident sunlight is intense and collimated due to the lack of an atmosphere to scatter the light	Long periods of darkness are a challenge for solar-powered missions, particularly at the equator. Polar regions can have areas of near-continuous sunlight
Dust (regolith)	The surface is covered with a layer of finely powdered, glass-like rock fragments called regolith, which is highly abrasive and electrostatically charged	Poses a significant hazard to equipment, optical systems, and human health
Micrometeoroids	The Moon is constantly bombarded by small, high-velocity meteoroids	Can damage equipment and surfaces exposed to space, necessitating protective shielding

16.4.1 Cementitious Planetary Additive Construction Applications

A summary of potential lunar applications for cementitious materials is presented in Figure 16.14 (Netti and Bisharat 2022). While not exhaustive, it provides the reader a good summary of the types of structures that will be required on the lunar surface. The needs are many and lend themselves to varying material property requirements (also a function of selected geometry) and fabrication technologies. Figure 16.15 (Shimizu Corp 2024) shows an artist's rendering of a Lunar Outpost with a manufacturing facility and ISRU-based concrete modules.

In November 1969, Apollo 12 landed about 197 yards from the Surveyor 3 spacecraft, which had landed in April 1967. The point was to remove some hardware elements from the Surveyor and bring them back to Earth for analysis. They were surprised to find that all surfaces of the Surveyor facing Apollo 12 had been "sandblasted" with lunar regolith accelerated by the plume of Apollo 12 as it approached the surface.

Metzger (2011), Morris (2012), and others characterized this phenomenon and identified that approximately 1 ton of regolith had been displaced under each Apollo

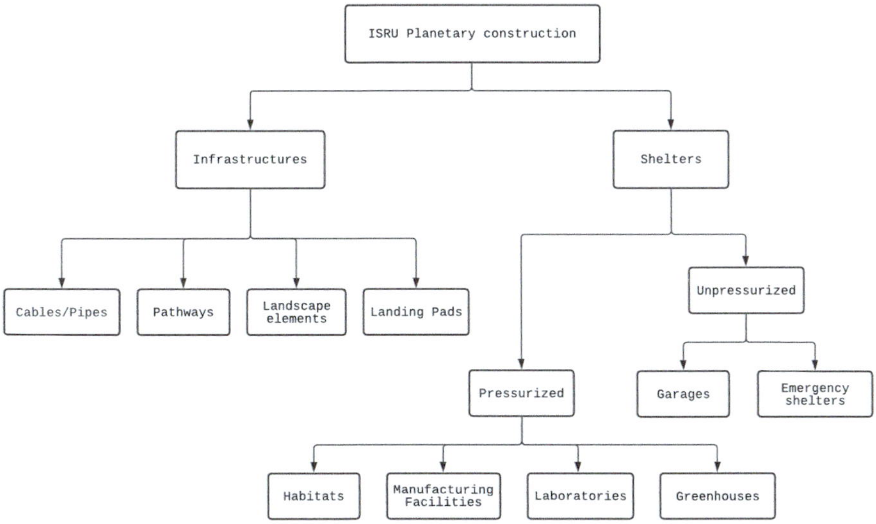

FIGURE 16.14 ISRU-based planetary outpost applications. Image Credit: V. Netti.

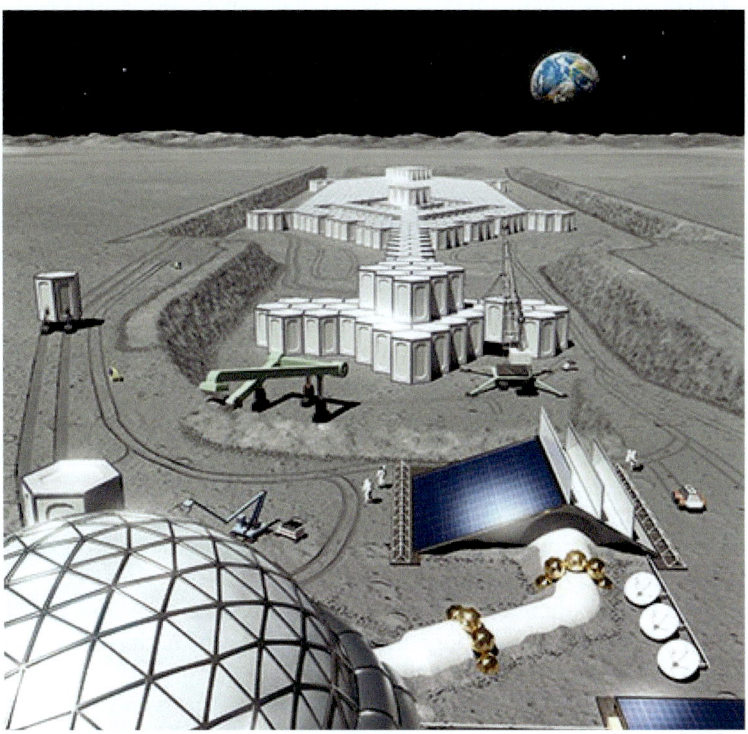

FIGURE 16.15 Shimizu Corporation design concept for a lunar outpost showing a manufacturing facility and concrete modules. Image Credit: Shimizu Corp.

lander during the approach to the lunar surface. This regolith is accelerated via plume surface interaction to speeds that could result in significant damage to nearby surface infrastructure, as well as regolith being placed into the lunar orbit, resulting in potential damage to other assets (i.e., Gateway). These early works defined the growing understanding that a lunar launch/landing pad should be one of the first ISRU-based surface structures constructed on the lunar surface.

16.4.2 Cementitious Planetary Additive Construction Materials

There are a number of processing materials and technologies being examined for use in lunar surface construction, many of which utilize lunar regolith in its natural state, and a number that rely on binders of some sort to use with lunar regolith as a cementitious material. As has been mentioned, it is not surprising that the development of lunar and Martian concretes began with an assessment of Earth-based concretes. These included ordinary portland cement (OPC) binder-based concretes, concretes based on calcium sulfoaluminate or magnesium oxide-based cements, sulfur concrete, geopolymer concretes, and polymer concretes.

Lunar concrete (also known as lunarcrete and mooncrete) is an idea first proposed by Larry A. Beyer of the University of Pittsburgh in 1985. Lunarcrete was proposed as an ISRU-based construction material, similar to concrete, formed from lunar regolith and cement, that would theoretically reduce the construction costs of building on the Moon (Beyer 1988).

Some of the other earliest suggestions for the development of lunar concrete were made by T.D. Lin in 1985 (Lin 1985), in 1987 (Lin 1987), and again in 1996 (Lin et al. 1996). The first experiments on the development of concrete using lunar regolith and OPC were performed by Lin in 1987 (Lin et al. 1996) using 40 gm of Apollo 16 regolith samples. The concrete was cured by using steam on a dry aggregate/cement mixture (also known as direct steam injection, DSI). Lin proposed that the water for such steam could be produced by mixing hydrogen with lunar ilmenite at 800°C, to produce titanium oxide, iron, and water. It was capable of withstanding compressive pressures of 75 MPa and lost only 20% of that strength after repeated exposure to vacuum. Based on recent Lunar Crater Observation and Sensing Satellite (LCROSS) and Lunar Reconnaissance Orbiter (LRO) (LCROSS 2022; Shadows 2012) results, the requirement for water as part of a cement binder could be accomplished by as simply as extracting water from icy regolith in the permanently shadowed regions of South Pole craters.

However, the development of portland cement on the Moon is significantly more challenging because the major raw material for the clinker making on Earth is typically limestone ($CaCO_3$, not available on the lunar surface) and because of the requirement to calcine the clinker above about 600°C and then fuse the clinker at about 1,450°C. Because, as mentioned earlier, the cost of taking 1 kg of material from Earth to the surface of the Moon is estimated at ~$1 million/kg, it is completely impractical to bring this material from Earth. This is also the primary reason that polymer concretes are not practical on the Moon, as all of the polymer constituents would have to be transported from Earth.

These constraints led to an analysis of cementitious materials that could be fabricated from 100% ISRU materials. Early analyses by a number of authors (Toutanji 2005; Fiske and Bodiford 2006; Montes 2015; Casanova 1997; Collins 2023; Lee and Van Riessen 2022; Mobili et al. 2017; Thibodeau 2017) resulted in the identification of the following classes of planetary cementitious materials:

- Sulfur concrete
- Geopolymer concretes
- Magnesium oxide-based concretes
- Calcium aluminate-based concretes

16.4.2.1 Sulfur Concrete

Sulfur is predominantly found on the Moon as troilite (FeS). While in low concentrations within the overall material composition, it can be extracted with a 90% yield through heating at 1,000°C. Additionally, sulfur serves as a by-product of oxygen extraction, a critical process for supporting human life on lunar bases, with a mass ratio of approximately 10% relative to oxygen mass. However, the technical feasibility of utilizing sulfur concrete for construction is challenged by the melting point of sulfur at around 120°C, particularly in the context of the lunar environment temperature range (−171 to 120°C). Sublimation of sulfur in the lunar vacuum over time has been identified as an issue as well, particularly as a function of temperature. However, recent work has shown that the sublimation rates are not as high at the lunar South Pole because of the reduced average temperature compared to the lunar equator (Giwa 2022).

16.4.2.2 Geopolymer Concrete

Recent studies suggest promising prospects for using lunar regolith simulant-based geopolymers to protect astronauts from the harsh lunar conditions. Geopolymers are artificially synthesized amorphous aluminosilicate material that acts as an inorganic binding agent. The most commonly studied to date for lunar applications is based on sodium hydroxide and sodium silicate. Its production involves mixing aluminosilicate source materials with alkali activators, followed by curing at either ambient or low temperatures (which can make curing 3D-printed structures problematic). Ongoing research in geopolymer development for lunar-based construction aims to enable extended human habitation on the Moon (Collins 2023; Lee and Van Riessen 2022).

16.4.2.3 Magnesium Oxide Concrete

The best studied magnesium oxide-based cement for lunar applications is magnesium oxy-sulfate (MOS). MOS is a type of air-dried cementitious binder. The general properties of MOS binder include its relative lightweight, high-temperature resistance, excellent fire resistance, and low thermal conductivity. These properties make it a promising material for several applications including producing lightweight insulators, electrically insulating materials and fire-proof materials (Mohan and Gosh 1991). Compared with magnesium oxychloride cement, which has drawn much research interest recently (Chau and Li 2008), little commercial or industrial attention has been paid to MOS mainly because of its low mechanical strength (Wu et al. 2015).

MOS binder is known to have a high range of its fresh and hardened properties such as its flow, initial setting time, final setting time, and the mechanical properties (strength and ductility). The wide range of these properties makes the MOS an excellent candidate to be utilized for 3D printing applications. Typically, set retardants are used in low quantities to control set time. Boric acid is a common set retardant for MOS concretes.

16.4.2.4 Calcium Aluminate Concrete

The most studied calcium aluminate cement for lunar applications is calcium sulfoaluminate (CSA). The main mineral components in CSA cement are anhydrous CSA ($4CaO{\cdot}3Al_2O_3{\cdot}CaSO_4$), dicalcium silicate ($2CaO{\cdot}SiO_2$) and gypsum ($CaSO_4{\cdot}2H_2O$). The lime in CSA cement is bonded and not free, so its alkali is lower.

CSA cements are based on alumina (Al_2O_3) and silica (SiO_2) with high mechanical strength and high impermeability. CSA cement requires less energy to produce, emits far less (or no) CO_2, and exhibits reduced shrinkage than portland cement production (Mobili et al. 2017). Alumina and silica are readily extracted from anorthite using HCl leaching (Thibodeau et al. 2024) as follows:

$$CaAl_2SiO_8 + 8HCl + 2H_2O \rightarrow CaCl_2 + 2AlCl_3{\cdot}6H_2O + SiO_2$$

$$AlCl_3{\cdot}6H_2O \rightarrow Al(OH)_3 + 3HCl + H_2O \text{ at } 100^\circ C$$

$$2Al(OH)_3 \rightarrow Al_2O_3 + 3H_2O \text{ at } 400^\circ C$$

Anorthite, a type of calcium-rich plagioclase feldspar, is a common mineral found on the Moon. It is primarily located in the lunar highlands. The obtained alumina (Al_2O_3) and silica (SiO_2) will be mixed at a ratio of 5:1. This mixture will then be sintered at 1,500°C using a liquid aluminum dihydrogen phosphate ($Al(H_2PO_4)_3 \cdot xH_2O$) binder. The resulting product is a high-shear-strength adhesive with excellent high-temperature tolerance (Fiske and Bodiford 2006). Typically, set retardants are used in low quantities to control set time. Citric acid is a common set retardant for CSA concretes, although there are also inorganic varieties.

16.4.3 Planetary Additive Construction Processing Technologies

As previously mentioned, there are a number of processing technologies being examined for use in lunar surface construction, many of which utilize lunar regolith in its natural state, and a number that rely on binders of some sort to use with lunar regolith as a cementitious material. This section will address the two primary technologies involved, namely, (i) thermally processed regolith and (ii) cementitious materials.

In the case of thermally processed regolith, there are two primary technologies used. Those include sintering processes that use laser, solar, or microwave heating sources (among others), and melting processes using laser, solar, induction, microwave, or resistance heating. In both of these thermal processes, the heat can be applied to the surface, or materials can be heated in a furnace and cast, or extruded using a 3D printing process.

In the sintering processes, the material is heated to a temperature where the lower-melting-point materials begin to melt and bond the remaining solid material together. In this scenario, it is difficult to obtain uniform volumetric heating with a narrow temperature range, depending on the heating source. Therefore, furnace sintering (usually of tiles or pavers) is often used for uniform heating. For lunar applications, despite recent work focusing on laser (Suhaizan eet al. 2024) or microwave (Shulman et al. 2023) sintering, issues such as relatively slow processing speed, high power requirements, and porosity due to gas evolution remain.

In contrast to sintering, melting followed by extrusion casting or layered deposition of lunar regolith heats the material to above the liquidus temperature and can produce materials that are fully dense, very strong, and abrasion resistant. Materials produced by melting and extrusion or casting are often classified as glass, glass-ceramics, depending on the degree of crystallization of the end product (Farries 2021). ICON's laser vitreous material transformation (VMX) process is a powder-bed fusion (PBF) method of additively constructing horizontal and vertical structures. Thin layers of regolith are progressively deposited and selectively lased – the lasing sequence applies a thermal schedule such that the material is pre-heated, sintered, melted/adhered to the previous layer, and crystallized (Jensen 2024). Issues with melting technologies also include relatively slow processing speed and high power requirements.

On the other hand, cementitious materials are construction materials utilizing binders extracted from ISRU materials (lunar water ice and regolith) that are used with regolith as the aggregate, and with or without reinforcement materials. In general, these materials use less energy and are typically faster during construction operations but do require extraction of elements from the regolith to formulate the binder material.

As early as 1989, the Precast/Prestressed Concrete Institute (PCI) performed a study to evaluate the construction of a lunar base habitat using these materials (Lindbergh 1989). A conceptual design for such a habitat is shown in Figure 16.16. In this study, it was proposed to use steel rod reinforcement manufactured by extracting the iron from ilmenite and adding carbon brought from Earth. This is one opportunity where the small amount of needed material from Earth (~0.02% C to be added to iron to make steel) can "buy" its way on to a lunar rocket.

When this PCI report was released, it had only been speculated that mixing, transport, pumping, casting, and extrusion in the lunar vacuum would cause porosity in water-based cementitious materials. Most recently, curing of several concretes at varying pressure levels were explored in terms of the effect of cure pressure on compressive strength. Materials tested have included CSA concrete (Fiske and Ellery 2025), geopolymer concrete (Collins 2022), and sulfur concrete (although this is actually a "waterless" concrete) (Fiske 2024). In all cases, the compressive strength of the material degrades with decreasing atmospheric pressure over the material as it cures. The degradation rate of strength with pressure appears to be unique to each material, but in all cases, the compressive strength is reduced by ~10% of

FIGURE 16.16 Proposed precast, prestressed concrete three-level lunar base, 1989. Image Credit: C. Lindbergh.

ambient-cured concrete strength at between 0.5 and 0.8 atm and continues to degrade at lower pressures.

These efforts suggest that a permanent, pressurized manufacturing facility on the lunar surface would benefit a number of these processes and materials. Such a facility could incorporate a large airlock such that segments of habitats or other structures could be cast or printed inside this facility. They could be rolled out through a large airlock and placed at its desired location on the lunar surface, where the final assembly would occur. Such a concept is not new. Figure 16.17 (Drake 1991) shows a rendering of such a facility, which included a regolith overburden to aid in micrometeorite and radiation protection, and in interior thermal control because of the insulating nature of lunar regolith, and the mass of the regolith could help offset some of the tensile loads in the structure as a result of being pressurized. A more recent perspective of a pressurized lunar manufacturing facility (LMF) with various incorporated manufacturing technologies is presented in Figures 16.18 and 16.19 (Hoe 2025).

Structures that could be precast/prestressed include any unpressurized or pressurized shelters/garages/habitats or laboratories, as well as large block structures, or large tile segments that could be part of a landing pad apron or containment wall/blast shield.

As the interest in 3D printing of concrete, or additive construction, grew through the 1990s and 2000s, it was only natural that the application of 3D print hardware to lunar concrete materials would be explored as well. As previously mentioned, Dr. Behrokh Khoshnevis was one of the first to envision and demonstrate the use of

FIGURE 16.17 Conceptual design of a large pressurized manufacturing facility on the lunar surface. Image Credit: R. Drake.

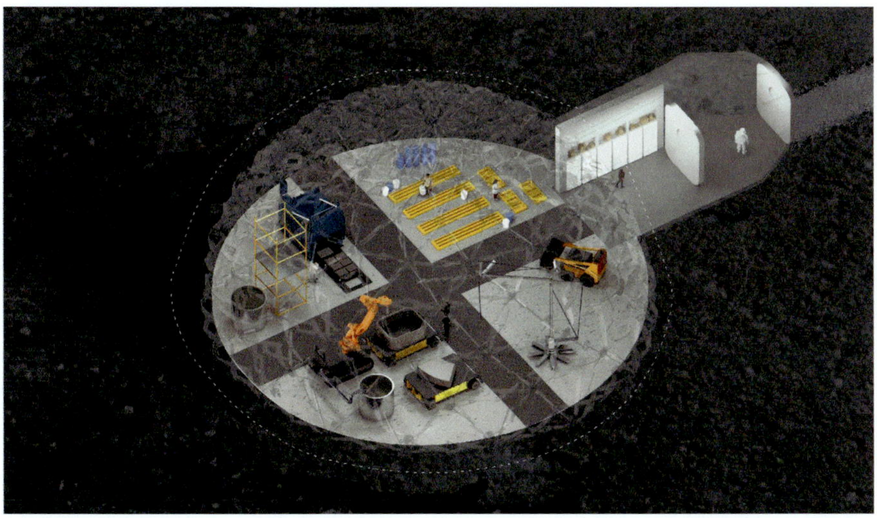

FIGURE 16.18 Lunar manufacturing facility design concept. Image Credit: A Hoe (2025).

a gantry printer or robotic arm to 3D print sulfur concrete on a planetary surface. Now, over 20 years later, the Moon to Mars Planetary Autonomous Construction Technology (MMPACT) project at NASA/MSFC recently completed (spring, 2024) thermal vacuum testing of an ICON-developed 14-ft robotic-arm prototype for potential use on an upcoming unmanned, CLPS lander mission to demonstrate a laser-based VMX process that ICON has also developed. However, this arm has been designed to support various printhead designs and other construction technologies,

FIGURE 16.19 Lunar manufacturing facility design concept interior. Image Credit: A. Hoe (2025).

such as microwave sintering of regolith and extrusion of a CAS lunar concrete. These processes/materials have also been identified as back-up, or "toolbox" technologies, as successful deployment of a long-duration human presence on the Moon will require a set of technologies for the construction of various infrastructures.

16.5 SUMMARY

NASA has identified six major development areas associated with sustainable living and working farther from Earth, including sustainable power and surface utilities, ISRU, survival in extreme environments, autonomous excavation, construction and outfitting, advanced robotics and autonomy, and low technology readiness level (TRL) habitation systems.

The MMPACT project at NASA's Marshall Space Flight Center addresses four of these areas. Development of autonomous surface construction technologies based on ISRU materials, and maximizing robotics while minimizing launch upmass associated with raw materials, is the focus of the MMPACT project.

MMPACT has been working on developing 3D printing construction systems for the Moon and Mars. The process of building landing pads, habitats, and roads on the Moon will likely look different from the common construction site on Earth. Excavation robots, for one, will need to be lightweight yet capable of digging in reduced gravity. A large-scale construction system could be autonomous and equipped to work without astronauts' help. As part of the Artemis program, NASA has a concept for the core surface elements needed to establish a sustained presence on the Moon, which emphasizes mobility to allow astronauts to explore more and conduct more science. NASA is considering putting in place a lunar terrain vehicle, habitable mobility platform or lunar RV, and surface habitat on the Moon by the end of the decade. The agency is investing in advanced manufacturing – one of five industries of the future to enable space exploration and improve life on Earth – including technologies that could find and use available resources on the Moon and Mars to build out future infrastructure.

Work is underway internationally as well in many countries. China has announced a plan to team with up to 50 countries to build a lunar base. Construction specifics are not available, but a clear interest has been expressed in launch and landing infrastructure (Collins 2023). All of the current development effort had its beginnings in a series of seemingly unrelated development activities that started with Urschel, included Pegna, and came together in Khoshnevis's lab in the late 1990s. The planetary concrete 3D printing community is grateful to these pioneers for their ground-breaking efforts.

REFERENCES

Beyer, L.A., "UND Engineers Would Like to Follow the Lunarcrete Road." *Grand Forks Herald, North Dakota*, February 28, 1988.

Buswell, R.A., et al., "3D Printing Using Concrete Extrusion: A Roadmap for Research." *Cement and Concrete Research* 112 (2018): 37–49.

Casanova, I., "Feasibility Study and Applications of Sulfur Concrete for Lunar Base Development: A Preliminary Study." *Proceedings of the 28th Annual Lunar and Planetary Science Conference*, Houston, TX, March 17–21, 1997, p. 209.

Chau, C.K., and Li, Z., "Microstructures of Magnesium Oxychloride." *Materials and Structures* 41, no. 5 (2008): 853–862.

Cohen, A., "Report of the 90-Day Study on Human Exploration of the Moon and Mars, NASA-TM-102999." 1989. https://share.google/knt5C7hRgQEEm9zG1.

Collins, P., "Towards Extraterrestrial Construction: Lunar Concrete for In-Situ Infrastructure on the Moon." PhD Dissertation, Civil Engineering, The Pennsylvania State University, May 31, 2023. https://etda.libraries.psu.edu/catalog/2334lpjc348.

Collins, P., et al., "Geopolymer Lunar Concrete Under Reduced-Pressure Curing and Vacuum Exposure." *73rd International Astronautical Conference*, Paris, France, September 2022.

Cooper, K.G., and Griffin, M.R., "Microgravity Manufacturing via Fused Deposition." NASA/TM-2003-212636, NASA Marshall Space Flight Center, July 2003.

Crawford, I., "Lunar Resources: A Review." *Progress in Physical Geography* 39 (2015): 137–167.

Curth, A., "3-D Printed Concrete." *Structure Magazine*, July 2022. https://www.structuremag.org/?p=20944.

Drake, R.M. "Production of Lunar Concrete in a Large Pressurized Assembly Facility." *Lunar Concrete*, Published by American Concrete Institute, SP 125-13, 1991, pp. 220–221.

Enrico Dini, "Pioneering 3D Construction Printing." *D-Shape*. www.d-shape.com/enrico-dini/. Accessed October 10, 2020.

Farries, K.W., et al., "Sintered or Melted Regolith for Lunar Construction: State-of-the-Art Review and Future Research Directions." *Construction and Building Materials* 296 (August 16, 2021).

Fiske, M., "Low Pressure Studies on Sulfur Concrete Curing." September–October 2024 (Unpublished).

Fiske, M., and Ellery, A., "On the Development of an ISRU-Based Calcium Sulfo-Aluminate (CSA) Concrete for 3-D Printed and Cast Lunar Surface Infrastructure Applications." *2025 AIAA Sci-Tech Conference*, Paper No. AIAA-2025–1863, Orlando, FL, January 6–10, 2025.

Fiske, M.R., and Bodiford, M.P., "Materials, Design and Construction of Lunar Structures Based Primarily on In-Situ Materials: An Overview." Poster presented at *ASCE Habitation Conference*, Orlando, FL, February 6–8, 2006.

Giwa, I., et al., "Planetary Construction 3D Printing Using Lunar and Martian In Situ Materials," *Earth and Space* (2022): 817–831. https://doi.org/10.1061/9780784484470.069.

Hoe, A., "Lunar Manufacturing Facility (LMF) Unpublished Design Work at NASA/MSFC." Summer 2025. https://www.esa.int/ESA_Multimedia/Images/2013/01/D-Shape_printer, Accessed October 27, 2024.

Jensen, E., "Project Olympus: Off-Planet Construction." *Lunar Surface Innovation Consortium (LSIC) Monthly Meeting*, January 2024.

Kennedy, K., "The Vernacular of Space Architecture." *AIAA Space Architecture Symposium*, Houston, TX, October 2002. https://doi.org/10.2514/6.2002-6102.

Khoshnevis, B., "Innovative Rapid Prototyping Process Makes Large Sized, Smooth Surfaced Complex Shapes in a Wide Variety of Materials." *Materials Technology* 13, no. 2 (1998): 52–63.

Khoshnevis, B., "Automated Construction by Contour Crafting – Related Robotics and Information Technologies." *Automation in Construction* 13 (2004): 5–19.

Khoshnevis, B., and Bekey, G., "Automated Construction Using Contour Crafting – Applications on Earth and Beyond." *Proceedings of the 19th International Symposium on Automation and Robotics in Construction, National Institute of Standards and Technology Special Publication Vol. 989*, U.S. Department of Commerce, Gaithersburg, MD, 2002, pp. 489–494.

Khoshnevis, B., et al., "Lunar Contour Crafting – A Novel Technique for ISRU-Based Habitat Development." *43rd AIAA Aerospace Sciences Meeting and Exhibit*, Paper No. AIAA 2005-538, Reno, NV, January 10–13, 2005.

Khoshnevis, B., et al., "ISRU-Based Robotic Construction Technologies for Lunar and Martian Infrastructures." *NIAC Phase II Final Report*, Report Number: HQ-E-DAA-TN41353, December 2020.

"LCROSS Mission Finds Water – Planetary News | the Planetary Society." www.planetary.org. Archived from the original on January 22, 2010. Accessed January 12, 2022.

Lee, S., and Van Riessen, A., "A Review on Geopolymer Technology for Lunar Base Construction." *Materials* 15, no. 13 (2022): 4516.

Lim, S., et al., "Developments in Construction-Scale Additive Manufacturing Processes." *Automation in Construction* 21, no. 1 (2012): 262–268.

Lim, S., et al., "Extra-terrestrial Construction Processes – Advancements, Opportunities, and Challenges." *Advances in Space Research* (2017). http://dx.doi.org/10.1016/j.asr.2017.06.038.

Lin, T.D., "Concrete for Lunar Base Construction." In *Lunar Bases and Space Activities of the 21st Century*, edited by W.W. Mendell, Lunar and Planetary Institute, Houston, 1985, p. 381–390.

Lin, T.D., "Concrete for Lunar Base Construction." *Concrete International* 9 (1987).

Lin, T.D., et al.,"Lunar Concrete Made with the Dry-Mix/Steam-Injection Method," *Engineering, Construction, and Operations in Space V* (1996): 592–599.

Lindbergh, C., et al., "A Feasibility Study of Prestressed Concrete Construction on the Moon," *PCI Journal* (July–August 1989): 60–75.

Mansouri, A., et al., "Thermal Modeling of the Convective Heat Transfer in the Large Air Cavities of the 3D Printed Walls." *Cogent Engineering* 9 (2022): 2130203. https://doi.org/10.1080/23311916.2022.2130203.

Metzger, P., et al., "Phenomology of Soil Erosion Due to Rocket Exhaust on the Moon and the Mauna Kea Lunar Test Site." *Journal of Geophysical Research* 116 (2011): E06005. https://doi.org/10.1029/2010JE003745.

Mobili, A., Belli, A., Giosue, C., Telesca, A., Marroccoli, M., and Tittarelli, F., "Calcium Sulfoaluminate, Geopolymeric and Cementitious Mortars for Structural Applications." *Environments* 4 (2017): article 64.

Mohan, L., and Gosh, S.N. "Advances in Some Special and Newer Cements." *Progress in Cement and Concrete, SN Ghosh (Edr.), ABI, New Delhi* 1, no. Part I (1991): 253–313.

Montes, C., et al., "Evaluation of Lunar Regolith Geopolymer Binder as a Radioactive Shielding Material for Space Exploration Applications," *Advanced Space Research* 56 (2015): 1212–1221.

Morris, A., "Simulation of Rocket Plume Impingement and Dust Dispersal on the Lunar Surface." PhD Dissertation, University of Texas, Austin, 2012.

National Inventors Hall of Fame (NIHF), "Who Invented 3D Printing? Meet Inductee Charles Hull." *NIHF*, June 6, 2005. https://www.invent.org/blog/inventors/3D-printing-chuck-hull.

Netti, V., and Bisharat, T., "A Review of Additive Manufacturing Technologies for Planetary Construction." *Proceedings of ASCE Earth & Space Conference*, April 2022.

Owens, A., Do, S., Kurtz, A., and de Weck, O., "Benefits of Additive Manufacturing for Human Exploration of Mars." *45th International Conference on Environmental Systems, No. ICES-2015-287, International Conference on Environmental Systems*, Bellevue, WA, 2015.

Pegna, J., "Exploratory Investigation of Solid Freeform Construction." *Automation in Construction* 5, no. 5 (1997): 427–437.

Prater, T., et al., "In-Space Manufacturing at NASA Marshall Space Flight Center: A Portfolio of Fabrication and Recycling Technology Development for the International Space Station." *Proceedings of the AIAA SPACE Forum*, Orlando, FL, 2018.

Prater, T., et al., "3D Printing in Zero G Technology Demonstration Mission: Complete Experimental Results and Summary of Related Materials Modeling Efforts." *The International Journal of Advanced Manufacturing Technology* 101 (2019): 391–417.

"Shadows of the Moon Hide 'Fluffy' Dirt & Water Ice." *Space.com*, January 19, 2012.

Shimizu Corporation, "Lunar Base." https://www.shimz.co.jp/en/topics/dream/content05/ . Accessed October 6, 2024.

Shulman, H., et al., "Microwave Sintering for Lunar Base Construction." *NASA Technology Report.* https://ntrs.nasa.gov/api/citations/20230014145/downloads/MST%20Invited091923.pdf.

Suhaizan, M.S., et al., "Regolith Sintering and 3D Printing for Lunar Construction: An Extensive Review on Recent Progress." *Progress in Additive Manufacturing* 9 (2024): 1715–1736. https://doi.org/10.1007/s40964-023-00537-1.

Thibodeau, B., Walls, X., Ellery, A., Cousens, B., and Marczenko, K., "Extraction of Silica and Alumina from Lunar Highland Simulant." *Proceedings of ASCE Earth & Space Conference*, Paper 6962, 2024.

Toutanji, H., et al., "Strength and Durability Performance of Waterless Lunar Concrete," *Proceedings of the 43rd AIAA Aerospace Sciences Meeting and Exhibit*, AIAA Paper No. AIAA-2005-1436, January 11, 2005.

Wu, C., Yu, H., Zhang, H., Dong, J., Wen, J., and Tan, Y., "Effects of Phosphoric Acid and Phosphates on Magnesium Oxysulfate Cement." *Materials and Structures* 48, no. 4 (2015): 907–917.

Index

Note: *Italic* and **Bold** page numbers refer to *figures* and **tables**

T

For Product Safety Concerns and Information please contact our EU representative GPSR@taylorandfrancis.com Taylor & Francis Verlag GmbH, Kaufingerstraße 24, 80331 München, Germany